T0364932

# TWENTIETH-CENTURY BUILDING MATERIALS

## HISTORY AND CONSERVATION

EDITED WITH A NEW PREFACE BY THOMAS C. JESTER

GETTY CONSERVATION INSTITUTE, LOS ANGELES

THIS BOOK IS DEDICATED TO THE MEMORY OF
HENRY WARD JANDL, 1946–1995



The Getty Conservation Institute works to advance conservation practice in the visual arts, broadly interpreted to include objects, collections, architecture, and sites. It serves the conservation community through scientific research, education and training, model field projects, and the broad dissemination of the results of both its own work and the work of others in the field. And in all its endeavors, it focuses on the creation and dissemination of knowledge that will benefit professionals and organizations responsible for the conservation of the world's cultural heritage.



First published in 1995 as "Twentieth-Century Building Materials: History and Conservation" by Technical Preservation Services, National Park Service, U.S. Department of the Interior. An illustrated version was published in 1995 by McGraw Hill in cooperation with the National Park Service. Development and packaging of the 1995 edition was produced by Archetype Press, Inc.: Diane Maddex, project director; Robert L. Wiser, designer; and Gretchen Smith Mui, editor.

Published by the Getty Conservation Institute

Getty Publications
1200 Getty Center Drive, Suite 500
Los Angeles, California 90049-1682
www.getty.edu/publications/

Affiliations of individuals and addresses of institutions mentioned herein were current at the time of this book's initial publication in 1995.

Tevvy Ball, *Project Editor*
Catherine Lorenz, *Designer*
Elizabeth Kahn, *Production Coordinator*

Front cover: The Alcoa Building, Pittsburgh, Pa. See fig. 1.3.

Page i: Standard-sized building materials with established properties, such as Celotex's 4-by-8-foot fiberboard, became a hallmark of the construction industry in the twentieth century. *Celotex Manual for Architects*, The Celotex Corporation, 1937.

Page ii: The Kresge Auditorium at the Massachusetts Institute of Technology, Cambridge, Mass. See plate 8.

Page viii: Cast aluminum elevator doors and a stepped over door grace the Cities Services Building (1932, Clinton and Russell, George and Holdon), New York City.

Back cover: The Salk Institute (Louis I. Kahn, architect), San Diego, Calif. Detail.

Printed in China

Library of Congress Cataloging-in-Publication Data

Twentieth-century building materials : history and conservation / edited with a new preface by Thomas C. Jester.
pages cm
First published in 1995 as "Twentieth-Century Building Materials: History and Conservation" by Technical Preservation Services, National Park Service, U.S. Department of the Interior. An illustrated version was published in 1995 by McGraw Hill in cooperation with the National Park Service.—ECIP galley.
Includes bibliographical references and index.
ISBN 978-1-60606-325-5 (pbk.)
1. Building materials—Conservation—United States. 2. Building materials— United States—History—20th century. I. Jester, Thomas C., editor. II. Tomlan, Michael A. Building modern America. III. Getty Conservation Institute, issuing body. IV. Title: 20th century building materials.
TA403.6.T87 2014
691.0973'0904—dc23
2014013583

# CONTENTS


FOREWORD TO THE 2014 EDITION vii
Timothy P. Whalen

PREFACE TO THE 2014 EDITION ix
Thomas C. Jester

PREFACE TO THE FIRST EDITION xi
Thomas C. Jester

ACKNOWLEDGMENTS xiii

TWENTIETH-CENTURY BUILDING MATERIALS: A PHOTOGRAPHIC ESSAY xv

INTRODUCTION: BUILDING MODERN AMERICA: AN ERA OF STANDARDIZATION AND EXPERIMENTATION 1
Michael A. Tomlan

PART I METALS

1 ALUMINUM 13
Stephen J. Kelley

2 MONEL 19
Derek H. Trelstad

3 NICKEL SILVER 25
Adrienne B. Cowden

4 STAINLESS STEEL 31
Robert Score and Irene J. Cohen

5 WEATHERING STEEL 39
John C. Scott and Carolyn L. Searls

PART II CONCRETE

6 CONCRETE BLOCK 47
Pamela H. Simpson; Harry J. Hunderman and Deborah Slaton

7 CAST STONE 53
Adrienne B. Cowden; David P. Wessel

8 REINFORCED CONCRETE 61
Amy E. Slaton; Paul E. Gaudette, William G. Hime, and James D. Connolly

9 SHOTCRETE 71
Anne T. Sullivan

10 ARCHITECTURAL PRECAST CONCRETE 77
Sidney Freedman

11 PRESTRESSED CONCRETE 83
Howard Newlon, Jr.

PART III WOOD AND PLASTICS

12 FIBERBOARD 89
Carol S. Gould, Kimberly A. Konrad, Kathleen Catalano Milley, and Rebecca Gallagher

13 DECORATIVE PLASTIC LAMINATES 95
Anthony J. T. Walker, Kimberly A. Konrad, and Nicole L. Stull

14 PLYWOOD 101
Thomas C. Jester

15 GLUED LAMINATED TIMBER 105
Andrew McNall; David C. Fischetti

16 FIBER REINFORCED PLASTIC 111
Anthony J. T. Walker

PART IV MASONRY

17 STRUCTURAL CLAY TILE 119
Conrad Paulson

18 TERRA COTTA 125
Deborah Slaton and Harry J. Hunderman

19 GYPSUM BLOCK AND TILE 131
Susan M. Escherich

20 THIN STONE VENEER 137
Michael J. Scheffler and Edward A. Gerns

21 SIMULATED MASONRY 143
Ann Milkovich McKee

PART V GLASS

22 PLATE GLASS 151
Kimberly A. Konrad and Kenneth M. Wilson; William J. Nugent and Flora A. Calabrese

23 PRISMATIC GLASS 157
Dietrich Neumann

24 GLASS BLOCK 163
Dietrich Neumann; Jerry G. Stockbridge and Bruce S. Kaskel

25 STRUCTURAL GLASS 169
Carol J. Dyson

26 SPANDREL GLASS 175
Robert W. McKinley

PART VI FLOORING

27 LINOLEUM 183
Bonnie Wehle Parks Snyder

28 RUBBER TILE 191
Sharon C. Park

29 CORK TILE 197
Anne E. Grimmer

30 TERRAZZO 203
Walker C. Johnson

31 VINYL TILE 209
Kimberly A. Konrad; Paul D. Kofoed

PART VII ROOFING, SIDING, AND WALLS

32 ASPHALT SHINGLES 217
Mike Jackson

33 PORCELAIN ENAMEL 223
Thomas C. Jester

34 ACOUSTICAL MATERIALS 231
Anne E. Weber

35 GYPSUM BOARD 237
Kimberly A. Konrad and Michael A. Tomlan

36 BUILDING SEALANTS 241
Michael J. Scheffler and James D. Connolly

NOTES 245

BIBLIOGRAPHY 293

SOURCES FOR RESEARCH 301

AUTHORS AND CONTRIBUTORS 313

INDEX 315

ILLUSTRATION CREDITS 320

# FOREWORD TO THE 2014 EDITION

Over the past few decades it has become increasingly apparent that modern buildings present a unique series of conservation challenges. Many of these buildings have not aged well, and conservation efforts have been hampered by a lack of information about the materials and technologies used in their construction.

The Getty Conservation Institute launched its Conserving Modern Architecture Initiative in 2012 with the goal of advancing the practice of conserving twentieth-century heritage, with a focus on modern architecture. The Initiative is intended to help address the needs of the field through research and investigation, including the development of practical conservation solutions, and the creation and distribution of information through training programs and publications.

In our early discussions about publications and the dissemination of critical literature, Thomas Jester, the original editor of *Twentieth-Century Building Materials*, suggested we revise or re-issue this landmark volume, which had been out of print for several years. Originally published under the aegis of the United States Department of the Interior's National Park Service's Preservation Assistance Division, *Twentieth-Century Building Materials* is considered the first in-depth survey of important construction materials used primarily in North America since 1900. Highly regarded by architects, architectural historians, and conservation professionals, it is still a relevant and valuable resource for professionals from various disciplines working in the field of modern heritage.

We thank Thomas for editing the original 1995 edition of *Twentieth-Century Building Materials* and for his assistance in this re-issue. We are also grateful to the National Park Service for supporting the development of this important project, and to Susan Macdonald and Kyle Normandin, who through the Conserving Modern Architecture Initiative have helped give new life to this important publication.

We hope that this book will serve conservation professionals, including architects, engineers, conservators, and material scientists, as well as students in those fields, as a resource for the conservation of building materials of the twentieth century—from everyday structures to iconic monumental architecture—and help ensure that these buildings are appropriately safeguarded in the future.

Timothy P. Whalen
Director
The Getty Conservation Institute

811210.
WURTS BROS.
PHOTO N.Y.

# PREFACE TO THE 2014 EDITION

The importance of preserving both modern architecture and the recent past is today clearly acknowledged as part of the mainstream in the field of heritage conservation. That wasn't the case in 1995, however, when *Twentieth-Century Building Materials: History and Conservation* was first published. Twenty years ago very little information was available on the repair and conservation of modern building materials and assemblies. *Twentieth-Century Building Materials* was conceived and developed by the National Park Service's Technical Preservation Services Branch, which was an early leader in advancing the practice of historic preservation through its technical publications and national conferences. As an introduction to a conservation perspective on the world of modern building materials, the book was groundbreaking for its time. It brought attention to this emerging area of preservation practice and effected a sea change in the understanding of midcentury construction. A tremendous collaborative undertaking, *Twentieth-Century Building Materials* would not have been possible without the significant contributions made by more than forty authors and the commitment of the National Park Service.

Extensive literature has been published over the past twenty-five years about the conservation and repair challenges associated with modern materials and assemblies. Beginning in the mid-1990s, the literature on preservation technology shifted from theoretical discussion about conservation issues to project-specific case studies, as buildings from the modern era began to undergo interventions. The Getty Conservation Institute's bibliography *Conserving Twentieth-Century Built Heritage, A Bibliography, Second Edition* is testimony to the worldwide interest in this specialty area of conservation practice, an interest that continues today. The Association for Preservation Technology International and DOCOMOMO International have been at the forefront of efforts to advance the technical understanding of modern heritage through conferences, training workshops, and publications, which have addressed topics ranging from exposed reinforced concrete and curtain walls to stone cladding, modern wood, and glass.

Despite the real advances in our ability to address the unique technical and philosophical challenges associated with modern building materials and components, there is still a great need for more advanced information on modes of deterioration, diagnostics, and repair and conservation treatments. Advanced and widely accepted technical solutions still do not exist for many of the most common materials, including reinforced and precast concrete, curtain wall systems, glass, and metals. Although the number of modern preservation projects continues to grow—and with it the associated knowledge base—decisions about specific materials are most frequently based on unique circumstances and conditions rather than on application of time-tested conservation approaches and methods. In part this is explained by the sheer number of proprietary materials and assemblies used in the twentieth century, but it is also because we simply do not know as much as we should about modern materials.

Put simply, not enough advanced information is available to conservation professionals. More conservation research is needed, including more partnerships among industry, academia, and government agencies. One could look for inspiration at the related fields of museum and objects conservation. Objects conservators have embraced the challenges of modern materials in their collections with research programs on topics such as modern metals, paints, and plastics. Similar research programs are needed for the most important conservation issues associated with immovable cultural property.

So, if preservation technology for modern buildings has advanced in the past twenty years, why republish *Twentieth-Century Building Materials*?

First, there is still a great need for basic information, and no other reference provides readers with an introduction to the history of modern materials and associated repair and conservation issues. The book was intended as a primer, and readers needing more detailed historical and repair information were then—and are now—encouraged to consult additional sources. Since its publication in 1995, more detailed technical literature has been published on many of the materials. In many other cases, however, the history sections of these essays are as current today as they were twenty years ago, because no additional scholarship has advanced our understanding of the history of a particular building technology.

Second, more and more modern buildings are at an age when renewal is required, and practitioners are faced with decisions about repair and replacement. Some modern materials were experimental; others simply lack the durability of more traditional materials or do not meet today's performance requirements. The need to replace certain materials and assemblies, most prominently curtain walls, for example, has become more widely accepted than in the past. Addressing this kind of technical obsolescence nonetheless remains a challenge for practitioners, who must find a balance between the need for performance and respect for historic values in significant buildings. Indeed, the starting point for designing an intervention should be not what must be replaced but rather what can be preserved. This requires a thorough understanding of the materials from a historical and conservation perspective. Finally, the book contains valuable bibliographic references and resource information that remain useful to students of the history of building technology as well as to conservation professionals.

For all these reasons, *Twentieth-Century Building Materials* remains a valuable introduction to its subject. It is also hoped that republication of this book will once again emphasize the importance of understanding the history of modern materials and associated conservation and repair challenges and serve as a bridge until more advanced technical literature becomes available. The preface to the 1995 edition noted that the book was intended to identify many of the conservation challenges that lie ahead and "sound a call" for more research into such subjects as the failure mechanisms afflicting modern materials. Despite the advances made in the past twenty years, those points remain valid today. The Getty's Conserving Modern Architecture Initiative (CMAI) has assumed a leadership role in much the same way that the National Park Service did in the 1990s. We hope that the CMAI will spur, conduct, and support research on the most ubiquitous modern materials and facilitate advances both in our understanding of how modern materials deteriorate and in the development of new conservation approaches and repair techniques. While this important work is being done, for anyone wishing to understand the most widely used construction materials of the last century, *Twentieth-Century Building Materials* is once again available as both resource and reference.

Thomas C. Jester, AIA, FAPT
January 2014

# PREFACE TO THE FIRST EDITION

*Twentieth-Century Building Materials* explores the history of many of the materials used to build modern America. Some of these materials, such as reinforced concrete and terra cotta, were nineteenth-century materials that found new uses after the turn of the century. Others, such as glass block, stainless steel, porcelain enamel, and decorative plastic laminates, did not appear until after 1900. Still others were not widely used until after World War II. Such materials can be found in buildings on Main Streets and in cities across the country.

Building materials not only make possible the structure and form of a building; they also reflect the industries that defined the machine age. Curved structural glass, extruded aluminum, glass block, and linoleum are examples of mass-produced materials that gave buildings a distinctively modern appearance. Building products from this century are also reminders of improvements in construction technology—prismatic glass, for example, improved lighting in buildings, and acoustical materials provided sound control. Other products, like gypsum board, signified the growing trend toward simplified drywall construction. Regardless of importance, building materials have shaped both architectural masterworks as well as vernacular buildings.

This book was inspired in part by two earlier books that focused on historic building materials and technology and underscored the importance of understanding materials and technology as the basis of sound historic preservation practice. More than twenty years ago Charles F. Peterson, then with the National Park Service, organized a symposium to celebrate the 250th anniversary of the Carpenters Company of the City and County of Philadelphia and focus attention on building practices before 1860. Until then the subject of building technology had largely been ignored by historians. The culmination of this effort was the publication of *Building Early America*, edited by Peterson. The symposium's success led the Association for Preservation Technology to undertake a research program nearly ten years later. Under the direction of Hugh Miller and Lee H. Nelson, both historical architects at the National Park Service, the association's foundation compiled and in 1983 published a collection of essays on late-nineteenth-century building technology and materials. H. Ward Jandl served as editor of this volume, entitled *The Technology of Historic American Buildings*. *Twentieth-Century Building Materials*, likewise, is the culmination of a tremendous collaborative effort, involving professionals and scholars.

The National Park Service is directed by Congress to conduct a range of activities to assist states and territories, federal agencies, and the general public in the treatment and long-term protection of historic properties listed in the National Register of Historic Places. The Preservation Assistance Division, one of the National Park Service's cultural divisions, sets professional standards and guidelines for conservation and rehabilitation work, sponsors and carries out training workshops and conferences, and produces and distributes technical information through a variety of formats, including the well-known Preservation Briefs series, Preservation Tech Notes, and copublished works, such as this book.

There is now an urgent need to address the relative dearth of historical and practical conservation information on twentieth-century building materials. Many buildings constructed in the first half of the century are now being designated as historic landmarks, and they are being rehabilitated and restored or are prime candidates for renewal. This volume is intended to promote awareness of the history of a range of building materials that can be found in twentieth-century buildings. The conservation sections will help readers identify common materials, explain how they deteriorate, and suggest conservation, repair, and replacement approaches.

Mass-produced materials used in this century are already beginning to pose conservation challenges for the current generation of architectural conservators, architects, and engineers involved in historic preservation. Contrary to popular perception, modern building materials are neither maintenance free nor more durable than traditional materials, from brick and timber to iron and stone. Just like handcrafted materials, modern materials are subject to environmental degradation and the effects of human use.

Building materials from this century are in some respects more complex than their predecessors, and for this reason we need to expand our understanding of historic manufacturing processes, standards, and testing methods. Many twentieth-century products are typically manufactured using large-scale and mechanized processes; in addition, they are often composite materials. Intricate alloys, materials based on patented formulas, complex plastics, and laminated materials are only a few examples of composite materials. Learning more about how materials in composites behave and interact will be a challenging task for historic preservation professionals in the coming years.

Even though few time-tested conservation techniques exist for many twentieth-century materials, the amount of current technical literature on the performance and durability of these materials is far larger than that for materials of earlier centuries. To avoid outdated, potentially damaging approaches to materials evaluation, testing, and repair and replication, preservationists must study the historical record documenting the development and early testing of these building materials. Research into early technical documents will be useful—if not essential—in guarding the integrity of twentieth-century architecture and limiting insensitive material repairs and replacement.

The introductory essay by Michael A. Tomlan, director of the Historic Preservation Planning Program at Cornell University, provides a historical overview of building material developments between the late nineteenth and mid-twentieth centuries. Subsequent sections cover specific materials in the following general categories: metals; concrete; wood and plastics; masonry; glass; flooring; and roofing, siding, and walls. Entries on materials, written by more than forty contributors, begin with the name of the author or authors. When multiple authors are listed and separated by a semicolon, the first author or authors listed prepared the history section, while those listed second prepared the conservation section. Dates of production refer specifically to U.S. production and principally to architectural applications. The decision to include trade names also deserves explanation. A broad definition of the term *trade name* has been adopted for this book to include the names of both trademarked and proprietary products, as well as names commonly used in the construction industry.

This book is intended to sound a call for more investigation into the deterioration and conservation of twentieth-century materials. For a few materials discussed in this volume, no conservation section has been included. Further research is necessary on these topics and others.

The sheer number of materials used in this century suggests that research should focus on prevalent products or specific companies' product lines. A positive beginning using this approach is a National Center for Preservation Technology and Training grant made to the National Council for Preservation Education in 1994 to support master's thesis research on the history and conservation of twentieth-century materials. This grant will, for example, enable a graduate student studying Formica to develop state-of-the-art techniques for conserving this well-known decorative plastic laminate.

Other topics that deserve immediate study include asbestos products, paint coatings and other finishes, ceramic veneers and clay tile panels, more recent flat glass products, modern woods, such as plywood, and a host of products made of laminated sandwich construction. Research on these and other topics can be developed with assistance from professionals in related fields, including conservators facing similar challenges with twentieth-century artifacts made of plastics, metal alloys, and laminated woods. Materials scientists at universities and private laboratories are another relatively untapped source, as indicated in the appendix.

With the publication of this introduction to the subject of twentieth-century building materials, the National Park Service seeks to promote awareness of the history of building materials from this century, present current information on known repair techniques, and identify many of the conservation challenges that lie ahead.

Thomas C. Jester
Editor

# ACKNOWLEDGMENTS

Edited works are by their very nature collaborative efforts. Thus, this book would not have been possible without the invaluable assistance of many people.

Shortly after joining the National Park Service's Preservation Assistance Division in 1991, H. Ward Jandl shared with me a conceptual outline for a guide to twentieth-century building materials. Aside from his interest in modern buildings, Ward recognized the importance of this subject to the future of historic preservation. Ward's support of my efforts to make this book a reality was unwavering, and I am deeply saddened that his death in March 1995 prevented him from seeing its completion.

I would like to thank each contributor for participating in this project and for each's assistance and patience during the editorial process. Each author's passion for his or her topic was evident and exciting, even when that passion translated into longer essays that did not fit into the allotted space. Without the expertise of the contributors and the willingness of some individuals to research specific topics, this book would not have been possible. Thanks are due to Flora A. Calabrese; Irene J. Cohen; James D. Connolly; Adrienne B. Cowden; Carol J. Dyson; Susan M. Escherich; David C. Fischetti; Sidney Freedman; Rebecca Gallagher; Paul E. Gaudette; Edward A. Gerns; Carol S. Gould; Anne E. Grimmer; William G. Hime; Harry J. Hunderman; Mike Jackson; Walker C. Johnson; Bruce S. Kaskel; Stephen J. Kelley; Paul D. Kofoed; Kimberly A. Konrad; Ann Milkovich McKee; Robert W. McKinley; Andrew McNall; Kathleen Catalano Milley; Dietrich Neumann; Howard Newlon, Jr.; William J. Nugent; Sharon C. Park; Conrad Paulson; Michael J. Scheffler; Robert Score; John C. Scott; Carolyn L. Searls; Pamela H. Simpson; Amy E. Slaton; Deborah Slaton; Bonnie Wehle Parks Snyder; Jerry G. Stockbridge; Nicole L. Stull; Anne T. Sullivan; Michael A. Tomlan; Derek H. Trelstad; Anthony J. T. Walker; Anne E. Weber; David P. Wessel; and Kenneth M. Wilson.

I would also like to acknowledge the peer reviewers who provided comments on portions of the manuscript and ensured the technical accuracy of the manuscript. Peer reviewers included Robert Alexander, Arthur B. Dodge, Jr., William Coney, Steve Gebler, Leon Glassgold, Anne E. Grimmer, Paul Hollister, Stephen Kelley, Jeffrey Meikle, Doug Knuth, Bruce Mitchell, Christian Overland, Brian Parkyn, Percy Reboul, Andreas Jordahl Rhude, Maurice Rhude, William Scarlet, Amy E. Slaton, Pamela Simpson, Julie Sloan, Michael A. Tomlan, Derek H. Trelstad, Anne E. Weber, Bob Wessel, and David P. Wessel.

I am indebted to a number of professional colleagues who made significant contributions. First and foremost, I must thank Michael Tomlan, who worked closely with me from the project's inception, providing advice on the book format and content and reviewing many of the essays. In addition to providing this invaluable assistance, Michael managed to find time to contribute the historical overview. I am also grateful for the considerable advice, encouragement, and editorial assistance I received from Michael Auer and Carol Gould, two colleagues at the National Park Service, and from Derek Trelstad and Deborah Slaton.

My sincere appreciation is extended to Diane Maddex, president of Archetype Press, who initially expressed interest in the publication, identified a publisher, and patiently assisted me in bringing this book to fruition. I am truly indebted to my editor at Archetype Press, Gretchen Smith Mui, with whom it was a pleasure to work. Gretchen skillfully edited the manuscript, worked with me to select illustrations, and was my alter ego at times. The publication's striking design and distinctive typeface were developed by Robert Wiser. Working with such a talented graphic designer was a delightful experience. Thanks also to Kristi Flis and Christina Hamme for their assistance preparing the manuscript. I am also grateful to Joel Stein, our editor at McGraw-Hill, for his commitment to this project.

The National Park Service permitted me to devote an extraordinary amount of time and resources to this book project, for which I am thankful. In addition to financial assistance for research, the project benefited immensely from a number of interns who made important contributions. My personal appreciation is extended to Karin Link, Adrienne Cowden, Rebecca Gallagher, Kimberly Konrad, and Nicole L. Stull. Dahlia Hernandez is also to be recognized for typing and checking the accuracy of portions of the manuscript.

I am grateful to Richard Cheek for making his impressive personal collection available as a source for illustrations. Richard and his wife, Betsy, welcomed me into their home on several occasions. Their hospitality and good humor made these research trips enjoyable as well as rewarding. Richard Longstreth also generously allowed me to use trade catalogues in his personal collection. I would also like to thank Jack Boucher of the Historic American Building Survey for contributing a number of important color and black-and-white photographs, and Carol Highsmith for working with me to take the photograph used on the book jacket.

I wish to thank the following individuals for helping identify illustrations for the book: Eileen Flanagan, Chicago Historical Society; John Ferry, Buckminster Fuller Institute; Amy Richert and Susan Wilkerson, National Building Museum; C. Ford Peatross, Prints and Photographs Division, Library of Congress; Robert Alexander and William Scarlet, William H. Scarlet and Associates; Georgette Wilson, Historic American Building Survey; Daniel Zilka, American Diner Project; James Roan, Museum of American History Branch, Smithsonian Libraries, National Museum of American History; Susan Lewin, Formica Corporation; Harry Warren, Masonite Corporation; Barb Bezat and Alan Lathrop, Northwest Architectural Archives, University of Minnesota; Rose Fronczak, Armstrong World Industries; Andreas Jordahl Rhude and Maurice Rhude, Sentinel Structures; Michelle Justice, Forest History Society; Katherine Hamilton-Smith and the staff of the Curt Teich Postcard Archives, Lake County Museum; Pete Peterson, Gladding McBean; Lawrence Difilippo, Venice Art Terrazzo Company; Kate Kelley, White Castle Systems; Barbara Floyd, Ward M. Canaday Center, University of Toledo; Laura Harmon, Linda Hall Library; George Nasser of the Precast/Prestressed Concrete Institute; Robert A. Bell Architects; Kristin and Corry Kenner; Derek H. Trelstad, *Building Renovation Magazine*; Robert Armbruster, Baha'i House of Worship; Eldon Davis, Armet and Davis; Don Chase, library, Department of the Interior; David Degrosso, Kentile; Theo Prudon; Randy Juster; Bob Barrett; Eldon Hambel; Mary Swisher; Laura Culberson; Dick Ryan; Jack Denzer; John C. Scott; David Wessel; Balthazar Korab; Kimberly A. Konrad; Conrad Paulson; Deborah Slaton; David C. Fischetti; Scott Bader; Sharon Park; David Bell; Michael Scheffler; Carolyn Searls; Dietrich Neumann; Carol Dyson; Bonnie Wehle Parks Snyder; and Mike Jackson.

Professional colleagues who provided encouragement, answered many questions, made suggestions, and shared information with me include Donald Albrecht, Bruce Barton, Tim Barton, Maribel Beas, Joan Brierton, Heather Burnham, Gordon Bock, Dinu Bumbaru, Michael Calafati, Woodrow Carpenter, David Chase, Blaine Cliver, Joel Davidson, Albert G. H. Dietz, Judith Fagan, Charles Fisher, Milan Galland, Robert Gutman, T. Gunny Harboe, Elizabeth Harris, Blake Hayes, Judy Hayward, Elizabeth Igleheart, Larry Karr, John Lauber, Antoinette J. Lee, Frederick Lindstrom, Michael Lynch, Frank Matero, Robert Mitchell, Al O'Bright, John Oliver, Kathleen Randall, Lori Plavin Salganicoff, Tim Samuelson, Leslie Schwartz, Eric Schatzberg, Rebecca Shiffer, William Smith, Sally Sims Stokes, Diane Tepfer, Emily Thompson, Thayer Tolles, Kay Weeks, Sara Wermeil, and Scott Zimmerman.

We are all molded by past mentors, and I would be remiss if I did not mention Charles Bassett and David Lubin, two inspiring American studies professors at Colby College; Earle G. Shettleworth, Jr., director of the Maine Historic Preservation Commission; and David De Long and Sam Harris, professors in the Graduate Program in Historic Preservation at the University of Pennsylvania.

During the past two years family and friends took an interest in my work and provided encouragement, for which I am grateful. I especially thank my parents for their love, support, and advice along the way. This book was to have been published at about the same time my daughter was born, but books have a way of taking longer than expected, and Hallie is now taking her first unassisted steps. I would not have survived the past two years without my wife, Jen Giblin. Through all the emotional highs and lows, she was supportive, encouraging, and always good-humored, especially when indulging my frequent requests to pull off the road so I could photograph a building. This book is a tribute to Jen and Hallie, who are my greatest sources of joy from one day to the next.

# TWENTIETH-CENTURY BUILDING MATERIALS: A PHOTOGRAPHIC ESSAY

**Plate 1.** Built for the 1939 New York World's Fair, Demonstration Home no. 2 (Lawrence Kocher) highlighted plywood's potential use in residential construction (opposite). Douglas fir plywood was used for the exterior walls, interior walls, and roof sheathing. Lithograph by Burland Printing Company, 1939.

# the town of Tomorrow

DEMONSTRATION HOME No. 2

NEW YORK WORLD'S FAIR 1939

PLYWOOD HOUSE

TRAY
UTILITY RM. 7'-0" x 7'-3" HEAT
SINK
KITCHEN 7'-8" x 9'-0"
RANGE
DINETTE 8'-0" x 7'-0"
CLOS.
COUNTER CASES
CUPBD.
BOOK SHELVES
PORCH SCREENED 7'-0" x 16'-0"
LIVING ROOM 19'-4" x 16'-2"
DESK
COATS
TERRACE
DRESSER
CLOSET
LINEN
CLOS.
CLOS.
BED ROOM 10'-0" x 9'-8"
BATH 5'-8" x 7'-0"
BED ROOM 10'-0" x 11'-3"
DRESS. T.

No. 2

## THE HOUSE OF PLYWOOD

*Sponsored by*

**Douglas Fir Plywood Association**

Crane Co.
General Electric Company
Johns-Manville
Johnson Wax Products
National Better Light-Better Sight Bureau
New York Telephone Company
Orange Screen Company
Truscon Steel Company
Wall Paper Institute

**Furnishings by Modernage**

*Architect:* A. Lawrence Kocher, 4 Park End Place, Forest Hills, N. Y.

**Plate 2.** The Paramount Theater (1931, Timothy Pflueger and James Miller), Oakland, Calif., was a billboard itself. The two 20-by-120-foot murals of the god and goddess of cinema, separated by an aluminum sign, were executed in glazed mosaic terra cotta tiles.

**Plate 3.** Perma-Stone's resemblance to quarried stone lent ordinary houses an aura of permanence (above). Perma-Stone was sold and applied by licensed dealers nationwide in a range of colors created with mineral pigments.

**Plate 4.** Glazed structural clay tiles, generally produced in 5-by-8- and 5-by-12-inch sizes, gained favor in the early 1930s for building exteriors and interiors (right). *Sweet's Architectural Catalogues,* 1932.

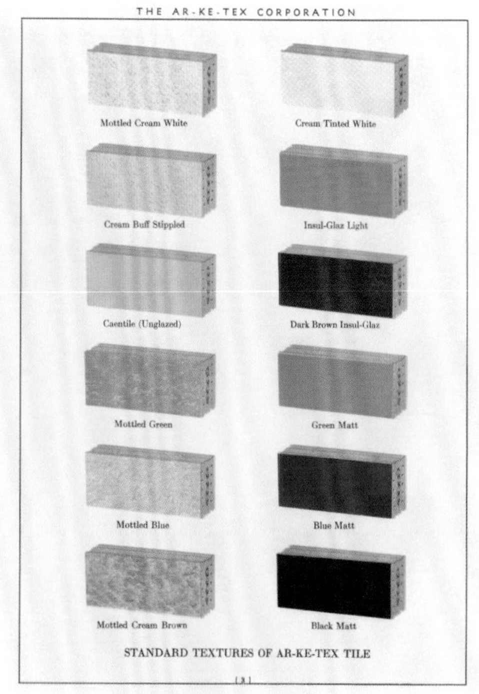

THE AR-KE-TEX CORPORATION

Mottled Cream White

Cream Tinted White

Cream Buff Stippled

Insul-Glaz Light

Caentile (Unglazed)

Dark Brown Insul-Glaz

Mottled Green

Green Matt

Mottled Blue

Blue Matt

Mottled Cream Brown

Black Matt

STANDARD TEXTURES OF AR-KE-TEX TILE

**Plate 5.** An intricate stainless steel sculpture, "Spirit of Light," adorns the Niagara-Hudson Building (1930, Clayton Frye and Albert Rumschik), Syracuse, N.Y. (left).

**Plate 6.** The cast metal door panels of the apartment building at 3 East 84th Street (1928, Howells and Hood), New York City, were made of Monel, a nickel-copper alloy manufactured by the International Nickel Company (bottom left).

**Plate 7.** Cast aluminum grilles with floral details were used for the entrance bay of the Seattle Art Museum (1932, Charles Bebb and Carl Gould) (bottom right).

**Plate 8.** Eero Saarinen's inventive use of modern materials—a thin reinforced concrete shell and aluminum-framed glass panels—contributes to the dramatic form of the Kresge Auditorium (1955) at the Massachusetts Institute of Technology, Cambridge, Mass.

**Plate 9.** Streamlined interior designs using modern materials were popular for restaurants during the 1930s. Metalwork was often featured prominently, as in this stainless steel bar front in Whitey's Cafe, East Grand Forks, Minn.

**Plate 10.** Walter Gropius selected mass-produced materials, including cork tile, glass block, and steel windows, when he designed his Lincoln, Mass., house (1938, with Marcel Breuer). When the house was restored in 1986, the deteriorated cork tile floor was replaced with new tile that matched the original.

MAKE **MODERN**

BATHROOMS OUT OF OLD

One thing that so often gives away the age of a house is the bathroom. Yet with two simple, inexpensive changes, any bathroom can be made really attractive. First hide the old worn floor under a fresh sparkling pattern of Armstrong's Linoleum. Then, right over the old plaster or wallboard, install Armstrong's Linowall. This new washable wall material (see pages 22 and 23 for the story and patterns) has all the good housekeeping advantages of Armstrong's Linoleum. It's easily installed, easy to clean, and long lasting. Join the wall and floor with Armstrong's Cove and Base, and presto! you have a new bathroom—and a much more livable house. In the room above, Linowall No. 721 covers the old plaster, while Armstrong's Straight Line Inlaid Linoleum No. 247 transforms the floor.

TEN

**Plate 11.** A geometric terrazzo floor complements the Art Deco showroom of the Auburn Automobile Company administration building (1930, A. M. Strauss), Auburn, Ind. (top left).

**Plate 12.** Lively terrazzo designs in building entrances were commonplace during the 1930s. The abstract terrazzo design of this San Francisco store entrance combines beige, green, blue, red, and yellow sections (top right).

**Plate 13.** Linoleum was touted as fashionable not only for kitchens and bathrooms but also for living rooms and bedrooms (left). Armstrong helped its customers plan new and remodeled rooms using linoleum. *Floors That Keep Homes in Fashion,* Armstrong Cork Company, 1937.

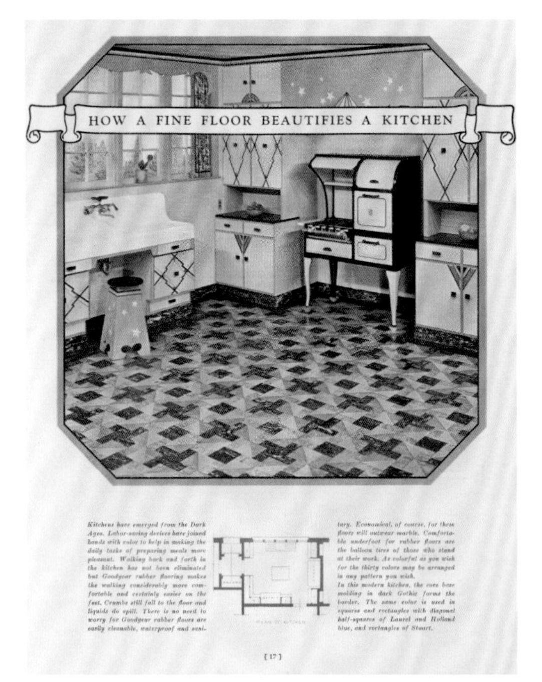
HOW A FINE FLOOR BEAUTIFIES A KITCHEN

*Kitchens have emerged from the Dark Ages. Labor-saving devices have joined hands with color to help in making the daily tasks of preparing meals more pleasant. Walking back and forth in the kitchen has not been eliminated but Goodyear rubber flooring makes the walking considerably more comfortable and certainly easier on the feet. Crumbs still fall to the floor and liquids do spill. There is no need to worry for Goodyear rubber floors are easily cleanable, waterproof and sanitary. Economical, of course, for these floors will outwear marble. Comfortable underfoot for rubber floors are the balloon tires of those who stand at their work. As colorful as you wish for the thirty colors may be arranged in any pattern you wish.*

*In this modern kitchen, the cove base molding in dark Gothic forms the border. The same color is used in squares and rectangles with diagonal half-squares of Laurel and Holland blue, and rectangles of Stuart.*

[ 17 ]

**Plate 14.** Rubber tile was marketed as a sanitary, economical, and colorful flooring material for kitchens (left). *Goodyear Rubber Floors: Architect's Reference Book,* Goodyear Rubber Company, 1929.

**Plate 15.** Amtico Permalife tile, a vinyl composition tile manufactured by the Biltrite Rubber Company, Trenton, N.J., was composed of vinyl chloride resin, plasticizers, abrasion-resistant fillers, and nonfading pigments (bottom left).

**Plate 16.** The wide variety of marbleized rubber tiles available in the late 1920s allowed architects to combine tiles to create unusual floor patterns and color schemes (bottom right). *Goodyear Rubber Floors: Architect's Reference Book,* Goodyear Rubber Company, 1929.

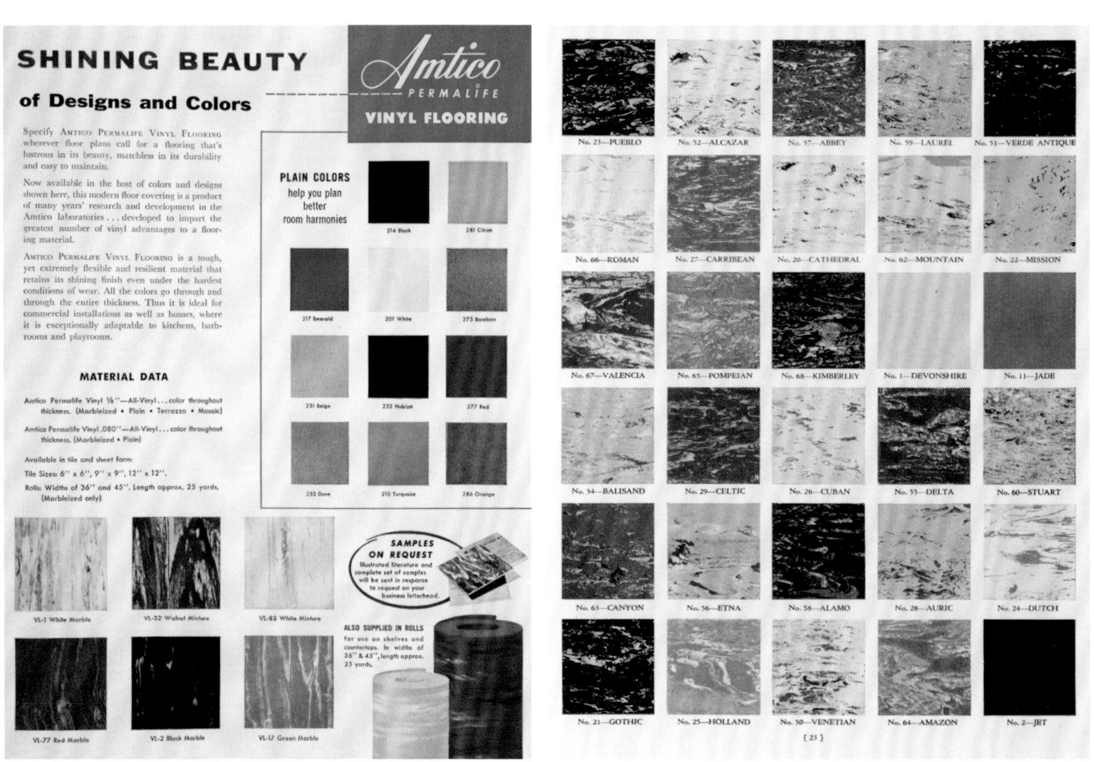
SHINING BEAUTY

of Designs and Colors

Amtico PERMALIFE

VINYL FLOORING

Specify AMTICO PERMALIFE VINYL FLOORING wherever floor plans call for a flooring that's lustrous in its beauty, matchless in its durability and easy to maintain.

Now available in the host of colors and designs shown here, this modern floor covering is a product of many years' research and development in the Amtico laboratories . . . developed to impart the greatest number of vinyl advantages to a flooring material.

AMTICO PERMALIFE VINYL FLOORING is a tough, yet extremely flexible and resilient material that retains its shining finish even under the hardest conditions of wear. All the colors go through and through the entire thickness. Thus it is ideal for commercial installations as well as homes, where it is exceptionally adaptable to kitchens, bathrooms and playrooms.

MATERIAL DATA

Amtico Permalife Vinyl ⅛"—All-Vinyl . . . color throughout thickness. (Marbleized • Plain • Terrazzo • Mosaic)

Amtico Permalife Vinyl .080"—All-Vinyl . . . color throughout thickness. (Marbleized • Plain)

Available in tile and sheet form:

Tile Sizes: 6" x 6", 9" x 9", 12" x 12".

Rolls: Widths of 36" and 45". Length approx. 25 yards. (Marbleized only)

PLAIN COLORS help you plan better room harmonies

214 Black, 281 Citron, 217 Emerald, 201 White, 275 Bonbon, 231 Beige, 232 Nubian, 277 Red, 233 Dove, 210 Turquoise, 286 Orange

VL-1 White Marble, VL-32 Walnut Mixture, VL-83 White Mixture, VL-77 Red Marble, VL-2 Black Marble, VL-17 Green Marble

SAMPLES ON REQUEST Illustrated literature and complete set of samples will be sent in response to request on your business letterhead.

ALSO SUPPLIED IN ROLLS for use on shelves and countertops. In widths of 36" & 45", length approx. 25 yards.

No. 23—PUEBLO, No. 52—ALCAZAR, No. 57—ABBEY, No. 59—LAUREL, No. 51—VERDE ANTIQUE

No. 66—ROMAN, No. 27—CARRIBEAN, No. 20—CATHEDRAL, No. 62—MOUNTAIN, No. 22—MISSION

No. 67—VALENCIA, No. 65—POMPEIAN, No. 68—KIMBERLEY, No. 1—DEVONSHIRE, No. 11—JADE

No. 54—BALISAND, No. 29—CELTIC, No. 26—CUBAN, No. 55—DELTA, No. 60—STUART

No. 63—CANYON, No. 56—ETNA, No. 58—ALAMO, No. 28—AURIC, No. 24—DUTCH

No. 21—GOTHIC, No. 25—HOLLAND, No. 50—VENETIAN, No. 64—AMAZON, No. 2—JET

[ 23 ]

**Plate 17.** Goodyear rubber tile could be custom ordered in odd sizes and strips to create floor patterns. Some patterns were simple enough to be installed by homeowners, while complicated designs were executed according to measured drawings. *Sweet's Architectural Catalogues,* 1931.

**Plate 18.** The machine-age aesthetic exemplified by the Hecht Company Warehouse (1937, Abbott, Merkt), Washington, D.C., derived largely from the extensive use of Insulux glass block, produced by the Owens-Illinois Company (top).

**Plate 19.** Plate glass, the highest quality glazing, was a necessity for display windows. Blumer's Bakery, Chicago, was distinguished by its curved plate glass windows and mottled porcelain enamel panels (bottom).

**Plate 20.** The Greyhound Bus Depot (1939, George D. Brown), Columbia, S.C., is a tour-de-force of twentieth-century building materials. On top of the reinforced concrete foundation, bands of glass block alternate with stainless steel. The mirrored blue and ivory Vitrolite structural glass color scheme above the aluminum marquee was developed for Greyhound by the industrial designer Raymond Loewy.

**Plate 21** The polychromatic lobby ceiling (1934, John J. Earley) of the Department of Justice building, Washington, D.C., features a series of thin architectural precast concrete panels (above).

**Plate 22.** Celotex fiberboard products, made from bagasse (sugar-cane fibers), included sheathing, wallboard, hardboard, insulating lath for plaster, and acoustical tile (left). *Celotex Manual for Architects,* The Celotex Corporation, 1937.

**Plate 23.** Formica covers the counters and booth tables at the Modern Diner (1940), Pawtucket, R.I., one of the Sterling Streamliners built by the Judkins Company (top).

**Plate 24.** In his design for the Beth Sholom Synagogue (1954), Elkins Park, Pa., Frank Lloyd Wright explored the translucent qualities of plastic and glass (bottom). Free of internal supports, the main sanctuary has a tripod frame covered by a double layer of translucent wire glass outside and corrugated, fiber-reinforced plastic inside.

**Plate 25.** Sheetrock, the United States Gypsum Company's trademarked gypsum board, was economical to install and provided a fireproof enclosure for wood walls. *Sheetrock: The Fireproof Wallboard,* United States Gypsum Company, 1937.

Suggested Stencils for Pre-decorated Acousti-Celotex

1 2 3 4

5 6 7 8

9 10 11 12

13 14 15 16

17 18 19 20

43

**Plate 26.** The design for this prefabricated Lustron house (1950, Morris Beckman), Chesterton, Ind.—number 2,329 off the production line—featured 2-foot-square porcelain enamel panels screwed to wall studs (above). All the components needed to construct a Lustron house were delivered on one tractor trailer.

**Plate 27.** Painted and stenciled Acousti-Celotex tiles, offered by the Celotex Corporation in the late 1920s, afforded greater light reflection than unpainted acoustical tile without decreasing sound absorption (left). *Acousti-Celotex,* The Celotex Corporation, 1927.

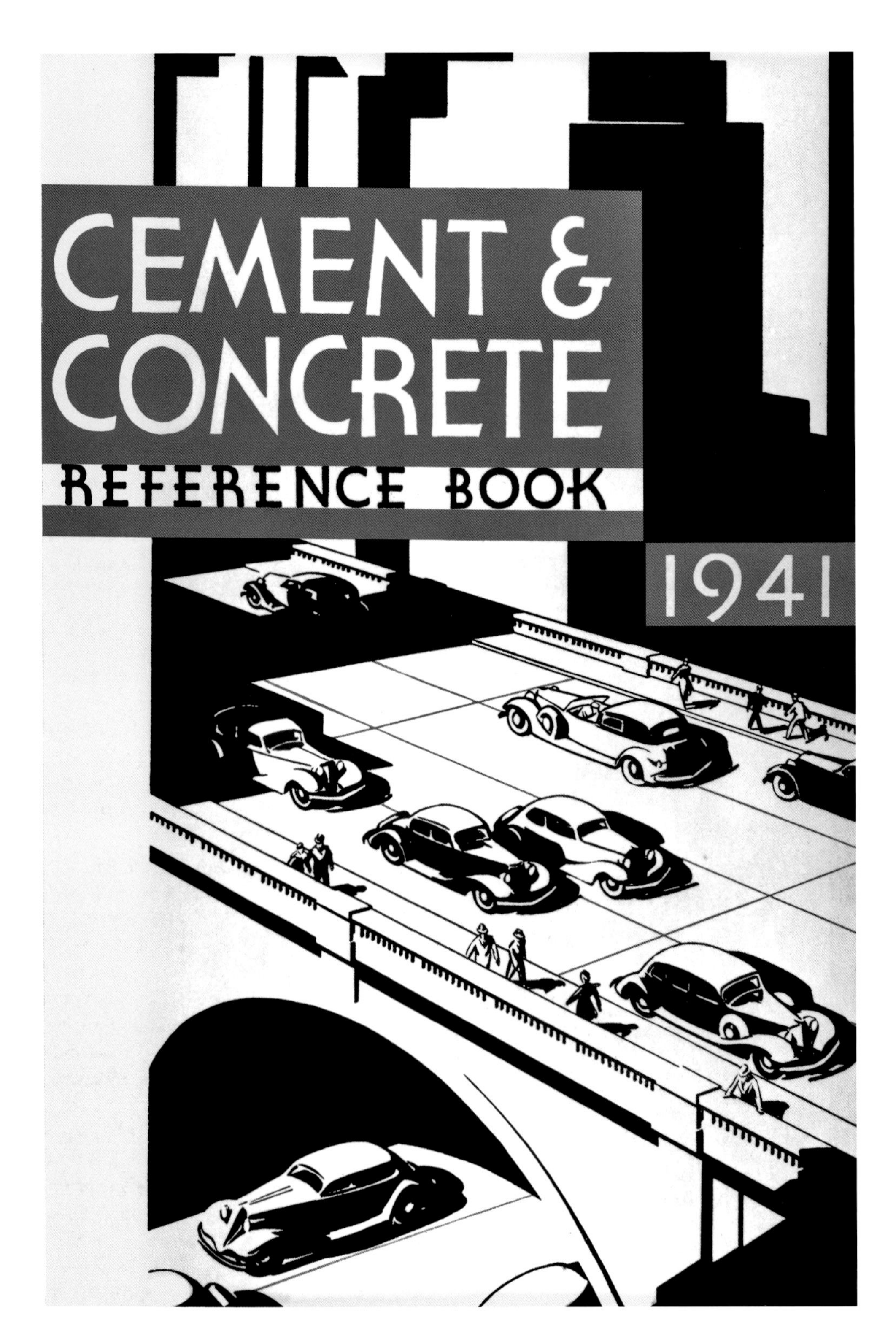


**Figure I.1.** The proliferation of technical literature and manuals on building materials in the twentieth century is unparalleled. The *Cement and Concrete Reference Book,* first published by the Portland Cement Association in the late 1920s, provided architects and engineers with the most up-to-date information available. *Cement and Concrete Reference Book,* 1941.

MICHAEL A. TOMLAN

# INTRODUCTION

## BUILDING MODERN AMERICA: AN ERA OF STANDARDIZATION AND EXPERIMENTATION

Historians often overlook the fact, but a country's use and development of building materials is a historic and cultural characteristic as distinctive as any other. In the United States the most commonly used building materials are wood and brick. In fact, 86 percent of all buildings are wood frame and are mostly residential structures, while about 4 percent are unreinforced brick buildings devoted primarily to commercial, institutional, and industrial use. A few other industrialized countries have used these materials in roughly the same percentages, but America's distinctive character becomes obvious when a third type, the metal structure, is considered.

The high-rise skeleton structure is the most significant contribution the United States has made to the world's building history. Although metal structures were introduced only in the last half of the nineteenth century, today they make up 7 percent of America's built environment and account for more than 11 percent of the square footage enclosed.[1] Both metal skeletons and balloon-frame structures became widespread because their parts were standardized and they were easy to assemble.

The development of building materials may be considered evolutionary rather than revolutionary. The gradual industrialization of the trades was possible because mass-produced building elements cost less and, thanks to the railroad, were distributed to ever more distant markets. The scarcity of labor and the low cost of materials in the United States continually tempted builders to try the latest, most efficient technological developments. Because a new material was often presented as a substitute for a traditional one, almost every experimental substance was given at least a trial installation. Substitution was so commonplace and change orders were so frequent that they were seen as characteristic of the American building industry.

Faith in science and technology paralleled America's belief in capitalism. The idea that large corporations could be more efficient producers than small businesses was widely accepted in manufacturing and set the tone for the building industry. The success of Carnegie Steel and Standard Oil in harnessing science to produce better products was hard to ignore. Perhaps the best known pacemaker in modern research and development was the DuPont Company, whose leadership in munitions depended on a laboratory that employed hundreds of people who had earned doctorates in chemistry.[2] The university laboratory was already being linked to the production line, and such linkage became a hallmark of large corporations.

### DAWN OF THE CENTURY: AN ERA OF EXPERIMENTATION

The twentieth century made use of traditional building materials—wood, brick, and stone—but their physical properties were often inadequate for modern needs. Traditional materials were subject chiefly to mechanical manipulation—cutting, turning, planing, and finishing. Increasingly steam was used for more than powering machinery; it helped form the product. Early-twentieth-century building products were often composed of more than one substance and were ground, mixed, heated, and pressed before being finished. In some cases several materials were sandwiched with adhesives; the combination was intended as a substitute, a better and more economical product than that in use.

America's abundant natural resources, both above and below ground, continued to provide the basis for building products. As these raw materials were exploited and trade expanded, advanced steel, concrete, and glass products quickly entered the market.

As architects dreamed of ever higher structures, engineers repeatedly turned to modern metallurgy. Whereas once only four or five kinds of steel were produced, by the 1920s a whole range of alloys were available.[3] These allowed the engineer to meet ever-widening needs more accurately and economically. The specialized treatment of metal provided a model for the development of other materials. Metal trade associations set voluntary

**Figure I.2.** To keep ahead of competitors, the Armstrong Cork Company designed and installed a new rotary machine to manufacture straight-line linoleum in 1920. Machinery used to produce building materials became increasingly mechanized in the early twentieth century, lowering the cost of many.

product standards with the help of professional societies, which lent time and expertise, particularly the influential American Society of Mechanical Engineers. The need for standardization was driven partly by commercial imperatives. For example, the American Society for Testing and Materials (ASTM), founded in 1898, came into being largely because of the need to standardize the quality of steel rails.[4] Commerce stimulated engineering advances, and the number of civil, mechanical, electrical, and chemical engineers boomed accordingly.[5]

One almost wholly twentieth-century phenomenon is the development of portland cement. The U.S. production of natural hydraulic cement peaked in 1899, and thereafter it was increasingly supplanted by portland cement. In 1902 the first ASTM standards for gray portland cement were published, and when white portland cement came on the market a few years later, the visual character of concrete was broadened considerably. The early use of lightweight aggregates, especially cinder ash, the introduction of aeration, the use of shale in the 1920s and expanded slag in the 1930s—all increased the range and application of concrete. Empirical testing was increasingly superseded by scientific formulation. Although portland cement was deemed acceptable for storage structures, chiefly silos and water tanks, it was not considered an appropriate material to be cast in place for walls until the second half of the century, when a general change in aesthetic sensibilities occurred.

## WORLD WAR I: HARNESSING RESOURCES

Although the United States was not directly involved in armed intervention during the first three years of the war, its transition to a wartime economy stimulated production. Supplying the Allies in Europe soon affected prices. The number of contracts dramatically increased in 1915, as steel mills were converted to produce armaments and munitions. In 1916 a survey of more than 18,000 industrial plants revealed America's manufacturing strengths and weaknesses, and an advisory council

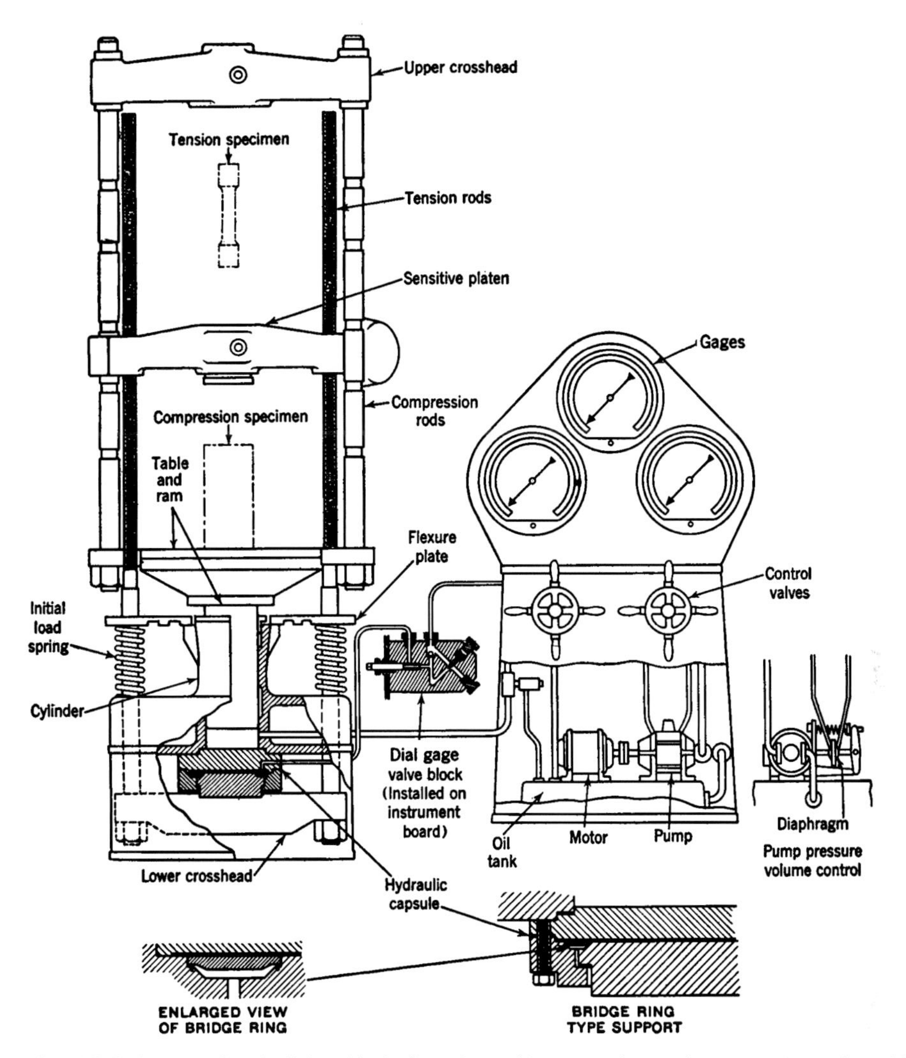


**Figure I.3.** Equipment such as the Universal hydraulic testing machine was used to test the compressive, tensile, and flexural strength of construction materials. Testing conducted by engineering departments in universities as well as in government and private laboratories expanded knowledge about materials and helped verify standard performance requirements.

was created as the liaison between government and industry.[6] This led to the establishment of the War Industries Board on July 28, 1917, whose duty was to stimulate production of industries making essential materials while protecting and restraining the others. Iron, steel, brass, and copper were especially valuable for the war effort, as were a number of industrial chemicals necessary for the production of munitions and explosives. In addition, fuel supplies and the use of railroad cars came under close scrutiny.

The nationwide shift in priorities was achieved primarily through voluntary cooperation. While some shortages in trained personnel were obvious, the fact that the United States produced most of the material to meet its domestic building necessities meant that there was no noticeable dearth of lumber or building products, despite the need for billions of feet of lumber to build housing, hospitals, and warehouses for training camps and airplanes.[7] When the board's building materials division was formed in March 1918, it began to monitor the needs for portland cement, brick, tile, and gypsum products. Only with gypsum products—wallboard and plasterboard—did the government allocate the industry's entire output for the construction of fireproof housing and storage facilities.[8]

The building materials division's chief contribution was that it advanced conservation by establishing schedules of standard specifications for war building projects.

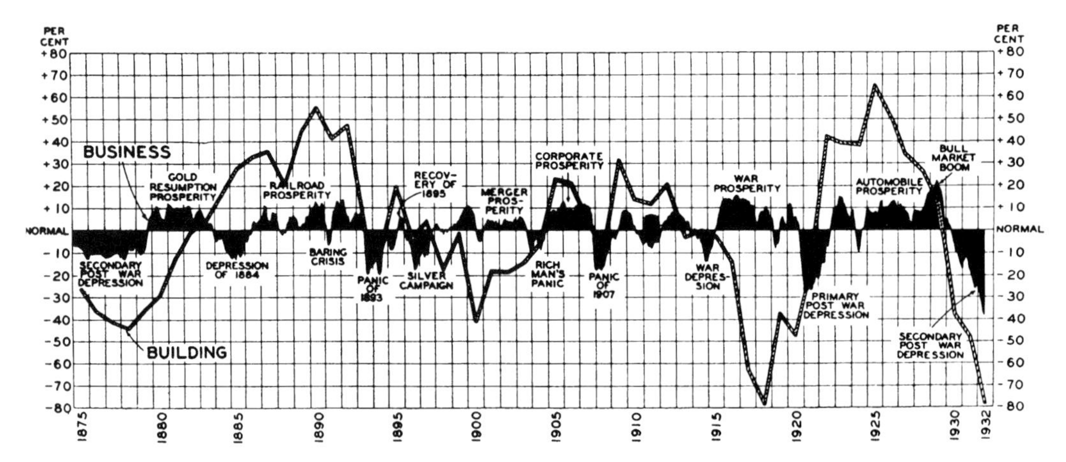


**Figure I.4.** Building activity has been characterized by cyclical peaks and valleys, as this graph of U.S. building cycles from 1875 to 1932 shows. Many twentieth-century building material advances occurred during periods of prosperity.

Standardization was accomplished in carpentry and millwork, roofing, fiberboard, hardware, plumbing, gas fitting, heating, electric wiring, lighting, painting, hollow building tile, stucco work, and fire prevention and protection. Although in many instances the division merely followed industry's prescribed procedures, the federal government's recommendations had an increasingly important role in the postwar economy.[9]

The materials developments of World War I found their way into peacetime use almost immediately afterward. Plywood was first recognized as anything more than a paneled veneer only during the closing days of the war, when airplane fuselages were made of this forest product. It gradually found general acceptance in the furniture industry and later as sheathing for house construction. Laminated glass, first developed to strengthen military vehicles' windshields against shrapnel bullets and rifle shells, was adapted for automobile windows and doors.[10] Micarta, the Westinghouse trade name for its phenolic resin laminate, was first used in the airplane industry for lightweight pulleys for the control wires that linked the cockpit to the rudder, ailerons, and throttles.[11] Micarta and Formica were used in electrical devices and gears before being developed as decorative veneers.

The high production levels of wartime goods inevitably led to the problem of disposing of government surplus after the war. The need for gypsum dropped considerably in 1918 and 1919, although it rose precipitately during the subsequent unprecedented building boom.[12] Meanwhile, tradition continued unbroken in the plastering trade. Well into the 1930s plasterers' apprentices continued to be taught that it was necessary to apply coats of wet plaster over boards.

## THE ROARING TWENTIES: EXPANDING PRODUCT LINES

The U.S. economy's tremendous expansion during the 1920s boosted the use and development of building materials. Before this period the study of construction in architectural education included a traditional litany of materials: cementitious binders (plaster, lime, and cement); concrete, stone, brick, and clay products; ferrous and nonferrous metals; timber; and mechanical fabrics (rope).[13] During the 1920s more scientific concerns—thermal insulation, acoustical control, and lighting—were added, although the nature of the materials was constantly evolving.

The use of the steel frame accelerated the need for exterior cladding that not only was securely fixed but also provided effective insulation. Even in domestic construction during the 1920s, the widespread use of thermal insulation reflected a new concern about environmental control. Slag wool, developed in the nineteenth century, was superseded by asbestos, promoted as the best alternative by heating engineers who dealt with the control and handling of steam.[14] Where refrigeration was needed, cork was preferred. Although fiberboard continued to be touted as the most cost-effective insulation, shredded and powdered fillings became more common after World War I, and batts were introduced, some with reflective foils of aluminum and copper.[15]

A scientific understanding of sound was necessary to create the best acoustical materials. Although Wallace Sabine made major theoretical advances early in the century and helped develop porous clay tile, not until the 1920s did the rising concern for noise abatement in urban settings lead acoustical scientists to address the problem and design perforated tiles. Soon thereafter, acoustics took up another challenge—the projection of sound with amplifiers and speakers, which required further product development.[16]

Advances in thermal insulation and acoustics were less obvious than those in lighting. Whereas in 1910 only one home in seven had electricity, in 1930 seven in ten boasted it. Although its impact on the development of building materials may be comparatively slight, the widespread use of electricity changed their manufacture and installation remarkably. The introduction of the small electric power saw, drill, sander, and associated appliances allowed contractors and home handymen to work faster anywhere and any time a light bulb was available.[17]

As might be expected, building materials development during the 1920s benefited from the enormous increase in industrial research. Figures compiled by the National Research Council indicate that in 1920 about 300 companies in this country employed 9,300 research personnel; by 1930, 1,625 companies employed 34,200 research personnel.[18]

The 1920s were also considered the golden age of standards, as the federal government joined with trade associations and professional groups, intensifying the effort to standardize products. In contrast to most countries, standard writing in the United States took place in the private sector. Trade associations were interested in standardization as a way to promote their products. Professional organizations also contributed to this effort as a means of educating their constituency and the public. Because most of the members were employed by large corporations, the standards often reflected the position of the firms that enjoyed a large share of the market.[19] Perhaps most visible, however, were the three nonprofit technical societies: the ASTM, the National Fire Protection Association (NFPA), and the Underwriters' Laboratory (UL). The ASTM made a considerable contribution to the dissemination of knowledge by publishing its *Standards,* containing thousands of pages devoted to all aspects of the building fabric. The NFPA's concern for fire prevention led it to sponsor standards in construction, many of which were adopted in federal, state, and local legislation.

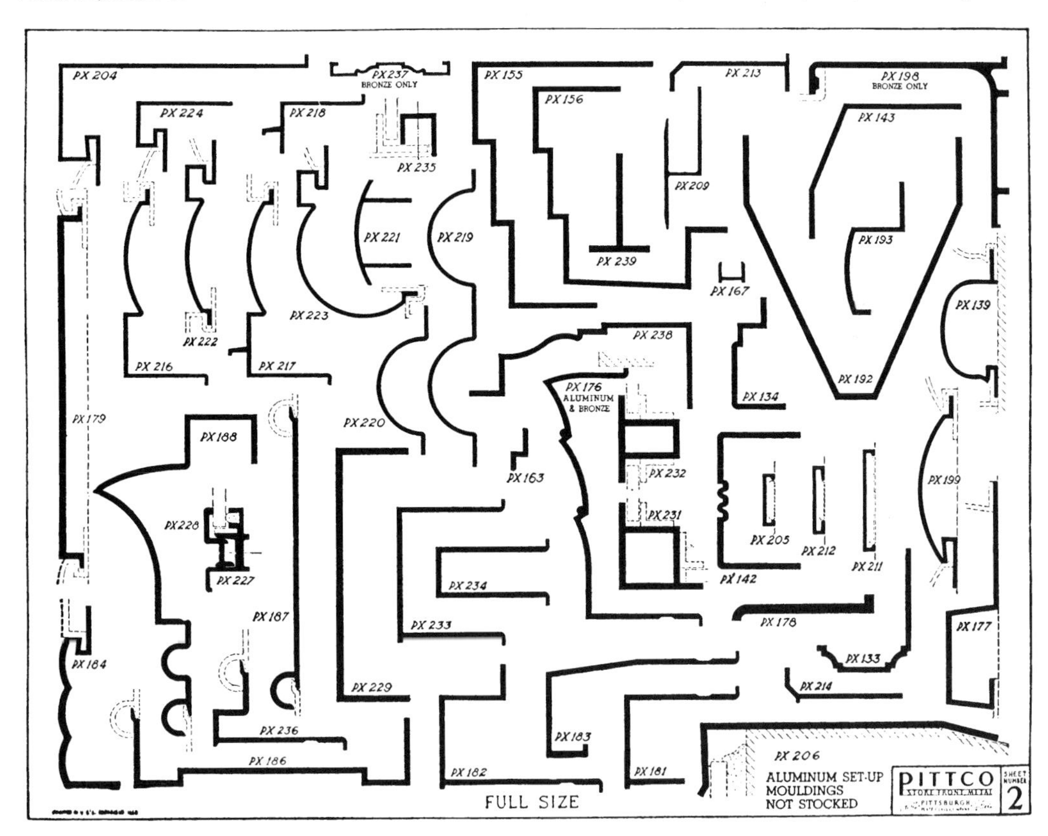


**Figure I.5.** The Pittsburgh Plate Glass Company offered industrially produced aluminite and architectural bronze moldings for storefronts under the trade name Pittco. Architects and builders could choose from a nearly endless array of shapes from a catalogue to create "modern" storefronts.

The UL, meanwhile, maintained in Chicago one of the country's oldest certification laboratories, noted for its role in preventing electrical fires.[20]

### THE GREAT DEPRESSION: SHRINKING MARKETS

Economic activity declined markedly from 1929 until 1933, the low point for the construction industry, and increased only slowly until after 1936, when building activity almost equaled what it had been in the 1910s. New housing was promoted as the mechanism to jump-start the economy, but the shift from large, elaborate houses to smaller ones with fewer rooms for ever-smaller families meant that the impact of any new housing would be comparatively modest.

Available financing was limited, and the U.S. government was slow to back building loans.[21] Once it became involved, its concerns for health, safety, and financial accountability began to make a difference in construction. The Federal Housing Administration set several materials standards to ensure the soundness of its projects. The Appraisal and Reconditioning Division of the Home Owners Loan Corporation was even more specific: It preferred gypsum, not fiberboard, in the rehabilitation of structures.[22] Meanwhile, the National Bureau of Standards, the fact-finding organization known for its *Journal of Research,* began to publish Building Materials and Structures Reports, a new series to further knowledge of the more technical aspects of materials, test procedures, and results.[23]

The dearth of building activity during the 1930s thinned out the number of raw materials suppliers and building product manufacturers. Those that survived expanded and consolidated, often by absorbing smaller firms. The National Gypsum Company, for example, expanded vigorously during the 1930s, growing from three plants to sixteen. It absorbed many of its smaller competitors and raw materials suppliers to produce gypsum board liner. In the face of declining new construction, the firm vigorously promoted its products as the answer for home improvement and commercial remodeling.[24]

Economic conditions themselves promoted standardization. Manufacturers reduced the number of raw materials needed in production, narrowed their product lines, and simplified their pricing structure. Acceptance of some new materials remained a problem. Although the *Sweet's* compilation of trade catalogue information assisted professionals, building products manufacturers redoubled their efforts to reach the public directly by lavishly advertising in popular magazines and hiring traveling salesmen. Even comparatively well-known new materials such as Masonite and Celotex had to be aggressively promoted to remain viable.[25]

Manufacturers also displayed their wares at expositions and fairs. New materials were presented as part of the average homeowner's dreams of a better tomorrow. The 1933 Century of Progress Exposition in Chicago, featuring the House of Tomorrow, and the 1939 New York World's Fair, showcasing the House That Chemistry Built, both provided the public with an idea of how science would transform the future.[26]

The changes were slow in coming, however. Although research activity increased in the late 1930s and a number of new materials were introduced, few were quickly adopted. Plexiglas, advanced by Rohm and Haas in 1936, had to prove itself in the aircraft industry before it was used as glazing in schools and industrial buildings, where impact resistance was a major concern.[27] Heat-absorbing glass found acceptance in medical circles but was not widely adopted in construction for another decade.[28]

By the end of the 1930s the tacit assumption on the part of many progressive thinkers in the construction industry was that scientific research coupled with mass-production techniques could create housing for all of America's citizens.[29] The introduction of the International Style, which emphasized twentieth-century materials and simplified forms, reinforced this thinking in the architectural community. These aesthetic preferences gained popular acceptance, but largely because many mass-produced building materials were soon needed for a completely different purpose.

### WORLD WAR II: CRITICAL NEEDS LEAD TO NEW MATERIALS

In comparison with its stance before entering World War I, the measures the U.S. government put in place before the Japanese attack on Pearl Harbor in December 1941 were much more deliberate in redirecting and accelerating production. By early 1940 foreign purchase of U.S. armaments and supplies was recognized as the United States became the "arsenal of democracy."[30] The building industry felt the shift to a wartime economy in June, when legislation was passed that empowered the Reconstruction Finance Corporation to purchase stock in any company producing strategic or critical materials and to aid plant construction and expansion.

Rubber and metals were the first materials to come under scrutiny and control. With shipping supplies endangered, the creation of synthetic rubber was a major national priority. The development of two new products

**Figure I.6.** A 1955 Owens-Illinois promotional photograph of glass block highlighted the translucent qualities as well as the allure of the company's product.

during the 1930s—Neoprene, by the DuPont Company, and Butyl, by the Standard Oil Company of New Jersey—provided the starting points for synthetic rubber.[31] The need for this product was so great that the government backed new research and production facilities built by the U.S. Rubber Corporation.[32] Contractors, lacking steel and copper, made adjustments. Concrete was poured without the benefit of much reinforcing, reflective foil on insulation was banned, and fiberglass was substituted for the more valued asbestos. The problem of spanning great distances was often met by using glued laminated timber.[33] Where wood paneling might be preferred, plywood would serve.

After the creation of the Supply Priorities and Production Board in 1941, construction unrelated to the war effort was prohibited. This ban was mainly symbolic, however, as the need for steel and copper in defense production was already so great that their use in domestic construction had come to a halt.[34] The War Production Board, a new council that superseded the Supply Priorities and Production Board, brought together representatives of the Departments of War, Navy, and Commerce with the leaders of industry. Cooperation was crucial, for it was widely accepted that mass production would be a major factor in winning the war.[35]

The manufacture of aircraft, ships, tanks, guns, munitions, and supplies created an unprecedented need for new structures. The United States, which had built about 30 airports before 1940, constructed another 1,284 during World War II, each with runways 3,500 feet or longer. While 9.4 million square feet of space were devoted to airplane assembly before the war, another 175 million square feet were constructed to meet the demand.[36]

The expansion of plants and facilities, although impressive, was only part of the story. The facilities that sprang up to train military personnel—in essence, small cities in locations where none existed before—pressed even further the need for building materials, which peaked in 1942. In the first nine months of the war, for example, fifty camps and cantonments were constructed, each with miles of roads flanked by barracks, mess halls, hospitals, shops, chapels, and theaters, all supported by self-contained systems to supply electrical power, heat, water, and sewerage.[37]

When the country began to turn its thoughts to peace, it was apparent that traditional building materials such as brick and stone were in short supply, while newer materials were available in abundance.[38] The two giants of the wallboard industry, the National Gypsum and United States Gypsum Companies, benefited from the need to enclose enormous war plants and provide housing. In fact, gypsum board completely eclipsed the use of metal lath as a plaster base during the war, not only because the prefabricated boards were faster to erect but also because the use of steel in construction was so severely constricted. Drywall construction put traditional plasterers out of work almost completely.[39]

The United States was always recognized as the largest consumer of asbestos.[40] During the war, however, the use of cement-asbestos siding and roofing rose to new levels, primarily as a result of the need to enclose munitions supplies with an easily assembled, inexpensive fireproof material. The Johns-Manville and Pabco Companies anticipated a rosy future; however, rising health concerns about the material in the early 1960s curtailed its use.[41]

Aluminum is perhaps the most obvious example of a material that was exploited to unprecedented levels during World War II. Although large-scale refining of bauxite was possible in the late nineteenth century, aluminum remained generally more costly than steel through the 1920s and 1930s. As a result, its use in buildings was restricted to trim or alloys. This period saw the production of laminates of two aluminum alloys in which the exterior coating would clad the core material to provide corrosion protection, a development that would spur similar research with other metals.[42] However, the need for aircraft during World War II led to increased production and the invention of sandwiched panels, providing a postwar surplus.[43] Aluminum manufacturers went in search of new markets, hailing the ease with which the metal could be fabricated. They entered the construction industry with extruded sections for door and window frames and aluminum siding as a substitute for asphalt and asbestos veneers. The Alcoa Building (1953, Harrison and Abramovitz) in Pittsburgh seemed to point to a future of high-rise buildings clad in light metal.[44]

## THE 1950S AND 1960S: NEW IDEAS, NEW CHALLENGES

The military and economic outcome of World War II left the United States as the dominant global power, although its rivalry with the Soviet Union often dominated the concerns of political decision makers. On the domestic front, after a brief recession, the economy began to expand and with it the building products industry. Savings accumulated during the war were used to purchase new domestic goods, the financial benefits granted to veterans stimulated a boom in suburbanization, and one of the most prosperous periods in American history began. Construction expenditures rose nearly every year from

**Figure I.7.** For many years one hallmark of *Architectural Record* was the cartoonist Alan Dunn's lampooning of architectural developments—here, the use of plastic building materials and unconventional styles.

1946 until 1969, the longest continuous period of growth in the nation's history.[45]

The belief prevailed that science and technology, when applied to mass production, would meet almost every need. In addition, it was widely held that government should take an active role in economic affairs and sponsor research and development, often in concert with universities.[46] Most noticeable were the federal government's new public housing, urban renewal, and transportation improvement programs. Whether for high-rise apartments, interstate highways, or airport runways, the material of choice was concrete. Research and development of building materials expanded, particularly in applications of prestressed concrete, as the nature of the matrix and the conditions under which it cured were better understood. In urban areas prefabricated concrete structures were often made of precast concrete panels lifted in place by tower cranes, first seen in the war-ravaged inner cities of Europe. And for the first time carpenters found that they were often able to make more money by leaving aside their traditional skills to provide the formwork for pouring walls and decks.[47]

The other technological alternative for high-rise construction was the steel-and-glass box, the expression of an aesthetic that valued transparency as much as form. Curtain wall construction developed rapidly throughout the 1960s, but the 1973 energy crisis created a groundswell of opinion against glass walls that was never reversed. Energy consumption became increasingly important with the widespread use of air-conditioning, developed before the war but largely restricted to office buildings and hospitals. Improved interior environmental controls called for a keener understanding of scientific principles, more attention to site, and improved insulation materials.

The daring experimental architecture of the 1950s often depended on improved structural and mechanical systems, as architects and engineers enclosed space with cantilevers, plate and shell structures, tents and hung roofs, geodesic domes, space frames, and inflatable forms. The roof often became the structure, covered with rubberized membranes or fiberglass-reinforced resins that contrasted starkly with the pitched roofs of previous decades.[48]

Meanwhile, national security continued to depend on the production and supply of critical materials. Although the focus was often on radioactive materials, perhaps the most crucial was oil. Research in petroleum chemistry provided an ever-widening number of plastics.[49] In the 1950s interior finishes often sported decorative laminates, a range of synthetic materials bonded with resins and sandwiched together under heat and pressure to complete the polymerization process. Injection blow-molding provided an even wider range of products, introducing polyvinyl chloride, polystyrene, and the polyolefins.[50] For several years the fear of fire and toxic gases led government agencies to restrict the use of plastics in the construction industry to surface coatings, only gradually allowing polyvinyl chloride piping for waste water.[51] Mastics and sealants were accepted more easily, as was the shift from polysulfides to urethanes and silicones at the end of the 1960s.[52]

The space age ushered in a new period of scientific development, driven by further military research and a higher set of expectations. At the same time materials science became recognized as a separate discipline, distinct from and yet dependent on all engineering sciences—chemistry, physics, metallurgy, mechanics, and crystallography. Building materials of the last half of the twentieth century are often composites or synthetics developed to respond to the needs of the project, not simply what nature provided and humans manipulated or treated in some basic fashion.

The evolution of the building materials industry will continue. The number of wood and brick structures will likely decrease while the number of metal and concrete structures will rise. Likewise, the principal influences on the use and development of building materials will be affected by military needs, while the distribution will be governed by the economics of supply and demand. Scientific experimentation will continue to expand the range of possibilities, while standardization will seem to limit the choices.

Given the numerous alternative materials and the press of economic imperatives, the temptation will be even greater to replace rather than repair or restore twentieth-century structures. Preserving these structures in the future will also be more difficult because of the number of building materials that are no longer manufactured. The integrity of twentieth-century structures must be protected. To do so we need to understand the history and development of the building materials of the recent past.

# PART I
# METALS

ALUMINUM

MONEL

NICKEL SILVER

STAINLESS STEEL

WEATHERING STEEL

December, 1930 THE ARCHITECTURAL FORUM 11

Alcoa Aluminum shows its adaptability to decoration

**Figure 1.1.** Elaborate cast, forged, and extruded aluminum elements were manufactured for the entrances, doors, and grilles of the New York Trust Company Building (1930, Cross and Cross), New York City. *Architectural Forum,* December 1930.

STEPHEN J. KELLEY

# 1 | ALUMINUM

**DATES OF PRODUCTION**
c. 1884 to the Present
**COMMON TRADE NAME**
Alclad

## HISTORY

One of the most abundant metallic elements near the earth's surface, aluminum occurs naturally as alum, a double salt (potassium aluminum sulfate dodecahydrate or potash alum). The use of alum for mordant dyeing of fabrics was described in Roman times. Aluminum is not found in a natural metallic state and remained undiscovered until it was obtained from bauxite ore.[1] Aluminum alloys were developed to achieve a harder and stronger material. Lightweight, workable, and resistant to corrosion, aluminum became popular in contemporary building construction for decorative metalwork, spandrel panels, curtain walls, windows and doors, architectural trim, and siding.

## ORIGINS AND DEVELOPMENT

In 1807 Sir Humphrey Davy, a British scientist, tried unsuccessfully to obtain aluminum from alum. Convinced that the base material was metal, he coined the term *aluminum* for this yet undiscovered element. Metallic aluminum was first isolated by Hans Christian Oersted of the University of Copenhagen in 1825 by heating a mixture of aluminum chloride and potassium amalgam and removing by-products by distillation. The product was a small amount of impure aluminum.[2] By 1852 the cost of aluminum was $545.00 per pound.[3]

In 1854 Henri Sainte-Claire Deville, a French scientist, and Robert Von Bunsen, a German scientist, working separately, obtained quantities of aluminum of 96 to 97 percent purity.[4] In 1884 a pyramidal shaped aluminum cap was cast for the Washington Monument and is considered the first recorded architectural application of aluminum in the United States.[5]

In 1886 Charles Martin Hall of Oberlin College discovered that metallic aluminum could be produced by dissolving alumina in molten cryolite and then passing an electric current through the solution. Paul T. Héroult of France also independently discovered this electrolytic process the same year.[6] The Hall-Heroult process made it possible to produce metallic aluminum easily in sufficient quantities and at reasonable prices. As a result of this discovery, the cost of aluminum dropped from $11.33 per pound in 1885 to $0.57 per pound by 1892.[7] The Pittsburgh Reduction Company, later known as the Aluminum Company of America (Alcoa), was formed in 1888 to promote the Hall-Heroult process, which is still used today.[8]

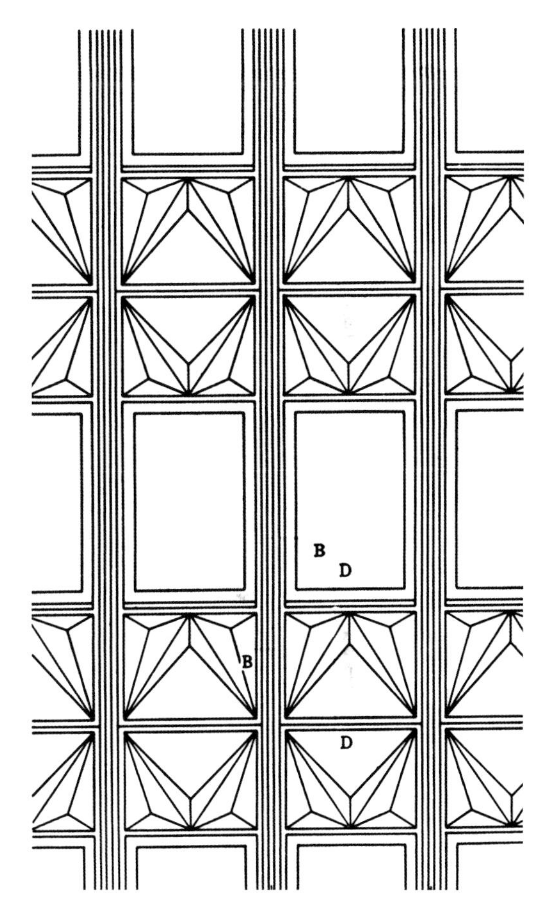


**Figure 1.2.** The stamped "prismatic" panels of the Republic National Bank (1955, Harrison and Abramovitz), Dallas, were insulated and sealed with aluminum foil.

Three Chicago buildings were among the earliest to use cast aluminum: the Venetian (1892, Holabird and Roche), Isabella (1892, Jenney and Mundie), and Monadnock (1893, Burnham and Root). In these structures cast aluminum was used for interior applications such as stairs, railings, and elevator metalwork.[9]

### MANUFACTURING PROCESS

Aluminum has been used in cast and wrought forms.[10] Of these processes, extruding has become the most widely used for aluminum. The initial concept of metal extrusion antedates the nineteenth century. By 1894 the American G. A. Dick had designed a 600-ton press for the hot extrusion of brass rod. Dick also extruded aluminum on an experimental basis. By 1900 the hydraulic extrusion press had been developed and was found capable of extruding copper, brass, aluminum, and other materials.[11] Aluminum extrusion processes were used to fabricate windows, doors, handrails, and the like, which appeared on the market and gained in popularity in the 1920s.

Aluminum alloys, which enhance properties such as strength, hardness, and workability, were first developed for use in the British Vickers Mayfly rigid airship in 1911. In 1917 the German Hugo Junkers developed an all-metal monoplane using an aluminum alloy for the tubular structure and corrugated sheet skin.[12]

Heat treating of aluminum alloys was found to increase strength, and heat-treated alloys were first manufactured in the United States during World War I to produce high-strength, lightweight structural material for aircraft.[13] Heat-treatable, intermediate-strength alloys with good extrudability were introduced after World War II.[14] Since the 1950s aluminum alloys have been classified as cast and wrought, and as heat-treatable and non-heat-treatable. Non-heat-treatable alloys contain trace amounts of manganese and magnesium and are used for roofing and cladding. Heat-treatable alloys contain magnesium, silicone, and sometimes copper and are used for windows, curtain walls, fasteners, and structural members.[15]

Aluminum finishes can be categorized as mechanical (grinding, buffing, blasting, and hammering), chemical (bright dips and frosted finishes), electrochemical (anodizing and electroplating), porcelain enamel coating, and organic coatings (lacquer, varnish, and paint).

Anodic coatings for aluminum, created by building up the natural aluminum oxide coating in an electrochemical bath, were first developed in the early 1920s. Almost simultaneously it was found that dyes could be used to color the coatings.[16] Anodizing was used mainly for electrical applications and did not become available for architectural purposes until after World War II. Anodized coatings became popular because of their outstanding resistance to atmospheric corrosion. Anodic coatings may be transparent or of varying degrees of silver, gray, or brown. They are also highly absorptive for dyes and paints.[17]

Other common postwar finishes for aluminum included porcelain enamels, which were first applied successfully around 1950, and baked-on enamel coatings, an outgrowth of the aluminum siding industry that developed in the early 1950s.[18]

### USES AND METHODS OF INSTALLATION

Following World War I aluminum and its alloys in cast, sheet, and extruded form were used for a wide variety of architectural purposes. Aluminum was used for doors, window sash, railings, trim, grilles, and signs and was particularly popular because of its color.[19] The 80-foot-high German Evangelical Church (1927, Henry Hornbustel) in Pittsburgh featured a spire reminiscent of European cast-iron church spires. The spire was constructed of cast aluminum bolted to a steel frame.[20]

Cast- or pressed-sheet aluminum spandrel panels were used in numerous structures during the 1920s and 1930s, including the Cathedral of Learning (1925, Day and Klauder) in Pittsburgh, the Chrysler Building (1930, William Van Alen), and the Empire State Building (1931, Shreve, Lamb, and Harmon), both in New York City. In these buildings, the spandrels are set into a masonry backup in a manner similar to stone or brick spandrels. Designers were beginning to realize the promise of a lightweight wall that could be easily fabricated and erected.[21]

One of the earliest buildings to be clad entirely in aluminum was the A. O. Smith Corporation Research and Engineering Building (1930, Holabird and Root) in Milwaukee. Sheet aluminum was used to face the walls, cornices, and parapets, while extruded aluminum was used for window and door frames and some exterior detailing. Aluminum was also used extensively for interior trim.[22]

A drop in consumption of aluminum in the late 1930s, a reflection of the Great Depression, was followed by a marked increase in consumption in 1939 and a continued upward trend through World War II.[23] During the war, aluminum production in the United States steadily increased to a peak of 835,000 metric tons—more than six times its prewar production. More than two hundred extrusion presses were in operation at this time to produce aluminum shapes for the war effort.[24]

Wartime research for the aircraft industry expanded knowledge about aluminum alloys and their properties.[25] Advances during World War II brought about new

**Figure 1.3.** The Alcoa Building (1953, Harrison and Abramovitz), Pittsburgh, was one of the first to use prefabricated curtain wall panels.

processes and techniques for fabricating and working aluminum and made unprecedented quantities of the material available for construction when the war ended.[26]

By 1952 aluminum production had surpassed wartime levels.[27] In addition to its use for storefronts, windows, and hardware, aluminum became an important component of the glass and metal curtain wall. The Equitable Building (1948, Pietro Belluschi) in Portland, Oregon, one of the first postwar skyscrapers, illustrated the aesthetic and technical potential of aluminum in combination with glass. Taking advantage of aluminum that had been stockpiled for World War II and assembly techniques derived from airplane assembly plants, the spandrel panels were sheet aluminum and the frames were extruded.[28]

The Alcoa Building (1953, Harrison and Abramovitz) in Pittsburgh was a reaction against the completely glazed curtain wall that had become popular and proclaimed the possibilities of an aluminum-clad curtain wall. The cladding consists of one-story-high prefabricated panels of pressed and anodized sheet aluminum penetrated by windows.[29] This type of curtain wall became prevalent in the 1960s.[30] The U.S. Air Force Academy (1962, Skidmore, Owings and Merrill) in Colorado Springs, Colorado, also used aluminum expressively.

By the 1950s aluminum had become a standard building material for a range of applications. And relatively new applications, such as residential siding and sun screens, gained momentum to complement existing uses.

**Figure 1.4.** Industrial aluminum roofing and siding for structures like this Tulsa, Okla., warehouse (top) were marketed by companies such as Kaiser Aluminum. Selling points included the material's low maintenance, light weight, and strength.

**Figure 1.5.** Aluminum curtain wall panels can be cleaned by hand, primed, and recoated with a custom metallic product to match the original appearance. The right panel shows deterioration; the left, the appearance after cleaning and recoating (bottom).

## CONSERVATION

Pure aluminum is extremely resistant to corrosion because of the transparent yet tenacious aluminum oxide film that forms on its surface. Aluminum alloys containing magnesium, manganese, chromium, or silicon are also highly resistant to corrosion. Alloys containing substantial proportions of copper are more susceptible to corrosion, depending markedly on the heat treatment.[31]

### DETERIORATION

Corrosive agents that break down the natural oxide film include hydrochloric acid, chlorides, sulfates, dirt, and high humidity.[32] Aluminum is also subject to corrosion when wet and in contact with certain alkalines, such as concrete, mortar, and plaster. Some forms of wood, wallboard, flooring, and insulation materials may also attack aluminum.[33] Most soils in direct contact with aluminum will cause corrosion. Acid-based chemicals used to clean facades and masonry surfaces should not be allowed to come into contact with aluminum.[34]

Once oxide films become porous, pollutants and corrosion products prevent access to oxygen so that the protective oxide film cannot form, accelerating corrosion.[35] This commonly occurs when unsealed joints in aluminum assemblies form crevices where moisture can penetrate and pool.[36]

Another common form of corrosion is galvanic corrosion. When two dissimilar metals are coupled in a conducting solution such as water, the metal having the most negative potential will be anodic to the other and an electric current will flow from it through the solution to the other metal. During galvanic action, the anode, typically the aluminum material, dissolves. Galvanic corrosion can be most readily seen in roofing applications where aluminum comes into contact with ferrous and cuprous metals. Aluminum and its various alloys can also be anodic or cathodic with each other.

As a ductile metal, aluminum and any protective oxide films or finishes are subject to erosion caused by airborne abrasives. This is a particular problem with aluminum roofing and high-rise installations.

## DIAGNOSTICS AND CONDITION ASSESSMENT

Because a variety of aluminum production methods, alloys, tempers, finishes, and treatments have been used historically, understanding these variables is important to develop sound conservation approaches. Available construction documents should be reviewed to understand the type of alloy and finish, and the building's maintenance record should also be researched.

Visual examination can help determine or confirm production methods and finishes. Castings are gray and uniform in surface appearance; hot-rolled products show a certain amount of discoloration or darkening, while cold-finished products have a whiter, brighter surface. Extruded products show traces of longitudinal striations caused by the extrusion process. Anodized finishes are generally clear compared to paint finishes. Laboratory analysis can be used to confirm alloys, tempers, and coatings but requires destructive testing.

## CONSERVATION TECHNIQUES

Common maintenance for aluminum includes cleaning the surface and applying a watertight coating. As a result of weathering and corrosion, unfinished aluminum naturally acquires a gray patina, an appearance that may or may not be acceptable. Periodic cleaning will prolong the service life of aluminum by retarding corrosion. Aluminum oxide formed on unprotected aluminum protects the underlying metal but also provides a base for accumulation of airborne dirt particles.[37]

Cleaning unfinished, weathered aluminum can be the most difficult because no cleaning procedure will remove the damage of significant corrosion caused by pitting or surface roughening. Only abrasive refinishing will restore the original surface appearance.[38] Aluminum with an applied surface treatment is easier to maintain and much easier to clean.

The effects of different chemical cleaners on aluminum are variable and depend on many factors, including chemical concentration and temperature, as well as the aluminum alloy type, production method, surface finish, and coatings. A cleaner's pH is of little value in predicting its effects on aluminum alloys.[39] However, strong alkaline and acidic cleaners should be avoided.

Mild cleaners (soaps, detergents, solvents, and emulsion cleaners) will remove loosely adhered soil but not tenacious soils, stains in the aluminum oxide film, or the patina. Moderate cleaners (more aggressive soaps and detergents) will remove most soils not removed by mild cleaners. Heavy-duty cleaners (abrasive cleaners and polishes and etching-chemical cleaners) can clean heavily soiled aluminum in the field but can, however, affect the appearance of aluminum alloys and damage surface treatments.[40] Mechanical cleaning, when other means are unsuccessful, can be considered. Plastic abrasives delivered at a low pressure (15 pounds per square inch), for instance, have been used to remove porous oxide layers without damaging the integrity of conversion coatings.[41]

Although steam alone has little effect on aluminum, it is sometimes used in combination with other cleaners to remove corrosion products.[42] Pitted or roughened surfaces can be renewed by wire brushing or blasting. These techniques may permanently alter the metal's surface or damage finishes and, therefore, are not typically recommended. The recommended method of determining the appropriate cleaning method is sample cleaning. Tests covering a range of concentrations should be implemented on the various alloys, finishes, and coatings represented on the building. After cleaning, the surfaces should be examined for evidence of damage. Cleaning methods that pit, stain, discolor, etch, or roughen the surface are unsuitable. Likewise, methods that alter or remove historically significant coatings should be avoided. The gentlest effective method should be selected.

Once cleaned, bare, anodized, or conversion-coated aluminum should be maintained with a wax or varnish. Waxing, however, can increase the buildup of surface dirt. Before coating, the surface should be thoroughly cleaned.[43] Aluminum that has been damaged by abrasion or corrosion can be protected from further damage by coatings of varnish or lacquer. Removable coatings help protect aluminum from alkaline materials, atmospheric corrosion, and abrasion. Satisfactory transparent coatings of this type should not noticeably discolor, peel, or chalk and will eventually wear away.[44] To protect aluminum from galvanic corrosion, it should be separated from dissimilar metals by nonconductive materials.

Restoring damaged anodized aluminum finishes acceptably in situ is difficult, if not impossible.[45] When aluminum is painted, product selection, surface preparation, and application method are critical. Surface preparation may be as basic as wiping the surface with a solvent-soaked rag and applying suitable primers. If preparation requires surface roughening, such as chemical etching or mechanical scarification, it may not be appropriate to apply a coating.

Applying paint to factory-painted aluminum can prove difficult and will require preparation and coordination with the paint manufacturer. Field painting of aluminum will increase the need for future maintenance. Painting surfaces that were not originally coated will alter the surface's appearance and is not generally recommended.

Where aluminum comes into contact with any material that may cause damage, such as masonry, concrete, or soil, it should be separated from these materials with compatible coatings that are resistant to these conditions. In all conditions where coatings are being considered, they should be coordinated with the coatings manufacturer and augmented by laboratory and field testing.

### REPLACEMENT

If damage is extensive, aluminum components may require in-kind replacement. Finding replacements for aluminum in cast and extruded forms is possible, although shapes no longer available may be expensive to recreate. Partial replacement of aluminum shapes in high-performance windows and curtain walls may affect the window performance.

In rare cases, aluminum sections may be repaired by welding. Pure aluminum and low-manganese alloys are weldable. Wrought and most cast aluminums can be welded, but die castings are difficult to weld. Alloys that are heat treated lose strength during welding.[46]

DEREK H. TRELSTAD

# 2 | MONEL

**DATES OF PRODUCTION**
1907 to the Present
**COMMON TRADE NAMES**
Monel®, Monel-Plymetyl

## HISTORY

Monel, an alloy of 68 to 70 percent nickel, 25 to 29 percent copper, and small percentages of iron, manganese, silicon, and carbon, was used in a wide range of architectural applications from 1909 to the mid-1950s.[1] One of several white metals popular in the first half of the century, Monel was stronger than steel, readily fabricated, and resistant to corrosion.[2] It also had a low coefficient of thermal expansion, which allowed it to be used in engineered assemblies in conjunction with steel and concrete without affecting tolerances or fit.[3] When first discovered, the major elements of this metal were present in the ore in the approximate proportion in which they were needed to form the alloy. Until this ore deposit was depleted, Monel was considered a natural alloy.[4] Today the metal, which is still available in sheet form for architectural applications, is alloyed from metallic nickel and copper.

### ORIGINS AND DEVELOPMENT

Monel originated in the mid-nineteenth-century nickel industry that developed around the ore deposits near Sudbury, Ontario, Canada.[5] That the large ore deposits in the southeastern reaches of the province contained nickel had been known since about 1863, but it was the iron in the ore that first interested investors.[6] Efforts to extract the iron were unsuccessful. By 1879 the metallurgy of nickel was more clearly understood, and the Orford Nickel and Copper Company had started shipping ore to Pennsylvania to be refined to metallic nickel. Orford's arrangement with the independent refiners in Pennsylvania soon proved unsatisfactory, and the company established a smelter in Bayonne, New Jersey.[7] In 1899 Robert Thompson, Orford's president, sought investors to expand the nickel operations, and on March 29, 1902, the International Nickel Company (Inco) was incorporated; its officers included Thompson and Ambrose Monell.[8]

In 1901 Robert Crooks Stanley joined the company.[9] Stanley was convinced that the major component metals of nickel silver (copper and nickel), which were both present in the ore Orford was extracting from its mines in Canada, need not be separated during the refining process or that a useful alloy could be extracted directly from the ores. To prove his hypothesis, Stanley ordered a boxcar load of low-sulfur Bessemer ore (a partially refined product) from Orford's Canadian operation. Stanley conducted several experiments on this ore and eventually stumbled on an alloy that was both malleable and tough.[10] The new metal was named Monel, in honor of the company's president.[11]

Inco began manufacturing sheet, plate, and mill goods. In 1907 the first experimental products were manufactured. In 1908 Monel sheet was installed as the roofing membrane on New York City's Pennsylvania Station (1910, McKim, Mead and White).[12] Despite this impressive first application, early sales of Monel were slow. Rolling mills, forges, and fabricators were loath to use the novel alloy, and the company's established distributors would not stock it.[13] In response, Inco formed the Bayonne Casting Company to provide an outlet for manufactured goods and began seeking other distributors. In 1918 Inco also established a rolling mill dedicated to producing Monel.[14]

In the 1920s, after Stanley had been promoted to president of Inco, he established the Development and Research Department to provide technical information to the trade press and to consumers of nickel products.[15] Advertising was integral to the company's efforts to market this metal.[16] Promotional campaigns stressed Monel's advantages: improved sanitation, corrosion resistance, low weight-to-strength ratio, and cost savings.[17] As a result, consumption figures for Monel rose from 3,000 tons in 1923 to 7,500 tons in 1925.[18]

Specifiers and manufacturers began integrating the metal into grocery coolers, countertops, sinks, laundry and food-preparation appliances, and roofing and flashing. To increase its domestic market share, Inco's advertisements highlighted "model kitchens" incorporating Monel sinks and Monel-trimmed appliances as a major factor in new home sales.[19] By 1929 the market for Monel

**Figure 2.1.** Monel sinks were heavily promoted as a sanitary, lightweight, and durable alternative to the copper, porcelain, and porcelain enamel sinks popular in the first decades of the twentieth century. Pantry sinks catalogue (Catalogue T), John Trageser Steam Copper Works, 1927.

and other nickel alloys had become so large that Inco and its main western competitor, the Mond Nickel Company, could no longer efficiently and independently extract ore from the Sudbury deposits.[20] In an effort to streamline their operations and maintain the steady expansion of the nickel market, the two companies merged in 1929. After the merger, Monel, a trademarked Inco alloy, was thrust into new markets under the Mond name.

In the 1950s Monel began to be displaced by stainless steel, which used a much lower percentage of costly nickel than Monel and could be fashioned into identical forms at a much lower cost. Nevertheless, many manufacturers who advertised in *Sweet's* catalogue offered Monel as an option long after Inco ceased advertising the material. Ultimately, Monel, which had pioneered some of the applications for "stainless" metals, was driven out of the market by a combination of steep competition with stainless steel and the U.S. government's interest in controlling the supply of nickel, which was considered a strategic metal.

### MANUFACTURING PROCESS

Monel can be formed using most common metalworking techniques except extrusion.[21] Sheet goods can be welded and brazed as well as soldered.[22] The same materials and techniques used for joining copper sheets can be used with Monel. As with other metals, the mechanical coupling between sheets—whether rivets, double-lock seams, or other joining methods—provides the greatest portion of the joint's strength.

A number of factory-applied finishes were available for Monel sheets and cast goods in the 1920s and 1930s. Monel was available as full-finished (hot-rolled) and cold-rolled sheets. Cold-rolled sheets, available in dead-soft, soft, skin-hard, quarter-hard, and half-hard tempers, among others, had a high roll-finished surface straight from the mill. This type was often specified for commercial cabinetry and equipment for hotels, restaurants, and hospitals.[23] A mirrorlike polish described by Inco as No. 5 was often used to contrast the satin finishes available on the full-finished sheets.

Full-finished sheets, unlike many of the cold-rolled sheets, were dead flat. A silver satin finish, known as No. 8, was often applied to these annealed sheets. Full-finished sheets were also factory polished, often with Monel wire brushes, to meet architects' finish specifications.

Castings also can be made readily. The alloys used for casting differ slightly from those used for sheet and bar goods.[24] Castings are first chipped and filed to remove all fins, risers, and other imperfections. The surface is then smoothed on a grinding wheel and a finish polish applied with a series of emery compounds, emery grease, Tripoli, and white finish buffs.[25]

Monel was easily forged and was used where iron or mild steel would typically be specified. It could not, however, be forge welded. New forgings had to be formed with care, using a low sulfur fuel.

Unlike iron and mild steel, no paint or lacquer was necessary to protect Monel from corrosive elements or to obtain two-tone effects. A tight, thin black oxide formed on the surface of Monel when it was heated; selectively polishing this surface brings out a two-tone finish.[26] Forged Monel can also be finished like bright nickel or with a pewterlike or satin finish.

Decorative finishes for Monel sheet, plate, rods and flats, angles, and castings included chemical etching, mechanical or hand engraving, and sandblasting. Polishing the sheets beyond the standard ground finishes was done with commercial polishing compounds.

### USES AND METHODS OF INSTALLATION

Architectural applications for Monel were diverse. Sheets were used for roofing. Both sheet and plate goods were used for heating and ventilation ducts, flashing, gutters and downspouts, countertops, letter boxes, mail chutes, laundry chutes, elevator fittings, lighting fixtures, and sinks and other appliances. Inco sold Monel Roofing Sheets specifically for roofing and provided customers with specifications for detailing and fastening the sheets. Monel's low coefficient of thermal expansion was an advantage in roofing applications; when used in skylights, this property meant fewer broken panes. When it was used as the primary roof membrane, the need for fewer expansion joints meant a tighter roof. After World War II the company introduced a crimped roofing sheet that was significantly more rigid and eliminated the need for expansion joints over long runs.[27] Monel sheet goods were also used for flashings and expansion joints: more than 16,000 lineal feet of Monel was used in expansion joints at Union Terminal (1930, Graham, Anderson, Probst and White) in Cleveland, Ohio. At the Lever House (1952, Gordon Bunshaft; Skidmore, Owings and Merrill) in New York City, more than 20 tons of Monel flashing sealed the facade.

Monel castings and forgings ranged from the purely ornamental to the pedestrian. The seven pairs of nickel silver doors at the United Nations General Assembly Building (1950, Wallace K. Harrison et al.) in New York City are decorated with cast Monel panels, designed by Ernest Cormier, and the lighting fixtures outside the entry of the church of St. Ignatius Loyola (1898, Maginnis

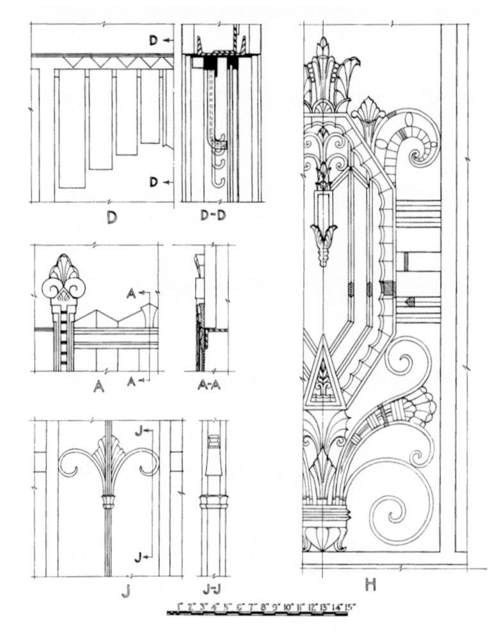

**Figure 2.2.** The Monel grilles of the Union Trust Building (1928, Smith, Hinchman and Grylls), Detroit, were created by etching and sandblasting (top).

**Figure 2.3.** Because Monel could not be extruded, decorative elements had to be built up from rolled stock, as was this ornament on the T. Eaton Department Store (1930, Ross and Macdonald, Sproat and Rolph), Toronto (bottom left).

**Figure 2.4.** Finishes and color for Monel elements varied from smooth, brightly polished surfaces to "hand-forged" black and two-toned finishes (bottom right).

and Walsh) in New York City are Monel. Castings were also used for rosettes, plaques, handrail fittings, lengths of molding, long thin castings (pilasters, mullions, and door jambs), and grilles. Monel forgings were used for hardware (bolts, hinges, locks, and ornamental trim) in bank vaults and safes, and on doors for handles and push plates.[28] Rod and bar stock can be found in bank screens (wickets), gates, directory boards, railings, and terrazzo floor strips.

Monel's unsurpassed corrosion resistance made it suitable for harsh environments and inaccessible locations. Monel Tie Wire was widely used in commercial structures to secure lathing for walls and suspended ceilings, fasten tile roofs, and anchor stone cladding to steel framework.

Where cost restrictions prohibited the use of solid-section Monel, a thin sheet of the metal could be laminated to a cheap backing material, gaining many of the advantages of Monel at a much lower cost. Monel-clad steel was one such material. Another was Monel-laminated plywood, sold in the United States under the trade name

Monel-Plymetyl. Monel-Plymetyl could be cut with a band saw, formed, and riveted, and was used commercially for doors, tabletops, paneling, shop faces, and bank vault linings.[29]

## CONSERVATION

The detrimental effects on Monel building elements caused by environmental pollutants are few. However, remedial treatments may be necessary for surface discoloration or more severe deterioration of the metal. It is important to select a treatment that is appropriately matched to the deterioration.

### DETERIORATION

In normal atmospheric conditions Monel develops a relatively stable patina. A light tarnish may be found on indoor installations of Monel, which can be removed with an occasional wiping, but often there will be no evidence of a patina. Outdoors, the metal will develop a patina that ranges from a light gray-green to a thin medium-brown film. In neutral environments where the metal is protected from rain, a coating of hygroscopic corrosion products will form. This coating will become moist on humid days. In marine environments, a smooth gray-green film develops.

The presence of large quantities of water, nitric oxides, and sulfur dioxide can have a profoundly corrosive effect on the alloy. Pitting may occur in stagnant saltwater exposures where marine organisms accumulate, but slows considerably after a rather rapid initial attack. Steam and hot water will not affect the metal in most instances; some combinations of steam, carbon dioxide, and air will cause significant corrosion.[30]

In concentrations found in the atmosphere, many common acids will not significantly affect Monel. The alloy's resistance to sulfurous acid varies considerably. In mild concentrations it resists attack by sulfurous acid. At room temperature Monel is resistant to solutions of hydrochloric acid that have a concentration of 10 percent or less. In the presence of oxidizing salts, corrosion can greatly exceed that caused by a 10 percent solution. In the absence of oxidizing salts, Monel is serviceably resistant to hydrofluoric acid. Stress corrosion cracking and accelerated surface corrosion may result if Monel is exposed to aerated hydrofluoric acid in moist conditions. Organic acids, such as acetic and fatty acids, have little effect on Monel. However, nitric and nitrous acids in solutions of less than 0.5 percent are severely corrosive to Monel at room temperature.[31]

Monel's resistance to alkalis is high; it resists corrosion by caustic soda (sodium hydroxide) at most concentrations. It is also resistant to anhydrous ammonia and ammonium hydroxide solutions up to a 3 percent concentration.

The acid salts zinc chloride, ammonium sulfate, aluminum sulfate, and ammonium chloride are only mildly corrosive—rates of 5 to 13 milli-inches per year are typical. Acid and alkaline oxidizing salts, however, are generally very corrosive. Ferric chloride, ferric sulfate, cupric chloride, stannic chloride, mercuric chloride, and silver nitrate all will corrode Monel, except in very dilute solutions.[32] Hypochlorites are significantly corrosive to the alloy; for continuous exposure, dilute solutions containing less than 500 parts per million available chlorine must be used. For intermittent exposure, concentrations as high as 3 grams per liter can be used, but should be followed by rinsing with an acidic solution.

Monel may be used in contact with brasses, bronzes, copper, and chromium alloys in indoor and outdoor installations without fear of galvanic action. It may also be used in contact with aluminum and its alloys, zinc, and coated iron, in indoor and most outdoor conditions without galvanic action. When exposed to severe conditions and when in contact with Monel, however, these metals will corrode galvanically and should not be used as fasteners or other critical elements when combined with Monel. Galvanic effects are more pronounced when Monel is exposed to salt solutions.[33]

### DIAGNOSTICS AND CONDITION ASSESSMENT

Because each white metal reacts differently to cleaning agents, it is essential to identify metals before undertaking conservation.[34] Some may be identified by visual inspection, but differentiating them through nondestructive means is difficult.[35] Identifying Monel is complicated by the variety of surface finishes and colors that were available.[36]

Start any investigation and conservation program by reviewing the original construction documents. Next, physical tests should be conducted.

Magnetic testing, one of the most accessible and inexpensive methods used to identify unknown metals, is not a reliable indicator for Monel, whose degree of magnetism varies with temperature.[37] Chemical testing is the most reliable way to identify Monel. Spot tests will often provide clear evidence of one or more elements through visible changes in the color or physical makeup of the reagents or sample.[38] Where simple spot tests are ineffective, other chemical and physical testing procedures can be used.[39]

## CONSERVATION TECHNIQUES

Inco's early literature emphasized the ease with which Monel could be cleaned but did not discuss methods for removing stubborn soiling or preserving existing patinas. No known treatments specifically address deterioration.

Cleaning processes range from washing and degreasing the metal surface, to removing and replacing a protective coating, to polishing to remove scratches or corrosion products. Washing and degreasing are typical for minor maintenance; all three steps can be used for major cleaning. Monel is rarely left unprotected by a lacquer or oil coating. Techniques used for bronze cleaning and conservation, such as hot wax treatments, are worth investigating for application to Monel.[40] The following procedures are typical of those used by commercial cleaning companies but are emphatically not recommended for historic metalwork with an original finish or patina.

The first step in cleaning Monel is to remove grease and dirt. To degrease the metal surface, use a neutral pH soap, warm water, and a lint-free cloth or natural bristle brush. Acid- or alkali-based solutions and abrasive agents are not recommended. Washing should be followed by a thorough rinsing with water and drying with another cloth to avoid spotting. The water used for the final wash should be distilled or de-ionized water. Removal of dirt and other elements from the metal surface discourages the formation of harmful galvanic cells.

To remove lacquer, oxides, or paint, a lacquer solvent, such as methylene chloride-based gels or acetone combined with ethyl acetate and methyl ethyl ketone, can be used. These substances are hazardous and should be used only by skilled mechanics.

Once the coating is removed, the surface should be washed with clean water and wiped with a lint-free rag. Stains should be removed with an aqueous slurry of 5 percent oxalic acid and powdered pumice applied with a brush and a rag coated with pumice. For heavy staining and polishing, plastic or nylon pads and abrasives can be used. After the stains are removed, rinse and wipe the surface with a clean rag. Apply a new coat of lacquer.

## REPLACEMENT

Monel is still manufactured by Inco Alloys International. Sheet goods are readily available, but rolled and cast forms may be prohibitively expensive to reproduce.

**Figure 2.5.** This door of the United Nations General Assembly Building (1953, Ernest Cormier), New York City, has developed a patina from its semimarine environment.

ADRIENNE B. COWDEN

# 3 | NICKEL SILVER

**DATES OF PRODUCTION**
c. 1835 to the Present
**COMMON TRADE NAMES**
German Silver, Liberty Silver, Nevada Silver, Queen's Metal, White Metal, Woldram Brass

## HISTORY

Nickel silver is a copper-nickel-zinc alloy that contains no silver and at most only 30 percent nickel. Its silver-white color, its ability to take a high polish, and early trade names such as German Silver and Liberty Silver, however, resulted in its present misleading name. Its composition depends on the production method as well as the mechanical and physical properties desired. Variants of this alloy are quite numerous, but most contain 50 to 80 percent copper, 5 to 30 percent nickel, and 10 to 35 percent zinc.[1] Small percentages of lead, tin, and manganese may also be present.

### ORIGINS AND DEVELOPMENT

Long before the discovery of nickel in its pure state or in meteoric iron, nickel alloys were used in coins and ornamental objects. One of the most ancient nickel alloys is nickel silver, which has existed in one form or another for more than two thousand years.[2] This alloy occurs naturally and was first used by the Chinese, who called it *pai-t'ung* (white copper). The reddish metal was cast into triangular rings and transported to Canton, where zinc was added.[3] This silver-white alloy, known as *paktong* (the Cantonese pronunciation of *pai-t'ung*), was the precursor of modern nickel silver.

The mineral niccolite, now known as nickel arsenide (NiAs), was first described in 1694 by Hiarni, who called it copper-nickel or false copper.[4] The reddish, fume-emitting ore resembled, but did not produce, copper, and as early as the mid-1600s lodes of this mineral ore were laid open at Schneeberg in Saxony.[5] When smelted, it produced something quite unlike ductile, malleable copper. The resulting white metal was extremely brittle and so tough it could not be hammered. Eventually the miners discarded the ore as worthless.

It was not until 1751 that Aksel Fredrik Cronstedt discovered that these ores contained nickel, which until then had remained unidentified. While studying silver ores at the Swedish Department of Mines, Cronstedt examined ore from the Los cobalt mines in Hälsingland, Sweden. He had noticed that when Los ore tarnished it resembled copper-nickel in color. Experimenting in the ore, Cronstedt discovered a new elemental metal, which he later named nickel.

Germans were the first to produce a satisfactory version of Chinese *paktong*. In 1824 Henninger Brothers of Berlin placed *Neusilber* (new silver) on the market, and shortly thereafter Maximilian Werner von Bausman Geitner of Schneeberg invented a similar alloy, which he named argentan. Geitner prepared argentan from cobalt speiss with an average content of 49 percent nickel, 37 percent arsenic, 7 percent sulfur, and small amounts of various other metals.[6] In contrast, Neusilber contained only copper, zinc, and metallic nickel. Neusilber was exported in 1830 to Sheffield, England, where it was called German Silver,[7] the name used until after World War I, when it was supplanted by the term *nickel silver*.[8]

Nickel silver products were first exported to the United States in the early nineteenth century. The first documented American nickel silver was produced in 1835 by Robert Wallace, a Connecticut spoon maker, who had purchased a formula for nickel silver consisting of 2 parts copper, 1 part zinc, and 1 part nickel.[9] Until the 1900s the production of nickel silver alloys was usually an adjunct of the brass rolling mill but made up only a small percentage of any company's output.[10] The alloy was used principally for tableware and plated articles.

Between 1830 and 1840 nickel silver began to supersede copper as a foundation metal for fused silver plating because it provided a harder, stronger, and more durable base than copper and was adaptable to hard soldering. The introduction of electroplating in 1840 by Elkington and Company in England and Rogers Brothers in 1847 in the United States further expanded the market for plated nickel silver tableware.[11]

### MANUFACTURING PROCESS

In the late nineteenth century and into the twentieth century, manufacturers typically used the German,

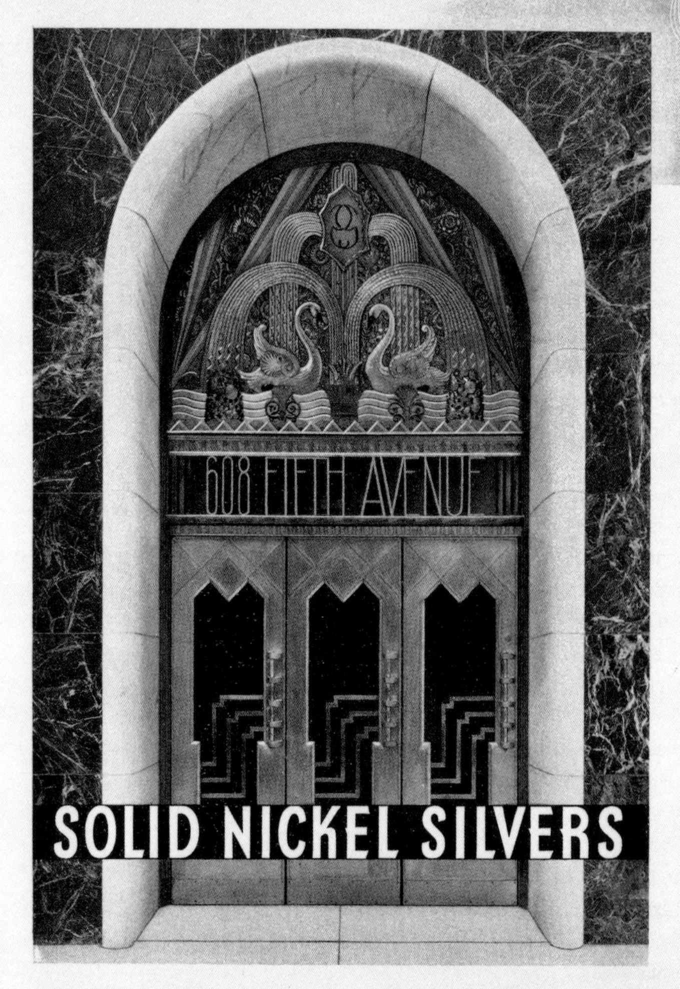


**Figure 3.1.** The International Nickel Company highlighted the polychromatic effects possible with nickel silver and other metals in this advertisement featuring the entrance to the Goelet Building (1932, E. H. Faile), New York City. *Architectural Record,* September 1938.

English, and American methods to produce nickel silver.[12] Barring differences in what order the metals were melted, the methods were essentially the same. In each case, carefully measured quantities of the three constituent metals were melted under a covering of charcoal or coke dust. Since nickel tends to produce gases when melted, producing a homogeneous alloy without blowholes was and is extremely difficult.[13] The charcoal or coke dust acted as both a protective covering and a deoxidizing flux to prevent porosity. Broken glass, soda ash, calcined borax, and other materials were also used as deoxidizers in addition to or instead of charcoal.[14]

The manufacture of nickel silver has remained essentially unchanged since the early twentieth century. Generally copper is melted first, followed by the nickel. Any scrap metals to be added are charged with zinc and mixed with the copper and nickel. The metals are then heated, deoxidized, and poured.[15]

Today most manufacturers produce only a bright white nickel silver. However, by varying the nickel content they can produce shades of pale yellow, green, pink, and blue.[16] If nickel silver is to be truly silver-white in color, it must contain at least 20 percent nickel and only a moderate amount of zinc. The specific physical and mechanical properties will vary with the percentages of copper, nickel, and zinc, but nickel silver is typically corrosion resistant, strong, ductile, and nonmagnetic. Copper is the most ductile component of nickel silver, and a high copper content is desirable if the metal is to be rolled, stamped, drawn, or spun.[17] Copper also improves nickel silver's corrosion resistance. Zinc lowers the melting point and slightly improves corrosion resistance. Increasing amounts of zinc also raise the yield point (the lowest stress in a material at which the material begins to exhibit plastic properties), maximum strength, and hardness but result in a corresponding loss of ductility.

Once formed, nickel silver may be wrought, cast, rolled, stamped, forged, drawn, extruded, and machined. Depending on the presence of lead, it can be joined by methods such as hard and soft soldering and spot-, arc-, and fusion-welding. These properties, in combination with its white color and receptivity to high-luster finishes, made nickel silver popular for a wide variety of products, including architectural hardware and ornamental metalwork.

## USES AND METHODS OF INSTALLATION

Although electroplating constituted one of the most important commercial applications of nickel silver, by 1914 the alloy was being used for an incredibly diverse array of items such as costume jewelry, keys, soda-fountain and bar equipment, cigarette cases, automobile radiators, and hubcaps.[18] It was also used for architectural hardware, but architects did not utilize it for decorative and structural elements until the late 1920s. Nickel silver was ideal for the streamlined edges, stylized motifs, and abstracted figures of the Art Deco style. It could be easily worked into any shape, was highly resistant to environmental corrosion, and could take a high-luster finish. Architects used it for decorative panels, doors, grilles, railings, plumbing fixtures, plaques, and trim. Strips of nickel silver were also used to enhance and separate different colors in terrazzo flooring.

By the early 1930s a variety of white metals had gained favor for building interiors and exteriors.[19] The Chicago Daily News Building (1929, Holabird and Root) contains nickel silver details from floor to ceiling throughout the concourse and main lobby. A highly polished nickel silver was used for radiator grilles, elevator doors, directory cases, stair railings, and trim. It was also incorporated in the terrazzo flooring and applied to the marble walls to give the illusion of fluted pilasters.[20] Nickel silver was used for the check desks, railings, and banking screens to accent the red, brown, and green color scheme of the National Bank of Commerce (c. 1929, Finn, Franzheim, and Carpenter) in Houston.[21] Architects also used different shades of nickel silver to relate to other building materials. In the Squibb Building (1930, Buchman and Kahn) in New York City, light pink nickel silver was used in the lobby to harmonize with the stone veneer.[22] The Waldorf-Astoria Hotel (1931, Schultz and Weaver) includes nickel silver in colors from bright white to a brassy yellow.[23]

Nickel silver was often used to enhance or contrast with other metals. The ornamental screen once located over the doors of the Goelet Building (1932, E. H. Faile) in New York City incorporated contrasting nickel silver, brass, bronze, and copper. The City Bank Farmers Trust Company Building (1931, Cross and Cross), also in New York City, used extensive amounts of rolled, drawn, extruded, and cast nickel silver alloys alone and in combination with various bronzes.[24]

Nickel silver continued to be used throughout the 1940s for trim and hardware, but it had almost entirely disappeared from architects' vocabulary by the mid-1950s. Its decline can be attributed largely to the development or increased usage of other materials. The onset of World War II also dramatically affected the nickel silver industry. Nickel and copper were diverted to the war effort, leaving little for the production of alloys. After the war, stainless steel and aluminum replaced nickel silver and other white metals in architectural designs because

Figure 3.2. A popular white metal for Art Deco designs, polished nickel silver was used extensively for the elevators, stair railings, directory cases, and radiator grilles in the Chicago Daily News Building (1929, Holabird and Root).

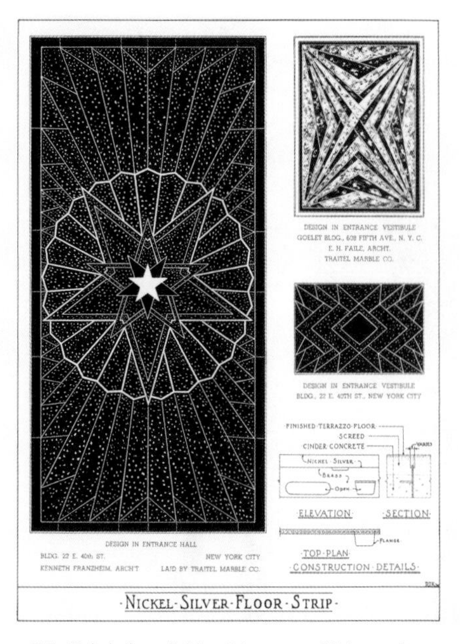


Figure 3.3. Nickel silver divider strips were widely used to separate sections in elaborate terrazzo floor designs, such as these for the entrance hall of 22 East 40th Street, New York City (left and right bottom), and the Goelet Building vestibule (top right).

of their lower production cost. Today nickel silver is used predominantly for industrial and electrical purposes. Cast and wrought forms of the alloy, however, are still used occasionally in building designs.

## CONSERVATION

Nickel silver is generally resistant to corrosion, and historically this has been cited as one of its most attractive features. It is also an extremely durable metal. It is not, however, impervious to chemical pollutants and mechanical strain. Unfortunately, there are currently no treatments that specifically address the conservation of nickel silver, and there is a corresponding dearth of information concerning its deterioration. The following is offered as a preliminary investigation of this metal—its deterioration and conservation.

### DETERIORATION

Nickel silver alloys are highly resistant to most atmospheric corrosion and to many mineral and organic acids. They will, however, tarnish slowly in the presence of sulfur because of the presence of zinc, which is particularly reactive with sulfur and acid pollutants.[25] Since it is a nonferrous metal, nickel silver forms a protective oxide when it comes into contact with oxygen. After prolonged exposure without cleaning, it acquires a compact and tightly adhering patina. This brownish-green patina protects the alloy against further corrosion and surface damage, such as pitting.

Nickel silvers may also be susceptible to stress corrosion cracking, which occurs when a metal is subjected to tensile stress and a corrosive environment. The stress may result from structural deficiencies that occurred during manufacture. Other variables affecting stress corrosion cracks are temperature, metal composition, and metal structure.

Nickel silver is an extremely hard alloy, and its abrasion resistance has made it popular for doorknobs, handrails, and push plates. It is not subject to creep. However, it may be scratched and dented, and nickel silver elements are also subject to mechanical deterioration such as fatigue.

### DIAGNOSTICS AND CONDITION ASSESSMENT

Before cleaning or repair, the metal must first be identified. Differentiating among various white metals can be extremely difficult. Nickel silver is often mistakenly identified as Monel, a binary alloy of copper and nickel. Nickel silver's brownish-green patina, which forms in highly corrosive environments, may also cause the metal to be identified as bronze or some other cupronickel alloy. A cupronickel metal is a copper-based alloy containing 10 to 30 percent nickel. Visual inspection alone will generally be inconclusive.

**Figure 3.4.** Abrasion resistance, workability, and bright surface finish made cast, drawn, and wrought forms of nickel silver suitable for plumbing and hardware in large buildings such as hotels, particularly during the 1930s.

**Figure 3.5.** The nickel silver statue of Alexander Hamilton (1940, Adolph A. Weinman) at the Museum of the City of New York has tarnished to a slate gray color, similar in tone to many bronze patinas, and is showing signs of deterioration.

Because most structures containing ornamental nickel silver date from the 1920s and 1930s, building records identifying the metals used may still exist. These documents sometimes note the foundry or the type of nickel silver used. But the existence of such records does not guarantee that the metals specified were used. Accordingly, metals should be identified by a metal conservator or metallurgist. Tests to identify nickel silver should be undertaken with extreme care. Its high copper content makes it extremely sensitive to strong oxidizing acids, such as chromic and nitric acid, used to identify white metals. Other methods such as x-ray diffraction and metallographic analysis may also be used to identify specific metals. These methods require a small sample and will generally provide an accurate analysis.

## CONSERVATION TECHNIQUES

Because of its high copper content, nickel silver has many of the same characteristics as bronze and brass. Both interior and exterior nickel silver metalwork, therefore, can usually be cleaned in the same manner as brass and bronze. Indoor applications generally require minimal attention. These elements can normally be cleaned with a mild non-ionic detergent and a wetting agent such as de-ionized water.[26] Well-maintained exterior metalwork and sculpture may require similarly minimal treatment. Natural-fiber bristle brushes can be used to scrub off loosened dirt; if greasy or oily residues have built up on the metal's surface, chemical compounds, such as ammonia with whiting or a 5 percent oxalic acid and water mixture, may be used as a degreaser.

Exterior metalwork with built-up corrosion and surface deposits, dirt, and graffiti may require greater intervention. Tightly adhering corrosive crusts and surface deposits can be removed through a number of abrasive procedures. Low-pressure mechanical blasting involves cleaning the surfaces with a stream of abrasives propelled through a special nozzle at a predetermined, low-pressure velocity. Highly abrasive materials, such as sand and glass beads, should not be used to clean nickel silver mechanically. Softer, organic media shot at a low pressure are generally considered to be the safest types of microabrasive treatment; walnut shells are currently favored by most conservators. Blasting must be carefully executed to limit removal of surface metal and intact patina. Low-grit sandpapers and plastic abrasive pads can also be used to remove thick crusts. These materials are used with clean, de-ionized water or a mineral oil as a lubricant. Under no circumstances should an element be scrubbed with steel-wire brushes, which will pit and scratch the surface. Graffiti and failed, or discolored, coatings may be removed with organic solvents. Because the properties of nickel silvers vary widely, test patches should be conducted on inconspicuous areas. A solvent appropriate for one variety of silver nickel may be less effective or even harmful for another.

Once nickel silver elements have been cleaned and all chemicals have been thoroughly washed off with deionized water, a protective coating should be applied. These coatings will limit further corrosion and temporarily protect the metal against abrasion, water, and environmental pollutants. Lacquers and synthetic waxes, or combinations of the two, are the most appropriate types of coatings.[27] Lanolin-based and oil coatings, which require more maintenance and remain sticky after application (and thus trap pollutants), are not recommended. Because of the high copper content of most nickel silvers, natural waxes are also not recommended, since they may cause acidic reactions with the metal in moist environments. Coatings are not permanent and should be cleaned and maintained on a regular schedule, depending on the type of coating. Failure in these protective barriers, caused by scratches or breaks, can result in selective corrosion.[28]

### REPLACEMENT

In some cases it may be necessary to replace damaged pieces or missing elements. Most companies that produce nickel silver today manufacture the alloy only in a silver white or a white with a slight yellow tint. However, custom orders can usually be made to match a nickel silver alloy and its finish.

ROBERT SCORE AND IRENE J. COHEN

# 4 | STAINLESS STEEL

**DATES OF PRODUCTION**
c. 1927 to the Present
**COMMON TRADE NAMES**
Allegheny Metal, Ascoloy, Enduro, Nirostametal, Rigid-Tex

## HISTORY

Stainless steel comprises a diverse group of metal alloys containing iron and at least 11 percent chromium. Chromium allows the alloy to form a tight film of iron-chromium oxide that resists corrosion and chemical attack.[1] Stainless steels fall into four classifications: (1) martensitic stainless steels—iron-chromium alloys hardened by heat treatment; (2) ferritic stainless steels—iron-chromium alloys that cannot be hardened by heat treating; (3) austenitic stainless steels—iron-chromium-nickel and iron-chromium-nickel-manganese alloys that are hardened only by cold working; and (4) precipitation-hardening stainless steels—iron-chromium-nickel alloys.[2]

### ORIGINS AND DEVELOPMENT

Chromium alloy steels were investigated as early as 1821 but were not discovered to be corrosion resistant. In fact, in 1892 a British metallurgist concluded that chromium impaired the corrosive resistance of steel.[3] This conclusion would not be disproved until 1908, when Philipp Monnartz of Germany discovered the role of carbon in obscuring the corrosion resistance of chromium steels.[4]

In 1913 Harry Brearley of Sheffield, England, discovered martensitic stainless steels. Brearley's alloy was approximately 1 percent chromium and 0.35 percent carbon.[5] An American patent for martensitic stainless steel was granted in 1916. This alloy was originally used for cutlery and munitions.[6]

In 1911 Christian Dantsizen of Germany developed ferritic stainless steel for the electric lightbulb filament. The low carbon content of ferritic stainless steel causes it to act like a stainless iron. In 1914 Dantsizen adopted the low-carbon alloy for turbine blades.[7]

Between 1909 and 1912 Eduard Maurer of Germany developed austenitic stainless steel, an alloy containing 18 percent chromium and 8 percent nickel (18–8 stainless steel). Austenitic iron-chromium-nickel alloys harden by cold working, which means that the metal will harden only through deformation. This quality makes austenitic ideal for deep drawing, a method of forming sheet and strip metal into cup-shaped articles or shells with a punch-and-die. Austenitic stainless steels are most frequently found in architectural uses as formed sheets.[8]

Precipitation-hardening stainless steels were first introduced in 1929 by the Germans William J. Kroll and R. Wasmuth, who added titanium, boron, or beryllium as hardeners. In 1946 E. Wyche, an American, patented a precipitation-hardenable austenitic stainless steel using titanium.[9]

### MANUFACTURING PROCESS

In the traditional process, stainless steel is made in three steps.[10] First, ordinary steel and iron ore or other oxides are melted. The oxygen from the iron ore reacts with the carbon to form carbon monoxide, thereby removing the harmful carbon. Nickel is added to the furnace at this stage. Next, silicon is added to reduce iron oxide. Finally, chromium in the form of ferrochromium is added. The molten metal is then ready to be poured into molds. After the stainless steel is formed into ingots, it is forged and can be rolled into billets and sheets. At this point surface defects are removed by grinding. In the final forming process billets are rolled into bars, rods, and tubing, while slabs are rolled into plates, strips, and sheets.[11] The manufacture of stainless steel products is similar to the process for carbon steel. Sheets can be worked on a stamp, break, and punch; they can be drilled, cut, ground, welded, brazed, and soldered.

As stainless steel began to be used for engineering and architectural purposes, many patents were issued for manufacturing techniques and finishes.[12] All mechanical finishing processes remove surface material to make the metal smoother, resulting in a higher shine or reflectance. Standardized names for the various finishes began to be developed around 1930, when the industry experimented with polishing methods. The No. 1 finish (hot-rolled, annealed, and pickled) was an early industry standard, but other finishes did not become standardized until

**Figure 4.1.** The ease with which stainless steel curtain wall panels could be installed made them a popular cladding material beginning in the 1950s. *Architectural Forum*, 1955.

**Figure 4.2.** Used for storefronts, Rigid-Tex patterned sheets of embossed and rolled finishes for stainless steel were produced by the Rigidized Metals Corporation.

**Figure 4.3.** Stainless steel sheets were manufactured for the dome (the top nine floors) of the Chrysler Building (1930, William Van Alen), New York City.

1934, when both Republic Steel and the American Rolling Mill Company adopted the same standard finishes. These finishes are still used on all types of stainless steel, but the appearance of each type differs slightly.[13]

Stainless steel can also be polished by an electrolytic process. In 1929 Pierre Jacquet, a French engineer, discovered the polishing effect of electrolytes on wire, leading others to experiment with electropolishing methods for stainless steel. Between 1939 and 1947 nearly one hundred American patents were issued for electropolishing processes.[14]

In 1947 the Sendzimer cold reduction process was developed. This process allowed for the cold rolling of thinner sheets and strips of stainless steel with greater precision and improved finish.[15] Stainless steel produced using this method has a slightly different finish from pre-1947 products. Other developments in stainless steel include the production of gold- and bronze-plated stainless steel and super-mirror finishes with clear coatings.[16]

## USES AND METHODS OF INSTALLATION

Architectural uses of stainless steel lagged behind engineering and manufacturing developments in the 1910s and 1920s. In 1921 *The American Architect* reported that Great Britain had been manufacturing stainless steel and iron for cutlery, stoves, motor cars, and other uses, and indicated that U.S. manufacturers were not marketing aggressively.[17] Architectural uses of stainless steel developed slowly over the next decade.

In 1927 the Allegheny Steel Company was the first to advertise stainless steel in *Sweet's* catalogue. It promoted its chromium-nickel-iron alloy, Ascoloy 22, for kitchen use and other equipment. In 1929 Allegheny Steel introduced Allegheny Metal, which became its staple product. Other early stainless steel manufacturers include Associated Alloy Steel Company, Republic Steel, American Rolling Mill Company, Electro Metallurgical Company, and U.S. Steel.[18]

In 1929 Allegheny stainless steels were available for elevator and entrance lobbies as well as exterior trim. The early marketing of stainless steel promoted it for

**Figure 4.4.** Throughout the 1940s and 1950s stamped stainless steel was ubiquitous in diners, where it was used for wall panels, food preparation surfaces, decorative trim, and other features. This diner was manufactured in 1939 by the Paramount Diners of Haledon, N.J.

every architectural use from structural to decorative applications. Through the 1930s the primary architectural products advertised in manufacturers' literature and architectural and trade journals included trim and ornament, railings, door hardware, elevator and entrance doors, light fixtures, furniture, signage, counters, and equipment. Some sculptured surfaces were highly decorative pieces of art, combining many manufacturing techniques in each piece.[19]

Manufacturers promoted stainless steel primarily for its corrosion resistance but also for its sanitary qualities and modern appearance. Although it was expensive, especially compared to aluminum, manufacturers pointed out that maintenance and replacement costs were negligible for both exterior and interior applications.[20]

Stainless steel was only one of several white metals that architects experimented with on Art Deco and Streamline Moderne buildings, and it had to compete with other corrosion-resistant metals such as aluminum, Monel, and nickel silver, which share many of these qualities. Aluminum was its strongest competitor because of its lower cost and light weight. Both stainless steel and aluminum were used extensively in combination with copper and bronze to create striking ornamental effects.[21]

Two influential early projects that incorporated stainless steel were the Chrysler Building (1930, William Van Alen) and the Empire State Building (1931, Shreve, Lamb, and Harmon), both in New York City. The Chrysler Building used about 55,000 square feet of Krupp-Nirosta stainless steel sheets for the dome, spire, gargoyles, storefronts, entrances, and doors, and the Empire State Building used 18–8 stainless steel for vertical trim.[22] Critics extolled both buildings as breakthrough projects for stainless steel, noting the low maintenance requirements.

Stainless steel can also be found in architecture as a cast metal in plaques, signs, and sculptural elements.[23] By 1925 metallurgists had already found methods using electric furnaces to cast stainless steel.[24] An early application was the cast Nirostametal door panel at 230 Park Avenue, New York.

During the Great Depression, stainless steel continued to grow in popularity. It was used with structural glass to

**Figure 4.5.** The Inland Steel Building (1957, Skidmore, Owings and Merrill), Chicago, an early application of stainless steel in modern architecture, juxtaposes polished stainless steel mullions and spandrel panels with tinted green glass.

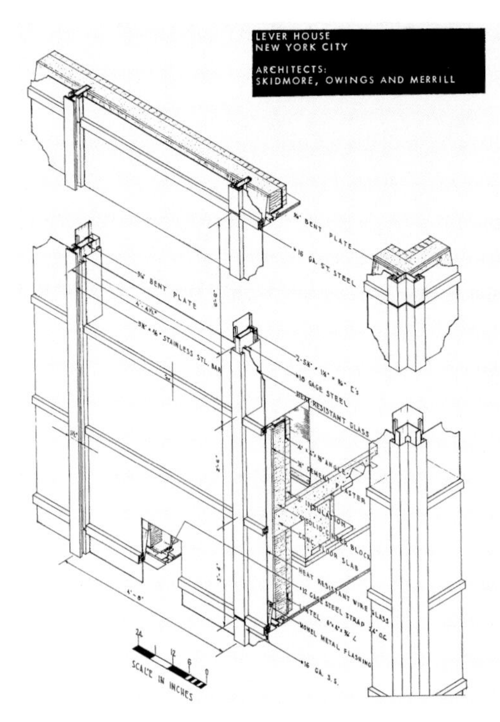


**Figure 4.6.** Thin, 16-gauge stainless steel mullions, combined with glass and Monel flashings, lent the Lever House (1952, Gordon Bunshaft, Skidmore, Owings and Merrill), New York City, a sleekness that symbolized machine-made building technology.

modernize storefronts and with marble in many new bar designs and soda fountains, principally for drain boards and glass-washing stations.[25]

By the 1940s stainless steel had established its place as a corrosion-resistant, low-maintenance, visually appealing material. Stainless steel products included many standard extruded shapes for storefronts and trim, as well as the earlier uses of equipment, hardware, lobbies, and entrances. Because of its high cost it was most often found in areas of high use and exposure; for example, it was bonded to plywood or composition material panels for standardized designs such as service stations.[26]

Through the late 1940s and 1950s curtain wall construction was developed using thin, lightweight panels hung on a structural frame to enclose a building. An early stainless steel curtain wall appeared on the General Electric Turbine Building (1948, Stone and Webster) in Schenectady, New York. As curtain wall construction was developed, more and more of the wall area was composed of glass, leaving only the mullions to be made of sheet metal. The Lever House (1952, Gordon Bunshaft, Skidmore, Owings and Merrill) in New York City, with its glass windows and opaque wire-reinforced glass spandrel panels set in stainless steel mullions, is an example of this construction.[27]

After World War II rigidized stainless steel, such as that produced by the Rigidized Metals Corporation in Buffalo, was introduced. Rigidized sheets have a rolled pattern that both increases the tensile strength of the sheet and visually breaks up the surface to eliminate distorted surface reflections (oil canning).[28] Quilted sheets, which have a grid pattern produced by lightly creasing the sheets on a break, were available by 1950 and were extensively used in diners for both exterior and interior wall panels. In the late 1950s and 1960s manufacturers also promoted stainless steel for gutters, roofing, and flashing and for resheathing existing masonry walls.

Stainless steel applications can appear in customized or standardized products, but certain methods are used for both. Trim is usually applied with screws, bolts, tacks, nails, or rivets, preferably of stainless steel, although some shapes simply snap into place. Trim can be directly attached to steel frames, anchors, or brackets. Joints can be formed with oxyacetylene or electric fusion welding, or occasionally bronze welding. Expansion joints must also be incorporated, since the 18–8 type of stainless steel has a coefficient of thermal expansion approximately 1.5 to 1.8 times that of low-carbon steel.[29]

## CONSERVATION

Under adverse conditions stainless steel will corrode and is often marred by scratches and dents. Its advantage is that it is more durable than most sheet metals; if it does become damaged, its homogeneity makes repair possible.

### DETERIORATION

All metals are susceptible to corrosion, and stainless steel is no exception. Its corrosion resistance comes from an invisible passive film of chromium oxide that protects the material in oxidizing environments, such as oxygen or nitric acid. The passive film does not protect the metal in reducing environments, such as hydrochloric acids. Stainless steel can also corrode if it is heated to about 900 to 1,500 degrees Fahrenheit for a short time—for example, during welding. The heat causes chromium to be removed, thus leaving these areas more prone to intergranular corrosion.[30] This condition is sometimes identified by a blue and orange halo (weld stain) around the area that has lost chromium.

Another form of corrosion is pitting. Pitting occurs in stainless steel when dirt builds up and keeps oxygen from reaching the metal's surface, preventing the metal

**Figure 4.7.** When exposed to a highly polluted atmosphere, stainless steels will corrode and are then often mistaken for aluminum. The gates at the Federal Trade Commission (1937, Bennett, Parsons, and Frost), Washington, D.C., have pitted because dirt on the surface has trapped moisture and created local electrolytic cells. Stainless steel should be cleaned periodically with water and a mild detergent.

from producing the passive film of chromium oxide that protects it. Corrosion then occurs in localized areas beneath the soil. Stainless steel, furthermore, is susceptible to galvanic corrosion when in contact with lead, nickel, copper, copper alloys, and graphite. In general, however, corrosion occurs only in areas of high pollution or where exposed to salts or curing mortar.

Other possible damage to stainless steel includes dents, scratches, and warping. Since stainless steel is commonly used in high-traffic areas, it is commonly damaged by abrasion and impact. Warping may be caused by thermal expansion or exposure to extreme heat. When warping occurs, it is necessary to determine if expansion control was provided in the original design.

## CONSERVATION TECHNIQUES

To prevent corrosion and maintain stainless steel's original finish, it is necessary to keep it clean.[31] For routine cleaning of soil and pollution, rinse the surface with warm water and a mild detergent. Steel wool or steel brushes will cause pitting and should not be used. Use only soft cloths, sponges, fibrous brushes, or plastic pads, taking care always to rub in the direction of the grain. Rubbing and wiping can scratch finishes and should be minimized. After cleaning, rinse surfaces thoroughly with fresh water.

Chlorides can be cleaned by rinsing with warm water. Fingerprints, grease, and oil should be rinsed with a combination of water, mild detergent, and mineral spirits. Cleaning should be followed by a thorough rinsing with clean water. Extensive, disfiguring iron stains from bolts, screws, and other elements are difficult to eliminate without removing the elements and immersing them in nitric acid; this approach probably would not be practical.

Cleaning techniques for graffiti depend on the type of ink or paint involved. For water-soluble inks, use warm water and a non-ionic detergent and then rinse with clean water. For paints and other inks, use a combination of water, non-ionic detergent, and mineral spirits. Rinse thoroughly with clean water afterward. Lead-pencil markings can be cleaned with an oily cleaner such as paste wax.

Weld stains can be cleaned by using a mild abrasive cleaner in paste form and water. Weld stains can be minimized by inserting thin-gauge aluminum between the welding tip and stainless steel on the side to be kept clean.

Corrosion products on stainless steel can usually be removed by using warm water, detergent, and plastic pads. However, when such finishes are severely corroded, mechanical methods, such as grinding or sandblasting, may be appropriate.[32] The removal of corrosion on stainless steels that were originally polished will require surface refinishing with fine abrasives to match the original finish.

When it is not possible to retain severely corroded stainless steel, new sections can be spliced in by welding and finishing the welds to match existing material or by reproducing the entire component.[33]

Most scratches are an aesthetic problem and, aside from trapping dirt, will not cause further damage to the metal. The likelihood of recurrence and the severity of the aesthetic damage should be evaluated because scratch repairs require the removal of material. The severity of the scratch will dictate the level of abrasive treatment needed to repair it. Minor scratches may require only a rubbing compound with a light abrasive, whereas heavier scratches will require grinding the surface down to an even base and refinishing with a series of finer abrasives to return the surface to its original finish. Deep gouges can be filled with weld and finished to match the existing finish.

**REPLACEMENT**

When replacing stainless steel, it is important to match existing finishes. Sheets and strips finished using the Sendzimer cold reduction process, developed in 1947, will not exactly match those produced before 1947. To match existing and original finishes, begin by comparing field samples to industry standard finishes available from steel manufacturers. The most common finishes are No. 2B, a bright cold-rolled finish; No. 4, a polished finish; No. 6, a dull satin finish; and No. 8, a mirror finish. If industry standards do not match the original material, it will be necessary to experiment with different abrasives and hand finishing to obtain an acceptable match. A test piece should always be made and compared to the original finish in the field before the repairs are performed to ensure a match. When reproducing cast stainless steel elements, it is necessary to make patterns larger to allow for shrinkage ($\frac{9}{32}$ inch per foot).[34]

JOHN C. SCOTT AND CAROLYN L. SEARLS

# 5 | WEATHERING STEEL

**DATES OF PRODUCTION**

1933 to the Present

**COMMON TRADE NAMES**

Cor-Ten, Cor-Ten A, Cor-Ten B, Mayari R, River-Ten, USS Cor-Ten

## HISTORY

Weathering steels make up the first and earliest of six current classes of high-strength, low-alloy steels. Weathering steels are low-carbon steel alloys, which achieve special properties by incorporating less carbon than plain low-carbon steels and adding small amounts of other alloying elements.[1] Weathering steels are considerably stronger and four to eight times more resistant to atmospheric corrosion than plain low-carbon steels.[2] A special feature of weathering steels is that as a result of exposure to normal outdoor conditions, they develop a thin, protective brown patina. This patina differs in appearance, structure, and composition from rusts that develop on most types of steel. On exposure to the atmosphere and repeated cycles of wetting and drying, clean weathering steel surfaces quickly rust to a bright orange color. After a few months of normal weathering, the surface becomes a reddish brown. Within a year the patina becomes darker brown and, after a few years of normal urban weathering, a deep rich brown. The patina becomes darker in urban than in rural locations. Under appropriate conditions weathering steel can be left outdoors unpainted.

### ORIGINS AND DEVELOPMENT

Iron-copper alloys have interested Western metallurgists since at least the early seventeenth century. By the late nineteenth century "copper steels," which contained from 1 to 2 percent copper in addition to carbon and other alloying elements, were in use. Early in the twentieth century, the mechanically strongest copper steels were found to be about 25 percent more corrosion resistant than most copper-free steel. And otherwise similar steels containing only about 0.20 to 0.25 percent copper were found to have even better resistance to atmospheric corrosion when exposed to wetting and drying.

Beginning in 1916 the American Society for Testing and Materials (ASTM) conducted tests that clarified the benefit of incorporating 0.2 to 0.5 percent copper and phosphorus into steel. In 1929 U.S. Steel started developing proprietary corrosion-resistant, low-alloy steels. In 1933 U.S. Steel introduced its low-alloy, high-tensile steel product, Cor-Ten A, later joined by Cor-Ten B. By the 1960s the term *USS Cor-Ten* was used, possibly to avoid loss of trademark, as this product name tended to be used generically. Today weathering steels are produced by several manufacturers. U.S. Steel promoted its new product for railroad equipment manufacture.

A 1937 promotional booklet on Cor-Ten steel sounded three themes that have proved persuasive to the present day: (1) exceptional strength and (2) significant maintenance reduction, both of which create economic benefits (cost savings are promised based on using less Cor-Ten in a given situation than carbon steel, with additional economy through the elimination of protective painting), and (3) the aesthetic value of a well-developed rust surface. When used properly, weathering steels are indeed economically beneficial because of their exceptional strength and minimal maintenance requirements. The dark, tightly adherent rust patina has proven compatible with a broad range of contemporary styles. Paint coatings last considerably longer on weathering steels than on plain low-carbon steels.

In the 1960s articles in professional journals began to warn of special design considerations required for weathering steel. Not until the late 1960s and 1970s, however, did industry literature feature special design constraints for unpainted use and recommend specific materials and methods.

### MANUFACTURING PROCESS

Weathering steel is manufactured much like other types of steel. High-carbon iron (containing 3 to 4 percent carbon) is obtained from iron ore in blast furnaces. To convert high-carbon iron into useful steels, excess carbon is first removed from molten metal by reaction with oxygen to form volatile carbon monoxide. Oxygen is added either by injecting the gas or by dissolving iron oxide into

**Figure 5.1.** Weathering steel's texture and color, which can range from light to dark to earth browns, appealed to some designers in the 1960s and 1970s. The Civic Center Building (1965, C. F. Murphy Associates) in Chicago was an early architectural application. Weathering steel catalogue, Bethlehem Steel, March 1978.

**Figure 5.2.** The Deere Company Administrative Center (1958, Eero Saarinen and Associates), Moline, Ill., the first architectural use of weathering steel, received the 1993 American Institute of Architects' Twenty-Five Year Award.

the melt. Next, alloying elements are added under vacuum melting conditions and then cooled. The solid steel is mill-rolled hot or while cooling under "controlled rolling," to produce plate and sheet or billets.

Weathering steels are produced in plate, bars, and shapes up to 8 inches thick. The alloy composition of the steel affects its strength, corrosion resistance, weathering characteristics, and appearance. The alloying element considered most important for corrosion resistance is copper, but high-phosphorus weathering steels, such as USS Cor-Ten A, may have up to twice the corrosion resistance of low-phosphorus weathering steels such as USS Cor-Ten B.[3]

### USES AND METHODS OF INSTALLATION

Weathering steels are used in architecture, civil and industrial engineering, and art. The most important factors affecting weathering steel structures are design, construction and fabrication, and site environment. To be successful, unpainted weathering steel structures must have regular cycles of wetting and complete drying. When weathering steel drains poorly or remains wet, it develops no protective rust, and rapid destructive corrosion proceeds.

Structural shapes such as channels and I-beams, surfacing shapes such as sheet and plate, and shaped components such as windows, ledges, columns, and light standards are fabricated and installed by welding or bolting. Welds and fasteners used in exposed structures must have compositions similar to the base steel. Special welding electrodes are available for weathering steels. Narrow welds will be enriched from adjoining metal during melting. High-strength bolts, nuts, and washers are available in weathering steel.[4] Bolt, washer, thread, and nut contacts should be tight and sealed to prevent crevice corrosion. Weathering steel sheet material has also been used for flat sheet siding, corrugated siding and roofing, and insulated panels.

The environment and microenvironment affect the corrosion rate as well as the color and texture of the patina. Weathering steels show the greatest superiority over normal steel in industrial atmospheres containing sulfur dioxide ($SO_2$). Great differences in corrosivity have been found at locations only a few miles apart.[5] So important is the environment that some early researchers recommended on-site corrosion and appearance testing of weathering steel before using it. Unpainted weathering steel is not recommended for marine environments.

Eero Saarinen and Associates' Deere Company Administrative Center (1958) in Moline, Illinois, was apparently the first architectural use in the United States of unpainted weathering steel.[6] The first high-rise building using unpainted weathering steel and located in an urban central business district was Chicago's Civic Center Building (1965, C. F. Murphy Associates).[7] U.S. Steel's Pittsburgh headquarters (1971, Harrison and Abramovitz) is a tour de force in the use of weathering steel.

**Figure 5.3.** Corrosion of weathering steel is accelerated when water pools on surfaces, joints, and crevices.

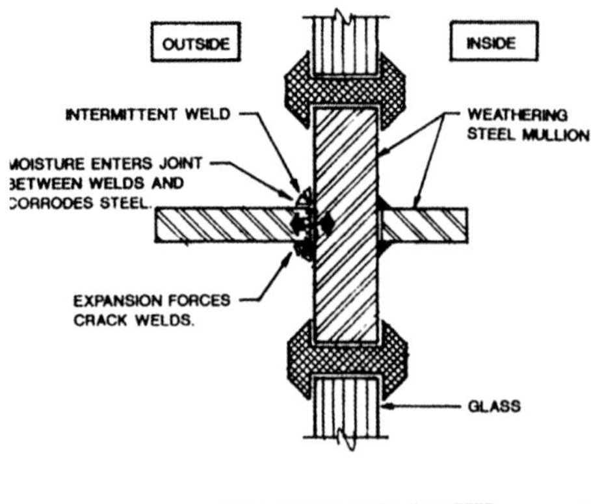


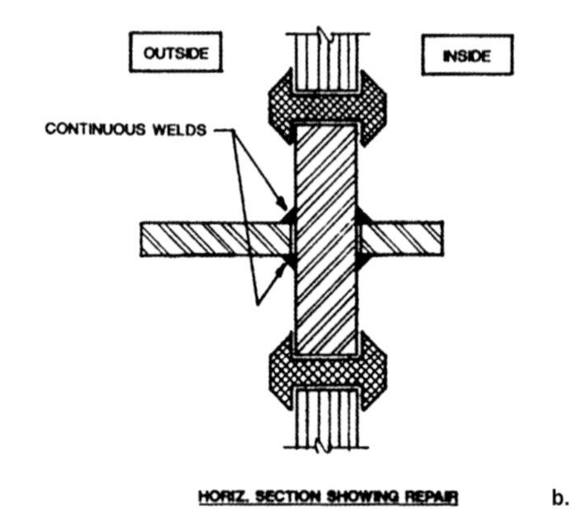


**Figure 5.4a–b. a.** Corrosion causes weathering steel window mullions to expand and welds to crack. **b.** Mullions should be continuously welded.

Many early weathering steel bridges were built in Michigan. Although most are in good condition, localized areas of accelerated attack are found on many structures as a result of salt deposits and nondraining elements.[8]

## CONSERVATION

Three primary factors affect the performance of weathering steel structures: the composition of the weathering steel, the environment in which it is exposed, and the structure's design and detailing. Conservation problems can usually be attributed to these factors. The cause of the observed deterioration should be well understood before applying conservation measures.

### DETERIORATION

The protective rust on weathering steels has a layered structure. A primary, inner layer consists of fine, tightly adhering amorphous and poorly crystallized ferric hydroxides such as goethite (alpha-FeOOH). This quasi-polymeric layer of ferric oxyhydroxide ($Fex[OH]_3$–2x) apparently keeps oxygen and corrosive species from the metal and prevents ferrous ions from dissolving. When the rust is wet, alloying elements, such as copper and phosphorus, dissolve at cracks in the inner layer and aid formation of additional goethite, which on drying fills crevices in the primary layer. This process is affected by the metal or oxidizer ions in solution and by the solution pH. The outer region of the protective inner layer is slowly converted into oxides and hydroxides of iron and alloying elements, which form more crystalline layers just below the rust surface.[9]

Thus, this protective rust is formed and maintained by frequent cycles of wetting and complete drying. When the steel remains dry, the process is slowed, since it requires aqueous solvent for the active ionic species. When the steel is wetted but not dried, the protective layer of somewhat soluble goethite is eroded, and corrosion proceeds.

Failure usually results from extensive corrosion due to extended time of wetness or from faulty design. As the protective rust deteriorates, flakes and sheets of expanding rust separate from the surface in exfoliating corrosion, and the underlying steel rapidly erodes and fails. If confined at joints and faying (tightly fitting) surfaces, the expansive corrosion products will exert tremendous force against opposing structures, which are often severely deformed by this corrosion jacking.

### DIAGNOSTICS AND CONDITION ASSESSMENT

Visual examination and mechanical probing of the structure are indispensable. Measurement of corrosion penetration and pit depth are made. Fatigue-sensitive elements should be examined for cracks. Study of architectural and construction shop drawings and specifications may also help. Some steel may require removal for laboratory analysis.

Undrained surfaces will corrode quickly. Such corrosion can occur where wide-flange beams are placed with the webs horizontal and flanges vertical, at cross-bracing connections, or open pockets. Failure due to water infiltration and corrosion at faying surfaces is common. Moisture will seep into a bolted or intermittently welded connection and corrode the steel. The expansion of the corrosion will force the steel sections apart, allowing more water infiltration and corrosion. These corrosion-jacking forces are strong enough to break welds and bolts.[10]

A common problem with glass installed in weathering steel is staining from corrosion rust runoff. Staining will become apparent when the iron oxide film and airborne dirt have dried on the glass. Mild abrasive cleaners may be required to remove this film.[11] When cleaning windows, care must be taken to rinse the adjacent steel with clean water to prevent discoloration. Also, glass breakage can occur when expansive corrosion puts extra pressure on the glass.[12]

Weathering steel skin panels (a single sheet) have been used as facing over masonry or as an industrial curtain wall.[13] In such situations, moisture can condense behind the skin. Failure can occur because of corrosion if the unexposed face is not painted. Corrosion will also occur if the space between the steel sheet and the backup is not well vented and drained. Failure has been observed at faying surfaces at panel lap joints and under battens used to hold panels in place.

Failure has also occurred in weathering steel buildings with insulated panels. Certain urethane foam insulation materials contain chlorides, which can become soluble on exposure to moisture or ultraviolet light; when in contact with the unprotected backs of insulated panels, they can corrode the panels from the inside out.[14] Sealant failure, coupled with negative air pressure drawing water into the wall, has caused insulated weathering steel panels to fail.[15]

During patina formation, the outer 2 to 3 millimeters of the steel surface corrodes. The corrosion product is washed from the structure onto adjacent materials, and unsightly staining can occur unless structures have been designed with drip edges to direct water away. Materials that experience severe staining and are difficult or impossible to clean include concrete, stucco, galvanized steel, matte porcelain enamels, stone, wood, and unglazed brick.[16]

Proper design of bases and footings is critically important. Foundations should be designed to reduce unsightly runoff staining as well as to protect the weathering steel from contact with soil. Concrete columns and foundations stained by runoff from weathering steel above are a common sight. If weathering steel is buried unprotected in the ground, it will corrode at the same rate as or faster than carbon steel.

## CONSERVATION TECHNIQUES

The design of a weathering steel structure—both the original and the retrofit of an existing building or sculpture—should minimize pockets and crevices where debris and water can accumulate. However, this is not always possible, especially in a retrofit design. In that case a periodic clean-up program should be instituted to keep all pockets and ledges free of exfoliating rust, leaves, and other trash.

Careful cleaning with a dry brush or broom, water blasting with low-to-medium pressure, and brushing and mopping with detergent solutions in water or organic solvents will often remove detritus buildup, marker graffiti, and paint, and these methods should always be tested first. Steam cleaning can be very effective. High- or ultra-high-pressure water blasting is effective in eliminating destructive rust, repair scars, graffiti, and old paint, but can disturb or remove weathered patina. If abrasion is required to expose white metal, sodium bicarbonate (baking soda) may be added into the water stream. Sandblasting should be avoided.

A number of measures can be undertaken to stabilize and improve the condition of weathering steel structures. Existing connections with faying surfaces should be sealed to prevent water infiltration into the space between the steel members. The most durable repair is to weld the edges of the faying surfaces. Sometimes welding is not practical because of the adjacent materials; in this case the faying surfaces can be protected with sealants (although these require maintenance). Surface preparation is critical to sealant installation.[17]

Drainage enhancement is a common strategy for dealing with water pooling. Foundations must be well drained and have an effective moisture barrier between the foot of the structure and the surface below. In the structure itself, weep holes, or slots, allow water to drain away rather than pool. Surfaces that slope but do not drain may be retrofitted by drilling weep holes at the low end to drain water. Weep holes must not empty into an undrained area. They must be properly located and of correct size: large enough to avoid clogging, yet small and visually unobtrusive. Undrained elements that do not slope should be painted and maintained or repositioned to add slope. Corrosion-prone surfaces in pooling areas that cannot be drained may require sealing locally with a clear, water-excluding coating. When such areas are out of sight, aesthetic impact of coatings is minimal. Such coatings must be maintained in good condition.

Weathering steel structures are often painted or clear coated. However, painting surfaces intended to be unpainted is an ineffective conservation measure because it adds paint maintenance to the problems of weathering steel and because both painting and clear coating markedly alter surface texture and appearance. Drainage and sealing measures are usually more effective and appropriate than coating.

However, some structures are on unfortunate sites where local moisture persists, graffiti are a problem, or other destructive contamination occurs. In such cases unobtrusive physical barriers should be installed to protect the structure. Following conservation and after new patina has developed, it may be necessary to clear coat only the threatened surfaces as a moisture barrier or to eliminate frequent localized cleanings.

Identifying and repairing all deteriorated surface panels when corrosion is occurring from the inside out is nearly impossible. In fact, care should be taken when investigating these structures because the corrosion may be much worse than it appears. In these cases complete recladding or overcladding is sometimes the only option. Repair attempts using epoxy patching on insulated panels is often futile because the panels corrode next to the repaired area.[18] Structures with uninsulated panels can sometimes be retrofitted by removing the panels and improving the connections to the structure to eliminate faying surfaces and nondraining members. Glass installations must be retrofitted so that the glass is not subject to expansion forces from the adjacent steel.

## REPLACEMENT

Refabrication is often necessary to reinforce or replace deteriorated parts or entire structures, or to allow sealing or elimination of corrosion-prone features. If possible, determine the specific weathering steel originally used in the area being repaired, and use the same for the repair. If weathering steel sheet material is used as a panel replacement material, it should be at least ¼ inch thick.[19] The underside should be coated and the connection details carefully designed to shed water.

Mill scale must be removed before installing replacement pieces to allow proper development of the patina. Except in unwelded structures, arc welding is the appropriate method of repair joining and should follow manufacturers' specifications. Generally, single-pass and structural welds may be made using E70-group, mild-steel, covered electrodes, as long as the weld procedure ensures proper composition enrichment from adjacent steel. Different weld methods and fill materials must be used when the exposed weld is broad or must match the surrounding surface in color and corrosion resistance. Manufacturers' specifications should be followed.

Development of new, visually homogeneous rust on weathering steel structures is usually initiated by abrasive blast cleaning. Although the bright contrast with adjacent, unaffected mature patina is alarming at first, the appearance of properly welded areas improves quickly with time. The visual contrast of repairs may be mitigated by also blasting adjacent surfaces. The entire blasted area then weathers and regains its patina together, and the appearance of the welded area matches the rest of the surface. However, traditional abrasive blasting is not always possible or advisable.[20]

# PART II
# CONCRETE

CONCRETE BLOCK

CAST STONE

REINFORCED CONCRETE

SHOTCRETE

ARCHITECTURAL PRECAST CONCRETE

PRESTRESSED CONCRETE

**Figure 6.1.** Concrete masonry elements made with Haydite, a burned clay or shale aggregate, were produced in the 1930s. Wood trim and furring could be nailed directly into the blocks, used for backup and partitions. Catalogue, Hydraulic-Press Brick Company, St. Louis, undated.

PAMELA H. SIMPSON; HARRY J. HUNDERMAN AND DEBORAH SLATON

# 6 | CONCRETE BLOCK

**DATES OF PRODUCTION**

c. 1900 to the Present

**COMMON TRADE NAMES**

Celocrete, Haydite, Pottsco, Straublox, Waylite

## HISTORY

Concrete block is produced from a mixture of portland cement and aggregates.[1] Commonly manufactured in a nominal 8-by-8-by-16-inch size, concrete blocks, also known as concrete masonry units, can be solid or hollow (typically with two or three cores), and block ends can be flat or flanged. Block weights vary depending on the aggregates used in the manufacturing process.[2] Most concrete blocks manufactured today have a core area of 40 to 50 percent and are defined as hollow.[3] The blocks' configuration determines two key characteristics: compressive strength and fire resistance.[4]

### ORIGINS AND DEVELOPMENT

The mass production of hollow concrete blocks is a twentieth-century phenomenon. Throughout the nineteenth century attempts had been made to cast concrete into building blocks, but the modern industry began with Harmon S. Palmer's invention of a cast-iron block machine, which he patented in 1900.[5]

The first phase of the industry, the period 1900–1920, was characterized by the Palmer-type hand-operated metal machines that made single blocks. Palmer founded the Hollow Building Block Company in 1902 to manufacture his machines, and by 1904 his company was producing about four hundred machines a year. Competitors soon began flooding the market with similar machines.[6] Palmer won a number of patent infringement lawsuits, but new machines continued to be produced.[7]

The growth in popularity of concrete block in the first two decades of the twentieth century was phenomenal.[8] One reason for this overnight success was the growing availability of an improved and reliable portland cement. Even though the appropriateness of ornamental faces was debated in the building press in the 1910s and 1920s, the concrete block industry was rapidly developing new machines, mixtures, and curing methods to improve the product.[9]

The industry also began to organize. The Concrete Block Machine Manufacturers Association was founded in 1905, the Concrete Producers Association in 1918, and the Concrete Block Manufacturers Association in 1919.[10] One significant result was the creation of standard block sizes. In the industry's early years, sizes varied, but in 1924 the associations agreed on standard sizes. By 1930 the 8-by-8-by-16-inch unit had become the most common block size. Another development was the increased use of testing to improve the blocks' reliability and durability.

These organizations also worked to promote the use of concrete blocks through trade magazines, catalogues, and books addressed to the home-building public. Such publications as *Plans for Concrete Houses* and *Concrete Garages: The Fireproof Home for the Automobile* appeared in the 1920s. The success of these promotional campaigns is reflected in industry production levels. In 1919, 50 million concrete blocks were produced in the United States; in 1928 that number had increased to 387 million. A brief downturn during the Great Depression was followed by an upsurge during World War II. By 1951, 1.6 billion blocks were being produced in the United States. Meanwhile, the number of plants was decreasing. In 1920 there were 7,000 manufacturers; in 1928 there were only 4,140. The industry was consolidating into large manufacturers that became increasingly connected with ready-mix concrete companies.[11]

A major change in the industry was the introduction of lightweight aggregates to correct the problem of weight. In 1917 F. J. Straub patented cinder blocks and in 1919 established a plant in Lancaster, Pennsylvania.[12] In its first year it produced 25,000 cinder blocks, but by 1926 Straub was making more than 70 million blocks annually. Straub promoted cinder blocks as an affordable, lightweight, structural material that could be faced with more traditional materials such as brick.[13] Cinder blocks were strong, could be nailed into, and were easy to lay. Like Palmer, Straub fought to prevent imitators from infringing

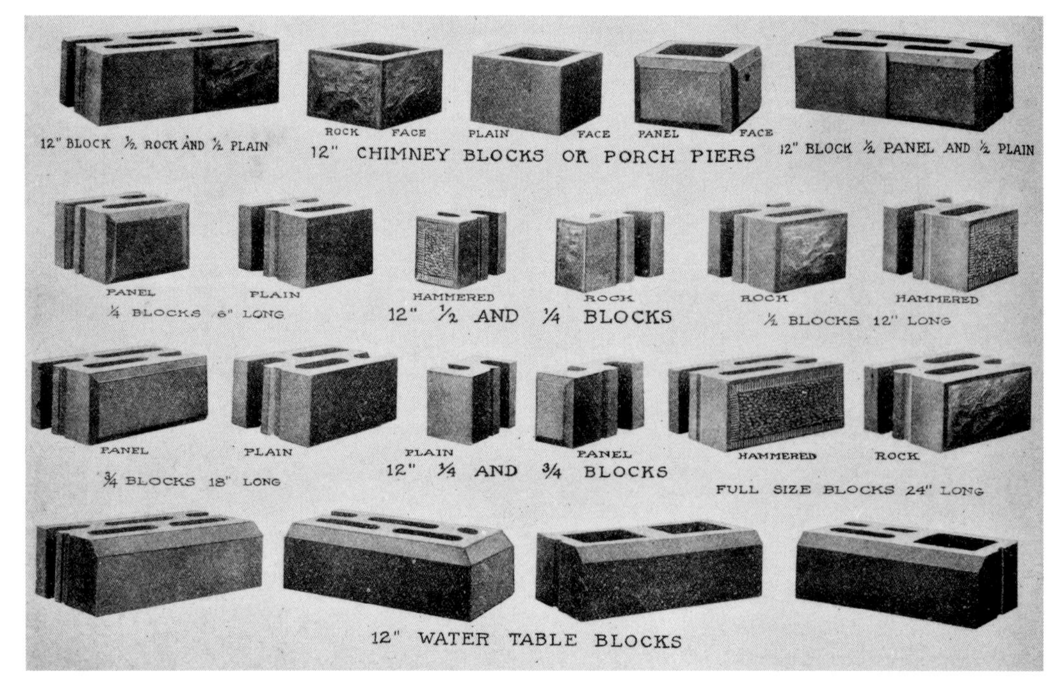


**Figure 6.2.** Frost- and moisture-proof because of their staggered air cavities, Miracle Blocks, patented in 1903 by the Miracle Pressed Stone Company of Minneapolis, were used for chimney flues, porch piers, and exterior walls.

on his patent rights and licensed production throughout the 1920s and 1930s.

In the 1930s, and especially in the 1940s, however, many other lightweight aggregates were introduced. Some, like pumice, were natural; some, like cinders and slag, were by-products; and still others, like expanded shale, clay, and slate, were manufactured. Expanded shale, clay, and slate aggregates, which have been particularly popular, are made by a process of rapid firing that bloats the raw material with expanded gases. The material is then quickly cooled to retain a cellular structure. Haydite, one of the earliest expanded shale products, was patented in 1919 and first used for blocks in 1923. Pottsco (later called Celocrete), a blast furnace slag treated with water, was introduced around 1930. In the late 1930s Waylite, a slag expanded with steam, was introduced. Many lightweight aggregates had pleasing light colors.[14]

## MANUFACTURING PROCESS

Most early-twentieth-century machines had a metal frame and mold box with a hand-release lever that allowed for removal of the sides and cores. The two main types of machines were downface and sideface, named according to the location of the faceplate. A fairly dry mixture of portland cement, water, sand, and stone or gravel aggregate was shoveled into the machine and tamped down to compress the mix and prevent voids.[15] Once released from the machine, blocks were stored on a pallet or board and dried.

An alternative way of making blocks was the slush method, which used the Zagelmeyer cast stone block machine. A wet mix was poured into metal molds that sat on open railcars, which were subsequently rolled into curing sheds. Each car could hold up to twenty molds, but the blocks had to dry in the mold before they could be released and the molds used again.

In the 1920s power tamping and stripper machines replaced the earlier hand-tamped, downface forms. The stripper had rigid sides that extruded the blocks from the molds rather than releasing them through movable sides. Most of the process, including the tamping, was automated. Two major changes in the 1930s were the introduction of automatic vibrators, which replaced tamping, and machines that could make multiple blocks.

In 1934 Louis Gelbman and the Stearns Manufacturing Company of Adrian, Michigan, developed the Joltcrete machine, which made three standard blocks simultaneously and nine blocks a minute. The vibrating machines also gave impetus to the use of lightweight aggregates because they could handle the mixture better than the tamp machines could. By 1940 most aspects of making blocks, from mixing to curing, had been automated as

**Figure 6.3.** In the early 1900s concrete blocks were manufactured in molds filled by hand (above).

**Figure 6.4.** Concrete blocks were occasionally produced in colors. Rose-colored blocks were used for this foursquare house (1906) in Devil's Lake, N.D. (left).

well. The use of steam, first suggested in 1908 as a way to cure blocks quickly, became standard in the 1930s.[16]

### USES AND METHODS OF INSTALLATION

The advantages claimed for hollow concrete blocks were that they were inexpensive and that they could be installed faster than traditional materials, such as brick. They were also fireproof, needed little care, and could be ornamental. The early block machines came with a variety of faceplates that imitated cobblestone, brick, and ashlar, as well as scrolls, wreaths, and roping. But the most popular face—the one that was standard on all machines in the period 1900–1930—was rockfaced, the rough-cut surface that looked like quarried stone.[17] Thousands of buildings were constructed of rockfaced hollow concrete blocks in this early period. Examples range from twenty houses in Oakherst Place, the 1906 middle-class suburban development in Saint Louis, Missouri, to fifty houses in the 1909 working-class company town of Mineville, New York.[18]

Of the concrete blocks manufactured before 1915, nearly 75 percent were used for foundation and basement walls or as partition walls. While the foundation blocks were often rockfaced, the partition walls, which were later covered, were always plainface block. Plainface blocks covered with stucco were also popular for exterior walls.[19]

Changes in taste and technology in the 1930s led the industry to shift into greater production of the more utilitarian and less decorative plainface blocks. When they were used as the finished surface, coloring pigments could be added.[20] However, the majority of blocks were used as backup and for cavity-wall construction.

**Figure 6.5.** Often relegated to foundation and backup use, concrete blocks could also be decorative. The concrete ashlar walls of Chicago's Will Rogers Theater (1935, Z. Erroll Smith) were elaborately painted.

## CONSERVATION

Concrete block is a durable construction material when properly manufactured, installed, and maintained. New repair technologies such as improved patch mixes, corrosion-resistant anchorage systems, and breathable coatings are available to extend the service life of concrete block.

### DETERIORATION

Like other concrete products, concrete blocks shrink as they cure and change when exposed to thermal and moisture changes.[21] Cracking is most often related to shrinkage of the concrete or movement of the blocks and the wall. Other common forms of deterioration include spalling and efflorescence. Spalling may be caused by problems in the block mix, water infiltration, or mechanical damage. Efflorescence is the accumulation of salts on the surface, generally carried through the masonry by the movement of water. Staining and dirt accumulation can also occur on concrete blocks.

### DIAGNOSTICS AND CONDITION ASSESSMENT

A visual survey can identify symptoms of distress—patterns of cracking, water movement, spalling, or other deterioration—and thus help determine the extent and causes of distress. Step cracking is usually related to differential settlement. Water movement through a wall results in efflorescence, while water entering from the ground at the base of a wall may result in rising damp. Localized water staining may indicate deficiencies in drainage and flashing systems. Particularly with unpainted concrete blocks, a visual survey during rainfall may determine patterns of water infiltration. Water penetration testing can monitor water infiltration and movement.

Laboratory testing is sometimes required to determine specific causes of deterioration. Causes of spalling that are related to improper mix design can be identified through petrographic and chemical analysis of samples. Mix problems may also be the cause of cracking. Use of inappropriate coatings or sealers that trap water in the wall can be diagnosed through a combination of field survey and laboratory testing. Laboratory structural testing to evaluate compressive strength may be required if the blocks appear to be failing structurally.

### CONSERVATION TECHNIQUES

Typical maintenance requirements include repointing joints occasionally and recoating the surface if paints or clear sealers are used. However, intervention may be required to address problems such as cracking or spalling, efflorescence, staining and dirt deposits, and water infiltration.

The selection of a cleaning method for concrete block construction depends on the condition and finish of the blocks and the degree and type of soiling. Unpainted concrete blocks may be cleaned with low-pressure water (400 pounds per square inch or less) and a mild, non-ionic detergent. When soiling of unpainted blocks is more severe, chemicals or detergents can be used in liquid or poultice form. Poultices can be effective on deep-set stains. Selection of a cleaner should be based on trial cleaning samples; the mildest effective one should be chosen.[22] For painted concrete blocks, test samples will reveal whether cleaning is likely to damage or remove the paint finish.

Efflorescence can be limited by preventing moisture from entering the walls by proper pointing, installing flashings and copings, and using appropriate vapor barriers. Paint or other surface coatings may be used on some masonry walls to prevent moisture infiltration. Efflorescence can generally be washed away with water; however, the source of water penetration must be located, or it will recur.

Before repairing cracking, the problem of movement must be solved, or the wall will crack again after repairs. In new or rebuilt concrete block walls, using concrete blocks manufactured to limit shrinkage and moisture movement will mitigate future cracking. Reinforcement can also help resist cracking.[23]

Cracks that extend through the joints between blocks should be repaired by repointing. If cracks are hairline or of fine width, a surface coating may be sufficient to prevent water infiltration, and further repairs may not be required. Wider cracks can be repaired by cutting the crack out to ½ inch wide at the surface and undercutting to ½ to ¾ inch wide at the back and ½ inch deep. The crack should be clean, free of dust, and dry before repairs are begun. Then the surfaces of the crack should be dampened and the crack filled with mortar.[24]

In existing construction, control joints (vertical separations constructed in the wall to accommodate movement) may be installed if such modification is not visually intrusive. Spacing of the joints depends on expected movement, the wall's resistance to horizontal tensile stress, and the number and location of wall openings. Control joints are typically located at changes in wall height, thickness, or orientation; at construction joints; and at the sides of wall openings. If concrete blocks are used as a backup material, the control joints should extend through the facing material as well as the blocks, unless the bond between the facing material and backup is flexible. When installing control joints is not possible, wider cracks may be repaired by cutting out the crack and installing a sealant, prepared according to the manufacturer's recommendations.[25]

Spalls may be repaired by installing a patch mix similar in nature to the concrete blocks. A typical patch mix consists of dry masonry sand, cement, and an admixture to improve bond. The sand should be selected to match the appearance of the adjacent block. Stainless steel pins, set in epoxy and covered by the patch surface, can be used to anchor large or deep patches to the concrete block substrate.

Proper pointing of joints between concrete blocks is an important factor in keeping walls watertight.[26] The mortar should be compatible with the blocks and, in an existing wall, with the historic mortar. Samples of existing mortar can be taken for laboratory analysis to develop a mix matching the composition and appearance of the original.[27]

If improper mix design or water infiltration is involved, overall repair of the wall and its water protection system as well as patching may be required. Another important means of preventing water-related deterioration of concrete blocks is incorporating flashing into the wall system. Flashing should be included in the design of new or replacement block walls and may be installed during repair of an existing wall.[28] Flashing should be used at all wall penetrations to direct any water that enters the wall to the exterior. Weep holes may also be used to expedite water removal. Flashings can be made from a variety of sheet metal or flexible membrane materials.

While many concrete block walls do not require coating or waterproofing, coatings can be used with concrete blocks to prevent water penetration or improve aesthetics. In some cases, a coating may be necessary to prevent water from penetrating or even passing through the wall.[29] The finish coat will need to be repainted at infrequent intervals as part of regular maintenance.[30]

Coatings for concrete block include portland cement paints, latex paints, oil-based coatings, rubber-based coatings, epoxy coatings, alkyd paints, urethanes, and silicones. The selection of a coating depends on the conditions to which the wall will be exposed and the special characteristics the coating provides, such as breathability, alkali resistance, color or colorlessness, ultraviolet resistance, and resistance to water absorption.[31]

#### REPLACEMENT

If cracking is severe or extends across entire blocks, replacement of an entire block may be necessary.[32] A summary of American Society for Testing and Materials (ASTM) specification requirements for concrete unit masonry is provided in the National Concrete Masonry Association's pamphlet "ASTM Specifications for Concrete Masonry Units" (TEK 36A).[33] Custom blocks may be manufactured in special shapes—with curved and beveled corners, for example—to meet particular replacement needs. Molds can be used to form replacement blocks matching the texture and shape of historic blocks. Laboratory studies to determine the composition can be useful in determining an appropriate mix for replacement blocks.

*Bas-relief ornamentation even of intricate pattern, is easily possible in cast stone at moderate expense.*

# *Modern Ornamentation*

CAST STONE—concrete moulded in sections to definite architectural designs—opens new possibilities of beauty in modern commercial buildings.

The furniture warehouse, the storage garage, the automobile sales and service depot, the wholesale grocery, the small manufactory—once bleak, wholly utilitarian structures—today may be buildings of dignity and character.

Cast stone affords wide latitude to the architect in achieving simple and effective design. It affords, too, a complete control of color. Owner and architect may plan a building of lightest hue, secure in the knowledge that it will resist weather, smoke and soot stain. When backed by reinforced concrete construction, cast stone permits buildings to be fire-safe and storm-proof, with economy.

Business executives planning on new buildings and architects specializing in the design of commercial structures are invited to write for complete information. The beauty and distinction economically attained through the new technique in concrete are well worth careful consideration.

*Building of the New York Telephone Co., Syracuse, New York. Vorhees, Gmelin and Walker, New York City, Architects.*

**PORTLAND CEMENT ASSOCIATION ~ Chicago**

**Figure 7.1.** Intricate ornamentation could be achieved with cast stone. This Portland Cement Association advertisement highlights the New York Telephone Company Building (1928, Vorhees, Gmelin and Walker), Syracuse, N.Y. *Architectural Forum,* April 1929.

ADRIENNE B. COWDEN; DAVID P. WESSEL

# 7 | CAST STONE

**DATES OF PRODUCTION**
c. 1869 to the Present
**COMMON TRADE NAMES**
Arnold Stone, Benedict Stone, Chicago Art Marble, Dextone, Edmunds Art Stone, Instone, Litholite

## HISTORY

Cast stone is a highly refined form of concrete made from portland cement, fine and coarse aggregates, and water.[1] Executed as veneer, block, or ornament, cast stone typically simulates evenly veined and colored stones.[2] Its manufacture—in custom molds—is labor intensive and site specific. Used extensively during the late nineteenth and early twentieth centuries, cast stone was produced in a wide range of colors and textures. By the late 1920s numerous formulations of varying compressive strengths were used to simulate natural stone.[3]

### ORIGINS AND DEVELOPMENT

Although cast stone dates to the Middle Ages, not until the nineteenth century did innovators seriously explore the possibility of using concrete as an economical substitute for natural stone. U.S. production of cast stone before the 1860s was limited to isolated, unrelated experiments in building construction.[4] After the Civil War, however, the demand for concrete increased dramatically.[5] One of the first patents for cast stone was granted in 1868 to George Frear of Chicago.[6] Frear Stone contained only hydraulic cement, sand, and an alkaline solution of gum shellac.[7] The Pacific Stone and Concrete Company, based in San Francisco, was also active by 1868. This enterprise was specifically organized to manufacture cast stone under the 1856 patent granted to Frederick Ransome,[8] whose Siliceous Stone contained a mixture of silicate of soda, sand, gravel, flints, chalk, limestone, caustic soda, calcium chloride, and water.[9]

The majority of early cast stone companies based their operations on European practices and relied on imported natural and portland cements.[10] However, as domestic portland cement became more easily available and its price fell, the number of cast stone manufacturers began to grow. By the mid-twentieth century, numerous patents had been granted for artificial stone formulas and finishes.[11] Initially white portland cement was prohibitively expensive, but as the cost decreased, the production of cast stone increased and manufacturers could offer a wider variety of colors and production increased. During this period cast stone was increasingly based on scientific formulations. Crushed aggregates, cements, and mineral coloring agents were carefully selected, graded, and proportioned.[12]

The need for quality control became apparent as the industry expanded. Periodicals such as *Fireproof* and *Brickbuilder* reported the collapse of cast stone structures, and each failure damaged the industry's reputation. Published technical data, however, was meager, and manufacturers rarely tested the compressive strength and absorption of concrete samples; often a minimal (and inadequate) compressive strength of 1,500 pounds per square inch was specified.[13]

In 1927, thirty-four leading manufacturers formed the Association of Cast Stone Manufacturers to improve the quality of cast stone and disseminate information regarding its characteristics and use.[14] This association, which changed its name to the Cast Stone Institute in 1929, regularly tested the physical and mechanical properties of cast stone.[15] The results of these tests and those undertaken by the American Concrete Institute laid the foundation for standard specification P3-A-29T, adopted by both associations in 1929.[16] This specification dictated a minimum compressive strength of 5,000 pounds per square inch and a permissible absorption of 7 percent.[17]

In 1931 the Federal Specifications Board followed suit, adopting a purchase specification for cast architectural stone.[18] Three years later the U.S. Department of Commerce's Division of Commercial Standards, in cooperation with the Cast Stone Institute, adopted standard colors and finishes. Standard specification TS-2126 included colors and finishes most frequently used by cast stone companies; purchasers could then easily specify a cast stone product by number.[19] These early specifications provided manufacturing guidelines and helped define the characteristics and limitations of cast stone.[20]

**Figure 7.2a–b.** The Delaware and Hudson Railroad Company Building (1918, Marcus T. Reynolds), Albany, N.Y., is richly ornamented with cast stone tracery trim (**a**), executed from these drawings (**b**).

## MANUFACTURING PROCESS

Cast stone manufacture begins with the preparation of working drawings that delineate section, bed, and face templates for the casting molds. Ostensibly molds can be assembled out of any material that can withstand both the weight of concrete as well as the tamping necessary to consolidate the raw mixture. Molding materials include wood, plaster, glue, sand, sheet metal, and gelatin.

Cast stone is produced by either a dry-tamp or a wet-cast process. The dry-tamp method requires a low-slump mixture that must be tamped into the mold.[21] Dry-tamped stone is often cast in two layers. An inner core is poured first and can be either dry-tamped or poured full of concrete. A decorative outer layer, referred to as the facing, is then cast around the core. For economic reasons, costly aggregates, pigments, and cement were generally used only in the facing layer.

A variety of aggregates—from granites and marbles to blast-furnace slag—have been used to create cast stone elements that require little finishing on removal from

**Figure 7.3.** Large cast stone operations employed modelers and carvers for elaborate details. At the Benedict Stone Company's plant in Chicago, ornamental components are in various stages of production.

the mold. The wet-cast method involves a more plastic, integrally colored mix that contains enough water for the material to flow easily into the mold. The method of manufacture determines the type of mold used and the number of castings possible in a working day. Although molds for dry-tamp mixtures can be used many times per day, the high water content of wet-cast mixtures generally allows only one piece of stone to be cast in a single mold each day.

The casting method also determines how a piece is cured. For both dry-tamp and wet-cast products to cure properly and attain sufficient compressive strength, the newly made stone must not dry out too rapidly. After it is removed from the mold, a dry-tamp product should be placed in a warm, moist environment; it is standard practice to cure dry-tamped stone with steam, a common practice by 1920. With wet-cast products the danger of the stone drying out is minimized by the excess moisture in the mixture. Since an element cast in this manner generally cannot be removed from the mold for at least a day, drying is also retarded by the mold itself. Wet-cast stone does not have to be steam cured; however, it should be kept damp for at least five days and placed in a warm environment to harden.

To create details such as veining, early manufacturers placed dye-soaked strings or thin strips of wood in the mold. The strings were removed before casting, leaving dye to be absorbed into the concrete mixture.[22] Veins were also simulated by applying color or dye with a fine brush directly on the surface of the newly cast stone or by adding pigment to the concrete mix but not thoroughly stirring it.[23] Typically, the finish selected for cast stone falls within one of three classes: surfaced cast stone, which includes hand-rubbed, brushed, and acid-washed (etched) finishes; cut cast stone, which includes machine-rubbed, planar-rubbed, bush-hammered, and machine- and hand-tooled finishes; and plain cast stone. Dry tamping results in a molded surface without a layer of cement and aggregate fines. In wet casting, manufacturers typically tool, abrade, or use an acid wash on the surface to remove the cement layer and create the desired finish.

**Figure 7.4.** Benedict Stone was used for Soldier Field (1924, Holabird and Roche), Chicago. The textured finish is visible in the steps.

## USES AND METHODS OF INSTALLATION

Since cast stone could be manufactured in virtually any shape or size, this material was not limited to any one architectural style or building type. It may be found in houses, banks, churches, schools, libraries, and office buildings. It was typically used for water tables, window-sills, steps, belt courses, chimney caps, spandrel panels, and even sculpture. Cast stone was also used for intricate ornamentation. The Doric colonnade and decorative details of Soldier Field (1924, Holabird and Roche) in Chicago were carved from Benedict Stone, a reinforced cast stone produced by Benedict Stone, Inc. Another notable example of ornamental cast stone work is the Delaware and Hudson Railroad Company Building (1918) in Albany, New York. The tracery and trim covering each facade was produced by the Emerson and Norris Company.[24]

Cast stone was and still is generally installed as a laid-in-place masonry product—that is, it is laid up with mortar exactly as if it were natural stone. Cast stone may also be attached to the building with metal anchors or similar fastening devices. Hooks and hangers to facilitate handling are placed in the elements while they are being cast.

During the Great Depression construction was limited, and although cast stone continued to be used throughout the 1930s, the demand for the product had dropped dramatically.[25] Some cast stone companies began to manufacture precast products such as blocks, floor slabs, joists, burial vaults, and manhole covers to stay in business, but few could compete with the increasingly mechanized manufacturers of concrete block and precast concrete products.[26] The maximum utilization of all building materials during World War II and the elimination of decorative elements during and after the war effort also hurt the cast stone industry. In addition, the lighter, precast concrete products were less expensive.[27] By the early 1950s most cast stone companies had been either dissolved or absorbed by the precast industry.

**Figure 7.5.** When built, the colonnade of Soldier Field was the largest in the world.

# CONSERVATION

The conservation of cast stone is often related to problems that begin with historic manufacturing techniques. Many causes of deterioration are related to techniques that used distinctive aggregates, varied grading profiles, and pigments to create specific surface colors and finishes. When proper standards for proportioning were ignored, cast stone products sometimes had unusually high porosity, which affected durability. Aggregates lacking suitable durability were also used frequently, and these can accelerate weathering processes.

## DETERIORATION

Crazing (hairline cracking) on cast stone surfaces is common, particularly for dry-tamp cast stone. Crazing can often be attributed to volumetric changes between the facing and backup mix, or to improper proportioning of the facing mix. Although proper proportioning of the cement, aggregate, and water mix was well understood by the twentieth century, mixes were often modified, deviating from accepted proportioning standards. The strongest, most durable cast stone was produced with aggregates of varying sizes. By grading the aggregate, smaller particles fit between larger pieces and remaining voids were filled with cement. To achieve a specific, uniform appearance, manufacturers often used a preponderance of one aggregate size, resulting in a more porous and less durable cast stone.[28]

In dry-tamp cast stone, delamination of the facing from the backup is a common occurrence. A condition often caused by inherent manufacturing flaws, delamination can also result from differing water absorption ratios in the layer, which is exacerbated by freeze-thaw cycles.

Cast stone, like other concrete products, is subject to deterioration caused by carbonation (loss of alkalinity), aggregate-alkali reactions, freeze-thaw cycling, and erosion. Freeze-thaw cycling may cause surface scaling if moisture penetrates a porous cast stone.[29] Because some aggregates and crushed stones used historically were not durable, acidic environments can be detrimental to some types of cast stone, such as calcareous stones like limestones, which were used to produce light-colored stone. Corrosion of metal reinforcement embedded in cast stone can cause cracking and spalling and can be detected by rust-colored staining.

**Figure 7.6.** Clarence B. McKay's design for a building entrance (1920s) typifies cast stone's ornamental possibilities.

## DIAGNOSTICS AND CONDITION ASSESSMENT

The type of cast stone (wet-cast or dry-tamp) should be identified. Often a broken cast stone feature will reveal a facing layer or integrally colored units. If visual examination is inconclusive, a sample can be removed for identification.

The original surface finish should also be identified before conservation work is begun. More subtle finishes may not be distinguishable from regular mortar and may require cleaning or paint removal. The presence of a coating is often a clue that the surface has been damaged or discolored. However, historically paint was applied for waterproofing and for hiding streaking of colored cementitious materials after rainfall.[30] Paint was also applied to dirty, stained, and patched areas.

Cast stone should be examined carefully to diagnose deterioration. Crazing appears as fine hairline cracks and indicates a high porosity. Delamination of dry-tamp stone can be detected by tapping the stone with a plastic hammer and noting hollow-sounding areas. Erosion is evidenced by surfaces where the aggregate and cement binder are weathered. Eroded surfaces appear sandy, pockmarked, or rough when the aggregate is exposed. Horizontal surfaces, such as the tops of parapet walls and cornices, are particularly prone to erosion.

The position of metal reinforcement can be determined using a metal detector. The depth of the reinforcement should also be measured. A simple, inexpensive method for determining depth is to drill a small hole using a $\frac{1}{16}$-inch masonry bit. When the drill bit touches the metal, mark the bit, remove it, and measure to the tip.

## CONSERVATION TECHNIQUES

Cast stone in relatively good condition may require cleaning only. Cleaning procedures should follow the general ap-

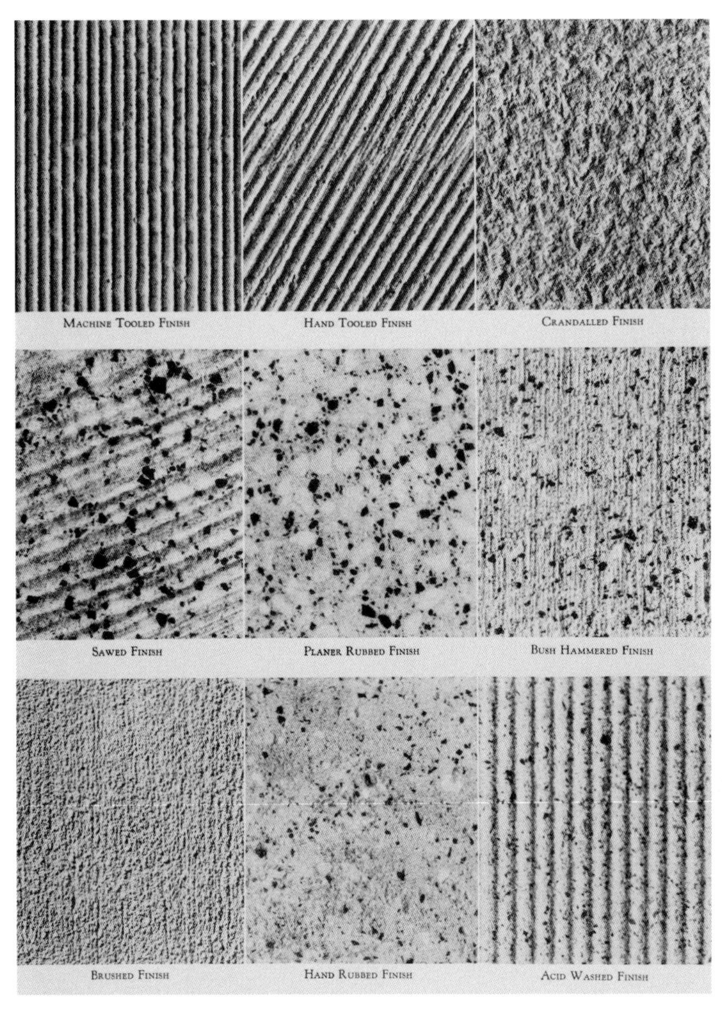


**Figure 7.7.** For decorative finishes, cast stone was mechanically tooled or chemically etched.

proach for cleaning all historic materials: using the gentlest means possible. Before cleaning cast stone surfaces, the exterior should be surveyed for cracks and weak spots. If a pressure water system is used, be careful to avoid damaging porous or crazed surfaces; the pounds per square inch should not exceed that used on soft masonry.[31] Acid-based cleaning systems that may react with decorative aggregates should be avoided. Paint removers to be used for cleaning graffiti should be tested carefully on small, inconspicuous areas, as some may lighten or damage cast stone.

Embedded metal that has corroded usually requires removal, particularly if the stone's structural integrity has been compromised. If it is necessary to install new metal reinforcement or use pins to reattach loose elements, stainless steel should be used.

Composite patching of damaged areas to match existing cast stone is a difficult repair procedure. It is important to develop a patching mix using the same or compatible materials, particularly for the aggregate. Lime, which was frequently used in the production of cast stone, could be added in the patching mortar to increase bond strength and workability. Patches should be slightly weaker than the cast stone to prevent stress cracking.

To determine the patching mortar components needed, testing similar to mortar analysis is required. At a minimum, a sample should be taken to separate the aggregate and binder. Obtaining suitable aggregates for patches and replacement cast stone is particularly challenging. Historical research on a building or cast stone manufacturer may uncover the original aggregate to complement physical samples. Manufacturers of texturing and coloring systems for concrete may also be suppliers of specialty aggregates or suggest other sources. Binder

**Figure 7.8.** A delaminated ornamental layer of cast stone on the Orpheum Theater (1925, B. Marcus Priteca), San Francisco, was reattached with epoxy.

colors can be matched using commercially available pigments to tint white or gray cement.

Larger patches may require mechanical anchorage. A variety of noncorroding materials, including stainless steel, copper, and polyester rods, are used for this purpose. The pins or rods are set in holes drilled in the area to be patched and secured with an adhesive, commonly epoxy.

Reestablishing the bond between delaminated facing and inner-core layers in dry-tamp cast stone is time-consuming and expensive. The selection of a repair material depends on many factors, including strength, adhesion, cohesion durability, and compatibility with the cast stone layers. The most compatible material is usually cementitious grout. However, penetration of a grout may not be possible if the separation between layers is small. In such cases, epoxy adhesives have been used to reestablish the bond between layers. Multiple delivery ports are inserted into the cast stone, and once the epoxy has been injected, the portholes are plugged with a patching compound. Any cracks must be blocked with a removable sealant to prevent the adhesive from leaking onto the cast stone.

Another method of repairing delaminated layers is mechanical anchoring—the insertion of threaded stainless steel rods into predrilled holes. Rods are subsequently secured with adhesive below the cast stone surface and patched. This technique can be used successfully only if the facing layer is sufficiently thick to accept the rod and adhesive.

Water repellents may be an effective way to reduce moisture infiltration and limit reinforcement corrosion in porous cast stone. Although repellents are usually viewed with suspicion, particularly for historic masonry, properly selected and tested repellents can extend the service life of deteriorating cast stone. Water repellents should be vapor permeable and not alter the visual qualities of the cast stone.

## REPLACEMENT

When cast stone components are beyond repair, replacement elements can be manufactured to match historic ones. Although some manufacturers can reproduce cast stone, matching the color and texture of the original material is extremely difficult. The process is expensive, partly because new molds have to be manufactured. When replacing cast stone, elements should be similar to those used in the original mix. Special attention must be paid to the aggregate, which in large part determines the material's color. New cast stone typically has a compressive strength of around 6,000 pounds per square inch and an absorption rate of less than 5 percent.

AMY E. SLATON; PAUL E. GAUDETTE, WILLIAM G. HIME, AND JAMES D. CONNOLLY

# 8 | REINFORCED CONCRETE

**DATES OF PRODUCTION**
c. 1885 to the Present
**COMMON TRADE NAMES**
Mushroom System, Ransome Unit System

## HISTORY

Concrete is composed of sand and stone (aggregate), water, and a cementitious material, almost always portland cement. Chemical reactions between cement components and water cause the mixture to harden as it dries, a process known as setting or curing. Well-cured concrete possesses great compressive strength but relatively little tensile strength. Reinforced concrete is concrete strengthened by the addition of another material, usually metal bars. The reinforcement, embedded in the concrete before it sets, can withstand tensile and shearing stresses, thus giving the material a much greater range of applications. The combination of concrete and metal allowed the construction of sturdy slabs, beams, columns, and pavements in the early twentieth century and eventually more complex architectural forms.

### ORIGINS AND DEVELOPMENT

S. T. Fowler patented a reinforced concrete wall in 1860, but widespread acceptance of the material took some time. In 1877 Thaddeus Hyatt, an American engineer, published his groundbreaking book, *An Account of Some Experiments with Portland Cement Concrete, Combined with Iron, as a Building Material*. Hyatt set forth the principle that iron reinforcement placed in the bottom of a beam would act in conjunction with concrete in the top, suggesting newly effective arrangements of reinforcing rods. He also advocated using bent and deformed reinforcing bars to achieve more controlled and more complete interactions between reinforcement and concrete than had been possible before.[1] Hyatt's publication of his theories did not, however, lead directly to commercial success.

The first practical commercial development of reinforced concrete construction in America is most often credited to Ernest Ransome, who promoted his building methods throughout the 1890s. Ransome refined procedures for casting girders, beams, and floor slabs as a unit on top of concrete columns, as well as designs by which load-bearing exterior walls could be replaced by expanses of windows (a significant development for factory design, in particular).[2] He marketed his services as a consultant and licensor, creating a successful business that greatly influenced the use of reinforced concrete in warehouses and factories. By 1900 competing techniques had emerged. Arrangements of mass-produced metal netting or fabric for reinforcing flat slabs appeared, and various bars or cables were designed for reinforcing beams and columns. The new technology was known by many names—armored concrete, ferro-concrete, concrete steel, steel concrete, and reinforced concrete—which became generally accepted about 1910. Many other methods were well known and widely marketed, including those developed by Albert and Julius Kahn and C. A. P. Turner.[3]

Significant advances were made in concrete technology during the twentieth century. Small quantities of admixtures or modifiers, such as air-entraining agents, superplasticizers, polymer modifiers, silica fume, and fiber reinforcements in the concrete dramatically improved its strength and durability.[4]

### USES AND METHODS OF INSTALLATION

The earliest reinforced concrete structures imitated the form of timber and steel buildings: reinforced concrete columns supported reinforced concrete girders, which in turn supported reinforced concrete joists. But soon after 1900 methods emerged for transferring the load-carrying capacity of beams and girders to floor slabs. This innovation saved overhead room (of particular importance in factories) and reduced floor-to-floor height and the expense of building elaborate formwork. This development is attributed to the Swiss engineer Robert Maillart and the Americans Orlando W. Norcross and C. A. P. Turner. In their systems, slabs rested directly on columns. Maillart developed a column capital that curved from shaft to ceiling; the American versions were straight and somewhat less expensive to construct. Reinforcement was

**Figure 8.1.** Reinforced concrete floor slabs, columns, and exterior walls were used for a Sears, Roebuck and Company building (1927, George C. Nimmons and Company) in Los Angeles. *Architectural Forum,* June 1928.

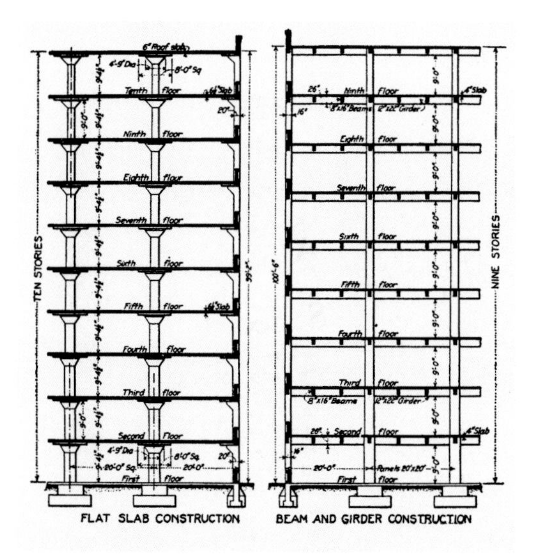


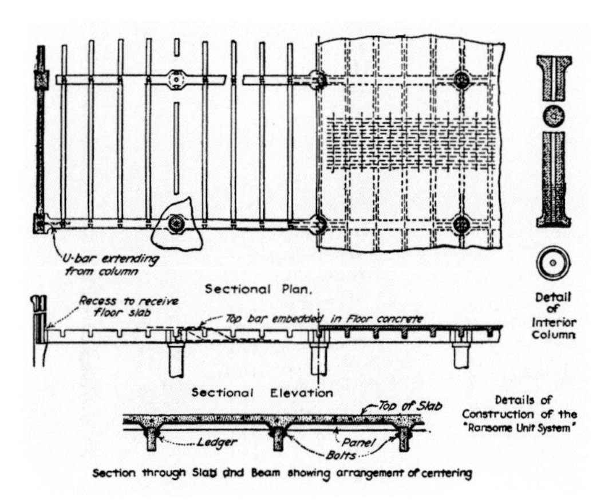


**Figure 8.2.** A comparison of flat slab and beam-and-girder construction (top) shows that use of reinforced concrete allows more floors in the same space.

**Figure 8.3.** In Ransome's unit system (bottom), the floor slab was poured in place after the girders, beams, and columns were erected.

concentrated in areas of highest tension (over the tops of columns). Shearing was counteracted by J-shaped stirrups. Julius Kahn patented a type of reinforcement that joined these features in a single unit. Column reinforcement at this time tended toward a spiral arrangement, after the work of the Frenchman Armand Considère.[5]

After 1900 reinforced concrete was rapidly adopted for industrial buildings of one or several stories. Reinforced concrete buildings could be built quickly, were fireproof, and could resist vibrations from heavy machinery. The first reinforced concrete skyscraper was the Ingalls Building (1903, Elzner and Anderson) in Cincinnati, constructed using the Ransome System. The Ingalls skyscraper adapted familiar techniques of steel-frame design to concrete elements, an approach soon imitated.

Reinforced concrete gained popularity as a material for buildings as well as other structures. French and German engineers adapted reinforcing methods for the construction of arch bridges, allowing for structures of much less mass than was traditional.[6] The system patented by Josef Melan in America in 1894 reduced the amount of steel needed in concrete bridges, and by 1897 a precursor of the concrete deck girder bridge had been achieved. Such bridges were more economical to construct than those involving elaborate arches. As their understanding of reinforcement improved, builders soon moved from using girders as projected beams to employing more sophisticated cantilevered girders. Maillart's elegant reinforced concrete bridges of the 1920s and 1930s are praised as the culmination of these design developments.[7]

At the same time, more unusual structures profited from the new understanding of concrete technologies. By the late 1890s two-way bar reinforcing developed by G. A. Wayss from Joseph Monier's later patents was used in grain elevators and storage tanks. Trade associations promoted reinforced concrete for smaller projects around the farm and home. Thomas Edison tried to create a system for rapidly molding reinforced concrete houses in a single pour into one huge form but was not successful.

The concept of using reinforced concrete for shells or domes emerged after 1910. The railroad station at Paris-Bercy (1910, Simon Boussiron) set a precedent for airplane hangars during World War I. In 1922 German engineers constructed a large dome by spraying concrete on a hemispherical metal framework. The concept of thin-shell construction as an inexpensive means of roofing auditoriums and markets spread quickly.[8]

Dramatic rib-and-arch designs of reinforced concrete began to appear in the 1930s, and in the Johnson Wax Administration Building (1939) Frank Lloyd Wright elevated the cantilevered slab-and-mushroom column to an aesthetic form.[9] For much of the first half of the century, frank expression of concrete technologies appeared only in the most utilitarian buildings. In all others, concrete elements were heavily disguised; the Ingalls Building, for example, featured marble, terra cotta, and brick cladding. But by the late 1950s, when the functionalist aesthetic gained favor among architects, the concrete frame asserted itself. Load-bearing reinforced concrete screen walls, of delicate vertical members joined to horizontal members at floor level, clearly revealed their material

**Figure 8.4.** In the Johnson Wax Administration Building (1939), Racine, Wisc., Frank Lloyd Wright experimented with unusual mushroom-shaped, reinforced concrete columns. Before construction began these were loaded with sandbags to test their structural capacity.

basis. In the 1970s and 1980s reinforced concrete buildings of extremely rough finish, even revealing the remnant of formwork, became popular.

The use of precast concrete elements has gained and lost favor over time. As early as 1905 slabs and girders could be purchased as precast components; an early example is the Ransome Unit System. Precasting eliminated on-site pouring, thus avoiding weather problems and labor shortages. But by the 1930s transit-mixed concrete came into use, and the demand for precast concrete diminished. In recent decades, with the refinement of prestressing techniques, it has regained some popularity for bridges, roads, and building projects.[10]

As the number of cement manufacturers increased and the use of reinforced concrete grew, several institutions emerged to establish standards, conduct research, and promote the product. Commercial materials producers and technical experts established the American Society for Testing and Materials (ASTM), the Portland Cement Association, and the American Concrete Institute between 1900 and 1910.[11]

Comparable agencies emerged in Germany, France, and England. The scientific study of concrete blossomed, shifting the understanding of reinforcement beyond the simple calculation of ratios of concrete to metal. Loading tests and laboratory analysis became commonplace, and regulations emerged alongside the new data. Governments issued legal guidelines for the use of concrete, and trade associations produced their own specifications. Ordinances for concrete construction grew partly because of several dramatic building failures between 1900 and 1910, but also as a favorable response to the performance of concrete buildings in fires during the same period.

## CONSERVATION

Reinforced concrete is commonly used in the construction of building frames, building facade elements, parking structures, bridges, dams, sculptures, and monuments. In many applications, reinforced concrete is exposed to the weather, and, as with masonry systems, the majority of deterioration is caused by moisture.

**Figure 8.5.** For the TWA Terminal (1962), New York City, Eero Saarinen used a more expressive form of reinforced concrete as well as thin-shell construction techniques. Saarinen's goal was "to design a building in which architecture itself would express the drama and . . . excitement of travel."

In the conservation of reinforced concrete, the key issue is whether the existing material can be repaired and conserved or whether it must be replaced. The concrete and the structure must be investigated to determine their condition, to determine whether the concrete can be retained, and to develop repair options and strategies. Laboratory analysis of the concrete to be repaired is an important part of the investigation.

## DETERIORATION

Concrete deterioration may occur for two principal reasons: corrosion of the embedded steel and degradation of the concrete itself. Concrete typically protects embedded reinforcing steel through its alkalinity. Corrosion occurs where embedded reinforcing steel is not protected by the concrete's normal alkaline environment and the steel is exposed to water or high relative humidity. For decades many instances of corrosion of steel in concrete were believed to be due to stray currents from nearby electric railroads and elevators. It was not until the 1950s that the two major causes of corrosion were determined: (1) the loss of passivation in the alkaline concrete due to the presence of chloride ions from seawater or deicing salts; and (2) the loss of the concrete's alkalinity due to penetration of atmospheric carbon dioxide and the consequent conversion of very alkaline components to less alkaline carbonates. Corroded steel expands significantly, resulting in expansive forces that cause the adjacent concrete to crack and spall. Concrete deterioration due to corrosion of embedded steel typically takes the form of cracking and delamination and rust staining at the location of embedded steel.

Carbonation, which results from the reaction of carbon dioxide and water with concrete, normally occurs only in the concrete's exposed surface but may extend to the level of the steel in poor-quality concrete. Once this occurs, the concrete offers no protection to the embedded reinforcing steel, and corrosion begins.

Corrosion of embedded reinforcing steel is often initiated and accelerated when calcium chloride is added to the concrete as a set accelerator during original construction or later by deicing salts used in northern climates. Seawater or other marine environments can also provide large amounts of chloride. When the chloride content of the concrete at the level of the reinforcing steel exceeds 0.2 percent by weight of cement, the normal passivating

**Figure 8.6.** Curved, petal-like balconies of reinforced concrete project from the twin towers of Marina City (1963, Bertrand Goldberg), Chicago.

**Figure 8.7.** At Fallingwater (1937, Frank Lloyd Wright), Mill Run, Pa., conservators in suspended scaffolding assess deterioration of reinforced concrete.

**Figure 8.8.** Cracking (top) can be caused by freeze–thaw cycles.

**Figure 8.9.** A worker measures the depth of concrete cover over reinforcement (bottom).

characteristic of the concrete is destroyed and corrosion can occur.

The concrete's exposed surface is also vulnerable to weathering, which may typically be observed as erosion of the cement paste. Especially in northern regions where precipitation has been found to be highly acidic, exposure has resulted in more significant erosion of the paste.

Damage due to cyclic freezing action results when concrete that is saturated with water freezes. This type of damage appears as surface degradation, including severe scaling and cracking, that extends into the concrete. It was accidentally discovered that portland cement concretes that incorporate microscopic air bubbles were resistant to damage from cyclic freezing and thawing. This air-entrainment provides "relief valves" that protect the concrete. Air-entraining agents are now usually added to mortar or concrete used in exposed applications in areas of the United States subject to subfreezing temperatures.

Alkali–aggregate reactions result when alkalis normally present in cement react with siliceous aggregates in concrete that is exposed to moisture. The reaction produces a toothpastelike gel that develops over years or decades until the forces created expand and crack the concrete. Most such deleterious aggregates can be detected by experience and testing, and low-alkali cements can be used in new construction to prevent significant reactions.

Sulfate attack is produced by the reaction of excessive amounts of sulfate salts with cement components that are exposed to moisture. The reaction leads to the development of expansive forces that eventually crack the concrete. The sulfate salts may come from the environment (e.g., sulfate waters or soils) or from one or more of the concrete constituents (e.g., aggregates, cement, or proprietary product providing fast set).

Deleterious porous aggregates can lead to pop-outs on concrete surfaces or D-cracking (fracturing of aggregate due to freezing water) of pavements. Concrete will also deteriorate if exposed to acidic soils or chemicals. Most of these conditions can be determined by laboratory analysis, primarily petrographic microscopy.

## DIAGNOSTICS AND CONDITION ASSESSMENT

A condition assessment of a concrete building or structure should begin with a review of all available construction documents. For a historic structure it is especially important to review historic photographs. The repair history of the concrete provides valuable information about its past performance.

A visual condition survey should always be conducted to evaluate the extent, types, and patterns of distress and deterioration. The American Concrete Institute has published several useful guides on how to perform a visual condition survey of concrete.[12] Generally, the condition survey documents all visible distress and deterioration.

Nondestructive field testing methods are used to evaluate the concrete's concealed condition. Sounding with a handheld hammer or chain helps identify areas of delamination. Impact-echo techniques locate voids or delaminations within the concrete by identifying vibrations and frequencies characteristic of the thickness of the concrete. Embedded reinforcing steel can be located with magnetic detection instruments, which can be calibrated to identify the size and depth of the reinforcement. Copper–copper sulfate half-cell tests are used to determine the probability of active corrosion of the reinforcing steel.

To further evaluate the condition of the concrete, appropriate samples may be removed for laboratory studies. Petrographic evaluation of the concrete samples is conducted in accordance with guidelines provided by ASTM.[13] The petrographic examination, performed with a microscope, is essentially a qualitative and semiquantitative visual survey of the concrete samples that identifies the original components of the concrete, information useful in developing a repair design, as well as evidence of damage due to various factors including cyclic freezing and thawing, alkali–aggregate reactivity, or sulfate attack. Laboratory studies can also include chemical analyses to determine chloride content, sulfate content, and alkali levels of the concrete, identify the presence of deleterious aggregates, and determine the depth of carbonation. Compressive-strength studies can be conducted to evaluate the strength of the existing structure and provide information for repair design.

## CONSERVATION TECHNIQUES

The concrete's durability or resistance to deterioration and its repair depends on its composition, design, and quality of workmanship. A mix design for durable replacement concrete should use materials similar to those of the original concrete mix and include air entrainment, and appropriate selection of aggregates and of cement and water contents. Good workmanship should address proper mix, placement, and curing procedures.

When repairs to existing concrete are designed, parameters based on visual evaluation and laboratory studies must be established to define the project's goals. A key concern is the aesthetics of the repair: it must match the existing concrete as closely as possible both visually and structurally. Another concern is to select repair interventions that retain as much of the original material as possible; however, an adequate amount of distressed concrete must be removed to provide a durable repair.

All repairs of existing concrete require proper preparation of the substrate to receive the repair material. This typically includes sandblasting, shotblasting, airblasting, mechanical scarification, and other appropriate means to provide a clean, sound surface to which the repair can adequately bond. Bonding agents are commonly used on the substrate surface to enhance the bond of the repair. Existing reinforcing steel that is exposed during the repair may require cleaning, priming, and painting with a rust-inhibitive coating.[14] In most cases the repair area should be reinforced and mechanically attached to the existing concrete. Reinforcement materials can include regular steel, epoxy-coated steel, or stainless steel, depending on the conditions.

Proper placement and finishing of the repair is important in achieving a match with the original concrete. Appropriate curing is essential for a durable repair. Wet curing is recommended to reduce the curing time and the potential for both surface cracking and cracking due to shrinkage.

Preparing trial repairs and mock-ups to refine the repair design and evaluate repair procedures is wise. Mock-ups also permit evaluation of the design's aesthetic acceptability.

Because concrete deterioration is heavily influenced by moisture penetration, rehabilitation may also entail application of a decorative surface coating or a clear penetrating sealer, if appropriate. These water-resistant coatings and sealers should be breathable and alkali resistant.[15]

Various repair methods and techniques available today reduce the rate of corrosion of embedded reinforcing and associated concrete deterioration. One method is cathodic protection, which uses an auxiliary anode so that the entire reinforcing bar is a cathode. Corrosion is an electrochemical process in which electrons flow between cathodic (positively charged) areas and anodic (negatively charged) areas on the metal surface; corrosion occurs at the anodes.[16] Cathodic protection is intended to reduce the rate of corrosion of embedded steel in concrete, which in turn reduces overall deterioration. Another technique to protect concrete is realkalization, which involves returning the concrete to its natural alkaline state by soaking the concrete with an alkaline solution, possibly speeded by passage of direct current. Other techniques include flooding with a corrosion inhibitor or, more aggressively, removing the concrete, cleaning the exposed steel coating with epoxy and covering the new concrete, and then sealing.

Careful evaluation of existing conditions, the causes and nature of distress, and environmental factors is essential before a protection method is selected and implemented. While cathodic protection may be effective in some cases, it is not the answer for every project.

## REPLACEMENT

If specific components in reinforced concrete structures are beyond repair, replacement components can be cast in place to match historic ones. As with surface repairs, placement and finishing will dictate how well replacement concrete matches the historic concrete. This usually requires mock-ups to ensure compatibility with adjacent areas.

**Figure 9.1.** Licensing agreements allowed companies to use the trademarked name Gunite, a dry-mix shotcrete. Today the term *Gunite* is nearly synonymous with shotcrete. Catalogue, Gunite Concrete and Construction Company, Kansas City, Mo., undated.

ANNE T. SULLIVAN

# 9 | SHOTCRETE

**DATES OF PRODUCTION**
c. 1910 to the Present
**COMMON TRADE NAMES**
Blocrete, Glastcrete, Guncrete, Gunite, Gun-Stone, Jetcrete, Nucrete, Pneucrete, Spraycrete

## HISTORY

*Shotcrete* is defined by the American Concrete Institute as "mortar or concrete pneumatically projected at high velocity onto a surface."[1] This term describes two separate application processes—pneumatically applied wet-mix and dry-mix mortar or concrete.[2] Although the methods of application differ, the result is similar for both procedures.[3] In the wet-mix process cement, sand, aggregate, admixtures, and water are combined before application. The mixture is carried from the delivery equipment through the hose to the nozzle, where air is added. Dry-mix shotcrete involves the mixture of cement, damp sand, and aggregate, which is delivered through the hose to the nozzle, where pressurized water and air are introduced.

### ORIGINS AND DEVELOPMENT

The development of wet- and dry-mix shotcrete is historically interrelated. The application of a concrete mixture using compressed air developed from early experiments using pneumatic machinery for grout and plaster application.[4] In 1908 Carl E. Akeley, a taxidermist at the Field Museum of Natural History in Chicago, applied for a patent for a machine that applied wet plaster mortar using the force of compressed air. Akeley obtained a second patent in 1911 for revised handling equipment that could mix and apply plastic adhesive materials.[5]

The commercial development of what we know as the dry-mix shotcrete industry began with the creation of the Cement Gun Company of Allentown, Pennsylvania, which acquired Akeley's machine in 1910. The machine, marketed commercially as the cement gun, was formally presented to the concrete industry at the second annual Cement Show held in New York City in 1912.[6]

Considerable experimentation with pneumatic application of both dry- and wet-mix shotcrete occurred between 1910 and the early 1920s.[7] Dry-mix shotcrete technology became more standardized in the 1920s, following documented research and development by several companies, including the Cement Gun Company.[8] The wet-mix shotcrete process was not as widespread, nor did it develop as rapidly, as dry-mix shotcrete. After about 1915 the dry-mix procedure surpassed the wet-mix, primarily because wet-mixes tended to clog the machinery. In the early 1950s interest in the wet-mix process returned with the marketing of the True-Gun, a pneumatic device for gunning a wet mortar mixture.[9]

The proprietary term *Gunite* was adopted by the Cement Gun Company to describe the "sand-cement product of the cement gun."[10] The company appears to have been the single supplier of the cement gun until it was succeeded by the Allentown Pneumatic Gun Company of Allentown, Pennsylvania.[11] In the early 1930s the American Railway Engineering Association introduced the term *shotcrete* for the dry-mix process.[12] This generic term covered all gunned applications of mortar and concrete materials, and continues to be the accepted term for the procedure today.

### MANUFACTURING PROCESS

Shotcrete is composed of the same materials used for poured-in-place concrete. Modern shotcrete mixes include aggregate, sand, and portland cement, typically mixed in the field before application.[13] Aggregates must be sized appropriately, and the amount of water in the mixture is carefully regulated. As long as the material is plastic enough to be pumpable, it can be projected into place as shotcrete.

Prepackaged factory mixes are available but are somewhat more expensive than mixing raw materials in the field. Proprietary mixes often contain admixtures such as pozzolans, synthetic fibers, air-entraining agents, water reducers, or accelerators.[14] Shotcrete admixtures are generally a post–World War II phenomenon, although there was some early experimentation with additives, such as iron fillings and calcium chloride.[15]

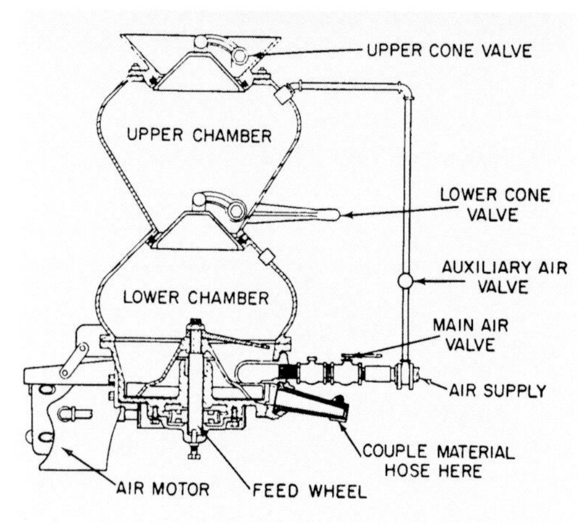


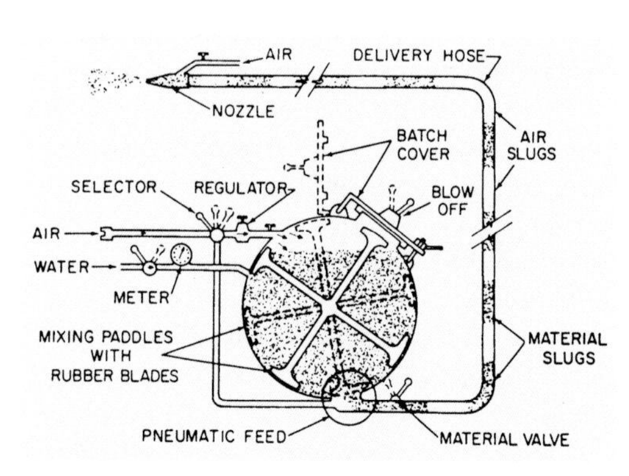


**Figure 9.2.** Carl Akeley's portable cement gun, developed about 1910, was the first for applying dry-mix shotcrete (top).

**Figure 9.3.** In dry-mix shotcrete systems, mixes move from the upper chamber to the delivery hose; pressurized water and air are added at the nozzle (middle).

**Figure 9.4.** In wet-mix shotcrete systems, the material is pneumatically discharged from the mixing chamber through the delivery hose (bottom).

## USES AND METHODS OF INSTALLATION

Shotcrete has been used in a number of industries, although it is best adapted to thin, often reinforced sections applied to a prepared base material. Industrial applications include coatings for mine shafts and tunnels and linings for refractories, furnaces, and boilers. Architectural shotcrete has been used as a coating over masonry materials, concrete, and steel, as well as for replacing or repairing deteriorated stucco, brick, and concrete.[16] Thin, reinforced sections of shotcrete have been used successfully to construct domes and thin shell roofs.

As early as 1911, it was suggested that shotcrete be used for coating building exteriors, both in new construction and on existing structures.[17] One of the earliest known applications of pneumatically placed mortar material was at the Field Museum of Natural History in Chicago (now the Museum of Science and Industry), constructed as the Palace of Fine Arts (Charles B. Atwood) for the 1893 World's Columbian Exposition. Around 1910 Carl Akeley's cement gun was used to repair the "staff," or lime stucco coating.[18] The use of shotcrete for fireproofing (it prevented buildings from burning quickly) aroused particular fervor in the early years of the industry. For example, it was used in 1910 as a fireproof coating for structural steel at Grand Central Terminal (1903–13, Reed and Stern, Warren and Wetmore) in New York City.[19]

By the early 1920s experimental applications gave way to more standard procedures. An article published in the proceedings of the American Concrete Institute's 1922 conference describes a system of construction utilizing Gun-Stone, a combination of Gunite and concrete, which appears to have been a form of dry-mix, large-aggregate shotcrete. The exterior walls of a house in Watertown, Massachusetts, were constructed by fastening expanded metal reinforcing over a structural concrete frame. Gunite was spray-applied to a depth of 1½ inches over the reinforcing.[20] The result resembled a stucco finish.

Another noteworthy early use of shotcrete was for the Hayden Planetarium dome (1935, Trowbridge and Livingston) of the American Museum of Natural History in New York City. The dome was designed using the German Zeiss Dywidag system of shell construction. A subdome, hung from the main reinforced concrete dome, was constructed of a reinforced concrete ring. Shotcrete was used to fill the segments of the inner dome.[21] While most modern shotcrete is used for engineering applications, it has also been used as an architectural material. Marcel Breuer designed a reinforced, folded, plate roof truss for

St. John's Abbey Church (1962) at St. John's University in Collegeville, Minnesota; the wet-mix shotcrete process was used.[22]

Shotcrete can be applied to vertical and overhead locations as well as horizontal surfaces. For successful adhesion it must be applied to a properly prepared substrate. Shotcrete can be applied by itself over the prepared surface or over reinforcing bars or mesh tied back into the substrate.[23] Reinforcement is recommended when an even, finished surface, rather than an uneven substrate, is desired, when the substrate cannot provide a proper bonding surface, or for crack repair. Shotcrete coatings are relatively thin, varying in depth from ⅛ inch to 4 inches. Greater thicknesses are required when reinforcing is to be covered.[24] Shotcrete is most often applied to surfaces without the use of formwork; however, either one- or two-sided forms can be used to aid in the formation of sharp corners or decorative elements.

The quality of the finished shotcrete surface is largely dependent on the skill of the applicator. Shotcrete is generally applied by one nozzleman with a partner handling the hose. In both dry- and wet-mix practice, the shotcrete is projected onto the receiving surface at a high velocity. As the shotcrete hits the surface, there is a certain degree of "rebound"; the larger aggregate and some sand bounce back off the receiving surface, leaving behind a rich cement paste. In cases where the shotcrete is being applied over reinforcing, care must be taken to prevent rebound aggregate from becoming trapped and leaving unfilled pockets behind reinforcing bars.[25]

A well finished shotcrete surface should appear compact, rich, and creamy rather than dry or sandy.[26] Textured finishes can be achieved by brooming the finished surface, applying a fine, sandy flash coat, or exposing the aggregate by removing surface cement with an acid wash. Other surface textures can be achieved using a wood or rubber float or steel trowel.[27] Like poured-in-place concrete, fresh shotcrete must have adequate protection to cure properly.

## CONSERVATION

Cracking and debonding (delamination) are the primary forms of shotcrete failure. A number of factors affect its long-term durability, including poor initial surface preparation, improper application, insufficient detailing, substrate flaws, and water infiltration. Thorough surface preparation is essential for achieving a secure coating, and a clean, stable substrate is necessary to assure proper bonding. In addition, shotcrete must be applied by a qualified applicator with proper equipment.

**Figure 9.5.** Unity Temple (1904, Frank Lloyd Wright), Oak Park, Ill., built of exposed aggregate reinforced concrete, was repaired with shotcrete in the 1970s.

### DETERIORATION

Cracks caused by shrinkage and movement are common. Hairline cracks, often less than $\frac{1}{80}$ inch and noncontinuous, can develop as a result of shrinkage during initial setting but generally do not give cause for concern because they are sometimes self-healing.[28] Continuous vertical cracks often indicate damage resulting from expansion and contraction of either the base concrete or the shotcrete surface. This condition occurs most frequently where large areas are coated with shotcrete without providing for material movement. Expansion cracks must be anticipated and details provided to allow for such movement. Failures in the shotcrete coat can also be caused by preexisting flaws in the substrate.

Cracking and debonding of shotcrete are often interrelated. Open cracks allow water infiltration, which can permit both carbonation and resultant corrosion of reinforcing bars.[29] Expansion as a result of such corrosion can cause shotcrete coatings to delaminate. Open cracks in either reinforced or nonreinforced areas can allow water to travel deep into the shotcrete coating, causing the cement-rich layer closest to the substrate to leach out of the cracks. Deterioration of this layer can eventually lead to delamination of the surface coat and spalling.

### DIAGNOSTICS AND CONDITION ASSESSMENT

Shotcrete deterioration is best investigated visually at close range or using binoculars. The pattern of cracking can be recorded on a "crack map." Similarly, visible cracks may be traced with chalk and photographed. Debonded areas are not as easy to detect. Areas suspected to be delaminating should be methodically sounded with a steel hammer; drummy (hollow) areas should be outlined with chalk and documented, either on the crack map or with photographs.

Some cracks, particularly those formed as a result of shrinkage during the initial set, are static. Other cracks may be active, such as those resulting from material expansion. Movement of such cracks may be tracked over time with the use of calibrated "tell-tale" crack monitors. These acrylic gauges can record horizontal and vertical movement to an accuracy of 1 millimeter.

Between 1965 and 1966 and again in 1989, durability tests were undertaken on shotcrete by Construction Testing Laboratories, a branch of the Portland Cement Association. The results indicated that, if installed properly, both non-air-entrained, dry-mix shotcrete and air-entrained, wet-mix shotcrete are extremely durable.[30] Until 1988 testing for shotcrete was based on the American Society for Testing and Materials (ASTM) standards for conventional concrete.[31] Since then, however, ASTM has published five shotcrete standards, and performance testing is ongoing.[32]

Core samples from representative areas, both bonded and debonded, should be analyzed by a qualified testing laboratory familiar with standards and procedures for testing shotcrete durability and failure. When extracting core samples at bonded areas, particular care must be taken not to break the bond between the shotcrete and the substrate. Core samples can be compared visually, examined microscopically, and analyzed chemically to establish the degree and depth of carbonation and determine the shotcrete constituents. Additional tests, derived from ASTM standards for concrete or developed for specific samples, may be performed by the testing laboratory.

## CONSERVATION TECHNIQUES

Maintenance procedures for shotcrete are similar to those for finished concrete. Surface soiling of shotcrete is dependent on the type and exposure of aggregate, the amount (if any) of lime in the mix, and the presence of nonoriginal coatings. As with early concrete, early shotcrete surfaces may have received waterproof coatings of linseed oil or of early proprietary products.

Generally, little is known about these coatings. Such coatings often leach to the surface over time, leaving a sticky residue that collects dirt. Solvent-based masonry cleaners are generally effective for removing such coatings, but test patches should be used to evaluate their effectiveness and compatibility. In some cases previous treatments may be too difficult to remove chemically and may have to be removed mechanically. Mechanical cleaning, such as sandblasting, will most certainly alter the appearance of the exposed shotcrete. The bond of exposed aggregate to the cement mixture may also be affected. Mechanical cleaning, which changes the integrity of historic shotcrete finishes, and acidic masonry cleaners should be avoided.

Sources of water infiltration should be identified. If water is allowed to collect behind the debonded shotcrete surfaces, the likelihood of damage due to freeze–thaw action is greatly increased, and cracking and debonding can occur. Water must be prevented from entering the shotcrete through larger cracks and washing away the cement in the mix, thus weakening the bond between the shotcrete and substrate. The application of breathable water repellents may help prevent water intrusion, but if cracks are greater than 1/80 inch, a repellent will not be effective.

Few techniques for repairing and conserving shotcrete surfaces are documented. Many modern shotcrete installations are on industrial structures. As these surfaces deteriorate, they are more frequently removed and replaced rather than conserved. Concrete consolidants may be applied to severely deteriorated shotcrete surfaces, but, depending on the application, it may be more cost-effective to selectively remove and replace the shotcrete coating. Deteriorated shotcrete surfaces are often removed, the base material is prepared, and the areas are shotcreted again.

Small repairs to shotcreted surfaces generally cannot be made using pneumatically applied concrete.[33] Shotcrete is designed to be applied as an even coating over a large area. Therefore, areas within a field of shotcrete requiring repair should be patched with a mortar mixture that matches the surrounding shotcrete as closely as possible. Ensuring visual compatibility of patch repairs with the surrounding shotcrete may be difficult. When a number of spalls have occurred, it may be necessary to remove a large panel of the material and reapply a shotcrete coating.[34]

An alternative to removal and replacement is reaffixing debonded material to the substrate using a structural adhesive. Epoxies can be injected, in situ, through ports placed strategically along the crack. The adhesive then spreads behind the delaminated material and reaffixes it to the substrate. Crack openings are dammed to prevent the epoxy from trickling out from behind. Injection pressure must be carefully monitored to prevent the debonded material from being forced off the substrate.

**Figure 9.6.** Wet-mix shotcrete, using pea-gravel aggregate from the original source, was applied to Unity Temple to match the original cast-in-place finish.

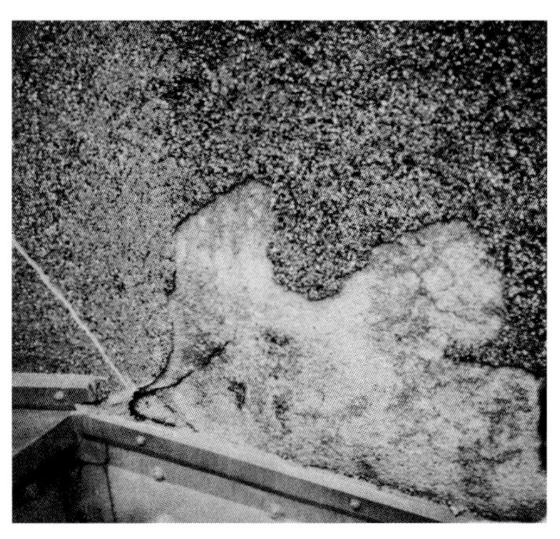

**Figure 9.7.** Expansion cracking caused the shotcrete surfaces to delaminate. Conservation of the deteriorated shotcrete is being planned.

## REPLACEMENT

In-kind replacement is the procedure used most frequently for the restoration of shotcreted surfaces in many engineering applications. Before application, test panels (at least 3 feet square) should be erected, especially when it is necessary to match existing surfaces. To ensure quality of material and workmanship, a testing program must be integrated into the specification and bidding process. Nozzlemen may also be prequalified during this trial period.

Hairline shrinkage cracks can be prevented through proper mix and application. Shotcrete panels should be detailed to allow for expansion and contraction and prevent continuous cracking due to material movement.

**Figure 10.1.** Member companies of the Mo-Sai Institute, following strict procedures, could produce panels 20 to 60 square feet and 2 inches thick with compressive strengths of 7,500 pounds per square inch. Precast concrete facing and curtain wall catalogue, Mo-Sai Institute, 1963.

SIDNEY FREEDMAN

# 10 | ARCHITECTURAL PRECAST CONCRETE

**DATES OF PRODUCTION**
c. 1920 to the Present
**COMMON TRADE NAMES**
Mo-Sai, Schokbeton

## HISTORY

Architectural precast concrete is any precast concrete element that through application, finish, shape, color, or texture contributes to a structure's architectural form and finished effect. Components may be standard or custom size, load-bearing or non-load-bearing, and conventionally reinforced or prestressed. Hardware for connection to the structure may consist of structural steel shapes, bolts, threaded rods, and reinforcing bars. Generally, connections are bolted or welded.[1] Architectural precast concrete gained prominence in the late 1950s and is still widely used today.

### ORIGINS AND DEVELOPMENT

The first documented modern use of architectural precast concrete was in the church of Notre Dame du Raincy (1923, Auguste Perret) in Le Raincy, France. In this building precast concrete was used only for screen walls and as infill in an otherwise cast-in-place concrete structure.[2] In the United States, the Great Depression years limited development of large-scale precast components. Not until World War II had ended did the architectural use of precast concrete begin to flourish.

Surface finishing techniques for architectural precast concrete, such as water washing and brushing, bush hammering, sandblasting, and acid etching, were initially developed for cast-in-place concrete, as was obtaining colored surfaces by means of pigments and special colored aggregates.[3] These techniques were also well established within the cast stone industry by the 1930s.

In 1932 John J. Earley and his associates at Earley Studio in Rosslyn, Virginia, began work on producing exposed aggregate ornamental elements for the Baha'i Temple (1920–53, Louis Bourgeois) in Wilmette, Illinois, one of the most beautiful and delicately detailed architectural precast concrete projects in the United States.[4] The panels are white concrete with exposed quartz aggregate. Lack of funds postponed completion of the exterior precast concrete until eleven years after the building was begun. Another early notable use of large precast panels was for the White Horse Barn (1937, G. P. Lagergren), constructed for the Minnesota State Fair. This structure had panels 15 feet long, 7 feet high, and 6 inches thick. Cast with a smooth finish and cured for seven days in steam-filled rooms, these panels were attached by the structural concrete around the edges of the panels.[5]

A pivotal development occurred in 1938, when administration buildings at the David W. Taylor Model Testing Basin (Ben Moreell, U.S. Navy Bureau of Yards and Docks) were built near Washington, D.C.[6] Panels 2½ inches thick and up to 10 feet by 8 feet were used as permanent cladding for cast-in-place walls. The project was significant as the first use of the Mo-Sai manufacturing technique, which permitted finishes with densely packed mineral aggregate and a minimum amount of cement and fine aggregate, produced by John Earley in collaboration with the Dextone Company of New Haven, Connecticut. It was also the first project in which large-area exposed aggregate panels were adapted to serve as exterior form and preinspected facing for reinforced concrete building construction.[7] Earley had patented the idea of using step (gap)-graded aggregate to achieve uniformity and color control for exposed aggregate work.[8]

Working from this background, the Dextone Company refined and obtained patents and copyrights in 1940 on the methods under which Mo-Sai Associates (later the Mo-Sai Institute, Inc.) operated.[9] The Mo-Sai Institute grew to include a number of licensed manufacturing firms in various parts of the United States. Its public relations and advertising activities, highlighting technical achievements, were a major factor underlying general acceptance of architectural precast concrete.

### MANUFACTURING PROCESS

Typical Mo-Sai panels were 2 inches thick and could be made up to 100 square feet, either as veneer for masonry construction or as form and facing for poured concrete.

All panels were cast facedown, pneumatically vibrated, and reinforced with galvanized welded mesh. The face mixes were always composed exclusively of coarse aggregate, either granite or quartz, and the fine aggregate was of the same material. Average mix proportions were approximately 1 part of fines to 7 parts of two or more sizes of coarse material. The aggregate-cement ratio was approximately 4:1, while the water-cement ratio in a very damp mix was as great as 5:1. Backup mixes were composed of washed and screened concrete sand plus crushed stone, and were placed integrally with the face material. Once cast, the entire plastic mass of the panel was pneumatically vibrated a second time to ensure compaction of the material and draw to the exposed back face any excess moisture, which was immediately evacuated by mechanical means and with hygroscopic materials.

After casting, panels were allowed to cure for 24 hours in the molds and were then removed and stacked vertically on easels for a further curing period of several days. Final finishing, usually acid etching, took place before shipment. Dextone was also producing polished concrete panels as early as 1949.

In 1958 a new panel-casting method was introduced in the United States under the name Schokbeton ("shocked concrete"), and a number of franchised plants were established.[10] The machinery used in this method was patented in Holland in 1932. The process is mainly a means of consolidating a no-slump concrete mixture by raising and dropping the form about $\frac{5}{16}$ inch approximately 250 times per minute. This contrasts with conventional methods of consolidation using high-frequency and low-amplitude vibration. Although the production of large precast panels by this method is relatively new, small concrete elements have been produced in the past on so-called drop tables, which follow the same technique without the refinement of modern machinery.[11]

## USES AND METHODS OF INSTALLATION

The use of architectural precast concrete was initially complicated by the lack of mobile cranes and other efficient materials-handling equipment. For this reason and because of competition from metal and glass curtain wall materials, the precast concrete industry was comparatively slow to develop. Its rise to parity with other materials—and even dominance in some locations—occurred in the early 1960s. Reasons for its expanding use included improved methods of production, better handling and erecting equipment, and development of new techniques and materials. Probably the greatest factor, however, was the realization that precast panels provided a pleasing variety of surface textures and patterns and exterior designs that generally could not be accomplished as economically in other materials.

A number of early postwar buildings constructed of architectural precast concrete were pioneering projects, although they were otherwise of limited significance. These included dormitories at the University of Connecticut at Storrs (1948, McKim, Mead and White), based on load-bearing precast concrete walls and partition sections; an eight-story office building (1949) in Columbia, South Carolina, with window wall-cladding panels installed by hand winch; and a six-story office building (1951) in Miami, where 4-inch-thick precast panels were suspended from the soffit of a cantilevered cast-in-place floor slab.[12]

The Hilton Hotel (1959, I. M. Pei and Associates) in Denver was one of the early significant uses of window wall panels fixed to a structural frame. One of the first major buildings to utilize the inherent structural characteristics of architectural precast concrete was the Police Administration Building (1962, Geddes Brecher Qualls Cunningham) in Philadelphia. It made history with its 5-foot-wide, 35-foot-high (three-story) exterior panels, which supported two upper floors and the roof, and was an early model for blending precasting and posttensioning techniques in one structure.[13]

## CONSERVATION

Current conservation techniques used for architectural precast concrete have application to historic structures. Field and laboratory investigations are necessary to determine the cause of the surface condition and the effect of the proposed remedial action on the concrete.

### DETERIORATION

Surface deterioration of architectural precast concrete may occur from the movement of water, chemical deposition, atmospheric staining, erosion, organic growths, cyclic freeze–thaw action, trapped moisture, scaling, joint deterioration, efflorescence, metal corrosion, and resultant rust stains, delamination, and spalling. In vertical wall units concrete seldom reaches the saturation point, which can create freeze–thaw expansion pressures. Therefore, the lack of air entrainment in concrete cast before 1940 should not be a major concern. However, in horizontal sections where water or snow can accumulate, freeze–thaw damage can occur. If the concrete retains moisture over time, metals with inadequate cover (distance from concrete surface to metal) may corrode and may cause

**Figure 10.2.** Mo-Sai anchoring systems fit various types of walls.

the concrete to spall. If water infiltrates architectural concrete cladding through panels or joints or if condensation is present, corrosion of the connections may occur.

## DIAGNOSTICS AND CONDITION ASSESSMENT

Before undertaking restoration, an evaluation of the structure's condition is often necessary to obtain information on the extent of deterioration and to establish its cause and significance. This information can be obtained only through a systematic review of service records and the structure's original design and construction details. Following such a review, a detailed field investigation should be planned. When records are incomplete or unavailable, intelligent observation and sound judgment should be applied in planning a field investigation program. The program should include visual examination and nondestructive testing to identify concrete, connection, and sealant failure and collect specimens for laboratory testing if necessary.[14]

## CONSERVATION TECHNIQUES

Cleaning should remove harmful and unsightly soilage and preserve the physical integrity of the precast concrete's surface and substrate.[15] Because a variety of materials have been used in concrete and each building is exposed to different conditions, both laboratory and field tests of proposed cleaning systems must be conducted to determine the effect on concrete surfaces. Before proceeding with the cleaning, a small inconspicuous area of at least 1 square yard should be cleaned and evaluated. The method's effectiveness on the sample area should not be judged until the surface has dried for at least one week.[16]

When developing a cleaning program to remove dirt, stains, and efflorescence, begin by using the least aggressive methods. Dry scrubbing with a stiff fiber (nylon) brush or wetting the surface with water and vigorously scrubbing the finish with the same type of brush and then thoroughly rinsing the surface with clean water may be effective. Low-pressure water spraying (water misting, either intermittent or continuous), moderate- to high-pressure water washing (300 to 1,500 pounds per square inch) with cold or heated water can also be considered. Chemical cleaning compounds, such as detergents, acids, and other commercial cleaners, should be used in accordance with the manufacturer's recommendations.[17] Dry or wet blasting using sand, ferrous aluminum silicate, industrial baking soda, and other abrasives that can dull, round, or erode the cement matrix should generally be avoided. Stain removal techniques vary according to the staining media.[18]

Structurally sound, durable, and aesthetically pleasing repairs can be made provided suitable mixtures are used, proper bonding is attained, and reasonable curing is possible.[19] Repair techniques and materials for restoring architectural precast concrete are selected on the basis of a number of factors, which include mix ingredients, final finish, size and location of the damaged area, temperature conditions, age of the component, and surface texture.[20]

Trial mixes are essential for determining exact quantities for the repair mix. This information is best determined by making a series of dark- to light-colored trial repairs on the project on the same day. Selection of the appropriate color-matched mix should begin after the trial repairs have been allowed to cure a minimum of 7 days but preferably 14 days, followed by normal drying to a total of 28 days. This length of time is important because curing and ultraviolet bleaching of the cement skin affect the finished color. Adequate curing methods for repairs should be implemented as soon as possible to ensure

**Figure 10.3a–c. a.** At the Baha'i House of Worship (1920–53, Louis Bourgeois), Wilmette, Ill., water infiltration led to (**b**) efflorescence and stalactite formation. **c.** Components were patched in place.

that the repair does not dry out too quickly, thus causing it to shrink away from the existing concrete. The repair area should be moist-cured for a minimum of 3 days, although 7 days is preferable.

Because repair areas have different densities, cure at a faster rate than the surrounding concrete, and are exposed to curing environments different from the existing component, it may be necessary to alter the color. To provide the same general color as adjacent areas, blending portland cements or altering the ratio of cement to aggregate of the original mix proportions may be necessary.

Bonding agents such as epoxies, polyesters, and water-resistant latexes may be desirable for large repairs. Latex bonding agents should generally not be used integrally with the repair material, because differences in texture, color, and appearance will be apparent when the repair material is wet; the repair may even change to an objectionable color. The bonding agent should be neatly applied to the dry surface of the repair area and allowed to dry to a tacky condition before the repair mix is applied.

If cracking has occurred and if repair is required for the restoration of structural integrity or aesthetics, cracks ranging in width from 0.003 to 0.250 inches and with a depth of less than 12 inches may be repaired by pressure injection of a low-viscosity, high-modulus, 100 percent solid, two-component epoxy. Type, grade, and class should be chosen to satisfy job conditions and requirements.[21] The epoxy should be capable of bonding to wet surfaces unless the crack is absolutely dry. Its color (amber, white, or gray) should match the concrete surface as closely as possible, and the crack line where it is injected should be unobtrusive.[22]

Completed repairs should be coated with a silane or dilute (2 to 3 percent solids) solution of an acrylic sealer to minimize moisture migration into the repair concrete. The sealer should not stain the concrete. Even with proper consolidation the repair is more porous than the original hardened concrete and will need this dilute sealer application.

To replace corroded connections, new galvanized, epoxy-coated, or stainless steel plates or angles can be installed with expansion or chemical anchors inserted into drilled holes in the hardened concrete.[23] The holes must be drilled straight, deep enough, and to the proper diameter and must be cleaned out. The bolts must be tightened to the recommended torque and may sometimes require pneumatic impact wrenches. The minimum distance from each bolt to the edge of the concrete and to other bolts should be based on the anchor manufacturer's recommendations. In placing the expansion an-

**Figure 10.4a–b.** The Police Administration Building (1962, Geddes Brecher Qualls Cunningham), Philadelphia, was constructed in sections.

chors in the predrilled holes, avoid expanding the anchor in the direction of the edge. A knowledgeable engineer should determine the movement capability of the new connections to avoid restraint and possible cracking.

In urban and industrial areas, sealers or clear surface coatings may be applied to improve weathering properties. Use of a sealer will also reduce the absorption of moisture, thereby minimizing the wet–dry cycle and the subsequent migration of water and salts to the surface. Proper application, following the manufacturer's instructions, depends on qualified operators and possibly expensive pretreatment of the precast concrete components. A sealer's application limitations related to timing, temperatures, the concrete's moisture content, and the method and rate of application should be fully investigated before selection.

Any sealer used should be guaranteed not to stain, soil, darken, or discolor the finish.[24] Also, some sealers may cause joint sealants to stain the panel surface or affect the sealant's bond. The manufacturers of both the sealant and the sealer should be consulted before application, or the materials specified should be pretested before application. Surface coatings applied onsite should not be applied until all repairs and cleaning have been completed.

## REPLACEMENT

Techniques using elastomeric materials are available to produce accurate form liners for replacement components. With the aid of a photogrammetric survey and original construction documents, the size and detailing of new components can be developed to create an accurate model.[25]

**Figure 11.1.** Applications of prestressed concrete range from industrial buildings to structures such as the monorail (1971) at Disney World in Orlando, Fla. *Reflections on the Beginnings of Prestressed Concrete in America,* 1981.

# 11 | PRESTRESSED CONCRETE

**DATES OF PRODUCTION**
1949 to the Present
**COMMON TRADE NAME**
Preload System

## HISTORY

Prestressed concrete is concrete that is internally stressed so that service loads are "counteracted to a designed degree."[1] Prestressed concrete is typically used in members subjected to bending or tension, such as beams, slabs, tanks, and piles. Linear prestressing, the most prevalent type of prestressing, is applied parallel to the long axis of the member to resist tensile stresses from bending.[2] Two methods are used to produce prestressed concrete: pretensioning and posttensioning.

### ORIGINS AND DEVELOPMENT

The potential benefits of prestressing reinforcement were recognized during the late nineteenth century when reinforced concrete was in its infancy. In 1886 P. H. Jackson of San Francisco obtained a patent for tightening steel tie rods in artificial stones or arches for use as floor slabs.[3] This and subsequent efforts during the early twentieth century failed primarily because the level of prestress was low and could not be maintained because of creep (shrinkage) in the concrete.

In 1923 Richard E. Dill of Alexandria, Nebraska, successfully pretensioned concrete to account for shrinkage, using coated steel for reinforcement.[4] Although Dill patented his system in 1925, it was not widely used. The French engineer Eugène Freyssinet, who began to use high-strength steel wires in 1928, is considered the father of modern prestressed concrete. Freyssinet's system was for posttensioning members. Methods that led to pretensioning were explored by Hoyer of Germany in the late 1930s.

The earliest true applications of prestressed concrete in the United States can also be traced to the 1930s, when the Preload Corporation began building large tanks and pipes using circular prestressing methods. This company used specially designed equipment to impose tension in the wires, which were wrapped around the tanks. However, the modern prestressed concrete industry in the United States is universally acknowledged to have

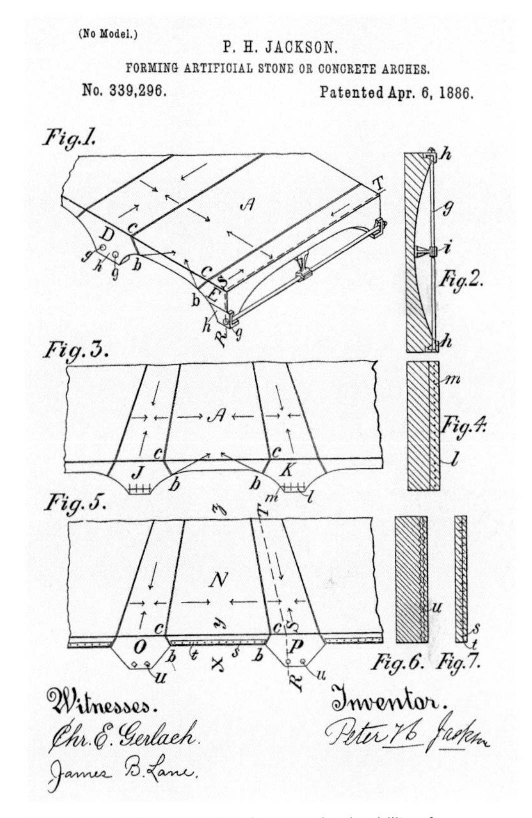


**Figure 11.2.** Early attempts to improve the durability of concrete by prestressing it, including P. H. Jackson's patented arch system, failed.

begun in 1949, the year the Walnut Lane Bridge in Philadelphia was designed. This bridge, the first major linearly prestressed structure, was opened to traffic in 1951.[5]

Success with posttensioned concrete depended on reliable and economical methods for stressing the steel and anchorages for locking the wire ends. Freyssinet developed conical wedges for end anchorage in 1939, and in 1940 the Belgian engineer Gustave Magnel developed an alternative anchorage system.[6] Magnel visited the United States in 1949 and delivered a number of lectures describing his research on reinforced concrete. Shortly after his visit, planning began for the Walnut Lane Bridge. The Preload Corporation, which fabricated the girders, hired Magnel as a design consultant.[7] The Walnut Lane Bridge used thirteen 160-foot-long posttensioned girders in the main span and seven 74-foot-long girders in both of the approach spans.[8]

The use of prestressed concrete expanded dramatically during the 1950s. The unprecedented construction activity following World War II, passage of the Interstate Highway Act in 1956, and steel shortages during the Korean War all contributed to the industry's growth. Important early efforts to promote prestressed concrete included a 1951 conference on the subject at the Massachusetts Institute of Technology and the formation of the Prestressed Concrete Institute in 1954.[9]

The material's acceptance hinged on the resolution of a number of technical problems, principally the production of high-strength, low-slump concrete, and methods for consolidating and curing the concrete to achieve the needed release strength.[10] After these problems were resolved, the range of prestressed members evolved rapidly.

## MANUFACTURING PROCESS

Prestressed concrete is produced by two techniques: pretensioning and posttensioning.[11] In pretensioning the steel is positioned in the forms, stretched to develop the required stress, and held in place; concrete is then applied in the same way as ordinary reinforced concrete. After the concrete has cured sufficiently to provide the necessary compressive resistance, the tension in the steel is released and the load is transferred to the concrete. The compression is transferred primarily through the bond between the concrete and the steel internally throughout the length of the member.

Posttensioned steel is installed in an unstressed condition and positioned in the concrete encased in conduits or sleeves so that the steel and concrete are not in contact. After the concrete achieves the necessary compressive strength, the steel, which is anchored at one end, is stretched by jacking to develop the necessary tension. When the required tension is achieved, the free end is locked in place and the force is transferred to the hardened concrete. This transfer takes place by concentrated loads at each end of the member so that end blocks and special anchorage devices are required to ensure that the compression is not lost. Grout is then injected into the conduits to prevent water from coming in contact with the metal. While most posttensioning is applied internally, it can also be applied externally to the member by an analogous process.

Posttensioning can also be used to compress blocks or segments of varied cross sections or with larger members to provide continuity over supports. Posttensioning is used extensively to join larger precast segments to span long distances, as in the case of modern segmental bridges.

## USES AND METHODS OF INSTALLATION

Before the construction of the Walnut Lane Bridge, prestressed concrete had been used only for industrial tanks and pipes. The Walnut Lane Bridge gave credibility to prestressed concrete as a viable construction material. Since 1950 prestressed concrete has been used in thousands of bridges. Another important early prestressed concrete bridge was the Tampa Bay Bridge (1951, William Dean et al.), a 17,500-foot-long trestle consisting of 46-foot-long spans.[12] This and other Florida bridges influenced the development of standard prestressed sections that are widely used today. Florida prestressed concrete companies, for instance, developed the widely used double-tee sections.[13]

In 1950, while the Walnut Lane Bridge was under construction, the Concrete Products Company of America, located in Pottstown, Pennsylvania, produced the first pretensioned bridge beam in the United States. This hollow box beam was formed with circular voids.[14] Other shapes were developed by the precasting industries.

The Doric Building (1951, Bryan and Dozier) in Nashville, Tennessee, is notable as the first true prestressed concrete building. The structural system was composed of slabs supported on posttensioned girders.[15]

During experimentation with pretensioning techniques, the seven-wire unit was developed in cooperation with the U.S. Steel Corporation.[16] This breakthrough system soon became the industry standard. Early attempts had used small piano wires less than 0.10 inch in diameter. Larger wires did not have sufficient surface area to provide the necessary bond. Initially the diameter of a twisted strand was 0.25 inch and was made up of seven smaller wires. Subsequently the strands were made larger, permitting hollow box beams as long as 100 feet to be produced.

The material's potential in structures requiring shorter spans and smaller members for use in light commercial buildings, schools, parking garages, and warehouses began in earnest in the mid-1950s. For these applications prefabricated and pretensioned members that could be transported to the construction site were manufactured.

**Figure 11.3.** The Walnut Lane Bridge (1950, Gustave Magnel et al.), Philadelphia, was the first linear prestressed concrete structure in the United States (top). Before construction a 160-foot girder identical to the type used in the bridge was tested to failure.

**Figure 11.4.** Rods for the Washburn Park Tank (1931, Harry W. Jones and W. D. Hewett) Minneapolis (bottom), were prestressed using the preload system, resulting in a higher compressive strength than that of conventional reinforced concrete.

Some were as small as 36 feet long, while the maximum length was limited only by transportation considerations.

One of the pioneering companies to use prestressed concrete for buildings was the Concrete Technology Corporation in Tacoma, Washington, founded in 1951. Concrete Technology's founder, Arthur Anderson, was convinced that successful prestressed concrete production required plant manufacture and control.[17] His firm designed and built the Boeing Company Development Center (1956) in Seattle, at the time the largest industrial, prestressed concrete building,[18] and the twenty-one-story Norton Building (1957), the first prestressed concrete structure higher than six stories.[19] These successes were followed by monorails at the Seattle World's Fair (1961) and Disney World (1971) in Orlando, Florida. Both are notable for their curved prestressed girders, which required special manufacturing techniques.

# PART III
# WOOD AND PLASTICS

FIBERBOARD

DECORATIVE PLASTIC LAMINATES

PLYWOOD

GLUED LAMINATED TIMBER

FIBER REINFORCED PLASTIC

**Figure 12.1.** The "Masonite Man" was featured regularly in Masonite advertisements. The company's 1938 catalogue gave specifications for Masonite sheathing, lath, interior finish board, and roof insulation, as well as insulating values. Catalogue, Masonite Corporation, 1938.

CAROL S. GOULD, KIMBERLY A. KONRAD,
KATHLEEN CATALANO MILLEY, AND REBECCA GALLAGHER

# 12 | FIBERBOARD

**DATES OF PRODUCTION**

1858 to the Present

**COMMON TRADE NAMES**

American Wallboard, Beaver Board, Cornell Board, C-X Boards, Feltex, Fir-Tex, Homasote, Insulite, J-M Board, Maftex, Masonite (Presdwood Quartrboard, Tempered Presdwood), Nu-Wood, Upson Board

## HISTORY

Fiberboard is a rigid sheet building material composed primarily of wood fiber or other vegetable fiber. Manufactured in various densities and thicknesses, this material has been used to insulate, sheathe, and finish building interiors and exteriors. Often referred to generally as wallboard, fiberboard falls into three basic categories: insulation board, medium-density fiberboard, and hardboard. These boards were either laminated or homogeneous compositions of grasses, reeds, rag, straw, bagasse (sugarcane waste), jute, flax, and hemp. Manufacturers also used waste materials such as sawdust, bark, oat hulls, spent hops, newspaper, and peanut shells, although most fiberboard was mechanically or chemically produced from wood pulp.

### ORIGINS AND DEVELOPMENT

The wallboard industry began in 1772 with Englishman Henry Clay's patent to produce a layered paper panel for use in rooms and for doors. The first U.S. patent was recorded in 1858.[1] This patent for "separating the fiber of wood and for the manufacture of paper . . . and other purposes" was granted to Azel Storrs Lyman of New York City.[2] W. E. Hale's 1870 patent for improved sheathing boards was the most explicit reference to the use of these boards in the building industry.[3] The following year Judd Cobb secured a patent for the manufacture of strawboard as a substitute for lath and plaster and as an exterior covering to be used in place of boards and siding.[4] Cobb's board was treated with a "cheap oil" to form the base of its waterproofing.

These patents suggest that the concept of using fiberboard for insulation, sheathing, and interior decoration was well established by the end of the nineteenth century.[5] However, in the early 1900s the patent engineer Joseph Rossman analyzed this industry and revealed that "only since 1906 [had] the first pulp board on the market [been] called 'wall board' with a view to take the place of lath, plaster and wood paneling."[6] The more than two hundred patents noted in Rossman's study indicate that numerous individuals were experimenting with variations of insulating boards and wallboards. However, fiberboard was probably not mass-produced or readily available until the 1910s.

One of the first companies to successfully mass-produce fiberboard was the Agasote Millboard Company in Trenton, New Jersey. Accounts suggest that the Trenton plant was established in 1909 and that by 1916 the company had introduced Homasote, its board product made of repulped newsprint.[7] In 1914 Carl G. Muench established a manufacturing operation at the Minnesota and Ontario Paper Company, where he patented Insulite, the first rigid insulation board, in 1915.[8] This plant at International Falls, Minnesota, could produce 60,000 square feet of Insulite per day.[9] The Celotex Corporation made the first board from bagasse in 1920.

The company most prolific in generating patents in this early period was the Beaver Board Company of Beaver Falls and later Buffalo, New York.[10] Between 1913 and 1920 people working on behalf of the company were awarded six patents for pulp-composite wallboard; however, corporate history and Buffalo city directories suggest that the company actually began producing wallboard in 1907.[11]

The first hardboard, a higher-density product with greater strength, was manufactured by the Mason Fiber Company (later the Masonite Corporation) in 1926. The Masonite Corporation's patented hardboard products, Presdwood and Tempered Presdwood, were widely used as interior finishing material as well as forming boards used in the construction of concrete structures.[12]

The early 1920s witnessed increased production of fiberboard of a homogeneous structure—that is, boards of a single-layer thickness made up of interlaced fibers. During the second half of the nineteenth century, only fifteen patents were granted for homogeneous, structured fiberboard, but there were four times that many by 1928. During the same period, the number of available laminated board products doubled.[13]

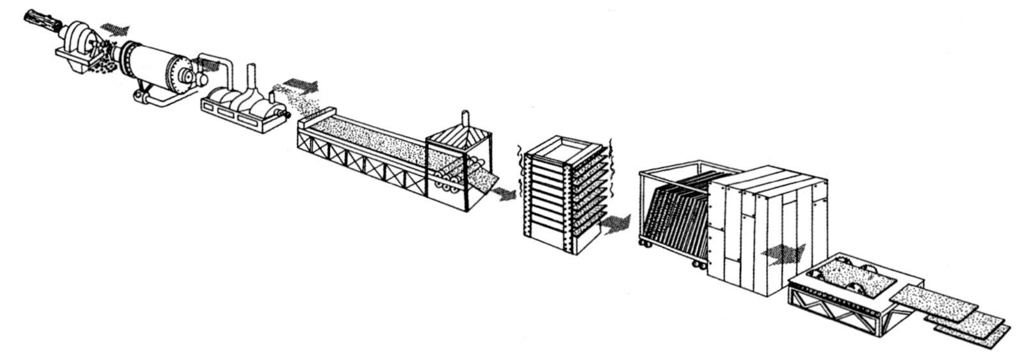

**Figure 12.2.** Steps in the manufacture of hardboard include (from left to right) chipping logs, defibering chips, refining fibers, compressing interlocking fibers into a continuous mat, treating with heat and hydraulic pressure to produce board sheets, adding moisture, and trimming for shipment.

Housing shortages in the 1930s drove the development of numerous insulation and wallboard products. Just before World War II, manufacturers such as Homasote developed specialized products for prefabricated housing, such as the Precision Built Home system, that allowed for the construction of 977 houses in Vallejo, California, in seventy-three days. By the 1940s war-related industries became the major consumer of these products, particularly hardboard, which was used extensively in constructing temporary military buildings. The continued explosion in the number of related patents, to more than six hundred by 1957, indicated the widespread application of this material in the American building industry as well as other markets, most notably the furniture industry, which became a major consumer of fiberboard after 1950.[14] Beginning in the 1960s, intense competition from plywood and particleboard had a grave effect on the production of fiberboard for the American building industry.[15]

## MANUFACTURING PROCESS

The majority of fiberboard was formed by mechanical processing. First, sawmill waste or logs were processed into pulp chips.[16] These were then sized and moved on to the grinder, where the wood cells were separated by frictional forces and steam pressure.[17] This energy-intensive pulping process involved high temperatures and a lot of water to soften the fiber bonds of the wood and permit better natural bonding in the consolidation stage.[18]

The pulp matter was allowed to flow in a current of water onto a screen, where heavy pressure was applied to remove excess water and form pulp sheets.[19] The sheets were then compressed between platens with a uniform force generated by hydraulic rams. Platens contained either steam or hot water and provided plain, smooth surfaces against which the fiberboard was molded. Pressure reduced the mass of wood fibers to a stiff, strong, dense board of interlocked fibers.[20] The last steps included drying, trimming, and fabricating to produce special finishes, colors, beveling, kerfing, laminating, and packaging.[21]

A relatively small percentage of fiberboard was manufactured by chemical or sulfite processing, a process in which wood cells are separated from one another primarily by dissolving and removing the natural bonding agent in a solution of sodium sulfate and caustic soda.[22] After the breakdown of wood pulp into fiber, the methods of pressing and drying were similar to those used in mechanical processing.

Both mechanically and chemically processed boards were treated in the pulping process with waterproofing material to prevent dry rot, fungal growths, and termites.[23] A variety of adhesives were used in the consolidation of the fibers, such as silicate of soda, flour paste, various glues, dextrin, and asphalt, or a mixture of water-glass and clay.[24] Other materials, including rosin, turpentine, paraffin wax, asphalt, asbestos, plaster of paris, and clay, were often incorporated to improve qualities such as tensile strength or resistance to moisture, fire, and vermin.[25]

Hardboard was manufactured to greater densities than insulation and medium-density fiberboard. This density was achieved through the application of greater pressure and higher temperatures. Hardboard was manufactured almost exclusively by the Masonite Corporation, which held a patent for the apparatus and process from 1928 until the early 1940s.[26]

Masonite's process was often referred to as a wet process because of the amount of water used in breaking down the pulp and distributing the fibrous materials, but it differed most significantly from other mechanical processes because it used steam pressure to explode

**Figure 12.3a–b. a.** The 1912 Beaver Companies catalogue showed how a carpenter could nail boards to wall studs and floor joists in any desired panel arrangement. **b.** Stained panel strips or battens used to cover nails created a paneled effect.

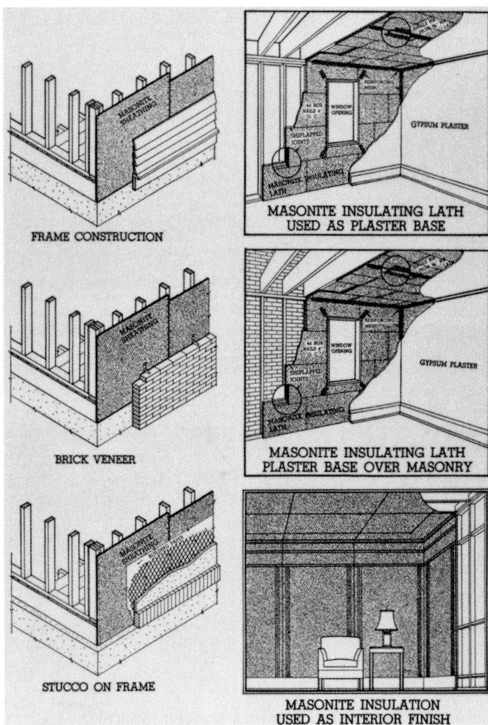


**Figure 12.4.** Masonite products could be used for both exterior sheathing (left) and as a plaster base for interior finish (right).

the wood chips. The advantage of this method was that it more effectively preserved the lignins (the natural cementing structure in the wood fiber) in the final product, giving the board greater strength.

A dry process for forming hardboard was developed in the early 1950s. The core fibrous mat was still produced by steam explosion and soaking in water; however, this mat was then dried completely before being consolidated into hardboard. Pressures used in the dry process ranged from 1,000 to 1,500 pounds per square inch. The resulting product was smooth on both sides; the fibers were refined to a higher degree, and waste was reduced to less than 3 percent.[27] Because this process did not entangle and form the ligneous bonds that the wet process did, synthetic resin adhesives were added to improve board strength.

The greatest changes in the manufacturing process of insulation board related to the speed with which boards could be produced. In 1910 insulation boards ½ inch thick needed 36 hours to dry; in 1947 they needed 50 minutes.[28] The post–World War II construction industry required mass production of insulation boards to meet American and international demand for this material. In turn, this encouraged continued research and

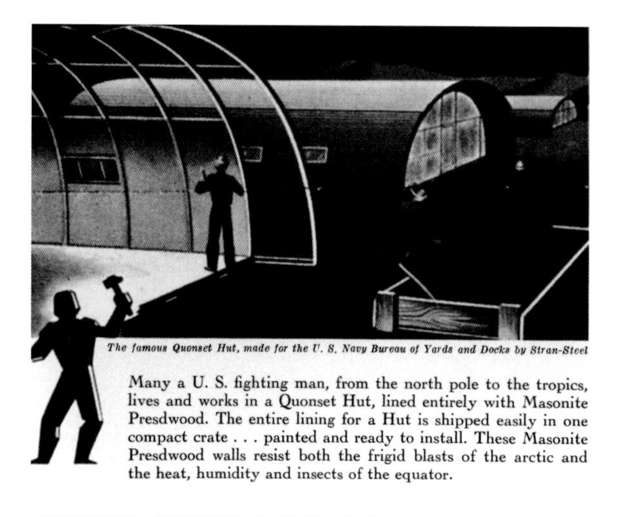

**Figure 12.5.** Masonite's Presdwood was used in the Quonset Hut, which revolutionized military housing (top).

**Figure 12.6.** The Masonite House at the 1933 Century of Progress Exposition in Chicago exemplified the use of hardboard in a modern home (bottom).

development of rapid production and finishing techniques, including application of paints, lacquer, plastics, and metals to make boards better suited for interior and exterior finishing of houses.

## USES AND METHODS OF INSTALLATION

Most early wallboard products were used either as insulation or sheathing beneath exterior cladding, or as a finish material for secondary spaces, such as attics and basements. Like so many new building materials, insulation board, medium-density fiberboard, and hardboard were featured at nationally recognized events, including world's fairs and Olympic Games, where they could be introduced to broad sections of the American public and colleagues in the building industry. As the versatility of all three types of material was recognized, fiberboard became popular for summer cottage and year-round home interiors, commercial displays, and partitioning of industrial facilities.

Low-density boards were generally of greater thickness and were ideal for heat- and sound-insulation purposes. Celotex products were used in the 1930s to insulate the roofs of the White House in Washington, D.C., and Rockefeller Center in New York City. Thinner medium-density fiberboards and hardboards possessed structural strength and rigidity suitable for sheathing but did not serve well as insulators. Products such as Celotex's Building Board, dubbed "structural insulation," were used as exterior finishes or as sheathing under roofing materials or wall veneers of brick, siding, wood shingles, or stucco.[29] New products in 1937 included Celotex's Cemesto, a fire-resistant insulation board surfaced on one or both sides with asbestos cement, which was used in low-cost housing, service stations, and industrial drying plants and for partitions in office and commercial buildings.

In general, fiberboard could be finished with paint, stencils, and baked-on enamels and glazes or with paper or fabric wall covering to make them suitable for use inside or out. Interiors might be finished with a pebble-surfaced Beaver Board, painted and decorated with stenciled designs instead of wallpaper, or, in more rustic settings, natural Celotex, which was used to finish the interior of the Lake Placid, New York, clubhouse (Distin and Wilson) for the 1932 winter Olympic Games.[30]

The thickness of insulation board ranged from 7⁄16 inch to 1 inch and in some cases up to 3 inches. The standard width was 4 feet, although board was available as wide as 8 feet; it could be 5 to 14 feet long, depending on the manufacturer. Standardized measurements for insulating lath (board used as the base for a plaster wall finish) were 18 or 24 inches wide and 48 inches long. Board used for roof insulation was 22 inches wide and 47 inches long. Medium-density fiberboard was generally 3⁄16 to 3⁄8 inch thick. Widths and lengths of this board were comparable to those of insulating board. Hardboard came in thicknesses of 1⁄8 to 3⁄16 inch, widths of 4 to 5 feet, and lengths of 6 to 16 feet. Thicker boards were occasionally produced by gluing together two pieces of board with a waterproof glue. Board finished to look like tile—typically used for ceiling finish—was produced with beveled edges in sizes from 6 by 12 inches to 23 by 48 inches and in thicknesses of 1⁄2 inch, 3⁄4 inch, and 1 inch.[31]

Contractors generally applied insulation boards to the studs in any frame construction on 12- or 16-inch centers with 2-by-4 headers inserted flush between the wood studs to provide a nailing surface for the wallboard and battens.[32] Some companies developed special mechanisms for fastening boards in place, such as Upson Company's self-clinching fasteners with "in-to-stay" nails, designed to cut installation time in half, eliminate nail holes at the center of panels, and avoid countersinking nails and filling holes.[33] Insulation board could be installed directly beneath the roofing material or between the structural framing members of the attic floor, or both. When used with flat roofs, insulation board was installed over the roof deck and under the roofing; however, in some instances it was installed in the ceiling.[34] When insulation board was used for sound conditioning, a suspended ceiling type of construction was recommended.

Numerous designs could be executed for interior panel arrangements with complementary ceiling treatment patterns. Use of all vertical panels was recommended to increase the perception of height in a room. Other arrangements included vertical panels with a continuous frieze, vertical panels above a continuous dado, or a design that included a combination of a dado, vertical panels, and a frieze. The latter design was recommended specifically for dining rooms, libraries, dens, offices, stores, schools, and theaters.[35] Decorative wood or panel strips were usually 1⁄4 to 1⁄2 inch thick and of widths to meet architectural requirements.[36]

Use of hardboard for exterior finishes was fairly limited before 1950. The earliest board was not well suited for exterior use because it did not resist moisture well enough. The Masonite Corporation introduced Tempered Presdwood in 1931 and promoted it for exterior use because it was water resistant. Masonite exhibited a house using its products, including Presdwood, at the 1933 Century of Progress Exposition in Chicago. Presdwood was marketed for storefronts, restaurants, beach cabins, barbecue stands, and many other business structures. The material was to be finished in the same manner as wood—either painted or stained.[37]

By the mid-twentieth century fiberboard had become a major industry. This reuse of a waste product was marketed to consumers as a material both for new construction and for remodeling because it was easily worked, light, and competitively priced.

**Figure 13.1.** Marketed as a man-made alternative to marble, Formica was a durable wall finish and countertop material for schools, banks, theaters, and restaurants. Color choices were wide, and designs could be incorporated into the material itself. *Architectural Record,* August 1943.

ANTHONY J. T. WALKER, KIMBERLY A. KONRAD, AND NICOLE L. STULL

# 13 | DECORATIVE PLASTIC LAMINATES

**DATES OF PRODUCTION**
c. 1911 to the Present
**COMMON TRADE NAMES**
Dilecto, Formica, Insurok, Lamicoid, Micarta, Panelyte, Parkwood, Roanoid

## HISTORY

Decorative plastic laminates typically consist of layers of kraft paper impregnated with a synthetic resin and cured under heat and pressure to form an insoluble, homogeneous piece.[1] Early laminates made from phenolic resins found wide application in both the electrical and automotive industries. Beginning in 1927, decorative laminated sheets were used for countertops, tables, bars, splash backs, interior paneling, doors, storefronts, and ornamental signs.[2]

### ORIGINS AND DEVELOPMENT

Laminates were first produced in 1907 by Leo Baekeland, who impregnated fibrous sheets with a phenol-formaldehyde resin. With the publication of his patent in 1909, the process for producing plastic laminates was adopted by many manufacturers.[3] The General Bakelite Company, established by Baekeland, sold resin to companies such as Westinghouse Electric and Manufacturing Company, which produced a phenolic laminate for electrical purposes by impregnating a heavy canvas with Bakelite resin.[4]

Two Westinghouse engineers, Daniel O'Conor and Herbert Faber, developed a process to make the laminate in sheet form. In 1913 they left Westinghouse to form the Formica Insulation Company of Cincinnati. The product name ("for mica") indicated that the laminate was a substitute for mica, an electrical insulator.[5]

With the expiration of the Baekeland patent in 1927, the market opened up considerably as many new competitors began producing plastic laminates.[6] In the early 1920s decorative laminates were used for the first time in radio cabinets. The dark color of the phenol-formaldehyde resin limited the product to dark colors and patterns of brown and black.[7] Early attempts to change the color with surface coloring caused problems because the dyes tended to rub off.[8] In 1927 Formica first used an opaque barrier sheet to block out the dark interior, allowing the production of light-colored, lithographed wood-grain laminates.[9]

Plastic laminates using clear urea- and thiourea-formaldehyde as resinous materials were introduced in 1927.[10] Because these resins were colorless, lighter-colored laminates, which were resistant to sunlight, were made possible. However, urea-formaldehyde resins tended to warp, absorbed water, and were less durable and much more expensive than the phenol resins.[11] The phenol and urea resins were not good conductors and, if subjected to an intense heat source such as a burning cigarette, they would discolor and ultimately char. To overcome this major disadvantage, in 1931 Formica patented a process to incorporate a very thin aluminum foil between the surface layer of the laminate and the core.

The introduction of melamine resins in 1938 allowed for more durable products.[12] When combined with formaldehyde they formed a laminate that was resistant to abrasion, heat, and moisture while allowing for a variety of color and surface patterns. The high cost of melamine prohibits its complete substitution for phenolic resins. Typically a colored melamine ply is used as a surface layer for phenolic laminates to provide a decorative finish, or a translucent melamine overlay is added to colored urea-formaldehyde plies to provide a tougher surface.

By 1948 a low-pressure polyester laminate had become commercially available. This material was flexible and could be used to produce curved or odd-shaped components.[13] Decorative applications included lighting fixtures and wall panels. Post-forming grades of phenolic laminates were also available using specially formulated resins that permitted the flat sheets to be reheated and pressed into simple curved forms for splash backs or counter edges. Polyester laminates have a high moisture resistance and good electrical properties; however, they have a low resistance to heat.[14]

World War II saw great expansion, experimentation, and development in the plastics industry.[15] Only industrial grades of laminates were produced for defense purposes. During the postwar construction boom, there was a dramatic growth in the use of laminates.[16] The application

**Figure 13.2.** Decorative plastic laminates are installed and trimmed to fit the backing material. A worker applies a laminate with an adhesive to a wood substrate on the panel of a curved counter.

of Formica and similar competitive products such as Roanoid and Micarta was seen in cinemas, diners, and kitchens, and by the 1950s a generation had grown familiar with laminated materials.[17]

## MANUFACTURING PROCESS

Laminates are made by first passing pieces of kraft paper through a bath of liquid resin.[18] The paper is then subjected to moderate heat to vaporize the solvents, leaving the dry paper fully impregnated. The sheets are then cut to the appropriate size and stacked in layers.[19] The heated plates of a hydraulic press apply temperatures of about 300 degrees Fahrenheit and 2,000 or more pounds of pressure per square inch.[20] The heat and pressure cause a chemical reaction in the resin in which the polymer molecules fuse the plies together, forming a solid piece. The laminate is then gradually cooled while still under pressure either by lowering the temperature of the water in the platens or by placing the laminate in an annealing oven.[21]

Gloss or matte surfaces are transferred by the press plates. A high gloss is achieved if the material is cured against a highly burnished metal plate; for a satin surface the sheet is buffed.[22] Decoration is achieved by applying a top surface sheet, which carries the design over the other layers of material, and curing the entire unit. Designs are etched on the surface of a cylindrical copper plate, which transfers the pattern onto a roll of paper.[23] A variety of design effects can be created using wood veneers or fabrics. Decorative inlays are built up on the surface of the laminate by cutting strips of paper or metal foil in the form of the desired pattern and laying them on the surface of the sheet.[24]

Although the process did not change appreciably, the development of new synthetic resins and the use of various filler sheets resulted in many decorative laminates. The fillers include kraft, alpha, or rag paper; cotton fabric; asbestos felt; and fiberglass cloth. Structural strength could be increased by adding thick gauges of metals, electrical properties could be enhanced with glass fiber, and fire resistance could be improved with aluminum foil or asbestos.[25]

## USES AND METHODS OF INSTALLATION

Decorative laminates range in thickness from 1/16 inch for countertops and veneering purposes to 5/16 or 5/32 inch for self-supporting panels and partitions.[26] A 3/8-inch laminate is typically used for exterior purposes.[27] Although the standard dimensions are 8 feet by 4 feet, larger sizes can be produced.[28] These veneer sheets are mounted on a backing of plywood, sheet metal, or asbestos. Adhesives commonly include casein and resin glues or bonding cement. Exterior laminates can also be attached with screws.

Laminates became very popular in the 1930s and 1940s as a material to modernize such spaces as theaters, diners, and stores.[29] The most common applications are storefronts, paneling and wainscoting, doors, countertops, and furniture.[30] Plastic laminates were also used in illuminated sign display.[31]

With constant research and development in the plastics industry, many experimental projects have used plastics. In the late 1930s Herbert Faber, president of Formica, built a house in Cincinnati using laminates for the windows, walls, light fixtures, and even the shower curtain.[32] In 1936 the ocean liner *Queen Mary* featured a fireproof plastic laminate for decorative paneling, tabletops, and furnishings, opening a new market for the material in marine applications.[33] The largest early specification of laminates was made by the architect of the Capitol in 1939 for the annex to the Library of Congress.[34]

At the 1964 World's Fair in Flushing Meadow, New York, Formica introduced a World's Fair House (Emil A. Schmidlin) that had laminates on the majority of its surfaces.[35] Using this house as a prototype, similar models

**Figure 13.3a–c.** Decorative plastic laminates took many forms: **(a.)** as the striking Formica wall panels in the lobby of the A. O. Smith Building (1930, Holabird and Root), Milwaukee, Wisc.; **(b.)** as countertops in innumerable diners and coffee shops, such as the Penguin Coffee Shop (1959, Armét and Davis), Santa Monica, Calif.; and **(c.)** as Micarta bathroom walls.

were built throughout the United States. In 1989 General Electric built an experimental house (Richardson Nagy Martin) entirely of plastics in Pittsfield, Massachusetts.

## CONSERVATION

A laminate's performance depends on its ability to function as a single homogeneous unit. The physical properties of laminates are determined by the degree of polymerization, the influence of the filler material, and the type of plasticizing resin used.[36] Although decorative laminates are characteristically hard and durable, they are subject to deterioration under normal operating conditions. Failure occurs when the plastic's polymer chain is broken in any way. Temperature, sunlight, moisture, stress, and chemicals are likely to cause deterioration. After prolonged exposure to these forces, deformation of the plastic assumes a constant rate and continues indefinitely.[37]

### DETERIORATION

When exposed to high temperatures, plastic materials are subject to physical and chemical changes that may cause degradation and a progressive change in physical properties.[38] Elevated temperatures will result in the eventual disintegration of the bond formed during polymerization between the resin and the laminate's base material.[39] Heat reduces the laminate by driving out volatiles such as moisture, solvents, or plasticizers. It also advances the cure of the thermosets and causes a change in the material's makeup.[40] Increased temperatures also decrease a laminate's resistance, making the material more susceptible to other types of degradation. Excessive cold temperatures will cause the laminate to become brittle. Exposure to heat may result in shrinkage or a change in dimension. Visual changes may include blistering, delamination, and cracking.

Prolonged exposure to moisture will also cause a laminate to deteriorate.[41] The rate of water absorption through each edge and surface may differ widely, but higher levels of absorption generally occur at cut edges. Water absorption results in a change of dimensional properties.[42] Laminates will swell with high humidity and become brittle when dried; this cycle can result in surface crazing.[43]

Light is a major factor in the surface deterioration of plastic laminates and causes a gradual change in their physical properties. Ultraviolet light is the most damaging part of the spectrum, but sunlight at any level can cause discoloration.[44] The change in color is a result of solar action on the surface dyes or the plastic base. Prolonged exposure to ultraviolet light causes the phenolic resin to gradually turn dark yellow.[45] Sunlight can also break down the surface sheen, giving the plastic an overall matte appearance.

Like other plastics, laminates are subject to deformation resulting from impact, which affects the material's internal structure.[46] Under direct tension and compression it takes a long time for appreciable laminate deformation to occur. The impact strength of laminated plastics depends largely on the direction of the load in relation to the plies of the material.[47] Every stress will produce a certain amount of deformation. When the load is removed, there is rapid, immediate elastic recovery.[48] A surface fracture or ply separation is characteristic of impact failure.

Although laminates are generally chemically inert, exposure to certain solvents may bring about rapid deterioration. Plastics may be affected by hydrochloric acid, sodium hydroxide, sulfuric acid, hydrogen peroxide, and acetone.[49] Evidence of chemical attack may include a change in weight or physical dimensions. Other signs of exposure to harmful chemical reagents are discoloration, delamination, warping, and shrinkage.[50]

Adhesive failure can also cause deterioration of laminated surfaces. Adhesives are highly resistant to heat and moisture, but the strength of the bond is affected by assembly conditions and material preparation.[51] These include the preparation of the base surface, mixing of the adhesive, the number of coats, the drying process, and assembly conditions, such as the application of pressure, open drying time, and room temperature.[52]

In general, laminates are very stable; attacks by bacteria, fungi, or insects present no serious problem. If the surface is cut or broken, however, the laminate becomes susceptible to biological attack because of its wood flour and cotton flock fillers.[53]

## DIAGNOSTICS AND CONDITION ASSESSMENT

Early detection of degradation and its likely cause is extremely important and can be achieved only if laminates are examined regularly and their condition noted. The surface condition—such as a change in hue, hazing, the loss of gloss, crazing, cracking, or delamination—may indicate degradation.[54]

A nondestructive test can be conducted to detect surface discontinuities such as cracks, seams, laps, isolated porosity, leaks, or lack of fusion. First, a liquid penetrant is applied evenly over the surface of the laminate and allowed to enter the discontinuities. After a dwell time (the time that the penetrant is in contact with the test surface), the penetrant is removed by wiping the surface dry. Next, a developer is applied, drawing the entrapped penetrant out of any open area and staining the developer. The surface is then examined to determine the presence or absence of failure.[55]

## CONSERVATION TECHNIQUES

Since environmental factors generally cannot be completely eliminated, the degree of exposure to extreme temperatures, humidity, and light must be controlled. Manufacturers recommend maintaining a temperature of about 70 degrees Fahrenheit to limit thermal effects on laminates.[56] Maintaining a constant level of humidity from 20 to 80 percent is also recommended to control the amount of moisture absorption.[57] Limiting exposure to light is an effective measure to prevent laminate deterioration.[58]

Decorative plastic laminates should be cleaned periodically to remove surface contamination that may have built up. Laminates can be washed with a soft brush and tepid water containing a small quantity of non-ionic detergent. After being washed, the material should be rinsed thoroughly with clean water to remove all the cleaning agent and immediately dried using an absorbent material.[59]

Methods of on-site laminate repair can be used in areas of limited damage; these are best used to repair minor surface failure, such as scratches or stains, but may also be useful in more serious repairs. A cellulose resin is used to fill in the damaged area and then recolored to

**Figure 13.4.** Decorative plastic laminates on the doors of the old Schine's Theater (1935, John Eberson), Auburn, N.Y., have faded from exposure to sunlight and have abraded from use.

obtain a suitable match. This method is most effective on plain-colored laminated surfaces without patterns.[60]

Further study is needed, however, on the early detection of failure in addition to visual inspection. More complete information is also needed on possible conservation methods to limit and arrest deterioration mechanisms.

## REPLACEMENT

When replacing a historic laminate, the deteriorated piece must first be removed by heating the surface to break the bond of the adhesive. Once the piece is removed, the base surface should be inspected. If the base shows signs of deterioration, it should be replaced.[61] The base surface must be cleaned to remove all foreign matter, traces of old glue, and cement. A bonding cement is applied evenly to the base surface and allowed to dry; another coat is then applied to the underside of the laminate and to the base surface.[62] The laminate should be fit into place and pressure applied over the entire surface. Wood clamps should be used to hold the laminate in place. The area should be protected from dust or dirt while drying.[63]

The main difficulty in replacing decorative laminates is that available patterns and styles change constantly. Recently, however, a number of historic laminate patterns have been reintroduced. Classic linen, boomerang, and wood patterns are once again available for replacement in historic buildings.

Damaged laminates may also be replaced by integrating original and new panels. If a pattern is no longer available, it may be necessary to select a solid color that is compatible with existing material. Integrating new laminates with historic ones is difficult, however, because the historic materials may have worn unevenly, depending on the original installation. Another problem is matching historic laminates that may have faded.

**Figure 14.1.** Plyshield, an exterior-grade Douglas fir plywood siding, was used as a structural skin for an unspecified home in this Douglas Fir Plywood Association advertisement. The modular plywood sections were covered with battens. *Progressive Architecture*, April 1950.

THOMAS C. JESTER

# 14 | PLYWOOD

**DATES OF PRODUCTION**

c. 1905 to the Present

**COMMON TRADE NAMES**

Haskelite, Plankweld, Plycrete, Plymetal, Plyshield, Super Hardobrd, Weldtex, Weldwood

## HISTORY

Plywood is an assembly of hardwood or softwood veneers (thin sheets of wood) bonded together with an adhesive. The grains of alternating sheets are perpendicular.[1] Noted for a high strength-to-weight ratio, dimensional stability, resistance to splitting, and ability to be molded into compound curves, plywood panels are used for both structural and decorative purposes.[2] In 1919 the term *plywood* was adopted by the Veneer Manufacturers Association, which changed its name to the Plywood Manufacturers Association the same year, in an effort to reduce the seemingly endless number of names—scale boards, pasted wood, and built-up wood—that were used to describe the material.[3]

### ORIGINS AND DEVELOPMENT

The earliest known plywood patent, granted to John K. Mayo in 1865, covered the manufacture of "scale boards." The second reissue of this patent referred to the material as useful for covering and lining structures. It states: "The invention consists in the formation of various structures used in Civil Engineering of a plurality of thin sheets or veneers of wood, cemented or otherwise firmly connected together, with the grain of the several scales or thicknesses crossed or diversified, so that they will afford to each other mutual strength, support, and protection against checking and splitting, shrinkage or swelling, expanding or contracting."[4] It is unlikely, however, that much material was produced under this patent.

Plywood panels were manufactured by producers of hardwood veneers in the late nineteenth century. These were custom ordered and used by furniture makers for drawer bottoms and other concealed parts. Plywood was also used in the late nineteenth century for pin planks in pianos, sewing machine covers, seating, and desk tops.[5] Lumber cores was also faced with veneers for furniture.

Plywood panels sold as a stock item were an outgrowth of the veneer industry on the West Coast. As early as 1890, hardwood plywood was used for panels in doors, replacing raised lumber panels.[6] The softwood plywood industry originated in 1905, when the Portland Manufacturing Company exhibited handmade Douglas fir plywood panels at the Lewis and Clark Centennial Exposition in Portland, Oregon.[7] The plywood industry developed slowly until World War I, when extensive tests were made to develop reliable plywood for airplane construction.[8]

In the 1920s manufacturers of Douglas fir plywood began selling it to automobile manufacturers to complement the existing door-panel market.[9] The Pacific Coast Manufacturers Association, which later became the Douglas Fir Plywood Association, was organized in 1924 to create a grading system and develop new markets for Douglas fir plywood, presciently recognizing the potential for plywood in house construction.[10]

The development of plywood as an engineered wood product with reliable properties hinged on the quality of the adhesive used to glue the veneers.[11] The earliest plywood panels were manufactured with hide glue. Vegetable glue made from cassava flour was introduced in 1905 by Frank G. Perkins, owner of the Perkins Glue Company. Although vegetable glue was the first inexpensive glue for plywood, was easy to manipulate, and could be stored for long periods of time, it was not water resistant.

Casein glue, made from milk and introduced in the United States around 1900, became the first important water-resistant glue for plywood and was widely used during World War I. In 1912 Henry Haskell perfected another water-resistant glue, blood albumin, which required hot pressing. Soybean glues also showed good water resistance and were first imported in the early 1920s. The Douglas fir industry began to use these glues more widely after I. F. Laucks introduced domestically produced soybean glue in 1927.[12]

German synthetic resin glue in sheet form, introduced in 1931, offered even greater water resistance. Resin-bonded plywoods opened new markets in construction.[13] Domestic production of synthetic resin adhesives—phenol-formaldehyde—began in 1935.[14]

As custom-ordered veneer core panels gained acceptance in the early 1900s, the number of plywood manufacturers grew. In 1919 U.S. Plywood was established; it would become the largest and most innovative plywood company to distribute and, after 1930, manufacture plywood.[15] By 1932 more than fifty manufacturers were producing plywood.[16] However, the greatest growth occurred after World War II, when the number of manufacturers grew to more than 150. Between 1939 and 1947 total industry output increased 380 percent.[17]

### MANUFACTURING PROCESS

Although manufacturing techniques have improved, principally in making the processes continuous and automated, plywood manufacture still follows the same basic steps to transform timber into panels that were used in the early nineteenth century.[18] Manufacture began with peeling bark off logs, then rotating the logs against a knife to create veneers to assemble plywood panels. Veneers were subsequently flattened, clipped to length, taped and patched if necessary, and dried.

Veneers were glued together using either a cold or a hot press.[19] In the cold process, the adhesive was applied in a liquid form by a mechanical spreader. The assembled plies were pressed together with either a hydraulic press or a hand- or motor-operated screw press. Unlike cold pressing, which required many hours of pressure, hot pressing required less time, and panels could be processed more immediately. Hot pressing, used for synthetic resin sheet and liquid-form adhesives, needed higher pressures but permitted thinner adhesive spreads. The final manufacturing step involved trimming the panels to an exact dimension and sanding.

Several manufacturing advances occurred during World War II. Electronic heating devices, which were based on radio-frequency technology, were used to replace convection systems as a means of rapidly polymerizing heat-reactive resins. This system enabled plywood adhesives to be cured at lower temperatures without changing the moisture content of the panels.[20] The war years also gave rise to molded plywood with compound curvatures. The curves were formed in a process called bag molding.[21]

### USES AND METHODS OF INSTALLATION

As a construction material, plywood has had a variety of nonstructural and structural uses. Nonstructural uses generally developed first and ranged from doors, paneling, and formwork for concrete to countertop bases for plastic laminates.[22] In the late 1920s the Forest Products Laboratory in Madison, Wisconsin, tested the strength of plywood for sheathing, and by the early 1930s house construction had become the next major market for plywood manufacturers.[23]

In 1934 the Harbor Plywood Corporation in Hoquiam, Washington, introduced the first waterproof plywood, Super Harbord, for exterior use.[24] The Forest Products Laboratory constructed a plywood house the same year, using exterior plywood and stressed-skin panels to carry part of the wall's load.[25]

Throughout the 1930s builders, architects, and foundations studying housing experimented with prefabricated plywood walls and partitions.[26] Foster Gunnison developed the first commercially prefabricated house with exterior plywood in 1936.[27] The Douglas Fir Plywood Association also offered a plywood house, purchasing the rights to the Dri-Bilt house system designed by Jacques Willis in 1938.[28] Dri-Bilt houses required no plaster and had plywood walls, subfloors, ceilings, and partitions.

Modernists too embraced the new material for exterior applications. In 1936 Richard Neutra designed a demonstration house with a plywood exterior, and Lawrence Kocher designed a plywood house for the Town of Tomorrow at the 1939 World's Fair in New York.[29]

Despite efforts to incorporate plywood in prefabricated houses and experimentation with Dri-Bilt construction, plywood found wider acceptance for sheathing and subflooring. As a commodity, it was logical to manufacture and sell standard plywood sheets, but at the construction site the material was easily cut as needed.

Plywood also made inroads in other areas for construction. Flush plywood doors, used for schools, public buildings, and residences, were available by 1930, and hollow-core doors were introduced in the late 1930s.[30] U.S. Plywood even manufactured a parquet plywood flooring.

Standards played an important part in plywood's success as a construction material. Building on its first standards adopted in the 1920s, the Douglas Fir Plywood Association developed new standards for quality of surface finish and structure in 1932 that were issued by the National Bureau of Standards in 1933. The grades were Good 2 Sides, Good 1 Side, Sound 2 Sides, Sound 1 Side, Wallboard, and Concrete Form Plywood.[31]

The earliest stock plywood panels were 3 feet by 6 feet. By the early 1930s the now-ubiquitous 4-by-8-foot panel was being produced. Common thicknesses for early three-ply panels were ⅜ inch and ½ inch, but the range of thicknesses expanded during the 1920s and 1930s.

Installation of structural plywood and paneling was usually based on nailing on 6-inch centers for panel perimeters and 10-inch centers for intermediate studs.[32] In some instances, paneling was glued and nailed to furring and studs. A range of joints, including flush, concealed with battens, V-joint, tongue-and-groove, and lap, have been used for structural and decorative plywood.[33]

Residential and commercial markets for plywood increased with the development of factory-prefinished panels. First introduced by the U.S. Plywood Corporation in the mid-1940s as Plankweld, prefinished hardwood panels were widely used during the 1950s.[34] Usually constructed of ¼-inch stock, the panels were coated with a roller-applied color compound and sealed with a lacquer or clear synthetic coating.[35] Other prefinished products included striated panels, such as U.S. Plywood's Weldtex, with V-grooves.

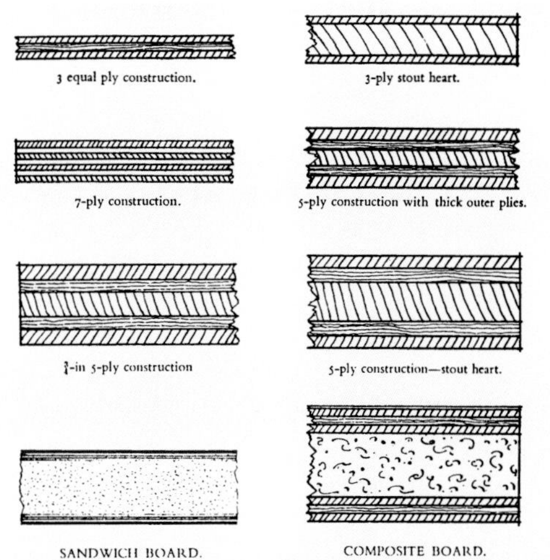

**Figure 14.2.** Plywood's strength and stability derived partly from the construction of odd numbers of plies—three, five, and seven—glued together perpendicular to one another, illustrated here by sections of various constructions (top).

**Figure 14.3.** In 1937 in Madison, Wisc., the U.S. Department of Agriculture's Forest Products Laboratory built the first completely prefabricated "low-cost" plywood house (bottom), based on methods used in another demonstration house in 1935.

**Figure 15.1.** Unit Structures, the first U.S. company to manufacture glued laminated timber, marketed laminated arches for gymnasiums, auditoriums, and recreation halls. Laminated Arches catalogue, Unit Structures, Peshtigo, Wisc., 1939.

ANDREW MCNALL; DAVID C. FISCHETTI

# 15 | GLUED LAMINATED TIMBER

**DATES OF PRODUCTION**
1934 to the Present
**COMMON TRADE NAME**
Unit Arches

## HISTORY

Glued laminated timber is any structural material composed of wood laminations glued together so that the grains of all laminations are longitudinally parallel.[1] Although laminations are typically 1 or 2 inches thick, they may be of any length and width. Douglas fir and southern pine are common species for use in glued laminated timber, which can be manufactured as straight timbers for columns and beams and truss components, or as curved laminations for arches and trusses. Arches have played an especially significant role in the development of glued laminated timber, permitting unique architectural effects and clear spans longer than solid timber.[2]

## ORIGINS AND DEVELOPMENT

The first recorded use of glued laminated arches occurred in 1893 in Basel, Switzerland.[3] However, the widespread use of glued laminated timber in Europe did not begin for at least another decade. In 1901 Otto Hetzer of Weimar, Germany, received the first patent for glued laminated beams, and in 1906 he received the first patent for curved laminated wood members.[4] Thereafter, glued laminated arches were known in Europe as the Hetzer construction method. During the next twenty years, Hetzer built structures such as railway stations, factories, workshops, and gymnasiums in Switzerland and Germany. This construction method was quite popular before World War I, but shortages of casein glue during the war curtailed its spread in Europe.[5]

While glued laminated timber was growing in popularity in Europe, another laminated wood product was being developed in the United States. Research on adhesives for plywood contributed to advances in glued laminated timber.[6] The McKeown Brothers Company of Chicago, for instance, used adhesive in addition to nails on bowstring-truss chords it manufactured in the early 1920s.[7] Wood laminating thus had some precedent in the United States when Max C. Hanisch, Sr., of Germany introduced glued laminated timber.

**Figure 15.2.** Workers at the U.S. Department of Agriculture's Forest Products Laboratory in Madison, Wisc., inspect one of the first glued laminated timbers.

Born in 1882, Hanisch received a degree in architectural engineering at age 21 and designed buildings with glued laminated timber for Otto Hetzer. Hanisch immigrated to the United States in 1923 but did not find conditions immediately suitable for introducing glued laminated timber.[8] In 1934 he founded Unit Structures in Peshtigo, Wisconsin, with the Thompson brothers, owners of a boatbuilding business. Unit Structures' first contract was in 1934 for the Peshtigo High School gymnasium (Max C. Hanisch, Sr.), the first building in the United States to use glued laminated timber.[9]

At about the same time, the Forest Products Laboratory of the U.S. Department of Agriculture in Madison, Wisconsin, began a comprehensive research program on glued laminated timber. In 1939 some of the findings were published. A second and more systematic study, published in 1954, included suggested design specifications.[10]

### MANUFACTURING PROCESS

To make glued laminated timber, manufacturers typically glue together strips of lumber 1 or 2 inches thick to form the desired shape. Lumber is selected on the basis of strength and appearance. In preparation for bonding, the lumber is surfaced (planed until smooth and flat), cleaned, and seasoned so that the moisture content is less than 16 percent at the time of bonding. Since glued laminated timber is custom manufactured, the number of laminations varies widely, as do the size, shape, and finish of the completed product.[11] To achieve the desired finished width, it may be necessary to place laminations edge to edge; no additional preparation is needed for these laminations. It is also usually necessary to join lumber end to end. Although a simple butt joint was sometimes used before World War II, this joint was superseded by stronger end joints, such as hooked, scarf, and finger joints. Once the timber is prepared, equipment uniformly distributes adhesive on the lumber faces. Laminations are then placed in a preset form and clamped together at a uniform pressure until the adhesive cures.[12]

Advances in adhesives have significantly influenced the development of glued laminated timber. The principal adhesive used before World War II was a water-resistant casein glue. This adhesive sufficed for building interiors but not exterior applications. Synthetic-resin glues, introduced in the early 1930s, improved the performance of glued laminated timber and expanded its use to bridges and other structures that would be exposed to the weather. Some resin adhesives used at least through the 1950s were urea-formaldehyde and phenol-formaldehyde. During this time resorcinol resin and phenol-resorcinol resin adhesives were also becoming popular.[13]

### USES AND METHODS OF INSTALLATION

Glued laminated timber met with resistance from skeptical engineers.[14] This skepticism led to the use of steel reinforcement in the Peshtigo gymnasium. However, engineers soon learned that reinforcement was unnecessary, as research proved that the adhesives in glued laminated timber provided bonds at least as strong as the wood itself.[15]

Tests on a packaging and storage building at the Forest Products Laboratory proved such claims. Built in 1935, this structure had seven arches, five of them three-hinge, glued laminated arches manufactured by Unit Structures and erected with a crane. The arches provided a clear span of 46 feet and an interior ceiling height of 19 feet at the apex. Tests included loading the center arch with sandbags totaling 31,500 pounds—exceeding the design specifications by 50 percent—for more than a year. During this time the weight deflected the roof peak 1¼ inches and spread the arch shoulders outward about ¼ inch each.[16]

In the period between this experimental building and the United States' entry into World War II, glued laminated timber was used in more than a hundred buildings, including auditoriums, barns, gyms, churches, garages, storage facilities, and warehouses. In 1936 the first church using glued laminated arches (St. Leonard Catholic Church, Max C. Hanisch, Sr.), was built in Laona, Wisconsin. Its three-hinge arches provided a clear span of 43 feet. Other types of buildings using glued laminated timber included highway-equipment buildings in Sturgeon Bay, Green Bay, Merrill, Altoona, and Oshkosh (all 1940) for the State of Wisconsin. The arches in these buildings provided a span of 80 feet, had a cross section of 7½ by 19½ inches, and were composed of twenty-four laminations of southern yellow pine.

Before World War II the use of glued laminated timber had not spread much beyond Wisconsin, and Unit Structures was still the dominant manufacturer.[17] The war, however, was an opportunity for this material to make a larger impact. Because steel was diverted for war equipment, the military turned to wood for roof construction. Timber Structures of Portland, Oregon, supplied the majority of the solid-sawn timber used in wood roof trusses, but Unit Structures also licensed this company to manufacture glued laminated timber.[18] In its first military contract, Unit Structures persuaded the military to use glued laminated timber, and several military drill halls, storage facilities, aircraft hangars, and factories were built with glued laminated timber, notably the eight drill halls at the U.S. Naval Training Station (1942) in Great Lakes, Illinois. Arches in these buildings provided a clear span of 115 feet and a height of 42 feet at the crown; in cross section they measured 8 inches wide by 27 inches thick.[19]

The notoriety glued laminated timber received during the war spurred new manufacturers to enter the market, and glued laminated timber met with increasing commercial success. Churches became one of the largest markets for the material; schools, hangars, supermarkets, auditoriums, factories, and warehouses were also built with glued laminated timber.[20]

To serve a growing industry, the American Institute of Timber Construction was organized in 1952.[21] Its first president, Ward Mayer, said that the institute would

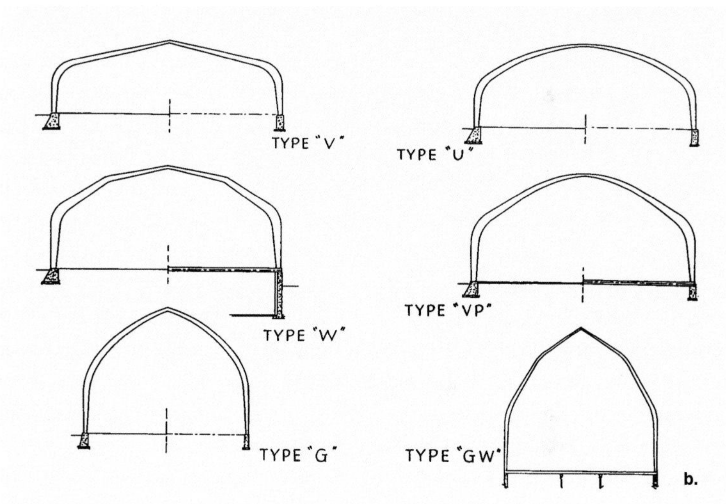


**Figure 15.3a–b.** The Peshtigo High School gymnasium (1934, Max C. Hanisch, Sr.), Peshtigo, Wisc., the first glued laminated building in the United States (**a**), used one of the typical arch shapes available in the 1930s, such as these manufactured by Unit Structures (**b**).

"attempt to do for the laminated timber industry what the American Institute of Steel Construction has done for the older industry," as well as develop quality standards for the product.[22] Between 1954 and 1963 the number of board feet of glued laminated timber produced in the United States more than doubled, increasing from 31,420,000 feet to 85,937,000 feet.[23] The commercial standards developed by the institute were adopted as official government standards in 1963.[24]

## CONSERVATION

Rehabilitating glued laminated timber structures requires the expertise of a structural engineer. Careful observation and measurement are required to determine how the building has performed during its service life. To develop an appropriate stabilization, repair, or replacement program for historic glued laminated timber structures, careful analysis is required to determine their structural capacity.

### DETERIORATION

Laminated timber deteriorates in the same manner as solid timber except for a few unique characteristics. Termites and fungi will destroy wood that has a moisture content of more than 20 percent.[25] In glued laminated timber, deterioration may follow certain laminations

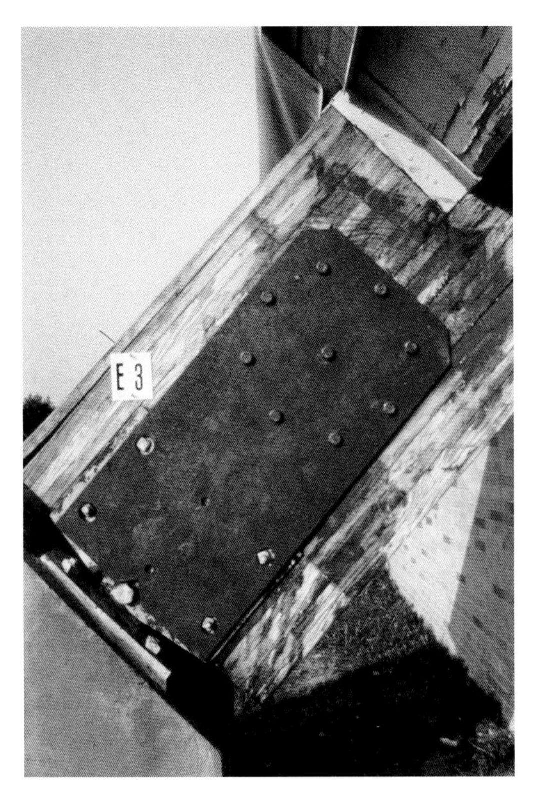


**Figure 15.4.** Shear plates were installed during repair of the Cary High School gymnasium, Cary, N.C.

because of moisture variations. The glue line, although thin, will act as a boundary; termites may be found in one or two laminations for a certain distance, while adjacent laminations may be free of infestation. Because laminated timber members are large, deterioration can be serious, requiring major intervention. Exposed ends of members, contact surfaces, and connection areas are particularly susceptible to moisture-related failure.[26]

As in other structural materials, glued laminated timber can fail because of deficiencies in design or manufacture, or changes in service conditions. Failure in glued laminated timber members can result from radial tension, bending or deflection, or horizontal shear, which may result in the splitting of the end of a member where wood fibers slide past one another. Horizontal shear usually occurs at connections, notched ends, or highly loaded beam ends. An inadequate face bond can cause a horizontal shear failure to occur at a glue joint.

The increase of weight during reroofing, or the ponding of rainwater on a flat roof, may cause beams to fail in bending. Shrinkage in wood around large bolted connections may cause splits because of overstresses in tension perpendicular to the wood grain. The misplacement of concealed pipe hanger connectors can cause a notched beam effect, resulting in excessive horizontal shear at a support.

## DIAGNOSTICS AND CONDITION ASSESSMENT

The first step in assessing a glued laminated timber structure is to determine the species of wood, design criteria, type of adhesive, and preservative treatments, if any. If the shop drawings survive, many of these questions can easily be answered.

Shop drawings provide such information as design loads, design values and lumber combinations, amount of camber (curvature), building geometry, member sizes and lengths, and the location of concealed or hidden connections. Shop drawings are important because they reflect what was actually built. Certain dimensions, such as the tangent depths of two-hinge Tudor arches, are very difficult to measure in the field.

The date of construction will establish the full range of design values associated with a particular lumber combination. Since the 1930s adjustments in design values or stresses have led to many changes in design. It is important to determine which set of design standards were used for a particular structure.

Shop drawings will also tell what adhesives and preservative treatments were used. Until the late 1960s waterproof and water-resistant adhesives were both available. Various preservative pressure treatments, applied either to individual laminations before gluing or the finished member after fabrication, have been used. The most prevalent pressure treatments for glued laminated timber have been pentachlorophenol (in a liquid petroleum gas) or methylene chloride. These treatments avoid the problems of warping or checking (splits caused by shrinkage due to drying) associated with the wetting and drying process used in water treatments such as copper chromated arsenate.

Waterproof adhesives usually are dark, whereas water-resistant adhesives are white or gray. The treatment type and the amount of retention may require that samples be laboratory tested.

Hollow cavities and weak areas of a deteriorated glued laminated timber can be identified by tapping with a carpenter's hammer. The extent of deterioration on the surface of a member can be easily recognized because of discoloration. Other signs of decay include swelling, brittle wood, and raised grain surfaces.[27] Often termite tubes or small pinholes caused by borer insects may be visible on the surface. Core samples, ranging from ⅛ inch to ⅜ inch in diameter, can be taken to evaluate decay, and ultrasonic devices may also be of value. A moisture meter can be used to determine areas likely to have deterioration. Members that have lost stiffness because of deterioration or failure can be identified by measuring deflection.

## CONSERVATION TECHNIQUES

Conservation techniques to repair decay or insect attack may include repair or reinforcement with steel or epoxy systems.[28] Many structural repairs require reinforcement of the decayed ends of arches or A-frames with moment resisting steel leg extensions connected with shear plates applied in the field. A repair such as this may consist of steel side plates welded to the original steel shoe and bolted, with shear plates, to both faces of the laminated timber member being spliced to transfer all bending, shear, and axial forces. In-place repair and conservation techniques require that the member be relieved of load by providing temporary jacking and shoring.

Shear reinforcing of other overstressed members that have not failed may include the use of lag screws to "stitch" the beam together. These are normally placed through the roof or floor deck, perpendicular to the axis of the beam. Lag screws are inserted into prebored lag lead holes that may extend almost the full depth of the beam. Lag screws with a diameter of ¾ inch are available up to 30 inches in length. Other shear reinforcing may include adding an exposed joist hanger to support the bottom of a notched beam, steel side plates, and steel- or fiber-reinforced plastic dowels inserted in vertical holes and epoxied in place. The zone around the bored hole may be strengthened with an epoxy consolidant before the dowel is inserted.

Radial tension, which is similar to horizontal shear, may occur in the curved portion of curved tapered beams and the knee area of Tudor arches. Again, the typical repair requires insertion of lag screws perpendicular to the failure lines.

If the failure is confined to a few tension laminations, damaged wood can be carefully cut away. New high-quality boards obtained from a laminator and planed to the actual thickness of the existing laminations can be glued in place using an epoxy applied in the field. This work requires the skill of an artisan and the cooperation of a manufacturer who can furnish long, finger-jointed replacement stock. For this type of repair to be successful, the cause of the failure must be removed; it may also be necessary to add one or more laminations to increase the depth of the member. After the individual laminations are replaced in the field by this process, known as segmental infill, the member can be planed, sanded, and finished to match the original's appearance.

Bending may require member reinforcement with steel side plates, bottom plates, or flitch plates (vertical steel plates usually placed between two or more timbers to act as beams, headers, or lintels). Any repair using steel requires that the load carried by the steel must balance that carried by the glued laminated timber, and the engineer must design the shear transfer between the two materials. Many engineers will ignore the contribution of the glued laminated timber and design the steel to carry the full load. In many glued laminated timber buildings, steel side plates have been painted to match the wood finish. If appearance is a concern, a steel flitch plate can be inserted into a kerf or rout cut into the center of the beam. Often these cuts are made with a chain saw attached to a guide. The flitch plate must then be securely attached to the wood with a suitable epoxy bonding agent. Determining plate thickness, height, length, and connection method requires a thorough structural analysis. Glued laminated timber beams with inadequate bending capacity can also be reinforced by adding exposed steel tie-rods or bottom plates, although this type of repair may be aesthetically unacceptable.

In glued laminated timber structures, the bases of Tudor or radial arches or of columns may become deteriorated if exposed to moisture. The connection of the laminated timber arch member to its supporting steel shoe is critical because the shoe transfers vertical forces and horizontal thrust through anchor bolts into the foundation. Repairing this type of connection often requires welding steel plates to the existing shoe and connecting them to the glued laminated timber member so that all forces are resisted. The connection of steel side plates to glued laminated timber may require the field installation of shear plates with a diameter of 2⅝ or 4 inches. Shear plates are inserted flush to the face of the member in a dap (notch) cut made with a special cutter.

## REPLACEMENT

Because laminated timber structures have many components, in-kind replacement of a damaged or deteriorated beam, purlin, column, or arch half span may be a more cost-efficient solution than an in-place structural repair or reinforcement. Laminators can usually make components to match historic ones in size, detail, and finish. Design values for replacement components should be made in accordance with applicable standards set by the American Institute for Timber Construction and the American Society for Testing and Materials.[29]

**Figure 16.1.** Probably the most celebrated plastic house ever built, the Monsanto House of the Future (1957, Albert G. H. Dietz, Hamilton and Goody), constructed at Disneyland, Anaheim, Calif., had modular fiber reinforced plastic walls filled with foam insulation. *Popular Science,* April 1956.

ANTHONY J. T. WALKER

# 16 | FIBER REINFORCED PLASTIC

**DATES OF PRODUCTION**
c. 1945 to the Present
**COMMON TRADE NAMES**
Alsynite, Corrulux, Filon, Indulux, Kalwall, Kalwood, Sanpan

## HISTORY

Fiber reinforced plastic (FRP) includes a wide range of polymers, such as acrylics, vinyls, polyolefins, phenolics, and polyesters, combined with reinforcing fibers, predominantly asbestos, carbon fibers, and glass fibers.[1] The most common combination for building components is glass fiber with unsaturated polyester resins.[2] Fillers, catalysts, stabilizers, and coloring agents are also added to produce rigid and semirigid materials that generally have a continuous mat reinforcement and varying physical and chemical properties.

### ORIGINS AND DEVELOPMENT

The first recorded display of woven glass fibers occurred in 1713, when René Réaumur submitted "glass cloth" to the Paris Academy of Science.[3] However, it was not until 1930 that fibers fine enough for modern use were produced.[4] By 1938 Owens-Corning Fiberglas was manufacturing fiberglass fabric woven from continuous filaments twisted together as a glass strand (yarn).[5]

The first polyester was made by the Swedish chemist Jöns Jacob Berzelius in 1847. It was followed by Daniel Vorländer's development of unsaturated resins.[6] Alkyd resins were first documented in England in 1901 by W. J. Smith.[7] In 1926 the General Electric Company introduced Glyptal, the first commercial alkyd resin. In 1933 Carleton Ellis, an American chemist, registered the first patent for using polyesters as a reinforced plastic. Ellis was also granted a patent in 1940 for copolymerizing polyester and styrene, the first molding resin. Allyl casting resins were introduced in 1941 as glass substitutes, and in 1942 a low-pressure laminating resin, allyl diglycol carbonate, was used to manufacture radomes with glass cloth and resin.[8]

Although reinforced plastics were used before World War II, the major techniques and resins available before 1940, such as phenolics, required greater heat and pressure for curing than the glass fibers could withstand.[9] With the introduction of cold low-pressure molding resin polyesters in 1941 and allyl resins in 1942, glass fibers

**Figure 16.2.** When the spray lay-up technique is used to fabricate components with fiber reinforced plastic, the resin and reinforcing materials are sprayed onto the mold simultaneously.

could be used for reinforcement. By 1944 FRP boats were being successfully manufactured by Winner Manufacturing, among others.[10] Growing interest in FRP during the war led to the creation in 1944 of a low-pressure division, later renamed the Reinforced Plastics Division, within the Society of the Plastics Industry, which incorporated in 1937 in New York.[11]

### MANUFACTURING PROCESS

The two main processes used to manufacture FRP components are contact molding, which uses an open mold, and machine molding, which requires matched molds.[12] Selection of the production method depends on the length of the run, the size of the component, and the quality of the surface required.

Contact molding, commonly used to fabricate building products, uses polyester resin's ability to cure without heat or pressure. Only one mold is required, female or male, with a smooth surface treated with a release agent. A gel coat is applied by brush or spray to a thickness of 4 to 5 millimeters. A synthetic surfacing material may be incorporated into the gel coat to provide a resin-rich surface and help balance the laminate; a fine glass tissue is often embedded in the gel coat to prevent the reinforcing fibers from popping through the surface.

Once the gel coat is sufficiently hardened, a resin coat is brushed over it. The glass fiber mat can be impregnated (usually referred to as "laid up"), either by hand or with a spray gun. A coat of resin is brushed over the gel coat, and the fiber-reinforcing mat is pressed into this and consolidated with a brush or roller until saturated. Further layers are built up as required, and stiffeners or blocking for fixings are built in.[13] The spray-up process allows the resin and chopped fiber to be deposited together by spray gun. Rolling is still required to consolidate the laminate. Although polyester resins cure at room temperature and pressure, heat can be used to accelerate the curing process. Moldings are trimmed and finished while the resin is still "green."

Machine molding permits better quality control and allows for finished surfaces on both faces. Sheeting is produced with this method by placing resin and reinforcement between two release films and rolled to remove air bubbles.[14] The consolidated sandwich then passes through an oven, in which rollers and dies form the corrugations and the material is cured. Patterns and colors can be incorporated in the sheets.[15] Automatic injection molding was introduced by Plaskon in 1946.[16]

## USES AND METHODS OF INSTALLATION

The building industry provided fertile ground for the suppliers and manufacturers of FRP seeking new markets after World War II.[17] Many suppliers began manufacturing car bodies or boats, and their practical understanding of the production techniques and the potential of three-dimensional forms was important in developing FRP products for construction.

Corrugated fiber-reinforced translucent sheets, which were to dominate the use of FRP in the construction industry, were introduced in the late 1940s.[18] Such sheets provided a high strength-to-material content ratio.[19] New techniques for strengthening and coloring sheets improved their quality, but few manufacturers could meet the highest standards for dimensional consistency.[20] By the mid-1960s the two leading products were Sanpan panels produced by Panel Structures in East Orange, New Jersey, and Kalwall panels produced by the Kalwall Corporation in Manchester, New Hampshire.

The Kalwall system was developed in the 1950s by the engineer Robert Keller. During World War II, Keller became familiar with FRP and started his company initially by using FRP supplied by the Alsynite Company and Filon Plastics Corporation.[21] Keller developed a self-supporting, light-transmitting building panel that could be used for large wall areas. Panels were made with two sheets of $\frac{1}{16}$-inch reinforced plastic sheeting bonded with a nonrigid polysulfide to an interlocking aluminum grid. An early example included a dramatic end wall and roof for a church by William H. Van Benshoten in Woodstock, New York.[22] At the 1958 Brussels World's Fair more than two thousand Kalwall panels were used for the roof of the U.S. pavilion, which covered 72,000 square feet of floor area with a clear span of 302 feet. Kalwall panels were also used for the windows of the Vehicle Assembly Building (1962, Max Urbahn and Anton Tedesko) at the John F. Kennedy Space Center.[23]

FRP use throughout the 1950s and 1960s followed two distinct trends. The material's plasticity and moldability were exploited, and sheet forms continued to develop, with a concentration on structural applications using sandwich construction principles. Following the tradition of the Vinylite House, built for the Century of Progress Exposition in 1933, a number of prototype houses and buildings highlighted the potential of plastic (and frequently FRP) for construction.[24]

Sandwich panels developed rapidly during the war as covers for radar equipment for which a single FRP skin was insufficient. Acorn Houses produced one of the earliest buildings with sandwich panels for the radar research laboratory at the Massachusetts Institute of Technology (MIT). Panels 1½ inches thick, consisting of impregnated honeycomb paper core, were bonded between skins consisting of a clear polyester reinforced with glass cloth.[25]

In 1949 Buckminster Fuller attempted to use FRP as a structural building material to produce a dome using triangular sections cast in fiberglass. However, the resin and chopped fiberglass mixture would not set in the plaster mold.[26] Fuller did not abandon FRP, however, and geodesic radomes 17 meters in diameter were developed for military use in 1956.

**Figure 16.3.** One of the most exciting buildings at the 1958 World's Fair in Brussels was the U.S. Pavilion (Edward Durell Stone) (top). The circular design had a clear span of 302 feet and was composed of approximately two thousand translucent Kalwall panels.

**Figure 16.4.** The first commercial radome (a dome housing radar equipment) made with fiber reinforced plastic was built in 1956 at Bell Laboratories, Whippany, N.J. (bottom). Buckminster Fuller was involved with the design.

**Figure 16.5.** Unsuitable gel coats or resin formulations of fiber reinforced plastic components can cause crazing, or fine surface cracks (top).

**Figure 16.6.** Unlike some plastics, fiber reinforced plastic can be repaired. Techniques range from resurfacing to cutting out and replacing the defective material. Translucent Kalwall panels that were damaged in a skylight installation have been resurfaced (bottom).

Shell structures were also very important for the development of FRP, as they took advantage of the natural rigidity of the curved or folded plate forms—the latter using a series of triangular folds to achieve stiffening with a thin sheet material. The importance of both forms was recognized by Albert G. H. Dietz, professor of building engineering and construction at MIT, who theorized that plastics could play a significant part by solving otherwise impossible designs.[27] In 1959 Dietz engineered a visitors' shelter using ninety parasol shells for the American National Exhibition in Moscow. The translucent tulip forms had additional reinforcement at the edges and ribs. Each tulip was composed of six petals ¹⁄₁₆-inch thick with edge ribs of ¼ inch and supported on a central hollow, tapered column with ¼-inch-thick walls.[28]

At the 1964 New York World's Fair several pavilions featured FRP enclosures. The 7-Up Pavilion (Becker and Becker) incorporated an open area sheltered by shallow, vaulted shells supported on columns at each corner, while the Marina Building (Peter Schladermundt) used hyperbolic paraboloids with infill glazing. Both were of sandwich construction using simple butt joints between the shells.[29]

Of the many sensational demonstration houses built, the Monsanto House of the Future, designed by Hamilton and Goody in conjunction with Dietz, had the widest influence.[30] Sponsored by the Monsanto Chemical Company and built at Disneyland, the house design used plastics demonstrably rather than as substitutes.[31] The plan, developed as a simple cross supported on a central core, enabled the house to be sited regardless of the terrain and required only two types of FRP floor and roof sections.[32] The Monsanto House proved to be enormously popular and had been visited by more than 30 million people by the time it was demolished in 1967.

Less dramatic uses of FRP also being explored include cladding for buildings with steel or concrete structural frames. Large-scale prefabrication included the development of complete rooms and even whole house units. In 1964 Holiday Inns developed a prefabricated system for motel units of four rooms using FRP for the outer wall cladding.[33] Cladding represented only a small proportion of FRP in buildings; other uses included roofing trim, gutters and flashing, corrugated sheeting, roof lights, and plastic forms for concrete.[34]

FRP installation hinged on effective joints, particularly because FRP has a high coefficient of thermal expansion that could produce significant movement stresses.[35] Often joints were detailed with another material overlapping or capping the joint, permitting movement. In other cases the inherent flexibility of the material was accepted.[36] Thin FRP sections often required tapered stiffeners bonded to the rear face.[37]

The need for FRP standards became apparent as the industry developed in the 1950s. Without performance data on fire resistance and flame spread, acceptance of the material was inhibited for walls and roofs under many building codes, even though special resins were developed with fire-retardant properties.[38] Special gel coats ultimately provided the necessary performance requirements.

## CONSERVATION

Unlike many plastics, FRP can be repaired. However, it is often difficult to match the surface appearance of opaque materials or achieve a similar degree of transparency for translucent FRP.

### DETERIORATION

The principal forms of polymer breakdown are bio-, photo-, atmospheric, and hydrolytic degradation.[39] Water and humidity contribute significantly to the deterioration of FRP.[40] Other causes are mechanical damage and manufacturing defects.

Yellowing from ultraviolet light can be delayed but not prevented by coatings and the use of ultraviolet inhibitors, which react to absorb the energy from the ultraviolet source. Sheets on the north face of a building or with heavy dirt deposits generally last longer than clean sheets facing the sun.

Weathering erodes the resin, exposing fibers on surface components. Water subsequently penetrates the laminate by capillary action (wicking) along the glass fibers and causes a bloom or whitening by attacking the resin–fiber interface.[41] Gel coats can prevent or limit such deterioration. However, additives used in the gel coat to meet fire-resistance requirements accelerate the surface breakdown, and coatings are sometimes used on fire-retardant FRP components to prevent excessive erosion.[42]

Impact damage before or after installation may enable water to reach the ends of the fibers through wicking, allowing organic growth and erosion of the reinforcement. In colored panels this appears as a series of dark lines; in translucent panels the breakdown of the glass–resin interface leads to diffraction of light passing through the panel and the appearance of a bloom that obscures light transmission.

In sandwich-panel construction the interface between the core and the facing material can be critical to the panel's performance. Small areas of trapped air are difficult to exclude during manufacture and when warmed by solar radiation may expand, causing a ripple effect on the surface.[43] Differential movement can take place between the outer skin and the insulation core or inner face; panel flexing can also produce a breakdown, especially in small corrugations caused by differential movement between the resin and fiber face.[44]

In addition to material degradation, failure in FRP components can be caused by manufacturing defects and flaws in component design and poor installation. Mechanical damage during manufacture or installation, such as scratching, can also lead to erosion. Environmental conditions can have a significant effect on the mechanical performance of FRP, particularly where there is prolonged tensile stress. Recent work has shown that failure from cracking can occur within a few hours when stresses are applied during installation.[45]

### CONSERVATION TECHNIQUES

FRP can be cleaned with soap and water to remove dirt and biological growth. A water rinse is recommended every four or five years.[46] When using abrasive cleaners care must be taken not to damage the gel coat. Stubborn dirt can be removed with rubbing compounds, such as those used for cellulose paint finishes. Polishes can be removed with acetone when recoating is necessary.

Although graffiti can be removed relatively easily from new FRP, a weathered surface provides a greater key and a route for media penetration. Graffiti on nonporous surfaces can be removed by applying solvent with a cotton swab, wiping it off after a very short time with water and detergent, and polishing dry with clean swabs. Hand pressure only should be used, and good ventilation is required. Translucent areas may be polished with a nonabrasive window cleaner and clean swabs. A small area should be tested first.[47]

Common repair techniques do not generally take aesthetics into consideration.[48] For most components the repaired area has to be blended in with the surrounding surface. Minor cosmetic damage is first cleaned with solvent; next a tapered edge is formed along the damaged area and any coating sanded off slightly beyond the damaged area. A resin filler is then applied, heat cured, and sanded to a smooth surface.[49] If structural damage has occurred, a backing strip and thicker laminate around the damaged area are necessary.

Holes in panels can be repaired by cutting out the damaged area, washing the adjacent areas with methyl ethyl ketone, and installing a new FRP section with a proprietary adhesive. Internal framing can sometimes be used to mask joint lines.[50]

Self-colored and transparent laminates in buildings present a different range of problems. Although repairs can be made, they are clearly visible unless complete panels are replaced or, in the case of opaque laminates, a colored coating is used.[51] Manufacturers can provide advice on suitable repair methods.[52]

Not all polymers are compatible; sandwich panels, in particular, have multiple layers with various resins, insulation, and coating systems. Primers may be used to separate finishings from the repair. In situ repairs may take some time to cure; a warm air heater or infrared lamp can be used to speed up the process. Such repairs generally require consultation with material manufacturers and other professionals knowledgeable about FRP.

### REPLACEMENT

Custom replacement units for building cladding can be produced with hand or spray lay-up in molds. Unfortunately, mold costs may prohibit the production of only small runs of replacement units. Color matching is also extremely difficult.

# PART IV
# MASONRY

STRUCTURAL CLAY TILE

TERRA COTTA

GYPSUM BLOCK AND TILE

THIN STONE VENEER

SIMULATED MASONRY

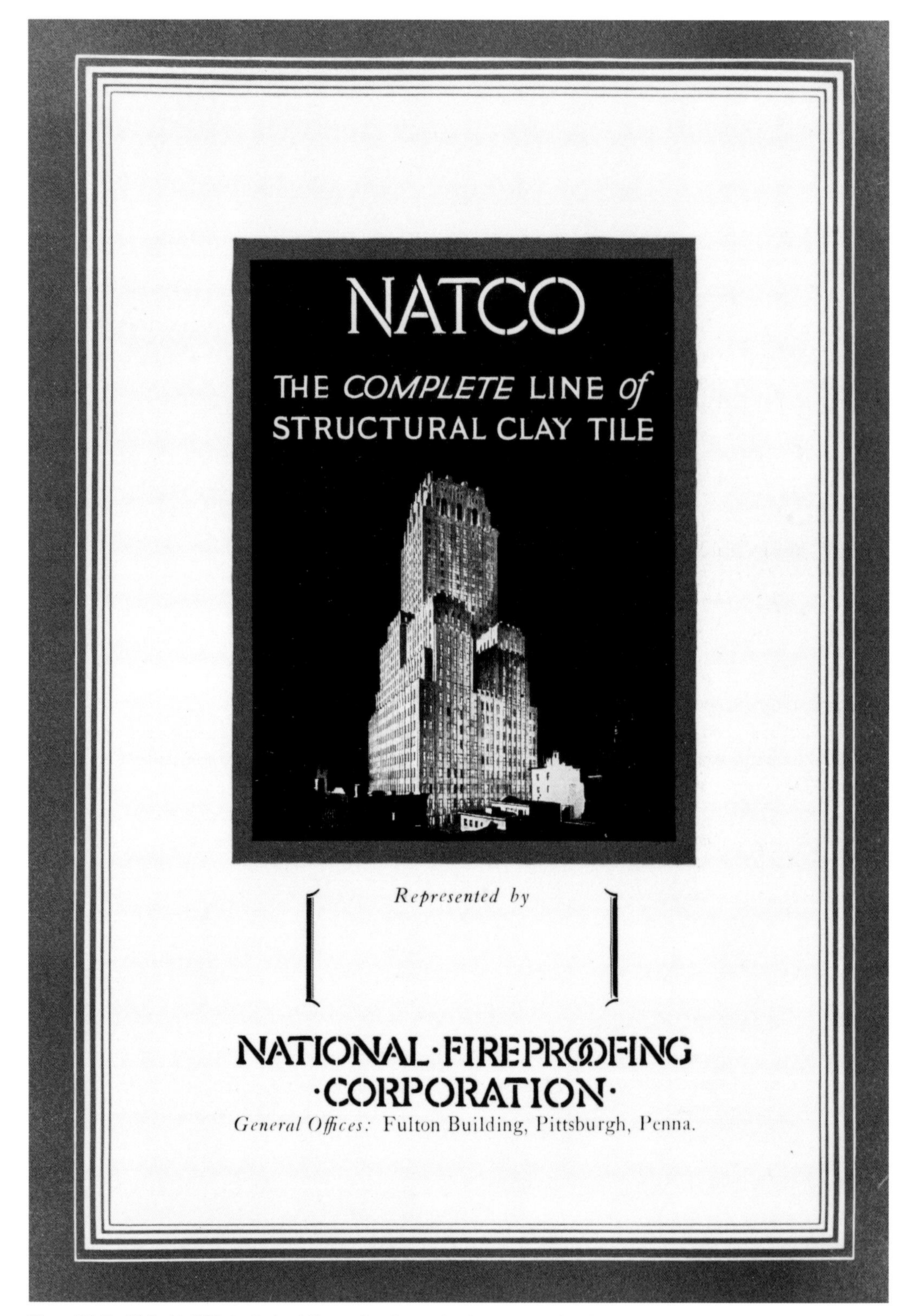


**Figure 17.1.** Established in 1889, the National Fireproofing Corporation became one of the largest producers of structural clay tile, selling tile for floor arches, interior walls, partitions, furring, backup, roofing, and column fireproofing. *Sweet's Architectural Catalogues*, 1932.

CONRAD PAULSON

# 17 | STRUCTURAL CLAY TILE

**DATES OF PRODUCTION**
c. 1870 to the Present
**COMMON TRADE NAMES**
Ar-ke-tex, Lee, Natco, Pioneer, Speedtile, Textile

## HISTORY

Structural clay tile includes a range of burned clay products used in both structural and nonstructural applications.[1] Typically hollow with parallel cells, clay tile was used widely for floor arches, fireproofing, partition walls, and furring in the last quarter of the nineteenth century. Load-bearing columns, pilasters, and backing for exterior walls became increasingly common after 1900. Structural tiles are classified into three primary grades—hard, semiporous, and porous—depending on the length of firing time during manufacture. This classification is important in determining whether a tile can be used in a particular application.

### ORIGINS AND DEVELOPMENT

The use of clay (terra cotta) forms as a lightweight structural material dates to eighteenth-century Italy, where clay pots lined with plaster of paris were set in concrete for the dome of the San Vitale Church in Ravenna.[2] In the late eighteenth century architects in Paris used a similar technique (with the addition of reinforcement), but only with the most expensive buildings.[3]

In the 1850s the American architect Fred A. Peterson developed a clay tile building method to span iron beams, which he used in the Cooper Union Building (1859) in New York City. However, these handmade tiles were never commercially produced.[4] In 1871 Balthasar Kreischer, a brick manufacturer, and George H. Johnson, a Chicago engineer, patented a similar method.[5]

The fires in Chicago and Boston in the late 1800s demonstrated that brick was one of the most fire-resistant building materials, and during the last quarter of the nineteenth century structural clay tile quickly gained acceptance for use in floors and for fireproofing iron.[6] Tile became a standard building material for high-quality buildings in New York, Chicago, and other major cities. The industry was also aided by requirements for fire-resistant materials in taller buildings.[7] Scientific testing of tiles began in earnest after 1910, and new products with greater compressive strength were developed; tile floor systems evolved and permitted longer floor spans.[8]

**Figure 17.2.** The blocks forming this combination floor arch were covered with concrete to stiffen the floor and distribute the load.

### MANUFACTURING PROCESS

Structural clay tile is manufactured today in much the same way as it was historically. Clay is kneaded until a proper consistency is achieved for molding, and then it is shaped or pressed into blocks and baked in kilns.[9] Tile was and is still classified according to density; other classifications included glazed and vitrified.

Hard or dense tile was burned the longest and therefore was the strongest and most impervious to moisture. It was used for heavy structural applications where the tile was to be exposed to weather. However, hard tile did not perform as well as porous tile in fireproofing applications and was more prone to crack in intense heat.

Semiporous tile, which was sufficiently hard burned for moderate strength and somewhat impervious to moisture, was commonly used in small buildings and houses. Porous tile was made from a modified clay mixed with ingredients such as sawdust and straw that were

**Figure 17.3.** Expanses of glazed structural clay tiles, applied over concrete masonry, and glass block give the Coral Court Motel (1941, Adolph L. Struebig), near St. Louis, Mo., its distinctive streamlined appearance.

consumed during the firing. Because of its numerous air spaces, porous tile was lighter and reduced structural dead loads. Porous tile was also more desirable for fireproofing applications because of its better performance under high heat.

The firing process for glazed tile produces a glass-like finish over the exposed surface of the tile by the introduction of common salt into the kiln when the material is nearly burned. Vitrified terra cotta is burned at a high temperature so that the material begins to fuse. Extremely hard throughout, vitrified tile is produced with the densest form of clay and is used as an exterior or interior facing.

## USES AND METHODS OF INSTALLATION

Structural applications for clay tile can be divided into three general categories: tile floor arches, gravity load-bearing elements, and shear walls. Different blocks and forms of construction are used in each application. Non-load-bearing applications include partition walls, fireproof casing for columns and beams, and furring.

Structural clay tiles were used extensively in various types of flat-arch floor construction from the 1870s through the 1930s. A wide variety of uniquely shaped blocks were manufactured for two general types of flat-arch construction: side construction and end construction.[10] At each end of the arch, a skew tile abutted the girder; intermediate tile and a key tile filled in the remainder of the arch. Floor arch tiles were centered on wood suspended on the beams with iron hangers. Tiles were set in mortar and filled over with concrete to distribute loads evenly.

In side construction, the earliest form of tile arch, the thrust of the arch was transmitted by bearing the side of one tile block against the side of another (thrust forces directed transverse to longitudinal direction of the tile cells). In end construction, thrust was transmitted by bearing through the ends of blocks oriented end-for-end (thrust forces directed parallel to the longitudinal direction of the tile cells). Historically, load tests have demonstrated that end construction is structurally more efficient and stronger than side construction.[11]

By the 1890s a system combining elements of both side and end construction had become the most popular form of flat-arch construction.[12] Nevertheless, as portland cement became more reliable in the early twentieth century, systems combining clay tiles with reinforced concrete began to supplant traditional flat- and segmented-arch systems, even though these continued to be produced well into the 1930s. In composite systems, structural clay tiles could be used as centering (essentially formwork) for reinforced concrete floors or as fillers in reinforced concrete slab floors.[13]

At the same time that structural clay tile for floors was developing, other structural clay tile products began to appear. Applications ranged from fireproofing for metal beams and columns to partitions to furring. Various cylindrical-shaped elements were used to cover columns, and "shoe" tiles were manufactured for girder flanges. For fireproofing, clay tiles were cased around beams and girders with steel hangers and mortar. Partition tiles gradually evolved into standardized 12-inch-square components; thicknesses varied from 2 to 12 inches.[14] Furring tiles, commonly a split version of the partition tile, were produced in similar sizes.

Structural clay tile blocks could also be used to form load-bearing walls and column or pilaster elements or used as backing for exterior walls. Buildings with load-bearing clay tile are most often single-story buildings (a load-bearing clay tile structure is almost never more than two or three stories tall). Load-bearing walls are often multi-wythe or constructed of heavy blocks. The load-bearing elements may be unreinforced with hollow cells, unreinforced with grouted cells, or with reinforced and grouted cells. Early load-bearing walls of this type were usually ungrouted. Long walls are often strengthened by pilaster elements. In some constructions the pilasters are grouted and also, in more recent construction since the 1950s, reinforced.

Clay tiles were widely used in the first half of the twentieth century for construction of infill walls in concrete and steel-framed structures. In many instances this infill provided lateral strength and stability to the structure for wind and seismic loads. In some structures, typically early high-rises from about 1880 to 1910, interior corridor partition walls served the same function. In these structures determination that clay tile walls served as shear walls was based on the absence of an explicitly designed lateral load system, such as diagonal bracing or rigid framing connections. In load-bearing wall construction, the wall supports gravity loads and inherently also acts as a shear wall for resisting lateral loads. These walls were designed explicitly for carrying both gravity loads and lateral loads.

As the industry evolved, new products were introduced to compete with related industries. A variety of facing tile (structural clay tile with a decorative finish) for interior and exterior walls was also available by the late 1920s.[15] Matte, glazed, and mottled finishes in a host of colors made facing tile an alternative to standard brick construction. Tiles were produced in sizes equivalent to two, three, and six common bricks and were also available with a colored finish on both faces. During the 1930s, for example, products such as Ar-ke-tex were used

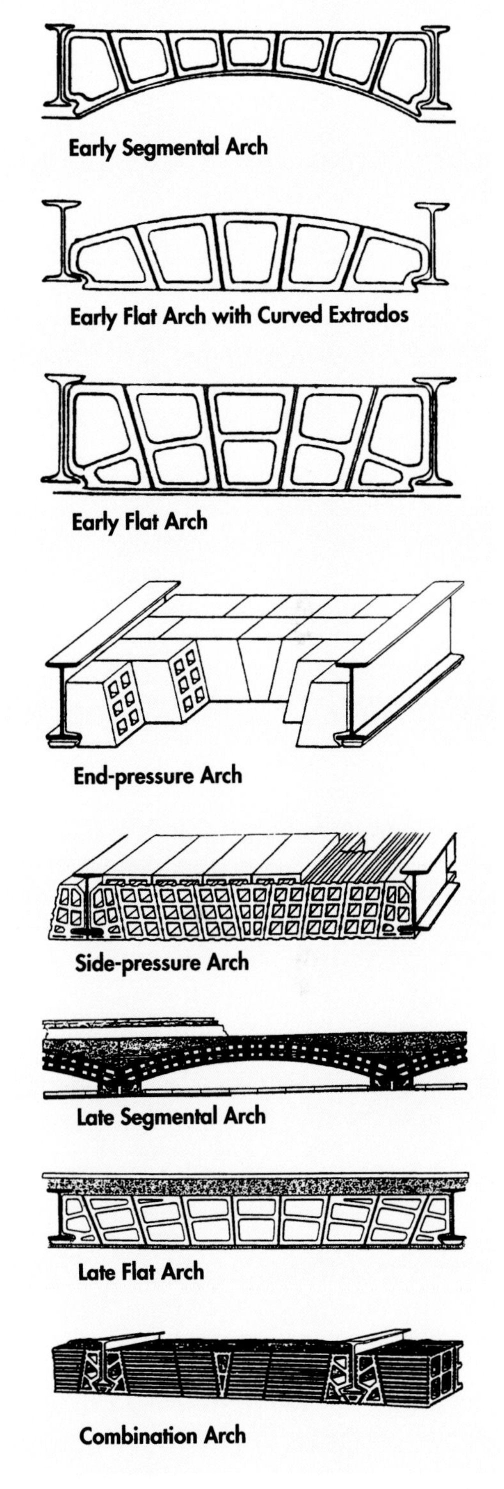


**Figure 17.4.** Tiles were used for a variety of flat and segmental arches. As spans lengthened, end-pressure tiles, stronger than side-pressure tiles, gained favor.

for corridors, auditoriums, and swimming pools and in many other interior and exterior applications where a sanitary, nonporous surface was required.[16] By 1932 the National Fireproofing Company of Pittsburgh, one of the largest producers of structural clay tile products, offered a complete range of tile types and was manufacturing more than one million tons of tile annually.[17]

Clay tile elements used in load-bearing applications for walls and for non-load-bearing applications, such as fireproofing, partitions, and furring, were installed, like ordinary masonry, with mortar. One advantage of clay tile was that it was lightweight. The Wheeling Tile Company in Fort Myers, Florida, even manufactured Speedtile, a patented load-bearing product with a handle to permit faster installation.[18]

After World War II reinforced concrete floor systems and composite metal decks replaced composite systems with tile components. These floor construction methods are still used today. Clay tile blocks remained popular for load-bearing wall construction until the early 1950s, when they were supplanted by concrete masonry units. During this period, steel and wood studs faced with plywood or gypsum wallboard replaced structural clay tile for infill and interior partitions.

## CONSERVATION

Environmental factors and structural movement can cause distress, particularly cracking, in structural clay tile, depending on where it is used in a building. Repair techniques should be selected only after determining the structural integrity of the tile through close visual examination and, in some instances, materials testing.

### DETERIORATION

Most deterioration of structural clay tile is caused by excessive moisture, water infiltration, and freeze–thaw action. The physical appearance of deteriorated clay tile (such as a soft, powdery, or flaky surface or the presence of efflorescence) is similar to that of common clay brick masonry that is deteriorating from these causes.

As with brick masonry, clay tile can be damaged by structural effects such as differential settlement, expansion and shrinkage, and structural overloading. These conditions usually result in crazing, cracking of clay tile blocks or the surrounding mortar joints, or bowing of wall panels constructed of clay tile.

### DIAGNOSTICS AND CONDITION ASSESSMENT

Visual observation is the primary means of diagnosing problems with structural clay tile systems. Most methods of observation, means of inspection, and tools for testing used with brick masonry are also applicable to structural clay tile.

The pattern of cracking helps determine the cause of the distress. Step cracking though mortar joints usually indicates differential settlement, diagonal cracks that extend through blocks suggest structural overloading, and extensive vertical cracks usually indicate restrained expansion or shrinkage. However, the nature of the cracks, including both pattern and location, should be closely studied to detect the exact cause.

The most reliable method of determining the strength of load-bearing structural elements is through in situ tests on the wall construction. Procedures commonly used on brick, such as bond-wrench tests, in-plane and out-of-plane shove tests, and in situ deformability testing (also called flat-jack testing), can be used on clay tile construction.[19]

Modern masonry codes can be used to establish the design strength of load-bearing tile elements.[20] Alternatively, older design handbooks can be used.[21] However, both these methods require knowledge of the strengths of each constituent material: clay tile, mortar, and grout. These properties are traditionally determined through materials testing. If tests are not conducted, use estimated strengths based on the type of material: grade of tile block (hard, semiporous, or porous) and type of mortar (ASTM Type M, N, or S).

Testing procedures based on strengths of constituent materials are less reliable than in situ strength tests on the completed assembly. The latter are preferred because they provide the most definitive information.

### CONSERVATION TECHNIQUES

Distress that is not too severe can be repaired by tuck-pointing, patching with mortar or grout, or perhaps by selective replacement of individual elements. Solidly filling existing individual hollow elements with grout may be appropriate at times when distress is attributed to structural overloading. However, the structural impact of the additional dead load due to grout fill must be evaluated.

Where there is a large extent of mild to moderate distress or where structural strengthening is necessary, solidly grouting existing hollow pilasters or wall panels may be a viable solution. Grouted cells should be reinforced whenever possible. Again, the effect on the structure of the additional dead load due to grout fill must be evaluated. If it is significant, strengthening of supporting structural elements and foundations may be necessary.

**Figure 17.5.** The soffit of this structural clay tile flat roof arch is severely deteriorated. Excessive moisture in the arch has caused efflorescence, and the tiles are spalling. A thorough structural assessment is required for tiles in this condition to determine whether replacement is necessary.

## REPLACEMENT

For a large extent of severely distressed tile, such as infill walls damaged by earthquake, replacement should be considered. In-kind replacement of wall systems is possible when a new structural tile system has adequate design capacity. Ordinary wall tile blocks are still commercially produced. Walls can be grouted and reinforced to achieve higher design strengths. Specialty tile blocks, such as decorative glazed units, will probably require custom production, which may not be readily available in some geographic regions.

Where structural strength cannot be achieved with new tile construction, replacement with other structural media, such as reinforced concrete, is necessary. Historic features requiring preservation could then be supported by the new structural system or, perhaps, carried by an independent system expressly designed to support them. Replacing the structural clay tile system with a contemporary system while preserving the visually obvious historic fabric may be possible, although it usually is expensive.

Severely damaged or severely deteriorated clay tile arch floor systems are never replaced in kind because tile blocks for floor systems are no longer manufactured, and artisans skilled in laying arch systems no longer exist. Most often, damaged tile arches can be grouted solid (with reinforcing steel added as necessary) and left in place. In extreme cases, the tile arch is removed and an entirely new structural floor is installed.

# THE NORTHWESTERN TERRA COTTA COMPANY

MAIN OFFICE AND CHICAGO PLANT

2525 Clybourn Avenue, Chicago, Ill.

OTHER OFFICES: Denver, Colo. St. Louis, Mo. Milwaukee, Wis. Cleveland, Ohio
OTHER PLANTS: Chicago Heights, Ill. Denver, Colo. St. Louis, Mo.

Sales Agencies in all Principal Cities

Terra Cotta Spandrel

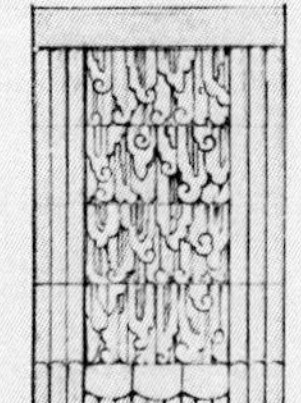

Terra Cotta Spandrel

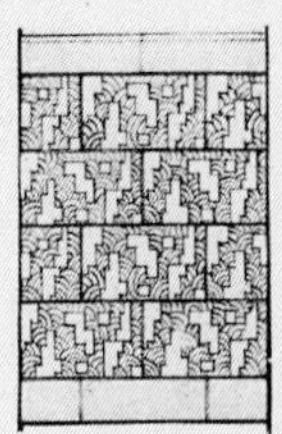

Terra Cotta Spandrel

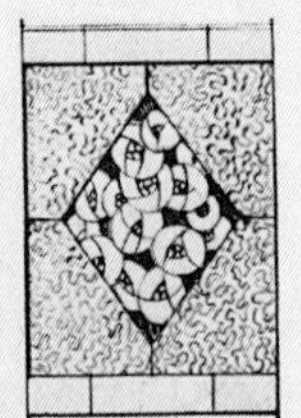

Terra Cotta Spandrel

Terra Cotta Spandrel

Terra Cotta Spandrel

Terra Cotta Spandrel

Terra Cotta Spandrel

**NORTHWESTERN PRODUCTS**

Architectural Terra Cotta for exteriors and interiors, in plain, ornamental or sculptured design, in all finishes, and in any desired color or combination of color.

Northwestern Terra Cotta in stock designs; Northwestern Multi-Unit Sills for steel sash windows; Northwestern Chimney Pots; Northwestern Garden Pottery; Northwestern Faience and Northwestern (Art-I-San) Wall Blocks.

**NORTHWESTERN SERVICE**

More than Fifty years of practical experience has equipped this Company with both personnel and plants that insure a service unsurpassed in the industry.

**NORTHWESTERN SPECIFICATION**

We suggest that architects write direct for information about our product — color treatment, mechanical texture, construction and design, for all types of buildings.

Terra Cotta Spandrel

Terra Cotta Spandrel

**Figure 18.1.** The Northwestern Terra Cotta Company offered stock and custom terra cotta elements with plain or mottled glazes to suit a variety of architectural applications. In the 1930s shallow relief patterns gained popularity for features such as spandrels. *Sweet's Architectural Catalogues*, 1932.

DEBORAH SLATON AND HARRY J. HUNDERMAN

# 18 | TERRA COTTA

**DATES OF PRODUCTION**
c. 1868 to the Present

## HISTORY

Terra cotta ("baked earth") is a fired clay product, usually glazed, used in construction as a structural component, fireproofing, or architectural cladding. For centuries artisans have used clay's plasticity to create formed, decorative objects. Ceramic veneer, a specific type of terra cotta, is a thin, machine-made pressed-clay product that is also usually glazed. Ceramic veneer elements, which were first produced in the 1930s, are usually simple in shape, although a variety of textures and finishes are manufactured.

### ORIGINS AND DEVELOPMENT

The use of fired clay for pottery and in construction has been known since ancient times. Terra cotta was manufactured for building construction in England by the early eighteenth century. This development was the origin of the American terra cotta production of the following century.[1]

Production of architectural terra cotta in the United States began in the late 1860s. Plants were located near large clay deposits, which tended to be sites where fired clay ware had been manufactured; for example, plants were located in Perth Amboy and South Amboy, New Jersey; Alfred, New York; Lincoln, California; and Chicago. Chicago was the center of the growing industry. The Chicago Terra Cotta Company, established in 1869, escaped the great fire of 1871 and provided materials for the rebuilding of the city. When the architect Sanford Loring, a former partner of William LeBaron Jenney, became president of the company, he expanded the range of its architectural products and also opened a plant in Boston in the 1870s.[2,3] Terra cotta was popularized in New York City during the 1870s by architects such as George B. Post and Francis H. Kimball.[4]

When the Chicago Terra Cotta Company failed in 1879, the Northwestern Terra Cotta Company took over many of its clients.[5] Northwestern produced terra cotta for the early commercial buildings designed by architects of the Chicago School and by 1900 had become the largest manufacturer of terra cotta in the country.[6]

### MANUFACTURING PROCESS

Production methods of hand-formed terra cotta have not changed significantly over the past several centuries. Terra cotta is made of a mixture of clay, grog (previously fired clay products), and water. These are mixed and kneaded in preparation for molding.

Manufacturers prepare full-scale shop drawings for the terra cotta elements based on architects' drawings. Models developed from the shop drawings are prepared for each type. Sculptors or artists may be commissioned to make the models for ornamental elements, typically using plaster and wood. The models are sized to accommodate shrinkage, as terra cotta shrinks approximately 7 to 11 percent during drying and firing. From the model, a plaster mold is made of individual pieces that are keyed together. The pieces of the mold are assembled and bound together with straps. Clay is pressed against the sculptured sides of the mold by hand to a thickness of approximately 1½ inches. Clay webs or stiffeners are added at regular intervals to provide stability during the drying and firing process. The mold is then removed, and the element is finished by hand to repair any surface defects. At this point surface texture, such as grooves or combing, may also be added. The terra cotta elements are dried for several days to several weeks, depending on size, and then are sprayed or brushed with a glaze and placed in a kiln for firing.

Terra cotta can also be extruded.[7] The clay is forced through a die as a continuous column and cut by wire into individual elements. A glaze or a slip coat of water and clay is applied to the exterior surface of the molded or extruded element, which is then fired. Typically, hand-pressed terra cotta uses a wetter clay mixture than machine-pressed terra cotta, and therefore more shrinkage may occur during drying of the hand-formed material.

Ceramic veneer was developed as an earthquake-resistant cladding on the West Coast in the 1930s.[8] Like

**Figure 18.2.** Modelers work at large easels at the Northwestern Terra Cotta Company plant in Chicago in the early twentieth century (top). Working from architects' drawings, skilled artisans created plaster models from which molds for the terra cotta were made.

**Figure 18.3.** The Pacific Telephone and Telegraph Building (1925, Miller and Pflueger; A. A. Cantin), San Francisco, exemplifies the use of terra cotta to emulate natural stone, an effect enhanced by the Granitex finish (bottom).

terra cotta, machine-pressed ceramic veneer is made from high-quality clays mixed with grog. It is typically made in relatively flat forms, no more than 1½ inches thick, and in standard sizes up to 4 square feet. Ceramic veneer is also used in panels for curtain wall applications. Multiple ceramic veneer elements are placed in a form, separated by strips of rubber, and brushed with grout. Reinforced concrete is placed into the form and partially cured, and the panels are removed from the form for final curing. These panels are then erected onto the building facade structure.[9]

## USES AND METHODS OF INSTALLATION

Terra cotta's fireproof quality made it initially attractive to the building trades in this country during the 1870s and 1880s, and its ability to be readily formed into a variety of shapes and its potential for ornamental use were rapidly recognized.[10] Multiple terra cotta elements could be produced from molds instead of having to be individually hand carved, as was the case with stone ornament, and terra cotta was also lighter and easier to handle and set than stone. Although terra cotta production was relatively labor intensive, it was considerably less costly than stonework.

The terra cotta facades of buildings in the United States in the 1880s typically incorporated terra cotta as an ornamental feature, as in the Rookery Building (1887, Burnham and Root) in Chicago, rather than as an overall cladding. In the 1890s changes in building construction technology provided the opportunity for skeleton-framed buildings with extensive window openings. The skeleton frames required a cladding, for which repeated units of plain or ornamental terra cotta provided ideal spandrels, mullions, belt courses, and cornices. Chicago's Reliance Building (1894, Burnham and Root) exemplifies this type of application.

Although terra cotta production was extensive before the turn of the century, standards to guide construction appeared only in the 1910s and 1920s. The National Terra Cotta Society, founded in 1911, published its first standards for construction and installation in 1914; these were revised in 1927.[11] In the 1910s the National Terra Cotta Society began a comprehensive research program in conjunction with the National Bureau of Standards. A summary report of their 1927 standards included "Tentative Recommendations for Improving Certain Qualities of Terra Cotta by Better Manufacture."[12]

In the early 1910s the use of stock terra cotta came to predominate the market.[13] Among the major suppliers from whom stock elements could be ordered was the Midland Terra Cotta Company, formed in 1910 near Chicago. Midland produced foliate ornaments inspired by the designs of Louis Sullivan that could be purchased to enhance the facades of even modest buildings.

In the 1910s and 1920s many new skeleton-framed structures were clad in terra cotta panels, foreshadowing the advent of ceramic veneer. Early-twentieth-century terra cotta facades tended to be glazed in white, ivory, or cream colors and feature detailing in neoclassical motifs. To some extent these features were probably a continued attempt to imitate stonework in a less expensive medium.[14] Noteworthy examples of fully terra cotta–clad facades include the Woolworth Building (1913, Cass Gilbert) in New York City and the Wrigley Building (1924, Graham, Anderson, Probst and White) in Chicago.

Following the 1925 Exposition Internationale des Arts Décoratifs et Industriels Modernes in Paris, terra cotta ornament and cladding in the Art Deco style appeared. The range of artistic expression available in terra cotta throughout the late 1920s and early 1930s ranged from individual elements on small apartment and commercial buildings to the complete cladding of high-rises.[15] The greater use of color encouraged technical improvements, including better quality control, a wider variety of glazes, and a better understanding of ceramic science.[16]

Terra cotta components were generally installed with masonry backup and were supported by steel lintels and angles set in the masonry wall. Before the turn of the century, bricks and mortar were used in the cavities of the terra cotta to support the masonry wall behind, but thereafter metal anchors or wires were increasingly used. The metal anchors were set into the masonry backup with mortar and placed into or through preformed holes in the webs of the terra cotta components. One common support configuration for projecting elements such as cornices was the use of a continuous rod through the webs of adjacent units, which in turn was supported by J-shaped bolts attached to a metal framework above. In contrast, ceramic veneer components were installed by two methods: individual elements were set in a mortar bed and adhered to a masonry or concrete backup, or they were held in place with metal anchors attached to the masonry backup.

The Great Depression took its toll on the terra cotta industry. After World War II ceramic veneer was often used instead of terra cotta in new construction. Ceramic veneer's manufacturing process was more highly mechanized and less labor intensive, and therefore veneer was less costly than terra cotta. Moreover, modern facade design no longer required terra cotta's ornamental capabilities. Masonry cladding took a back seat to new glass and metal panels in high-rise curtain wall construction.[17]

## CONSERVATION

A resurgence of interest in terra cotta has occurred because of an increasing number of preservation and repair projects on terra cotta–clad structures. The material is also being used again in new construction. If properly manufactured, installed, and maintained, terra cotta and ceramic veneer are durable and serviceable building materials. Buildings properly constructed with well-made terra cotta have survived intact and with little evidence of distress for more than a hundred years.

### DETERIORATION

The principal cause of failure in terra cotta is water penetration and associated cycles of wetting and drying and freezing and thawing. Symptoms of deterioration include cracking, crazing, and loss of the surface glaze; cracking and spalling of the body of the terra cotta; and fracturing and displacement of terra cotta components. The performance of terra cotta and ceramic veneer products depends on such factors as the manufacturing process, installation methods, and environmental conditions.

The terra cotta cladding of many buildings constructed in the 1910s and 1920s is supported by steel structural and anchorage elements. Water penetrating the terra cotta or mortar joints causes corrosion of the embedded metal anchorage, which can expand to up to ten times

**Figure 18.4.** Ceramic veneer clads the John Breuner Company Building (1931, Albert F. Roller), Oakland, Calif.

its original size. The expanding metal exerts pressure and causes cracking, fracturing of the terra cotta component, or displacement of adjacent terra cotta elements.

Terra cotta facades constructed without proper expansion joints to accommodate structural and thermal movement will experience cracking and displacement. Compressive stresses caused by moisture and temperature-related expansion in the terra cotta components can also cause cracking. Horizontal expansion and vertical stresses can cause vertical cracking, typically at corners and returns in the wall.[18] Within terra cotta elements, differential expansion and contraction of the glaze and body can result in crazing or cracking of the glazed surface.

In cold climates terra cotta may suffer severe distress from cyclic freezing and thawing. As water that has penetrated the wall freezes and expands within the terra cotta components or the masonry fill, pressure is exerted on the terra cotta, resulting in cracking or spalling.

The growth of molds, algae, or other organisms on the surface of terra cotta components or mortar joints can occur when the material is exposed to moisture over long periods of time. Such growth can hold water against the surface and encourage glaze deterioration. Excessive humidity can also encourage algae growth below the surface of the glaze, resulting in glaze spalling.

Although distress in terra cotta is more often related to exposure to severe weather, it may also be related to improper manufacturing. Poorly manufactured terra cotta may have high absorption, be vulnerable to water penetration, or be susceptible to glaze loss through incompatibility of the glaze and body. Distress may also result from inappropriate repairs, such as the application of sealant to the mortar joints, which can entrap water within the terra cotta. The inappropriate application of nonbreathable coatings to a terra cotta face can also lead to moisture entrapment and associated spalling of the glaze and body.

## DIAGNOSTICS AND CONDITION ASSESSMENT

In evaluating the condition of existing terra cotta on building facades, it is helpful to research available historic sources, including shop drawings, building construction documents such as drawings and specifications, and records of repairs. The investigator will often find that details found in situ are very similar to those illustrated in the 1914 and 1927 National Terra Cotta Society publications. This research may help limit the number of inspection openings needed to determine the causes of distress.

Investigation techniques used to evaluate existing terra cotta facades include nonintrusive methods such as using a borescope to evaluate concealed conditions, tapping the terra cotta with a wood or rubber mallet to identify delaminations within the components, or using a metal detector to locate hidden metal anchorage to correlate observed distress with the presence of metal elements.[19] Intrusive investigation methods such as removing terra cotta components or portions of these permit direct examination of concealed conditions and provide samples for laboratory testing.[20]

Laboratory testing is performed both on terra cotta and ceramic veneer removed from an existing building, and on new components manufactured for replacement or new construction. The purpose of laboratory testing

**Figure 18.5.** Terra cotta distress is often related to expansion corrosion of embedded metal anchorage elements.

of elements from an existing building is to evaluate the material's properties, determine causes of distress, and gather information for use in developing repairs or manufacturing replacement elements. The purpose of laboratory testing of new elements is to maintain quality control: to ensure that the new elements meet the required specifications and to maintain consistency in manufacturing. Testing of terra cotta and ceramic veneer typically includes petrographic examination and compressive strength tests. Laboratory tests are also performed to provide information about the composition of the terra cotta body and glaze, adhesion of the glaze, and quantitative data about characteristics of the particular terra cotta or ceramic veneer.[21]

Petrographic examination consists of a visual, qualitative inspection of the material. A stereomicroscope is used to evaluate the composition and density of the body, characteristics of the glaze, and nature of deterioration. Petrographic studies can also provide clues to problems with the original manufacturing process, such as inappropriate firing temperatures. These studies may be performed on both terra cotta and ceramic veneer.

Compressive strength tests are conducted by placing small prepared specimens of terra cotta in a testing machine that applies compression until the specimens fail. By determining the maximum load that the specimen will carry, the compressive strength of the terra cotta can be calculated.[22]

Terra cotta's relative absorption is closely related to durability. Dry samples of terra cotta are weighed before and after immersion in cold water for 24 hours, and before and after immersion in boiling water for 5 hours. Absorption is calculated as a proportion of the saturated weight to dry weight. The saturation coefficient, a ratio of the 24-hour absorption to the 5-hour absorption, is a relative measure of the material's resistance to freeze–thaw damage.

Other laboratory tests are used to evaluate thermal coefficients and permeability of the terra cotta body and glaze, adhesion of the glaze, moisture expansion, and firing temperature. To evaluate durability more directly, samples may be aged by exposure in an accelerated weathering chamber and tested for various physical characteristics before and after aging. Test results as well as archival research and a field inspection are used to develop conservation and maintenance procedures.

## CONSERVATION TECHNIQUES

Cleaning improves the appearance of the terra cotta, removes deleterious contaminants or inappropriate coatings, and helps identify needed repairs. Traditional abrasive methods are not appropriate for cleaning, as these will damage the glaze and sometimes the body of the terra cotta. Water-cleaning techniques may be effective on slightly to moderately soiled terra cotta. Heavily soiled terra cotta is effectively cleaned by chemicals such as two-part systems consisting of an alkali prewash and a diluted acid afterwash, preferably containing organic acids.[23] New techniques such as facade gommage and laser cleaning have not been extensively used or tested on terra cotta in this country.

Terra cotta components that are severely damaged may require replacement; however, they often can be repaired. Fractured units can be fastened together with stainless steel pins set in epoxy. Portions of fractured units can also be repaired in situ by installing a stainless steel pin through the face to anchor each portion to the backup construction. The pin is countersunk, and the pin location patched with a cementitious patch painted to match the adjacent terra cotta. This repair tends to be visually intrusive and should generally be avoided.

Nonmoving cracks are generally pointed with mortar that is compatible in strength and hardness with the terra cotta body. Spalls that extend through the glaze and into the body of the terra cotta are repaired by application of a cementitious patch, which is covered with a breathable masonry coating to match the adjacent glaze.[24] New coating technologies are providing more choices for colors and finishes to match the appearance of terra cotta.

## REPLACEMENT

When existing terra cotta or ceramic veneer components cannot be repaired, or when components are missing, new terra cotta elements can be manufactured. Specifications must take into account factors that affect the strength and particularly the durability of the finished product. The characteristics of both historic and new terra cotta or ceramic veneer can vary widely with production and location.[25] Carefully prepared specifications and quality control in manufacturing and installation are critical to the successful use of replacement elements.

SUSAN M. ESCHERICH

# 19 | GYPSUM BLOCK AND TILE

**DATES OF PRODUCTION**
c. 1890 to c. 1965
**COMMON TRADE NAMES**
Gypsite, Gypsteel, Pyrobar, Structolite, Unitrave

## HISTORY

Gypsum block and gypsum tile (the names were used interchangeably) were made of calcined gypsum (plaster of paris) with the addition of from 3 to 5 percent fiber, usually wood, and could be solid or cored.[1] They were produced principally for non-load-bearing partition walls, although precast gypsum tiles of the same formulation were also made for roofs and floors and sheathing for steel columns and girders. The American Society for Testing and Materials standards published in 1925 required manufacturers to stamp their product name on gypsum tiles shipped for resale.[2]

### ORIGINS AND DEVELOPMENT

Gypsum was discovered in the United States in 1792 in Camillus, Onondaga County, New York.[3] In the early decades of the nineteenth century it was used to fertilize fields. Deposits of gypsum were mined for agricultural use in Virginia in the 1830s and in Michigan and Ohio in the 1840s. By 1895 gypsum was being mined in thirteen northern and western states.

Commercial production of calcined gypsum began in New York City in the 1890s; plants for calcining were also built in Michigan and Iowa in that era. The process for making gypsum blocks from calcined gypsum was probably introduced from Germany in the 1890s.[4] Calcined gypsum was a component of precast tiles used for floors, partitions, and roofs, as well as of plaster used for finishing both exterior and interior walls.

Precast gypsum tiles first became popular for their fire-retardant properties. In 1903 the United States Gypsum Company's Genesee, New York, plant began production of gypsum tiles, which it marketed as a fireproof replacement for clay tile in non-load-bearing situations. Other companies soon followed suit.[5]

By 1915 there were sixty-eight gypsum mills in the United States. The largest producers at that time were the United States Gypsum Company of Chicago (22 mills in 12 states); the Acme Cement Plaster Company of St. Louis (12 mills in 9 states); and the American Cement Plaster Company of Chicago (7 mills in 6 states).

The 1920s saw record growth in the building industry, with gypsum products gaining in market share at the expense of other building materials, largely because of their fireproofing qualities. Between 1900 and 1926 production of gypsum rock in the United States increased from 594,462 tons to 5,635,441 tons; during the same period its value increased from $1.6 million to $46.7 million.[6]

The properties of gypsum continued to inspire new uses. By 1931 precast gypsum units as form molds for reinforced concrete beams and columns had been introduced. The process was used to pour walls, floors, and roofs. At least one of these houses was erected in a New York suburb; the design was touted as an example of affordable housing.[7] The following year a flooring system using a similar concept featured gypsum floor blocks laid on steel floor beams, which were then overpoured with concrete. Called the Unitrave system, it provided a smooth surface for a plaster ceiling below and support for poured concrete above.[8]

Competing technology grew slowly alongside that of gypsum partition tile. As early as 1920, the economic and technical reasons for switching from gypsum blocks to gypsum wallboard were becoming apparent.[9] Gypsum partition tile was being replaced by metal or wood partitions finished with gypsum wallboard. These materials could also be installed much faster than gypsum blocks, which had to be laid up and required drying time before finishing. Gypsum wallboard, furthermore, was as fireproof as the gypsum partition board.[10] In 1951 Certain-Teed Products Corporation's huge gypsum plant at Fort Dodge, Iowa, was still producing gypsum blocks but was concentrating on wallboard. The United States Gypsum Company, one of the biggest producers of gypsum tiles, ceased producing it in the 1960s, although it described Pyrobar (its trade name for gypsum blocks) and its uses in its 1965 *Red Book.* The Gypsum Association continued to give the fire-resistance rating of gypsum partition tiles through 1967.

134 PENCIL POINTS

## *Lighter* in weight!

Pyrobar Gypsum Tile for partitions and furring weigh 25% to 50% less per square foot than ordinary building tile of equal thickness. That's why Pyrobar Tile are so easy to handle and show such worth-while savings in structural steel. They are also fireproof, soundproof and permanent. We contract to erect Pyrobar. Write for booklet showing typical installations.

UNITED STATES GYPSUM COMPANY
*General Offices:* Dept. K 205 W. Monroe Street, Chicago, Illinois

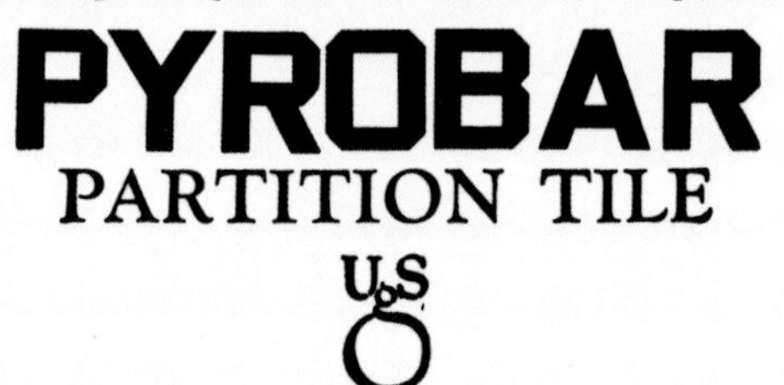


**Figure 19.1.** Pyrobar's lightness, which permitted rapid construction, and its fireproofing qualities were the biggest selling points for this gypsum partition tile, manufactured by the United States Gypsum Company. *Pencil Points,* October 1925.

## MANUFACTURING PROCESS

Gypsum tiles were manufactured largely by hand until the 1920s. A framework as high as the thickness of the finished tiles was set on a rubber mat. Tapered wood or metal plugs were placed through the frame lengthwise to form the hollow cores of the tiles. A thin mixture of calcined gypsum and wood fiber was poured into the frame and allowed to set. The plugs were removed, and the set tiles were then trucked to long sheds for drying, either naturally or by heat circulated by blowers.[11]

By 1925 standard shapes were more often made on either a belt or a circular machine.[12] Custom shapes, however, continued to be made by hand. Mills were situated beside open-pit mines. The gypsum was calcined, crushed, weighed, mixed with water and fiber, and then dispensed into molds placed on a small car. After the tiles were set, the car moved forward, releasing the core-forming units. The tiles were then unloaded by hand and put on another car, which rolled into the dryer. The tiles baked in the dryer for 12 hours at 160 degrees Fahrenheit, while a fan circulated the air. Certain-Teed's mill in Acme, Texas, for instance, had five tile machines capable of turning out 10,000 square feet of 3-inch tiles per day.[13]

## USES AND METHODS OF INSTALLATION

Gypsum blocks and tiles were sold as fireproof in themselves and as protection for steel structural members. Gypsum tiles were also publicized for their sound and thermal insulation properties, resistance to cracking from vibrations, workability, ability to be sawed to size, and low cost (because breakage was negligible). They were also touted as saving mortar and plaster because of their regularity of size and smoothness of finish, which provided a backing for finish plaster work. Finally, because they were lightweight, blocks could be made larger than bricks and stone, and thus a wall could be constructed in less time.

The United States Gypsum Company, one of the major producers, made the following recommendations for installation of its standard 30-by-12-inch tiles: "As a rule 3 or 4 inch tile is the thickness used for the average building where partitions do not exceed 15 ft. in height. . . . Two-inch solid Pyrobar Tile are used for partitions not exceeding 10 ft. in height and for covering columns. . . . Split and hollow Pyrobar furring are laid up against exterior walls and securely spiked every square yard. Joints should be broken the same as for partition work. Furring may also be fastened to the building wall by means of wall ties built into the masonry. The hollow portion of the tile should be placed against the walls, thus minimizing the contact area with the outside walls."[14] This type of tile was used in the Wanamaker Building (1902, D. H. Burnham and Company) in Philadelphia.

Specially finished blocks were developed for specialized purposes. United States Gypsum developed blocks that were smooth on one side to form a base for finish plastering or for the final wall surface in warehouses.[15] In dry areas, such as Douglas, Arizona, and Bighorn, Wyoming, gypsum blocks with exterior surfaces that resembled rough-dressed stone were used for exterior walls.

Several companies developed roofing systems using precast gypsum tiles covered with a waterproof coating. Gypsum roofing tiles were touted for their lightness and their insulating and fireproof qualities. In addition, they could cover large spans. The H. H. Robertson Company in Pittsburgh, for example, made precast roofing slabs as wide as 8½ feet. The company described the installation of the tiles as follows: "The slabs are molded in steel forms, into which the steel cables are placed and put into uniform deflection and tension by means of the same deflection-rods, restrained at the sides of the molds. These cables consist of 3⁄16" cold-drawn wire rods, spaced from 2 to 4½" apart, according to the spans and the loads to be carried. The cables are securely anchored to the molds at points beyond the ends of the slab, being removed from this anchorage when the gypsum is set, each cable emerging from the ends of the slab within ½" of the top surface and projecting about 2"."[16]

A typical installation for floors using precast gypsum blocks was advertised in the 1920s by the Structural Gypsum Corporation of New York City. Its Gypsteel precast gypsum floors consisted of rolled steel channels, spaced 30 inches on center, on which were laid precast gypsum floor slabs 1½ inches thick and from which were suspended ceiling slabs of the same material 2 inches thick. These ceiling slabs were reinforced and fireproofed the channels. Steel reinforcing bars ran through the slabs lengthwise and projected from either end. The Gypsteel floor slabs were 2½ inches thick, 2 feet wide, and 2½ feet long and were reinforced with wire rods. The company advertised that these products "eliminated the need for a plaster scratch coat, and could be plastered with brown and finish coats the same day they were installed, with no necessity of waiting for drying of the base material."[17]

In 1928 United States Gypsum advertised that its precast floors could be used with concrete joists for spans of up to 30 feet.[18] At the same time the company was advertising Pyrobar floor voids and roof tiles (the latter reinforced with electrically welded, galvanized steel mats) made of Structolite, a dense, specially prepared gypsum that came in the standard 12 by 30 inches as well as 19 by 18 inches.[19]

## CONSERVATION

Gypsum, an inert substance, is remarkably stable if kept dry. Since it was used for inside walls and flooring and protected by waterproof layers when used for roofing, it will generally be found in good condition.

### DETERIORATION

Water is the principal cause of deterioration in gypsum blocks.[20] Although gypsum loses compressive strength while saturated, it will regain that strength if allowed to dry out.[21] Moisture can cause corrosion of steel rods embedded in the gypsum. However, calcined gypsum is chemically inert and will not of itself corrode the steel bars. Examinations of steel rods embedded for fifteen years in gypsum have shown no evidence of progressive corrosion.[22]

### DIAGNOSTICS AND CONDITION ASSESSMENT

Because almost no gypsum blocks and tiles are ever exposed, evaluation may require destructive methods. Partition blocks can be evaluated with a borescope inserted into a small hole in a wall to determine whether any units are cracked or have sustained water damage.

Roofing and floor tiles are usually evaluated when a roofing or flooring system is being repaired or replaced. Corroded reinforcing bars indicate that water penetration has occurred, and the location and source should be identified. Water penetration is most likely to occur in roofing systems where flashing is inadequate or has failed.

Sound tiles and blocks can usually be retained, but severely deteriorated ones should be removed if their structural integrity is questionable. Fire code requirements should also be considered.

### CONSERVATION TECHNIQUES

If the plaster of the gypsum block becomes soft, porous, or powdery, it can be consolidated by treating it with a solution of 1 pint of household vinegar in 1 gallon of water. A spray bottle is good for applying the solution, and this treatment should be continued until the surface is hard.[23] Once dry, the consolidated area can be repaired using the technique described here.

Gypsum plaster should be used to repair gypsum blocks that have cracked or have been chipped, as neither portland cement plaster nor lime plaster will typically bond to the gypsum blocks.[24] Gypsum partition

**Figure 19.2.** Once cast in molds, tiles were dried in the open, in kilns, or in drying sheds, which were heated with air, steam coils, or direct furnace heat. Stacks of gypsum blocks and tiles await shipment to distributors around 1917 (top).

**Figure 19.3.** Quick-setting cement mortar was used to lay up United States Gypsum's Pyrobar tiles to create partitions (bottom). Tiles could easily be cut to fit various types of wall constructions and fireproof beams and girders, thus speeding construction and lowering costs.

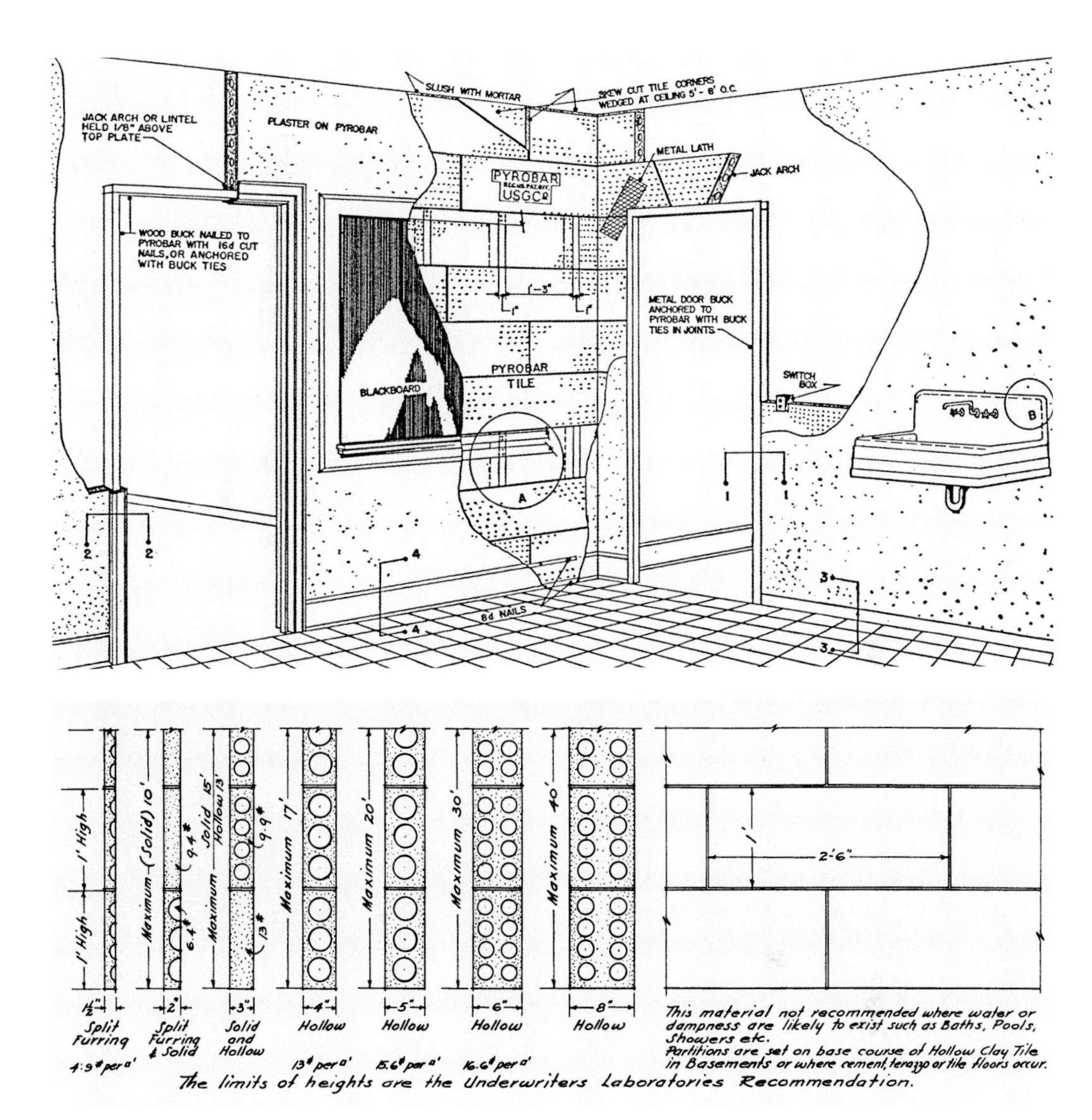


**Figure 19.4.** This cross section shows typical locations of Pyrobar installation (top). Lintels could be created with a jack arch or reinforced with metal lath. Door frames and baseboards were nailed directly into the gypsum tiles. Indented surfaces on partition tiles served as a key for plaster finishes.

**Figure 19.5.** Sections of typical furring and partition tiles show thicknesses and hollow cores (bottom). Recommended height limits for partition tile varied according to tile thickness.

tiles needing repair may be patched in a manner similar to other plaster. Thin plaster of paris can be forced into a narrow crack with a wide spackling tool; if the crack is large, vacuum out loose particles, wet down the area, spread a thin layer of plaster joint compound over the crack, and press down cloth or paper joint tape with a wide joint knife, as on modern drywall. The cracks should then be covered with a thin layer of joint compound, making it as smooth as possible. After this layer is dry, it should be wiped down with a damp sponge; another thin layer of joint compound can be applied and feathered out to blend with the surrounding area. This step can be repeated as necessary.

Larger holes can be undercut on the edges with a knife or bottle opener. After vacuuming, the area should be wet down and then filled with plaster of paris, spackle, or a modern cellulose filler. If the hole is deep, it can be filled in stages, with no layer more than ⅛ inch thick.

**Figure 19.6** During restoration of the Vinoy Park Hotel (1925, Henry L. Taylor), St. Petersburg, Fla., deteriorated gypsum furring tiles in the lobby were replaced with concrete blocks, visible in the columns.

Wadded newspaper can be forced into the hole to provide a temporary backing if the original is missing. Finally, the patch should be finished with thin layers of plaster.[25]

## REPLACEMENT

Gypsum blocks and tiles are no longer produced in the United States. When blocks and tiles are so severely deteriorated that they must be replaced, substitute materials must be considered. Block wall systems may be replaced by wood, concrete block, or metal studs and wallboard with veneer plaster.[26]

MICHAEL J. SCHEFFLER AND EDWARD A. GERNS

# 20 | THIN STONE VENEER

**DATES OF PRODUCTION**
c. 1900 to the Present

## HISTORY

Since about 1960 the term *thin stone veneer* has been defined as stone that is cut to less than 2 inches thick and applied to a building facade in a non-load-bearing manner.[1] Granite, marble, travertine, limestone, and slate are the most common stone materials used, and these are treated with a variety of surface finishes and colors to achieve differing architectural effects.[2]

### ORIGINS AND DEVELOPMENT

Before the 1890s building stone was commonly used in thick slabs and blocks for construction. The use of the term *veneer* to describe building stone can be traced to the 1890s, when hand-cut stone as thin as 4 inches was sometimes used to construct tall buildings.[3] For the Reliance Building (1895, Burnham and Root) in Chicago, stone veneer between 2 and 4 inches thick was used on the exterior of the first and second stories.

Thinner slabs of stone were also used in the early twentieth century for storefronts, bulkheads, and building interiors. By the early 1930s stone veneer was incorporated into curtain walls, and in the late 1930s thin stone veneer began to gain acceptance as cladding for entire facades. Early examples include the Rule-Page Building (1940, Hansen and Waggoner) in Mason City, Iowa, and the Federal Reserve Bank (1950, Smith, Hinchman and Grylls) in Detroit. Since the 1950s thin stone veneer has also been incorporated as a component in cladding systems.

Prominent buildings were not clad with thin stone veneer until the early 1960s. One early example is the John F. Kennedy Center for the Performing Arts (1964, Edward Durell Stone) in Washington, D.C., which is clad with marble veneer 1 inch thick. In recent decades technological developments have increased the body of knowledge about thin stone veneer behavior, and architectural preferences have resulted in a dramatic increase in use.

### MANUFACTURING PROCESS

Before 1900 almost all building stone blocks from quarries were finished by hand into thick slabs or blocks. In the early 1900s this process was mechanized with the introduction of multibladed frame saws featuring parallel blades that "simultaneously cut the blocks into smaller uniform slabs. Initially, water wheels were used to carry the water necessary to cool the blades and to take away the resultant abrasive sludge."[4] This process was later further refined with water pumps and machines to regulate the quantity of abrasive used.

In 1932 the National Building Granite Quarries Association described the gang-saw process used to make granite panel veneers as thin as 1 inch.[5] The process, by then "almost entirely by machine," included using three to seven long, reciprocating steel blades placed parallel in a frame that cut down through the stone block to produce thinner uniform slabs. These were then cut down to even thinner slabs with Carborundum wheels. Finishes available at that time included rockfaced, sawed, sandblasted, peen-hammered, pointed, bush-hammered, rubbed, honed, and polished finishes.[6] A honed finish was obtained using a fine-grained grinding wheel, and a polished finish was obtained using a heavy felt-coated wheel and jeweler's rouge. During this period thin stone veneer was typically used for interiors, street-level facades, storefronts, and bulkheads.

Developments in the 1970s included using diamond-studded cables to slab blocks, and in the 1980s the introduction of frame saws with metal abrasives allowed stone to be cut as thin as ⅛ inch for use in composite panels.[7]

### USES AND METHODS OF INSTALLATION

Veneer panels were typically laid up on mortar beds, and joints were finished with mortar in a manner similar to traditional masonry construction. Interior panels were tied to the backup material, brick or block, by wire anchors set in holes drilled into the panels and discrete mortar spots. Lateral anchoring of exterior panels

**Figure 20.1.** A design guide by the National Building Granite Quarries Association included details about anchoring "modern granite veneer," stone less than 4 inches thick. *Sweet's Architectural Catalogues*, 1932.

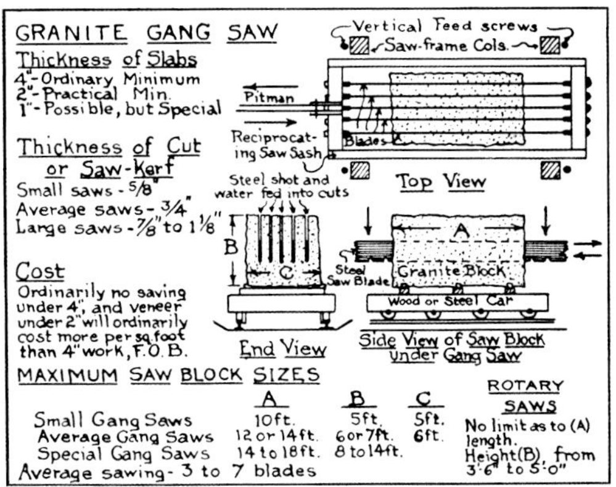


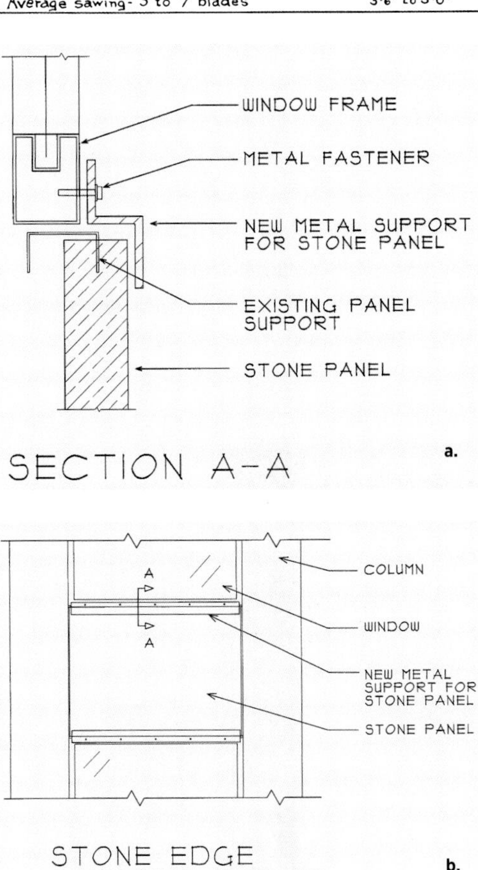


**Figure 20.2.** Thin stone veneer slabs were cut from large blocks of stone using reciprocating gang saws with three to seven steel blades (top). As blocks were moved back and forth on cars, steel shot and water were fed into the saw kerfs to aid cutting.

**Figure 20.3a–b.** One type of repair involves adding edge support for increased anchorage of panels, as shown in (**a.**) detail and (**b.**) cross section.

typically consisted of galvanized and asphalt-coated ¼-inch-diameter steel rods, which were bent and set into ⅜-inch-diameter holes drilled into the edges of the panels. The rods engaged the stone for a depth of between ⅜ inch (for panels 2 inches thick) and 1¼ inches (for panels 4 inches thick). Plaster of paris was recommended as a quick-setting spacer behind panels. Manufacturers typically recommended four anchors per panel, with two anchors on parallel sides. The anchor sides could be top to bottom or side to side.[8]

In 1940 the Cold Spring and North Star Granite Corporations promoted 1-inch-thick granite veneer.[9] In 1944 the Aberene Stone Company promoted slate veneer as thin as ⅞ inch. Manufacturers advised the use of a shelf angle at each floor to help carry the weight of the panels. Typically joints were still filled with mortar. Lateral anchors for thin stone were usually brass or steel rods set into holes filled with cement mortar. For stone more than 2 inches thick, steel straps 1¼ by ¼ inches were used. Four anchors per stone were generally recommended for stone up to 20 feet square.

In the late 1940s strap anchors became more prevalent for lateral support of thin stone veneer. For recladding, the anchor was twisted 90 degrees and mortared or plastered into a dovetail slot in the backup masonry. In new construction the straps were mechanically fastened into the backing material. The straps could be bent out of the way during initial setting of the veneer, then bent into a kerf (slot), providing easy installation. The straps were usually stainless steel and were 1 inch wide by 3⁄16 inch thick for thicker stones and 1⁄16 inch thick for thin stone veneer. Straps were turned down ¾ inch into thick stones and ½ inch into thin stone veneer.

In the early 1950s stone was typically produced in thicknesses of ⅞, 1¼, 1½, and 2 inches, although 1½ inch was usually recommended for veneer. Some manufacturers began to recommend that stone less than 1½ inches not be used, claiming that thin stone, such as ⅞ inch thick, was more expensive to produce. Industry standards listed tolerances as plus or minus ¼ inch, and joint spacing between panels was ¼ inch. Expansion joints were also recommended every 30 feet vertically and every second floor horizontally. A cavity of about 1 inch behind panels was still suggested. Filling the cavity with mortar and coating the backs of the panel with damp proofing was promoted in some instances. During this time veneer panels with overall dimensions generally between 3 by 3 feet and 4 by 4 feet were used.

The use of thin stone veneer increased in the 1950s because of the rise in popularity of the curtain wall and because it became less costly, a result of improved manufacturing efficiencies such as the introduction of diamond-bladed tools.

By the mid-1950s thin stone veneer construction was becoming much more refined. Horizontal joints between panels were sometimes sealed with elastomeric sealants rather than mortar. Shims of aluminum, plastic, or lead were placed in horizontal joints to transfer panel weight. Manufacturers' recommendations became more conservative, calling for an increase in the number of lateral anchors: for panels less than 12 square feet, four anchors were recommended; for panels between 12 and 20 square feet, six anchors were recommended; and for pieces larger than 20 square feet, one anchor was suggested for every 3 square feet. Many different types of lateral anchors, including stainless steel straps, brass dowel and wires, and disks were in use by this time. Gravity and lateral support could be incorporated in the same anchor with the development of the split-tail anchor; in some cases steel tabs were welded to relief angles.

By the late 1950s stone was being incorporated into composite building panels. Most common among these were precast concrete panels faced with a thin stone veneer. Typically, hairpin wire anchors or dowel anchors were preset into the back of the stone panel before casting. The precast concrete or thin stone veneer component was then formed by placing the concrete against the backside of the stone panel with the panel set facedown in a form. During this period a bond-breaker sheet or coating between concrete and stone was not usually incorporated before casting the concrete. This technique of separating materials was more commonly used in the later 1960s.

Construction of thin stone veneer become more standardized in the 1960s, with the publication of the *Marble Engineering Handbook* by the Marble Institute of America and *Marble-Faced Precast Panels* by the National Association of Marble Producers.[10] The demand for thin stone veneer grew throughout the 1970s and 1980s. An overall increase in building construction in the 1980s resulted in a 600 percent increase in the use of marble and a 1,735 percent increase in the use of granite between 1980 and 1985.[11]

## CONSERVATION

Conservation of thin stone veneer requires an understanding of the common causes of distress and deterioration and the basic techniques for diagnosing and assessing thin stone veneers. Conservation techniques can be wide ranging, and a thorough understanding of the various techniques available will aid in developing an appropriate and less intrusive repair.

### DETERIORATION

The vast majority of distress observed in stone cladding is associated with connections and support components. The major types of distress observed with these typical anchorage systems include cracks at or around anchor locations, spalls at anchor locations, spalls at shelf angle supports, cracks at panel corners, panel displacements, and rust staining. A number of factors can cause or contribute to these types of distress. Improperly designed anchors can lead to spalling and cracking of adjacent stone or displacement of panels. If anchors are missing, loads on existing anchors can exceed those for which the anchors were designed. This can cause failure of the anchors and distress to the stone. Water penetrating the cladding can corrode metal anchors and shims. Other factors include improper attachment of anchors to the building structure, improper positioning of anchors in the cladding, and a missing separator between concrete backup and panel.

Deterioration of stone veneer is not limited to the support systems. All natural building stone loses strength over time from the effects of cyclical heating and cooling, water penetration, cyclical freezing and thawing, and acid rain. Materials such as marble typically lose more strength than granites. The degree of strength loss is also related to panel thickness; the thinnest panels can be expected to be most affected by strength loss.

Stone curtain walls usually rely on the strength of the stone panel material to span connections. If the material is not strong enough, either from improper design or aging, it may evidence cracking between supports or bowing.

Generally, marbles are not volume stable; they experience permanent volume growth as a result of heating and cooling cycles. This growth reduces the strength of the material and can lead to warping and bowing, as the opposite sides of the panel experience differential thermal expansion. The other stone types typically are volume stable.

Certain visible characteristics of the stone have been found to correlate to strength loss. These include bowing, warping, saccharification (sugaring of the surface), exfoliation, existence of stylolites, and other variations in the surface appearance.[12] In extreme cases cracking may be related to strength loss. Significant veining, caused by mineral impurities, may or may not be related to the panel's planes of weakness or strength loss.

Even within one type of stone, composition can vary considerably, and components may behave differently. All stones, including granite, marble, limestone, and slate, have a natural bedding orientation. Depending on the direction in which the stone types are manufactured, strength and durability may vary. Even similar marbles manufactured in different orientations can behave differently because of the alignment of the crystals in the stone structure.

Distress in stone cladding components can also be related to building and wall movements, or differential movements not accommodated by the cladding design, or by the construction of the cladding system. If insufficiently sized joints are used to accommodate building or wall movements, cladding panels can come in contact with one another during thermal expansion or building movement. This could result in cracking and spalling near panel edges and corners and in failed or distorted sealant joints.

Staining and discoloration do not usually pose structural or safety concerns and may be related to preferential weathering of the stone's mineral components, corrosion of its embedded ferrous components, water runoff containing staining elements from other materials such as copper or iron, and improper cleaning or maintenance. Staining can also be related to excessive water leakage through joints or behind the cladding units, which may indicate associated corrosion of embedded metal elements and may create structural concerns. Some types of staining may be related to sealant materials in adjacent joints.

## DIAGNOSTICS AND CONDITION ASSESSMENT

Diagnosing and assessing thin stone veneer distress is best begun by reviewing original construction documents and design calculations, if available, to understand the overall construction. Often the only way to verify existing construction of thin stone veneer facades is by removing portions for inspection and obtaining samples for laboratory testing. Less intrusive techniques may be implemented first to preclude or mitigate the need for significant intrusion. These may include tapping or sounding with a hammer, using a metal detector, and using a field microscope.

Conservation of thin stone veneer usually involves documenting a number of specific conditions related to the veneer's attachment to the building. It is important to determine whether anchors are promoting distress, whether there are enough anchors, whether anchors are spaced properly, whether the anchors are properly sized, and whether they are correctly positioned and attached to the stone and properly fastened to the structure. It is also important to ascertain whether corrosion of materials is occurring and causing distress. Also, consider whether the shim material promotes distress; whether shims are properly used in slots and kerfs cut in stone,

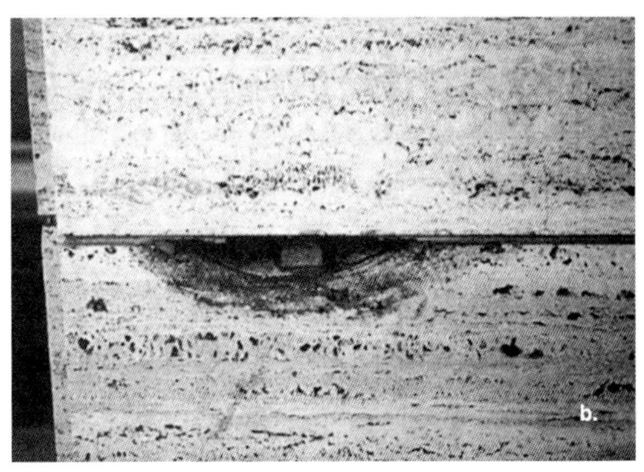

**Figure 20.4a–c.** The John F. Kennedy Center for the Performing Arts (1964, Edward Durell Stone), Washington, D.C. **a.** This was one of the first prominent buildings to be clad in thin stone veneer. **b.** Spalls, such as this one in a travertine panel, often occur at concealed metal anchors. **c.** Torque testing of a remedial expansion anchor installed in a precast concrete backup ensures that the panel is properly attached.

between panels, or between panel and shelf angle support; and whether shims inhibit proper engagement of slots in stone kerf connections.

Examine panels for movement, cracking, and unusual surface conditions. It is important to determine whether joints between veneer panels are properly constructed to accommodate building movement and whether joints are free from inappropriate materials. Often sealant in the joints provides important information relating to movement and distress.

## CONSERVATION TECHNIQUES

Repairs to stone cladding most often involve repair or replacement of connections and support components and may also involve selective or complete replacement of the cladding. If additional positive mechanical anchorage is required, through-bolt anchors covered with sealant, mortar, or a stone plug set in mortar are most commonly used when only a few components need repair. Expansion anchors can be attached to the precast backup, or dowels set in epoxy can be attached to the precast backup or adjacent stable panels. Perimeter edge support systems attached to adjacent building construction may also be used.

Often, cracked portions of panels adjacent to anchors can be reanchored. If a portion of the stone panel is severely distressed, that portion can be removed and a dutchman (a small panel piece) replacement installed in its place. Dutchmen typically are pinned or fastened to the existing adjacent stone or backup. In lieu of dutchmen, mortar patches, often reinforced with steel and matching the color and texture of the remaining panel portion, can be installed.

If distress has been caused by unaccommodated building and wall movements, repairs may entail grinding the joints to make them wider to allow movement. Extraneous materials must also be removed from the joints.

If distress is evidenced by staining and discoloration and the causes have been determined, it may be desirable to clean the stone cladding. Typical cleaners used on stone facades include water washes, mild detergents, and two-part chemical systems consisting of an alkali prewash and a mild acidic afterwash. Any proposed cleaning system should be thoroughly evaluated through laboratory and field studies before implementation. The gentlest effective system should be selected.

Field or laboratory testing of proposed repair methods is recommended before the repairs are implemented. Durability tests can be performed on the proposed repair or anchorage system to assess the expected life of the repairs.

## REPLACEMENT

In some cases removal and reinstallation or replacement of the stone cladding components can be considered. If significant loss of material strength has occurred and additional anchors will not provide adequate support, the panels will likely have to be replaced. In cases of widespread deterioration, replacement may be appropriate.

The performance characteristics of the new stone should be evaluated and a rational structural design undertaken. Testing should be performed on the stone selected, on anchorage components, and, in some instances, on whole assemblies.

All tests, including those to determine flexural strength, compressive strength, density, absorption, and aging characteristics, should be performed in accordance with standards established by the American Society for Testing and Materials.[13]

ANN MILKOVICH MCKEE

# 21 | SIMULATED MASONRY

**DATES OF PRODUCTION**

c. 1929 to the Present

**COMMON TRADE NAMES**

Bermuda Stone, Fieldstone, Formstone, Magnolia Stone, Modern Stone, Perma-Stone, Romanstone, Rostone, Silverstone, Terox

## HISTORY

The term *simulated masonry* covers a number of products manufactured to imitate the appearance and characteristics of stone. These products are made from various materials, including cement, minerals, epoxy, and fiberglass, among others. They can be cast in specific shapes or applied directly onto a building substrate, molded or shaped to resemble the texture of masonry, and struck to create mortar joints.[1]

### ORIGINS AND DEVELOPMENT

The attempt to imitate masonry and stone is not a modern phenomenon. Cast stone, which is often considered a form of simulated masonry, has been used in the United States since the last quarter of the nineteenth century. Another product, rockfaced concrete block, gained popularity in the early twentieth century. Simulated masonry is similar to both cast stone and concrete block in that it, too, mimics another material, but its construction technique made it a more flexible product. Simulated masonry was not a modular system cast off-site but was usually manufactured on-site and applied as a facing material. The process allowed for maximum flexibility to adapt to specific and sometimes unexpected site conditions. While simulated masonry products were marketed for new construction, they were also widely used on existing buildings. They were seen as an easy way to update a building or construct a new building without incurring the cost of actual stone construction while conveying a sense of permanence.[2]

Simulated masonry played a large role in the changing aesthetics of the American public beginning in the 1930s. Of the simulated masonries that could be applied directly to a building, probably the best known is Perma-Stone, which was touted as the "originator of moulded stone wall-facing."[3] Beginning in 1929, the Perma-Stone Company, based in Columbus, Ohio, sold and marketed its patented and trademarked product through the use of licensed and trained dealers.[4] The company provided the molds and materials (portland cement, aggregate, crushed quartz, mineral colors, and metallic hardeners) necessary for the job, but the dealers manufactured and installed the materials. The company's success spawned many competitors attempting to capture a share of the growing market for this type of remodeling process.

Another successful cement-based simulated masonry was Formstone, developed by the Lasting Products Company in Baltimore. Formstone was first available in 1937, the same year that Lasting Products obtained a patent for its process.[5] The company was responsible for the manufacture and distribution of the specific tools and materials necessary to complete a Formstone project. The actual on-site work was done by registered contractors who had been trained by the company.

Simulated masonry products were often hailed as thoroughly modern inventions. Rostone, made from pressurized shale, alkaline earths, quarry waste (lime), and water, was first produced in 1933 for the Century of Progress Exposition by the Rostone Company in Lafayette, Indiana.[6] Rostone was used to create the Wieboldt-Rostone House (1933, Walter Scholer), one of ten houses designed to exhibit modern building materials and innovative construction methods. Rostone was manufactured as prefabricated panels and shipped to the site for construction by trained contractors.

In addition to cement, many other materials have been used as the base for simulated masonry products. By 1960, for instance, fiber reinforced plastic panels were available. One product, Terox, was "moulded in dies cast from selected quarry stone."[7] As with other products, pigments were used to match the desired stone color.

### MANUFACTURING PROCESS

There are two categories of simulated stone: products manufactured off-site and those mixed on-site. Both types of simulated masonry can be applied to existing conditions or used as part of new construction on almost any building type.

**Figure 21.1.** Perma-Stone, a molded wall facing made of aggregates, cement, crushed quartz, mineral colors, and metallic hardeners, was suitable for both new construction and renovation. Perma-Stone catalogue, 1954.

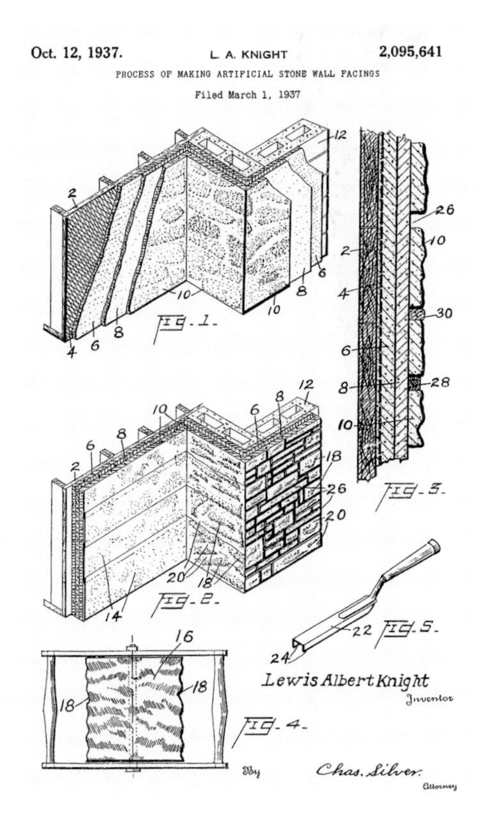


Figure 21.2. The finished surface of Formstone, patented in 1937 by Lewis A. Knight, was tooled and tinted to produce polychromatic effects.

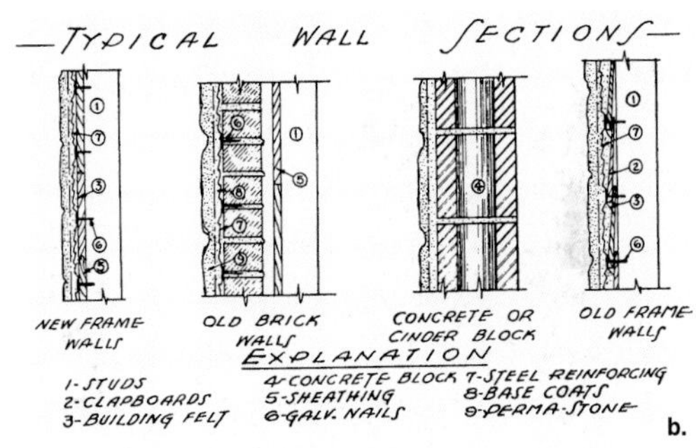


**Figure 21.3a–b. a.** A worker hand-presses a final Perma-Stone coat to a wall surface. **b.** Perma-Stone could be applied to many new and existing substrates.

Rostone simulated not only the appearance of stone; its manufacturing process closely recreated the formation of natural stone. No portland cement was used in its manufacture; rather, it was made from natural ingredients that underwent a chemical reaction to form a new material. The manufacturing process began with drying, pulverizing, and finely grinding shale. Lime and then water were added, and the resulting mixture was placed in molds or hand-formed into specified shapes. To induce the necessary chemical reaction, the molded mixture was hardened through exposure to heat. The material could be colored during the manufacturing process, and afterward the surface could be finished in a variety of textures.[8] On cooling, Rostone components were ready for shipment to the job site.[9]

## USES AND METHODS OF INSTALLATION

Simulated masonry was often used as a remodeling material, but it could also be used for new construction. Relying on the stereotypical perception of stone as signifying wealth, stability, and grandeur, these products were sold as the modern version of natural stone. They provided an inexpensive way for middle-class Americans to enjoy the prestige of a "stone" house. Many companies stressed the opportunity to update a house or building to a level, or class, that would never be possible without their product. Advertisements also routinely proclaimed that simulated masonry was maintenance free, fireproof, and energy efficient. All these properties appealed to buyers seeking an inexpensive way to modernize buildings by covering over deteriorating facades.

Perma-Stone was marketed for new construction, but the majority of its applications were in remodeling or renovation projects. Formstone, on the other hand, appears to have been used more widely for new construction. The Rostone Company initially marketed its simulated masonry as a "modernization" material for storefronts. Schemes for gas stations built of Rostone were also proposed, but whether any were actually built is unclear.[10] The company also targeted the residential market for the use of Rostone on exterior and interior walls, floors, roofs, and decorative elements such as door surrounds and fireplace mantels.

**Figure 21.4.** The Wieboldt-Rostone House (1933, Walter Scholer), Beverly Shores, Ind., used Rostone, a precast form of simulated masonry.

Rostone panels were produced as standardized 16-by-24-inch sheets either 1 inch or 1¼ inches thick; custom sizes and shapes were available at an additional cost. The panels could be finished in three surface textures: honed, a polished surface; natural, a slightly rough finish that mimicked natural stone; and shot blast, a moderately rough surface. Rostone could be made in any color, but earth tones were most popular. Greens, pinks, and reds were used as accent colors in carved designs and to convey a multidimensional look.[11]

In modernization and remodeling projects, Rostone panels were applied directly over the existing wall surface. When Rostone was used in new construction, the panels were attached to a steel frame with specially designed clips. The joints were filled with a mastic, thus providing an easily maintained system.[12]

Perma-Stone, a cementitious material, is produced on-site and applied directly to the building in a process similar to that of applying stucco.[13] Used extensively for remodeling, Perma-Stone provides a stonelike concrete veneer that can be "permanently" attached to wood or steel lath or applied directly onto concrete and masonry.

To install Perma-Stone, metal or wood lath is secured to the building to provide a base for attachment. The application process is a three-part procedure. A first coat, referred to as a brown coat, is applied over the lath. Before it sets, it is grooved to provide additional surface area for the next coat. Once dry, the second coat (the scratch coat) is applied. While the second coat is still wet, the finish coat is applied though the use of pressure molds designed to imitate the look of natural stone.[14] Minor hand finishing and application of a final membrane coating to serve as a water repellent complete the job.

In the direct-application technique, Perma-Stone can be applied to curved surfaces as well as broad flat ones. The "stone" wall can be laid up in random, broken, or coursed ashlar. Joints can be raked, beaded, or pointed. Because color is added directly into the product on-site, the color choice is unlimited and can vary. This provides the opportunity to develop interesting strata and varied

**Figure 21.5.** Across America in the 1940s and 1950s buildings like this bank were remodeled with Perma-Stone.

stone colors. The texture of the finished simulated stone is restricted only by the molds and the amount of hand finishing a mason is willing to do.

As a frontrunner in the simulated masonry industry, the Perma-Stone Company zealously protected its patented and trademarked product and did not hesitate to pursue court injunctions against those who tried to use it without permission.[15] Perma-Stone held several patents covering its product recipe, pressure-casting procedure and molds, and membrane-curing technique. To ensure quality control, only licensed dealers and contractors are permitted to use the process, molds, and materials.[16] Today the Perma-Stone Company still maintains a registered trademark for Perma-Stone products.[17]

The philosophy behind Formstone was to provide a process for making an artificial stone facing that used the tools of masons and cement finishers and could be readily carried out by them.[18] The procedure and finished installations share some similarities with Perma-Stone, for Formstone too is a cementitious material applied in a multilayer process. Where the two differ markedly is in the formulation of the "stones" and finishing procedures.

When applying Formstone to an existing building, the walls are covered by a perforated lath of wood or metal if they are not masonry (no lath is needed if the wall is masonry or stone). A layer of cement mortar $\frac{3}{8}$ to $\frac{3}{4}$ inch thick is applied over the lath, and the surface of this layer is scored before it sets and dries. A second layer, typically $\frac{1}{2}$ to $\frac{3}{8}$ inch thick, is applied. While this layer is still plastic, the finish layer is applied. The finish layer, also ranging from $\frac{3}{4}$ to $\frac{3}{8}$ inch thick, can be formed with two or more colored or shaded mortar cements that are distributed to produce the polychromatic effects desired to achieve the appearance of stone.

Before the material in the top two layers has set, a waxed paper or other nonadhering material is placed on the wall. A cast aluminum roller with a crinkled surface is passed over the waxed paper, creating a crinkled impression in the mortar. Several rollers of different sizes or textures could be used on the same project to achieve the desired effect. After the waxed paper is removed, the crinkled surface is scored with guide lines for the "mortar joints." Grooves are cut into the top layer with a chasing tool, which has two parallel cutting edges that create

**Figure 21.6.** In 1950 the Wieboldt-Rostone House was covered with Perma-Stone, shown in poor condition.

a mortar joint mimicking those found in natural stone construction. The groove may be either left unfinished or pointed with mortar.[19]

A variety of finishes can be created with different textures (using different rollers) and colors. Most tinting is created by adding color into the mortar mix, but surface color can also be achieved by dashing colored powdered materials such as "mica, oxide pigments, stone dust, slate dust or chips of mineral or artificial stone . . . on the outer layer. This produces a speckled surface, simulating particular natural rocks or stones." The powdered material is placed on the surface either before the waxed paper is applied or after the texture of the stone has been created and the wax paper removed.[20]

Formstone seems to have been marketed as a product to refurbish and modernize existing buildings of any type. It was promoted as a material that could solve problems of deteriorating masonry and stone structures and poor insulation. Purchasers received a twenty-year guarantee and assurances that the wall facing was "maintenance free." Baltimore, with its large number of brick buildings of an indigenous soft brick, became the "Formstone capital of the world."[21]

Simulated masonry products reached their zenith during the 1950s, but by the early 1980s interest in such products had nearly died. Other products such as vinyl and aluminum siding were mass-produced and more economically installed on both new and existing construction. More important, these new products appealed to the changing public aesthetics. However, both Perma-Stone and Formstone are still being produced in small quantities today. The countless examples of simulated masonry across America are reminders of the public's penchant for remodeling houses.

# PART V
# GLASS

PLATE GLASS

PRISMATIC GLASS

GLASS BLOCK

STRUCTURAL GLASS

SPANDREL GLASS

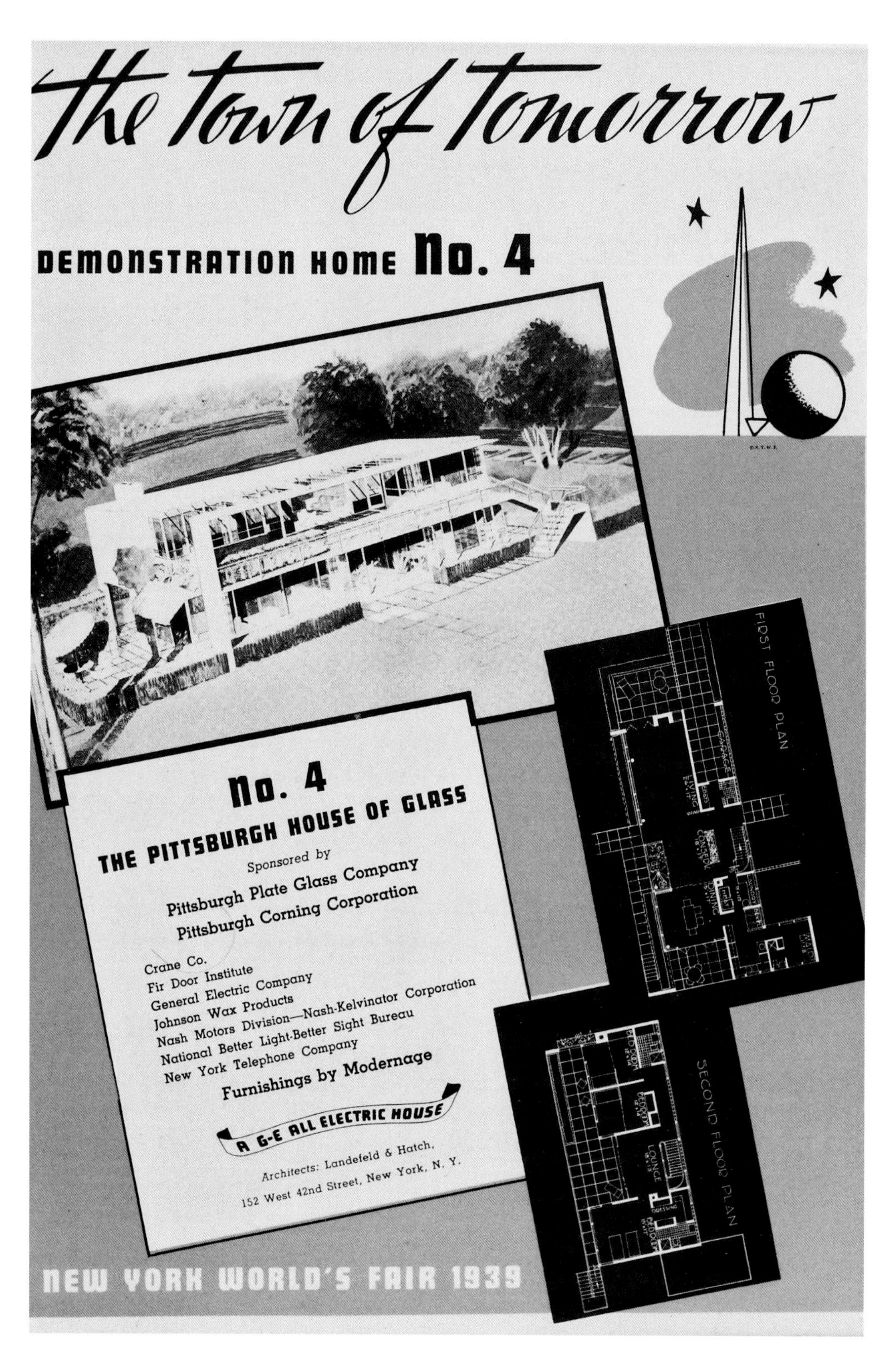


**Figure 22.1.** At the 1939 New York World's Fair, Demonstration Home No. 4 (Landefeld and Hatch), also known as the House of Glass, featured expanses of polished plate glass, which the Pittsburgh Plate Glass Company extolled as the "aristocrat of the flat glass family." Lithograph, Burland Printing Company, New York City.

KIMBERLY A. KONRAD AND KENNETH M. WILSON;
WILLIAM J. NUGENT AND FLORA A. CALABRESE

# 22 | PLATE GLASS

**DATES OF PRODUCTION**
c. 1850 to the Present
**COMMON TRADE NAMES**
Herculite, Solex, Thermopane, Tuf-Flex, Twindow

## HISTORY

Plate glass is transparent glass that is thicker and stronger than window glass and has little or no distortion. Until about 1960 plate glass was produced by casting and rolling large sheets that were then ground and polished.[1] Sizes typically ranged from ¼ inch to 1¼ inches thick and were as large as 14 feet wide and 20 feet long.[2] The introduction of the float process in 1959 eliminated the need to grind and polish plate glass and revolutionized plate glass manufacture. Strength, optical clarity, and the availability of large sheets made plate glass ideal for shop windows.[3]

### ORIGINS AND DEVELOPMENT

Until the mid-nineteenth century most plate glass used in the United States was imported, primarily from France.[4] Despite the availability of ample raw materials and fuels, there were few American manufacturers of plate glass.

Experimentation in the United States began in the 1850s and resulted for the most part in rough plate glass.[5] Manufacturers such as the Cheshire Glass Company, Cheshire, Massachusetts, and the Lenox Plate Glass Company, Berkshire County, Massachusetts, experimented with French and English methods of producing polished plate glass. In 1865, after visiting the Lenox works, John B. Ford opened a glass works in New Albany, Indiana, importing grinding and polishing machinery from England.[6] It was here that polished plate glass was first successfully and continuously manufactured in the United States. In 1882 Ford, whose firm became known as the Pittsburgh Plate Glass Company, constructed a plant in Creighton, Pennsylvania, near Pittsburgh.[7]

The plate glass industry grew rapidly in the last quarter of the nineteenth century. In 1860 Ford's factory produced $30,000 worth of rough plate; ten years later five plants had been established and were producing $355,250 of rough and polished plate glass per year.[8] By 1900 only 15 percent of plate glass for domestic needs was still being imported; by 1920 domestic production constituted 99 percent of total consumption.[9]

### MANUFACTURING PROCESS

The raw materials used in the production of plate glass are sand, limestone, and soda ash with additions of carbon, arsenic, and manganese. Plate glass manufacture was attempted in a number of locations across the country where the necessary materials and fuel supply were readily available.[10] Between 1880 and 1920 the methods of casting, annealing (cooling), grinding, and polishing remained virtually unchanged, aside from the ongoing effort to increase the size of plates cast.[11]

Plate glass manufacture depended largely on the use of mechanical appliances.[12] After the raw materials were melted in a furnace, a pot of molten glass was lifted from the furnace and poured onto a casting table. A large metal roller was drawn over it to create a sheet of glass. Once rolled and partially cooled, the glass was moved into the annealing lehr (furnace), where it was slowly cooled for four to five days.[13] The plates were then embedded in plaster of paris on large revolving tables, ground to the desired thickness using a series of ever finer abrasives, and finally polished with felt-covered blocks and rouge. The finished plates were then washed, inspected, and cut to size.

Two important innovations in the manufacturing process had been introduced by the early 1920s.[14] The first was a large power-driven crane to transport the pot of molten glass from the furnace to the casting table.[15] The second was the introduction of a continuous annealing lehr, consisting typically of multiple preliminary ovens and a long runway, which greatly reduced the time needed to anneal and cool the cast glass.[16] By 1923 the manufacturing process had been reduced from 10 days to 36 hours.[17]

Following World War I the automobile industry created a new market for plate glass. In 1918 the Ford Motor Company conceived the idea of manufacturing plate glass for automobiles by continuous methods.[18] The continuous process, introduced in 1923, automated the entire manufacturing process.[19] Molten glass in large tanks flowed

**Figure 22.2.** In the continuous grinding process perfected at the Ford Motor Company, Detroit, the annealed glass moved on railcars under the grinders, fed with progressively finer sands (top). Plates were then inverted, and the process was repeated.

**Figure 22.3.** The Hallidie Building (1918, Willis Polk), San Francisco, one of the earliest buildings with a glass facade, had large, operable window sashes placed several feet in front of the end of the floor beams and window sills only 3 inches thick (bottom).

through a discharge spout onto a movable table and under a roller, forming a flat sheet of a specified width and thickness. The glass then was moved through a long annealing lehr, after which it was ground, polished, cut, and inspected.

The new continuous process was suitable for making thinner plates—both a threat and an opportunity for window glass manufacturers. This invasion of plate and window glass in each other's markets rapidly blurred the distinction between the two industries.[20]

The advances in the casting and annealing process were quickly adopted, and attention turned to methods for finishing plate glass. In the early 1930s the Libbey-Owens-Ford Company installed twin grinders that would grind rough glass simultaneously on both sides.[21] About a decade later Pittsburgh Plate Glass equipped its newest plant in Cumberland, Maryland, with both twin grinders and twin polishers. However, in 1959 Pilkington Brothers of St. Helens, England, announced the development of the float process, which made twin grinders and polishers obsolete almost overnight.[22] In this process a continuous sheet of glass from the furnace is floated on a bed of molten tin, eliminating the need for grinding and polishing.[23] In 1962 Pittsburgh Plate Glass was the first American company to adopt the float process for plate glass production. By 1993, 90 percent of the world's flat glass was produced using this process.

## USES AND METHODS OF INSTALLATION

Until the early twentieth century plate glass was used principally for shop and store windows. However, the growing number of buildings constructed of iron and steel with non-load-bearing walls created new markets for plate glass. The greater thickness and availability in larger sheets made plate glass essential for glazing openings in tall buildings. The Woolworth Building (1913, Cass Gilbert) in New York City contained more than five thousand windows, each glazed with two panes of plate glass. The Empire State Building (1931, Shreve, Lamb, and Harmon) was also constructed using plate glass (manufactured by Libbey-Owens-Ford).

As the skyscraper developed, windows evolved from individual slits in thick walls to whole columns of windows. After World War II larger windows became walls of glass.[24] The all-glass facade on the Hallidie Building (1918, Willis Polk) in San Francisco was considered one of the boldest American attempts to create a glass wall. Usually no more than 50 percent of a commercial building's facade was glazed. Transparent walls of plate glass were also designed for residential applications, such as

Ludwig Mies van der Rohe's Farnsworth House (1950), in Plano, Illinois.

As the amount of glass used in curtain wall systems increased, manufacturers introduced plate glass with specific properties. Insulated plate glasses, such as Libbey-Owens-Ford's Thermopane and Pittsburgh Plate Glass's Twindow, reduced heat loss.[25] Heat-absorbing plate glass, such as Solex by Pittsburgh Plate Glass, limited interior heat levels.[26] Thermally toughened glass, also known as tempered plate glass, allowed even larger areas to be glazed. Two such products were Tuf-Flex and Herculite, made by Libbey-Owens-Ford and Pittsburgh Plate Glass respectively.[27] Such glasses made all-glass buildings more commonplace in the postwar period.

Figure 22.4. Plate glass was the material of choice for shop windows, where the highest quality glazing was a necessity for displays. This simple Kress Company store with curved plate glass is a typical example.

## CONSERVATION

Plate glass is noted for its transparency, outstanding impermeability, and durability.[28] The most common form of deterioration is loss of transparency due to staining or etching of the surface. Such damage can also reduce strength, although loss of strength is not a prevalent problem.

### DETERIORATION

In 1921 A. A. Griffith introduced the theory that the surface of glass has many invisibly small, inherent defects (Griffith flaws) that can lead to cracking when these defects coincide with high stress. The strength of glass depends on the number of Griffith flaws, their location, and the presence of chemicals, including water, that attack the strained atomic bonds in the flaws.[29]

The average tensile strength of plate glass is generally about 6,000 pounds per square inch.[30] Failure always results from fractures occurring under tensile stresses, since the compressive strength of glass greatly exceeds its tensile strength. The stress required to induce failure can vary widely for seemingly identical specimens. Glass strength can therefore be expressed only in statistical terms and is generally based on empirical data. Charts published by the major glass manufacturers show recommended glass thicknesses for various design conditions based on a given expected probability of breakage.[31]

Glass strength is also degraded by the presence of larger surface flaws, particularly near edges. Edge flaws most significantly affect the strength of glass under thermal loads. High-tensile stress can occur near the edges of a glass sheet when the central part is significantly warmer than the edges, a condition common in buildings with overhangs or framing that shades or insulates the glass edges.

Because plate glass is brittle, it deforms elastically when loaded until it reaches its fracture point. If loaded glass does not fracture, it returns to its original shape on removal of the load without permanent deformation. Because of its inability to deform plastically, glass is susceptible to local overstressing and is vulnerable to breakage.[32]

Exposure to alkaline solutions can damage glass and affect its appearance. The extent of deterioration depends on the period of contact. Mildly alkaline solutions used to clean glass are not detrimental if residues from the cleaning solution are thoroughly removed from the glass.[33]

Another source of alkalis is leachate from concrete or masonry components of a building facade. Rainwater running down the face of a building can leach alkalis from the concrete and deposit them on the glass. If permitted to remain on the surface, even for only a few days, these deposits can become difficult to remove. Deposits remaining for longer periods may actually stain or etch the glass.[34] Overspray from alkaline solutions used to clean masonry can cause similar damage to glass.

Prolonged contact with water can also damage glass. Through reaction with the glass, the water becomes alkaline and thereby more corrosive. Under normal exposure to rainwater, this alkali leaching does not readily take place. However, in cases where water is allowed to remain in contact with glass for long periods, such as when condensation cannot evaporate quickly or when excessive dirt traps moisture on the glass, surface deterioration can be significant. The result is hazing and roughening of the glass surface, which in turn makes removal of contaminants more difficult. The rate of attack increases as the ratio of the surface area to volume of liquid trapped on the surface increases, because the liquid becomes alkaline more rapidly. Accumulation of contaminants sufficient to trap moisture is more common on sloped glass than on vertical glass.

Plate glass is resistant to most acids. However, hydrofluoric and phosphoric acids, even in dilute concentrations, readily attack glass by decomposing and dissolving the silicates of which glass is principally composed.[35] These substances can sometimes be found in air pollution, especially in industrial environments. More commonly, these acids are found in chemical compounds used to clean metal framing and masonry.[36] If the glass is not protected during cleaning operations, etching can occur.

Impact from wind-borne debris, such as roof gravel or hail, can cause pitting of the glass surface. During renovation activities, welding or sandblasting can cause damage to glass that has not been sufficiently protected. This damage, in addition to being unsightly, can reduce the strength of the glass.

Corrosion of metal framing systems; deterioration and hardening of glazing compounds, sealants, or gaskets; and lateral movement ("walking") of the glass sheet within its frame can damage the glass edges or cause high local stresses to be applied to the edges. In either case the result can be fracture of the glass sheet. In annealed glass, fractures originating at a damaged edge usually evidence themselves as a single break line extending all or partway across the glass sheet. Fractures induced by high stress applied to the edge usually evidence themselves as multiple break lines extending across the sheet.

Fracture of annealed glass caused by impact to the surface away from an edge usually appears as a series of break lines radiating from the point of impact. When tempered glass fractures, it usually fragments and falls out of the window frame, making identification of the exact cause of failure difficult.

## DIAGNOSTICS AND CONDITION ASSESSMENT

It is advisable to identify any damaging agents and the extent of damage through chemical and microscopic analysis before proceeding with cleaning or restoration. If the presence of damaging agents is suspected, samples of the residue on the glass and adjacent materials (gaskets, sealants, paints, and so forth) should be obtained. Stain patterns may also be helpful in locating the source of damaging agents. In the laboratory the samples can be analyzed to identify the presence of harmful agents. The glass can be examined in place to determine the extent of damage. Once the damaging agents are identified, they should be removed, or contamination of the glass will recur.

The condition of gaskets, sealants, and glazing system members should also be evaluated for signs of damage, cracking, corrosion, or deformation. Corrosion of metal framing can create expansive forces that may cause glass failure. With the exception of material incompatibility, which can be identified by laboratory testing, the visual appearance of materials is generally a good indicator of their condition. The condition of gaskets and sealants can also be judged to some degree by their pliability. Hardened gaskets and sealants may impart undesirable stresses to the glass.

## CONSERVATION TECHNIQUES

Decisions on whether to restore scratched or hazed glass or replace it must take into account its historical significance. Since polished plate glass is no longer available, efforts should be made to preserve it. In all cases where an abrasive or chemical is proposed for restoring the surfaces of in-place glass, field tests on representative areas should be performed before undertaking the entire project. These tests will help establish the effectiveness of the proposed techniques as well as cost.

Cleaning of glass that does not have significant contaminants on its surface should be performed using a soft, grit-free cloth and a mild, nonabrasive, ammonia-based or similar cleaning solution. The surface should then be thoroughly rinsed with clean water and the excess removed. If significant accumulation of dirt has occurred, potentially abrasive particles must first be removed by flushing with large amounts of water to avoid scratching the glass surface. Organic solvents such as xylene or toluene may be necessary to remove grease or other substances not dissolved by water alone. Solutions containing a detergent such as trisodium phosphate have also proven to be effective in cleaning plate glass surfaces. With any cleaning method or solution, it is always advisable to test the procedure on a small area before proceeding to determine the possibility of damage to the glass or adjacent building materials.

Stains or light etching requires more specialized and time-consuming cleaning or restoration techniques.[37] To limit intervention minor stains or light etching in less prominent locations can be retained. Cleaners containing mineral acids such as dilute hydrofluoric acid should generally be avoided. Hydrofluoric acid decomposes the silicates in glass and thus removes stains by removing a thin layer of glass.

A final approach to removing various stains and light etching, short of replacing glass, is to polish with a cerium oxide paste, made by mixing cerium oxide powder with clean water, and a felt pad.[38] The felt pad is usually attached to a handheld, rotating mechanical polisher. An area of the glass is then polished using the cerium oxide slurry applied to the glass under moderate pressure for several minutes. If the stains are not removed after 10 minutes of polishing, the glass may be irreparable. Continually wet the paste and avoid using paste in direct sunlight since the cerium oxide can become overly abrasive if allowed to dry out.[39] To avoid further damage to the glass, special skill and care are required with this technique.

When the deterioration of the sealants, glazing compounds, or gaskets has significantly reduced the glass's structural support or has allowed the glass to move laterally, replacement of sealants and gaskets is generally the only feasible option. Choose replacement materials with adequate durability and take care to replicate the appearance of the existing glazing materials.

Where corrosion of the metal framing has resulted in significant section loss or where the corrosion products are causing high stress on the glass edges, the metal framing may be restored, supplemented, or selectively replaced, as necessary. A corroded metal frame can be restored by cleaning the frame to remove the rust products and coating it to retard future corrosion. If section loss is significant, it may be necessary to supplement or replace the framing with new materials to restore the glass's structural support.

### REPLACEMENT

When breakage has occurred or when cleaning techniques prove futile, glass replacement may be the only option. In the United States flat glass is produced by five companies: Pittsburgh Plate Glass Industries, Libbey-Owens-Ford Company, Guardian Industries, Ford Glass, and AFG Industries. Virtually all new glass is manufactured by the float process. Although its constituents are the same, the plate glass found in older or historic structures was produced by a different process. When existing plate glass is beyond repair, it can generally be matched with float glass.

As in the design of new glazing, the required transparency, color, and strength to withstand wind and thermal loads must be considered. Effects of building configuration and facade components must be taken into account when determining the design loads. Glass industry guidelines should be followed during installation.[40] Compatibility of sealants and other materials used in the glazing system must also be investigated during the design phase.

**Figure 23.1.** An early, evocative advertisement for the Luxfer Prism Company depicted its motto—Illustrous Obscuritas ("light from darkness")—and suggested that prismatic glass could shower the world in light.

DIETRICH NEUMANN

# 23 | PRISMATIC GLASS

**DATES OF PRODUCTION**

1896 to c. 1940

**COMMON TRADE NAMES**

American 3-Way Prism, Daylight Prism, Luminous Prism, Luxfer Prism, Searchlight Prism, Solar Prism

## HISTORY

Prismatic glass is characterized by small horizontal triangular ribs on the interior face that refract light rays deep into a room. Commonly used in pre-electric commercial buildings to improve lighting, this type of glass was originally produced in tile form and later in larger sheets.[1]

### ORIGINS AND DEVELOPMENT

Because prismatically shaped glass can redirect light, it seems to have been applied to the illumination of rooms as early as the eighteenth century.[2] Used at first to distribute light from above into the dark interiors of ships, this application was modified in the second half of the nineteenth century for pavement lights, used for lighting basements in urban centers in the United States and Europe. While the deck prism distributed light evenly, pavement lights redirected the light sideways into the basements underneath buildings.

Beginning in 1845 Thaddeus Hyatt, a pioneer of reinforced concrete construction, patented a number of pavement lights, which he produced after 1873 in his New York factory.[3] They were so successful that some building codes recommended Hyatt lights as a fireproof source of light in basements.[4]

Translucent, light-diffusing glasses such as rough plate glass ground on both sides or corrugated glass with horizontal grooves were used from the 1800s on to provide factories with an evenly distributed light. Corrugated glass had the additional effect of refracting a large part of the light deeper into the room and gained some success in New England's factories as "factory-ribbed glass."[5] The product that came closest to the development of the new prism glass was the so-called stallboard light, which had been developed in England in 1883 and quickly became a standard feature in British commercial architecture. It consisted of square glass tiles with horizontal V-shaped ribs on the inside; these were set in rows in a metal frame adjacent to the pavement lights at the bottom of the exterior wall or above the display window.[6]

Similar ideas were patented at the same time by a number of inventors in both Europe and the United States.[7] Among them was the British inventor James Pennycuick, who in 1882 filed a U.S. patent that was eventually to provide the basis for the Luxfer Company's success. He proposed window glass with prismatic ridges on the inside, which would "double the quantity of reflection or illumination of the plain window-glass of the same size."[8] After several unsuccessful attempts at finding commercial support for his "improvement in window glasses," Pennycuick founded the Radiating Light Company of Chicago in 1896 with a small group of entrepreneurs.[9] In April 1897 the company adopted the name Luxfer, derived from the Latin *lux* (light) and *ferre* (to carry).[10]

The company hired the prominent physics professor and spectroscopist Henry Crew of Northwestern University and his assistant Olin H. Basquin to develop the product further, explore its potential applications, and signal the scientific intent of its enterprise.[11] In contrast to the existing applications such as corrugated glass, stallboard lights, or Pennycuick's patent, Crew and Basquin aimed at a precisely predictable light refraction. They developed mathematical formulas to calculate specific, individual prescriptions for a building's lighting needs, similar to the way eyeglass prescriptions are precisely adjusted to a patient's imperfect vision.[12] Crew claimed that their products were directly inspired by Augustin-Jean Fresnel's mathematically exact system of prismatic lenses, which since 1821 had been the standard equipment for lighthouses all over the world.

In 1897 alone the Luxfer Prism Company submitted 162 patents for designs and technical details of the frames and machinery necessary to produce prisms and prismatic pavement lights, such as molds and grinding machines, as well as angle-measurement devices.

### MANUFACTURING PROCESS

To produce prismatic glass, glass was pressed in iron molds. Special dies were created to form the ribs in

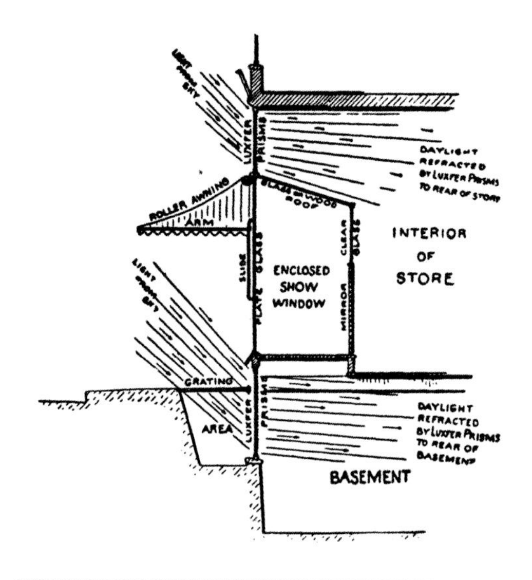


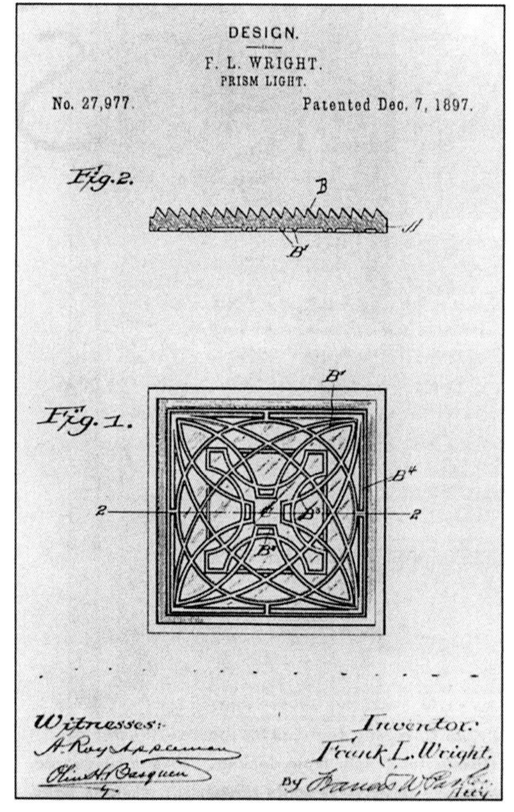


**Figure 23.2.** Prismatic glass refracted light, directing it to the rear of store interiors and basements, as shown in this diagram from a circa 1910 Luxfer catalogue (top).

**Figure 23.3.** Luxfer hired young Frank Lloyd Wright to design a prism pattern, which he patented in 1897. The only one of Wright's glass patterns ever mass-produced, it enlivened the glass's surface appearance but did not enhance its refractive qualities (bottom).

various degrees and any ornamental patterns on the opposite face. Because of the complicated manufacturing process, initially only small, 4-by-4 inch tiles with twenty-one triangular ribs could be produced; larger plates tended to crack when they were removed from the molds.

The tiles were supported by a grid of thin metal bars of soldered zinc caming and cement or the new, sophisticated, and very expensive methods of copper electroglazing, which had been invented in 1897 by William H. Winslow in Chicago, one of Luxfer's shareholders.[13] In this process the prisms were placed between thin copper ribbons (3⁄8 inch wide and 1⁄32 inch thick) and exposed to an electrolytic bath in copper sulfate (36 hours for vertical plates, 48 hours for canopies). During this procedure additional copper molecules would be deposited on the edges of the copper bands, thereby forming flanges that eventually firmly welded the materials together. The copper surface was sometimes plated with nickel or silver.

Prisms were generally about 3⁄32 inch thick with an additional 3⁄32 inch for the depth of the ribs. Luxfer provided nine angles of refraction and various qualities of prism glass: "cut," the best quality; "Iridian," with linear ornament; "commercial," which was untested; and "factory," with minor flaws.[14]

## USES AND METHODS OF INSTALLATION

Prismatic glass was used in window sash as a replacement for a sheet of window glass, in a separate frame in front of an existing window, or as a protruding canopy to capture direct zenith light in narrow streets or alleyways. Prismatic glass was used most frequently in commercial structures but was also installed in residences, schools, and hospitals.

Compound installations mixed glasses of different qualities and ornamental patterns in plates. Vertical plates, held in a wooden, brass, or iron frame, could be up to 15 square feet without any additional internal support. Luxfer guaranteed that the sash would be at least as stable and durable as ordinary window glass. Canopies needed intermediate supports in areas larger than 5 square feet and were not to exceed 15 square feet altogether.[15]

Long, complicated tables in Luxfer's handbook were used to calculate the right kind of prism in the appropriate combination and placement, leading to a specific prescription for a building's lighting requirements.[16] For architects who shied away from such complicated calculations, an average angle of 57 degrees was suggested. Luxfer also distributed color samples with the degree of light absorption, and formulas of how to respond with an additional amount of prismatic glass.[17]

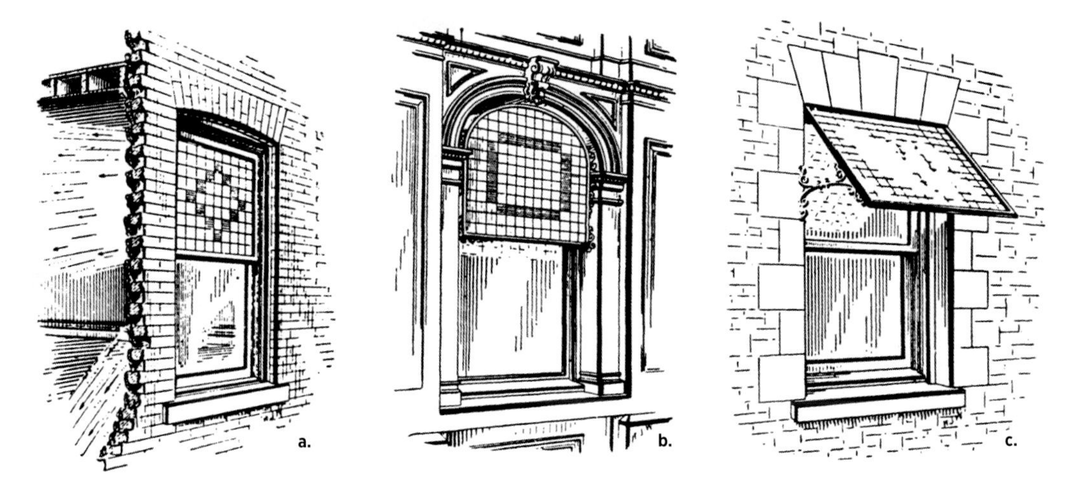


**Figure 23.4a–c.** Common types of prismatic glass installations included (**a.**) window sash filled with prismatic glass; (**b.**) separate screens set in front of window glass; and (**c.**) protruding canopies, often used for buildings on narrow streets or where tall buildings blocked natural light.

A number of independent tests between 1900 and 1913 reported that the lighting capacity of prismatic glass in the depth of a room was between five and fifty times that of ordinary glass.[18] However, its success depended strongly on specific local circumstances. Vertical prism installations reached their limits if the building opposite was higher than twice its distance from the facade. In such cases protruding canopies were recommended. They could deliver sufficient light for storage in rooms overshadowed by a building up to five times as high as the width of the street. Prism installations were not necessary when the street width was at least three times the height of the building on the other side (or, of course, if there was no opposite facade at all, as was often the case with the upper stories of tall buildings). Luxfer's calculation tables assumed the even brightness of an overcast sky. If sunlight struck the prism plates directly from above, it could cause a strong, unpleasant glare and increase the temperature in the back of the room.

The introduction of prismatic glass changed the architecture of office buildings and warehouses considerably. At a time when, especially in Chicago, facades were often reduced to little more than a simple post-and-beam construction with enormous windows, this new element assumed a prominent position: a horizontal checkerboard band, 2 to 4 feet high in the upper third of the storefront of the first one or two floors. It also occasionally influenced the floor plan, when existing light shafts were converted into additional floor space once prismatic glass had been installed. In newly planned buildings light shafts were left out entirely and replaced by prismatic glass in the facades.[19] Ceiling heights could also be reduced because light could reach deeper into spaces, eliminating the need for high ceilings.

Although the whole procedure of choosing the right angle of refraction, producing the tiles, and assembling them in frames was far more complicated and costly than the application of ordinary ribbed glass, Luxfer Prisms were an immediate commercial success. This was probably due to both the product's highly convincing performance and the company's sophisticated promotion campaign. It advertised extensively, published articles in the leading architectural magazines, sponsored a design competition, and edited a handbook, *Luxfer Prism*, with detailed descriptions of the product's qualities.[20] After less than one year on the market, the company had already equipped 296 buildings throughout the United States.[21] This number jumped to more than 12,000 by 1906.[22]

The Luxfer Prism Company emphasized additional advantages of prismatic glass. As a "new building material," it could be made part of the surface decoration of a building, competing with plate glass, which Luxfer considered a "necessary evil."[23] Luxfer urged its customers to go beyond the transom and fill entire openings with prismatic glass.

Hailed as the "Century's Triumph in Lighting," prismatic glass claimed to contribute to the rise of "good modern architecture."[24] Prominent architects such as Burnham and Root, Holabird and Roche, William Le Baron Jenney, and Ernest Flagg used it frequently in their buildings. In 1898 Louis Sullivan used prismatic glass in the facade of the Gage Building in Chicago.[25] Later he

**Figure 23.5.** The widespread use of prismatic glass panels in storefronts, usually incorporated as a horizontal band in the upper third of the storefront, changed the appearance as well as the design of stores. The 2- to 4-foot-high panels were generally fixed in steel frames with copper, zinc, or lead.

carefully integrated huge plates of prismatic glass into the wild flowering of his trademark ornament in the facades of the Carson, Pirie, Scott Department Store (1904) in Chicago. Frank Lloyd Wright, who in 1897–98 served as a product designer for Luxfer, produced a series of patterns for the outer surface of the prismatic tiles; at least one of them, an interwoven, linear pattern of circular segments and squares, was mass-produced.[26] Wright also designed a visionary office building, never executed, whose whole facade consisted entirely of Luxfer Prisms.[27]

Luxfer's success encouraged competition. By 1905, fourteen firms were offering prismatic glass under such names as American 3-Way Prism, Searchlight Prism, Daylight Prism, and Solar Prism. Some advertised products similar to Luxfer's, clearly the industry leader. The American 3-Way Prism Company advertised prismatic glass tiles in slightly different sizes (5 and 5½ inches square) and a newly developed prismatic wire glass for fire protection.[28] Sizes also increased. By 1910 Luxfer offered sheets as large as 36 by 84 inches for factories.[29] In the 1920s plates as large as 54 by 60 inches were available, and many large glass producers, including the Pittsburgh Plate Glass Company, offered prismatic glass in two or three prism angles.[30] Large plates were easier to clean and maximized the refracting surface.

The introduction of electricity as well as technical disadvantages contributed to the dwindling success of the prisms. Contrary to Luxfer's claims, the installations required a certain amount of continual maintenance. The horizontal ridges on the inside were much more difficult to clean than ordinary window glass; as a result they darkened, allowing in even less light than normal windows.[31] The soldered zinc bands did not age well, and leaks were frequent. None of the canopies survived the impact of inclement weather. The complicated procedure of choosing different angles of refraction according to local conditions and the assemblage of small glass tiles into larger frames was too costly to be continued at a time when electric lighting became generally affordable. In the 1930s fashions for storefront designs shifted toward simplified, larger forms and grander, illuminated store signs, often covering the transom area above the display windows. The introduction of air-conditioning required generally lower ceiling heights. When load-

**Figure 23.6.** Many prismatic glass installations from the early twentieth century, such as this one in Selinsgrove, Pa., combined tile patterns for a striking visual effect.

bearing, hollow glass blocks were introduced in 1935, a new, more efficient, and cheaper day-lighting device became available.

In 1926 the American Luxfer Prism Company merged with one of its competitors, the 3-Way Prism Company, and existed as late as 1936 as the American 3-Way Luxfer Prism Company in Cicero, Illinois. However, no manufacturer of prismatic glass seems to have survived the 1930s.

NEW INSULUX DECORATIVE DESIGNS

*Dramatize the Beauty of Glass*

**ABOVE:** In the Coffee Shop of the Commodore Perry Hotel, Toledo, exterior panels of Insulux "Circon" block bring diffused daylight . . . form a decorative feature of the room . . . exclude an objectionable view . . . and, having high insulation value, lower costs of heating and air conditioning.

**DESIGN No. 30**—The "Circon," so called because the design suggests concentric circles when the panel is completed. An excellent pattern for residential use. Available in 8 x 8-inch sizes.

**DESIGN No. 24**—A design by Walter Dorwin Teague. The waved ribs form a continuous pattern that gives the panel unusual beauty. Available in 8 x 8 and 12 x 12-inch sizes.

INSULUX Glass Block are basically a functional material, designed to transmit daylight, insulate effectively and help maintain better control of interior conditions.

Yet there are many places—in homes, theaters, hotels, shops, restaurants, etc.—where Insulux Glass Block are used mainly as a design element to add decorative beauty to interior and exterior. For such uses, Insulux offers you special decorative designs that take full advantage of the fluidity and translucence of glass . . . dramatize the natural beauty of the material.

It is impossible to catch in a photograph the full effect of these exclusive Insulux designs. May we suggest that you ask your Insulux distributor to show you samples of the decorative block pictured at the left. Owens-Illinois Glass Company, Insulux Division, Toledo, Ohio.

OWENS-ILLINOIS
INSULUX
*Glass Block*

THERE ARE PLACES IN EVERY BUILDING THAT *NEED* INSULUX

**Figure 24.1.** In the 1930s and 1940s glass blocks were used in both industrial and architectural applications. Appealing to designers was so important to Owens-Illinois that the company commissioned Walter Dorwin Teague to develop its No. 24 design. *Pencil Points*, April 1940.

DIETRICH NEUMANN; JERRY G. STOCKBRIDGE AND BRUCE S. KASKEL

# 24 | GLASS BLOCK

**DATES OF PRODUCTION**
1932 to the Present
**COMMON TRADE NAMES**
Insulux, PC Block

## HISTORY

Glass blocks, known historically as glass bricks or hollow glass tiles, are hollow assemblies manufactured by sealing, at high temperatures, two shallow rectangular cups along their open faces. Glass block structures are typically built using the techniques and materials of masonry construction. Partitions and curtain walls made of glass block are not load bearing, but blocks do have compressive strength ranging from 400 to 600 pounds per square inch. Glass blocks have excellent insulating properties and can reduce both thermal and sound transmission. They are available in a variety of patterns for directing or diffusing light. Although manufactured in several sizes since their introduction in the 1930s, glass blocks were typically either 8 or 12 inches square and 4 inches thick.

### ORIGINS AND DEVELOPMENT

Since the turn of the century the term *glass block* has frequently been used to describe different but related products and applications, including pavement lights, glass-concrete construction, prismatic glass, and hollow glass blocks. The terminology often differed from manufacturer to manufacturer and changed frequently.[1] A number of inventions and technical advances in the glass industry led to what is known today as hollow glass block.

The concept of hollow glass block as a building material is credited to the French engineer Gustave Falconnier, who in 1886 patented hexagonal or lozenge-shaped hollow glass blocks that were individually manufactured by being blown into a mold.[2] Falconnier's products were soon produced under patent agreements by several European manufacturers and exhibited at the 1893 World's Columbian Exposition in Chicago.[3] Despite problems with stability and condensation, these glass blocks enjoyed some success with young French architects at the turn of the century.[4] They were still on the market in the 1930s and occasionally even used in the United States.[5]

At the same time the search continued for more stable alternatives to Falconnier's products. In 1903 hollow, open glass bricks were introduced in Germany; these were manufactured by pressing the glass into a mold and remained open on the bottom. Although their stability was also insufficient, they were still available in Europe and the United States by the late 1920s.[6]

The use of heavy (sometimes also conical or prismatic) glass slabs for lighting dark interiors, basements, and cellars underneath sidewalks had been common in European and American urban centers throughout the nineteenth century. After 1890 these illuminating glasses were frequently applied in a reinforced concrete setting. In 1911 Friedrich Keppler, a German lighting engineer and head of the German Luxfer Company, developed a vertical version for partition walls and windows.[7] Solid square glass tiles, usually between 4 and 6 inches square, were cast in the section of a dumbbell. The edges contained the supporting reinforced concrete bars, covering them completely between their flanges and rendering them almost invisible because of internal reflection.[8] This system, as well as similar competing designs, enjoyed tremendous success during the rise of modern architecture in the 1920s, and was frequently used by such prominent architects as Walter Gropius and Le Corbusier.[9] Bruno Taut's famous glass pavilion at the 1914 Werkbund Exhibition in Cologne and Pierre Chareau's Maison de Verre (1932) in Paris were built with solid glass tiles in a concrete setting. To improve sound and heat insulation German producers had experimented from 1920 on with two layers of glass tiles using a tongue-and-groove system to fit the halves tightly onto each other and create a hollow space in between. Both parts would be embedded in concrete. Internal condensation remained a problem, however.

Several American glass producers experimented in the early 1930s with solid and partially evacuated blocks.[10] In 1929 the Structural Glass Corporation introduced several sealed hollow glass blocks, and the Owens-Illinois Glass Company introduced the first pressed glass block in 1932.

**Figure 24.2.** By pressing molten glass into molds, manufacturers were able to produce blocks with walls of uniform thickness and strength (top). The plunger creates the pattern on the interior face.

**Figure 24.3.** The Hecht Company Warehouse (1937, Abbott, Merkt), Washington, D.C., represented one of the earliest large-scale architectural applications of glass block (bottom). The sleekness of glass block made it a perfectly compatible material for Art Deco designs.

The Owens-Illinois block, a five-sided dish sealed with flat plate glass, was first used for a demonstration house at the Century of Progress Exposition in 1933. William Lescaze's own town house (1934) at 211 East 48th Street in New York City was equipped with solid square glass blocks manufactured by the MacBeth-Evans Glass Company.[11]

Advances in pressing techniques made by several manufacturers led to the development of functional hollow glass blocks. The decisive step in the new production process was the method by which the machinery pressed two identical concave halves together under heat and pressure, making glass blocks airtight. Machine-made hollow glass blocks had a higher degree of insulation and a more uniform look and were stronger and easier to install than their predecessors.

In 1935 Owens-Illinois introduced Insulux, the first widely used hollow glass block, which was sealed with lead. It was advertised and used for exterior windows and partition walls for factories, offices, and apartments. Owens-Illinois's initial rectangular No. 1 series had clearly been an attempt to match its blocks to brickwork. The blocks bonded with two courses of ordinary brickwork and could be laid in similar running bonds. Soon they were replaced by the Insulux 200, 300, and 400 series blocks, which were square, more economical, and available in nominal 6-, 8-, and 12-inch sizes respectively.

Challenged by Owens-Illinois's success, Pittsburgh Corning, an enterprise of the Corning Glass Works and the Pittsburgh Plate Glass Company, was formed in 1936 to develop a similar product.[12] Corning Glass Works had presented its Corning Steuben Block to the public in 1935, but this joint venture led to its perfection and mass production beginning in 1938.[13] This block, known as PC Block, was made from Corning's heat-resistant Pyrex, which reduced its expansion and contraction considerably. It was produced in nominal sizes of 6-, 8-, and 12-inch squares, all 4 inches thick.

In only a few years a new glass block aesthetic was firmly established. In 1938 *Architectural Forum* declared (on behalf of Pittsburgh Corning) that the glass age had finally arrived.[14] Two years later more than 20 million blocks had been sold. Never had "a new building product caught on so quickly."[15] After a decline in popularity in the late 1970s, which almost ended U.S. production, glass block is still produced in many of its original forms by the Pittsburgh Corning Company.

## MANUFACTURING PROCESS

The earliest glass blocks were either hand-blown or cast into molds.[16] To reduce the pressure on the corners, manufacturers designed hexagonal or lozenge-shaped forms that measured 8 by 5½ by 3⅞ inches. To create an airtight seal between the two halves, manufacturers initially used cementitious and solder-based sealants.[17]

The first step in the automated process for producing glass blocks is melting the glass in a tank furnace. The molten glass is then deposited in measured quantities into the pressing machine.[18] Once the appropriate amount of glass for half a block is released, it is cut and poured into a waiting mold on an automatic pressing machine. The pressing machine typically consists of a revolving table carrying twelve female molds. Each mold consists of two machined cast-iron parts: one establishes the size of the face, and the other determines the walls and thicknesses of the flange. Each mold is moved successively into position under a plunger equipped with a male mold machined to produce the configuration for the block's interior face. At each station air is blown into the mold and glass, cooling the glass. From the last station the block is moved to the sealing machine, where its two halves are sealed. The edges of the block are heated, brought together with only ¼ or ½ inch between them, allowing enough time for the air to escape, and fused. During the cooling process the air inside contracts, creating a vacuum.

The completed homogeneous piece of glass is then placed on a conveyor belt and passed through an annealing lehr (furnace). After inspection, early glass blocks were placed on revolving spindles as the side walls were covered with a resilient coating of synthetic resin and a fine spread of sharp-grained sand or marble dust.[19] Later, plastic coatings were used to improve bonding.

## USES AND METHODS OF INSTALLATION

The applications of glass block in the United States include industrial, commercial, and residential architecture. Glass blocks were laid by masons in similar fashion as bricks. Portland cement mortar was typically used, and walls required proper expansion joints at jambs and heads. For large areas of glass block, metal reinforcement was added to increase strength.

Owens-Illinois's Insulux block was advertised and used for exterior windows and partition walls in factories, offices, schools, and apartments. Among the many notable examples of new residential, institutional, and industrial structures featured by *Architectural Forum* in 1940 were Walter Gropius's home (1938) in Lincoln, Massachusetts, and the Museum of Modern Art (1939, Philip L. Goodwin and Edward Durell Stone) in New York City.[20]

Proponents of glass block also suggested that the material could be used to improve lighting conditions in the

**Figure 24.4a–b.** The dramatic plan of the Bruning Residence (1936, George Fred Keck), Wilmette, Ill., included a two-story staircase enclosure of glass block. Residential applications of glass block were numerous, ranging from sweeping curved walls to interior partitions to replacement for standard sash. The luminous quality of glass block, when lighted from behind, was both a selling point and an indication of the material's consummate modernity: it permitted light to enter but did not sacrifice privacy.

**Figure 24.5.** Replacement glass blocks may be difficult to find. Restoration of the Hecht Company Warehouse required 10,000 custom manufactured blocks.

workplace.[21] With its nonporous surface—a natural barrier to dirt, odors, grease, water, and air—glass block was also ideal for installation in areas requiring highly sanitary conditions.[22] Around 1939–40 glass block producers also adopted the principles of light refraction through prismatic ridges placed on the inside of both halves of the hollow glass blocks.[23] The light would be redirected up against the ceiling and then bounce back into the work spaces. Special no-glare blocks were introduced for the lower half of the installation. In the 1940s and 1950s many schools in the United States used glass blocks placed directly above the steel-framed windows. By 1958 ceramic-coated glass blocks in blue, green, yellow, and coral had been introduced.

## CONSERVATION

Glass block is a durable building material and generally weathers well. Cracked and fogged glass blocks may require replacement, depending on site conditions. Duplicating historic glass blocks, however, is expensive.

### DETERIORATION

In the mid-1930s glass blocks were produced by dipping the edges of the two halves into molten metal and then pressing them together to seal each unit.[24] Seal failure in blocks of this type is common. Once the seal breaks, moisture seeps into the block and minerals borne by the moisture cloud the inside faces.[25] Later manufacturing improvements allowed the two halves to be sealed by fusing the glass, eliminating the need for a metal seal.

Failure in glass block walls can also occur when the bond between the side walls and the mortar is inadequate. As a result, some early glass block panels did not have the structural integrity necessary to withstand exterior wind and climatic conditions, a disadvantage overcome by applying a sanded adhesive coating to the side walls during manufacture.[26] Further refinements in the 1960s incorporated corrugated sides with silica sand side walls to improve the bond. Today either paintlike or polymer coatings are often applied to increase bond adhesion.

As a non-load-bearing material, glass block is not intended to support the weight of walls, floors, or roofs. To prevent the blocks from carrying the weight of construction above and to accommodate thermal movement, glass block installations are typically isolated into small panels, which are subdivided with vertical and horizontal expansion joints. When a long, uninterrupted glass block wall panel lacks expansion or isolation joints, bowing, cracking, and crushing tend to occur.[27]

Perimeter joints and expansion joints rely on preformed metal edging that retains the glass blocks laterally while allowing differential movement in the wall plane. In the earliest applications joints between the block wall and the support structure were often caulked with oakum and sealed with mastic. Today these joints are typically sealed with backer rod and elastomeric sealants. If deteriorated, these materials should be replaced with contemporary sealant materials.

### DIAGNOSTICS AND CONDITION ASSESSMENT

To retain as many glass blocks as possible in a building, the condition of each block and the surrounding details should be inspected at close range. Bowing and cracking may be a sign of inadequate expansion joints or differential settlement.

Glass blocks in historic buildings may be fogged internally yet provide satisfactory performance. Blocks with lead seals lose their airtightness over time. Small chips and isolated cracks are common. Replacement should be confined to areas where blocks have lost significant structural strength, have cracks that have separated or a shard that might fall, or have chips or cracks that allow moisture to enter. Internal staining is normally the best indication of whether water is infiltrating a glass block.

### CONSERVATION TECHNIQUES

Typical maintenance of glass blocks should include repointing when joints are deteriorated or missing. Glass blocks may also have small chips or minor cracks. Such aging effects, like slight fogging, normally do not adversely affect the blocks' structural performance and are not generally grounds for replacement of the historic fabric. Blocks can usually be retained unless cracks have been judged to make them structurally unsound.

Cleaning should be undertaken after selective replacement of damaged blocks. On small walls and partitions, glass block can be cleaned with water and a mild detergent and wiped clean with a cotton cloth. Stubborn dirt and graffiti can be removed with solvents appropriate for use with glass. On larger areas a mist water spray can be used. Cotton cloths used for wiping should be washed out regularly to avoid scratching the block.

In walls that are experiencing bowing and extensive cracking, it may be necessary to incorporate expansion joints in areas that require replacement of blocks. These can be provided at panel jambs and heads.

## REPLACEMENT

Finding suitable replacements is often the most difficult problem in restoring glass block. It is very difficult to get a good match between old and new blocks because the old blocks may have a pattern that is no longer produced. Today only one manufacturer in the United States, Pittsburgh Corning, produces glass block, although it is imported from a number of manufacturers in Germany, Holland, and Japan. Custom dies can be cast for obsolete patterns, but these may be prohibitively expensive for all but the largest projects.

Even though it may be possible to find a source of old used glass blocks that match a pattern or to reuse blocks from other locations on the building, many building codes prohibit the use of salvaged glass blocks. Damage that occurs to the side wall material during removal and preparation for reuse can affect the new bond. Also, chips and other defects in old used glass blocks make them more prone to detrimental cracking after reuse.

The type and quality of mortar used with glass blocks ultimately affect the durability of the construction. Therefore, masonry construction for glass block should follow the same general guidelines as for other masonry construction. Type S (portland cement and lime) mortar is commonly used, mixed stiffer and dryer than for ordinary masonry. Any mortar for repointing or replacement work should match the original mortar in color but should not be stronger than the original mix.[28] Early glass blocks were sometimes laid with a mix of 1 part masonry cement to 2½ parts sand.

Conventional ladder-type joint reinforcement is usually used in glass block panels and is installed every 16 inches on-center vertically. Embedded metal panel anchors are often used to secure the glass block panel to adjoining construction while allowing for differential movement. While contemporary reinforcement products are not exact duplicates of their historic counterparts, they serve the same purpose and are suitable for replacement work and for new walls.

**Figure 25.1.** Structural glass was widely used during the 1930s for Moderne shop fronts, service stations, movie theaters, and other buildings. These designs show the range of decorative effects possible with this modern building material. Vitrolite catalogue, Libbey-Owens-Ford Company, 1937.

CAROL J. DYSON

# 25 | STRUCTURAL GLASS

**DATES OF PRODUCTION**
c. 1900 to c. 1963

**COMMON TRADE NAMES**
Argentine, Carrara, Glastone, Marbrunite, Novus, Nuralite, Opalite, Sani-Onyx, Vitrolite

## HISTORY

*Structural glass* is the general term for a variety of architectural glasses but usually refers to colored opaque glass slabs.[1] Structural glass is fused at high temperatures, rolled into slab form, slowly annealed, and mechanically polished. Its properties include resistance to abrasion and warping and an impervious surface. Sold in black, white, and a variety of colors and finishes, structural glass can be bent, carved, laminated, inlaid, and sandblasted to create patterns.[2]

### ORIGINS AND DEVELOPMENT

Opal glasses are a category of historic glass that range from semitranslucent to opaque. They are composed of small, clear particles held in a clear matrix. The particles reflect and scatter light so that transparency is reduced. Traditionally this translucent opal glass has been used for illuminating glassware.[3]

Colored opal glass was first developed in ancient Egypt. The ancient Romans used a translucent glass that imitated marble. By the sixteenth century the technology had resurfaced, and Venetian artisans were producing a semitranslucent glass by adding fluorides such as cryolite to the matrix. The Chinese also added cryolite to glass to produce an imitation porcelain.[4]

In the last decades of the nineteenth century the development of the regenerative furnace and discovery of natural gas reserves in Pennsylvania, West Virginia, Oklahoma, Arkansas, Texas, and Missouri led to rapid expansion of U.S. domestic flat glass production. This brought varied innovations in flat glass technology and products in the early twentieth century.[5]

Opaque structural glass slabs were first developed about 1900 as a sanitary alternative to white marble slabs for wainscoting and table surfaces. The first structural glass, Sani-Onyx, was created by the Marietta Manufacturing Company of Indianapolis, Indiana.[6] By 1906 the Pittsburgh Plate Glass Company was producing Carrara glass in white and black, and the Penn-American Plate Glass Company, also based in Pittsburgh, had begun production of Novus Sanitary Structural Glass.[7]

Eventually at least eight American firms made structural glass, but the two products that dominated the market were Pittsburgh Plate Glass's Carrara glass and Libbey-Owens-Ford's Vitrolite, first produced about 1916.[8] By 1929 U.S. production of opaque structural glass exceeded 5 million square feet.[9] As demand increased, new colors and finishes were offered.

By the early 1950s the market for structural glass was only a fraction of what it had been in the 1940s. Changing design tastes and competition from other materials, such as porcelain enamel, contributed to its declining use for storefronts. Advertisements during the 1950s again emphasized use in utilitarian interior spaces such as kitchens and bathrooms.[10]

### MANUFACTURING PROCESS

Opaque structural glass was composed of silica, feldspar, fluorspar, china clay, cryolite, manganese, and other materials vitrified with intense heat (about 3,000 degrees Fahrenheit) in pots or tanks.[11] Sheets were cast and then rolled to the desired thickness, much like plate glass.[12] The glass was annealed (cooled) much more slowly than modern plate glass, taking from three to five days.

Once annealed, the glass finish was fire polished.[13] Some applications made use of this soft finish without further polishing. To achieve a more glossy finished glass, the surface of the slabs was mechanically ground with fine sand and rollers and then polished to a mirrorlike finish with felt blocks and rouge (ferric oxide powder).[14] After polishing, the slabs were cut to size. Normally the material was cut and drilled and the edges finished in the factory.

The addition of fluorides produced the glass's opacity. Annealing caused the fluorides to precipitate, creating a dense mass of particles suspended in the clear matrix. The fluoride particles scattered, reflected, and trapped light until the glass was semitranslucent or completely

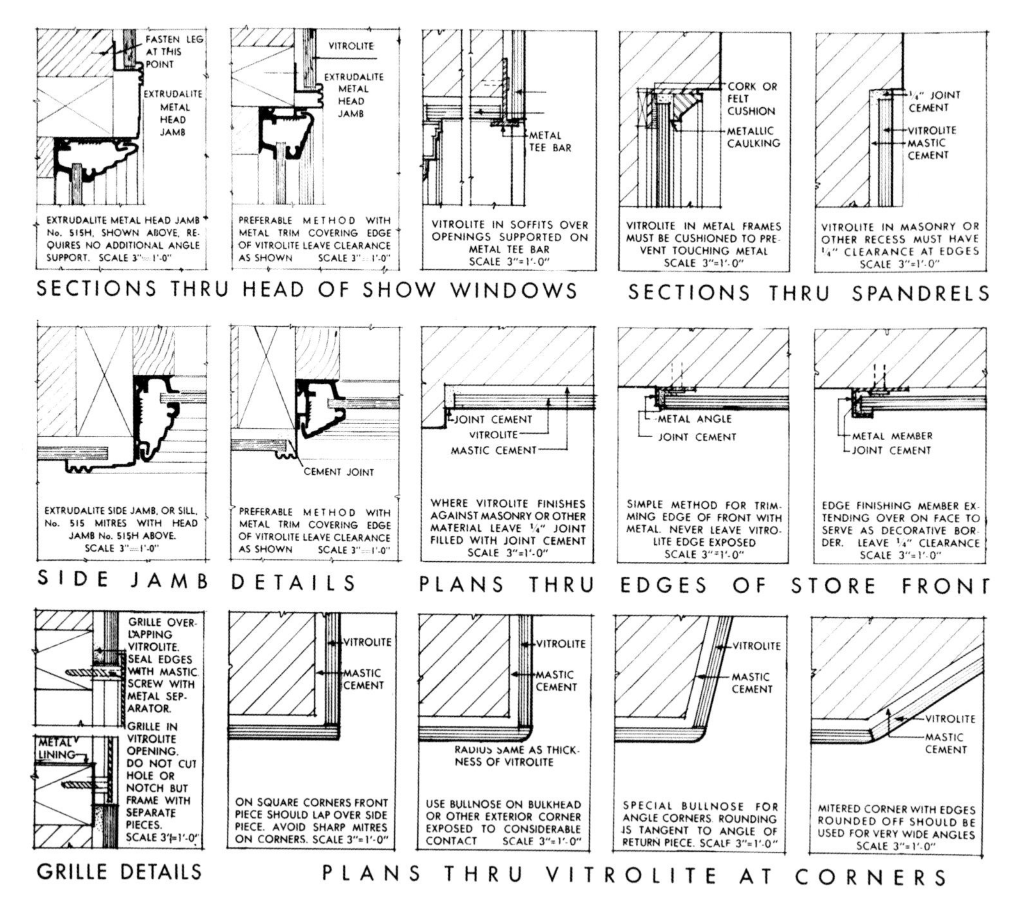


**Figure 25.2.** A brittle material, structural glass had to be properly installed to prevent breakage. These cross sections show typical late-1930s installation of such storefront details as window heads and jambs, spandrels, and corners. Edges were always separated by joint cement, cork tape, felt, or caulking.

opaque. Coloring oxides were added to the matrix before firing.[15]

Although the Penn-American Plate Glass Company produced its Novus Sanitary Structural Glass in several colors as early as 1906, the majority of structural glass was made only in shades of white and black until the late 1920s. In the 1930s it was produced by various manufacturers in a range of colors, and by the late 1930s it was available in more than thirty colors ranging from pastels, jewel tones, and solids to striated agate and dendritic patterns.[16]

In early applications the softer, fire-polished and satin finishes predominated. By the 1930s, however, glossy, colorful, mirrorlike finishes had become popular. In the late 1950s and early 1960s Pittsburgh Plate Glass marketed Carrara as spandrel glass, but its use in curtain wall systems was not widespread.[17]

## USES AND METHODS OF INSTALLATION

When first developed around 1900, structural glass was compared to statuary marble in appearance and marketed for its sanitary qualities. The fact that the glass was nonporous, homogeneous, and noncrazing and could be produced in large sheets made it more desirable than marble or tile for aseptic conditions such as in hospitals.[18]

During the first two decades of the twentieth century, structural glass was used principally in utilitarian locations requiring durable, nonstaining, and easily cleaned and maintained slab materials for wainscoting, flooring, refrigerator linings, lavatories, tabletops and countertops, bank coupon desks, and electrical switchboards. Because structural glass diffused light, it was used extensively for corridors, operating rooms, and laboratories.[19] In the early twentieth century when structural glass was used for storefronts, it was substituted for stone in bulkheads and dados.

Structural glass reached its zenith as a building material with the advent of new design aesthetics, including Art Deco, Art Moderne, and Modernism. Its variety of colors and versatility made it very popular during the 1930s and 1940s. The material could be bent, carved, inlaid, sand-blasted, and even factory painted with gold and silver. It was installed in sleek Moderne office-building lobbies, movie theaters, restaurants, confectioners, residential kitchens, and bathroom interiors.[20]

Structural glass also proved to be an ideal material for modernizing the exteriors of commercial buildings and was marketed extensively for this purpose.[21] New construction for storefronts, movie theaters, gas stations, and automobile dealerships was clad in gleaming structural glass set in aluminum framing. Opaque structural glass was no longer viewed as a substitute for stone; it was extremely popular in its own right.

Storefront panels were coordinated with aluminum glazing systems because the lustrous finishes were compatible with the white metal. To complement its Carrara structural glass, Pittsburgh Plate Glass produced Pittco-Carrara Glass Store Fronts, with metal window sash that overlapped the Carrara glass, protecting the edges. To promote the use of Vitrolite in new construction, Libbey-Owens-Ford offered a prefabricated Vitrolite-faced concrete masonry unit called Glastone.[22]

Structural glass was available in thicknesses ranging from ¼ inch to 1¼ inches. The panel sizes were determined by use. On exteriors the maximum size was 6 square feet if the panel was to be installed 15 feet or more above the sidewalk, and 10 square feet if installed less than 15 feet above it. Interior wall panels up to 15 square feet were available. Toilet partition panels could be produced up to 25 square feet and were created by laminating together two ⅞-inch slabs with bituminous adhesives.[23]

Structural glass's versatility was due partly to its tolerance of various substrates. It could be readily applied to most flat surfaces, including plaster on metal lath, concrete, or masonry (wood substrates, however, were discouraged because of their absorptive nature and potential to warp). The backing surface was prepared and sealed with a bonding coat supplied or approved by the glass manufacturer, and mechanical fasteners (nonferrous metal brackets, angles, or channels) were secured to the substrate. The panels were prefabricated to specifications and attached with an asphaltic mastic.

Mastic was applied to the back of the glass in 3-inch daubs covering 50 percent of the back of the panel. The

**Figure 25.3.** The Tic Toc Cafe (1941, Furbringer and Ehrman), Louisville, Ky., used Carrara glass, a popular type of structural glass commonly selected as a material for "modernizing" jobs (top).

**Figure 25.4.** The Club Moderne (1937, Frank Wallus), Anaconda, Mont.—with its polished mirror finish, curved glass corner, and bands of black, beige, and light blue structural glass—is a classic of streamlined design (bottom).

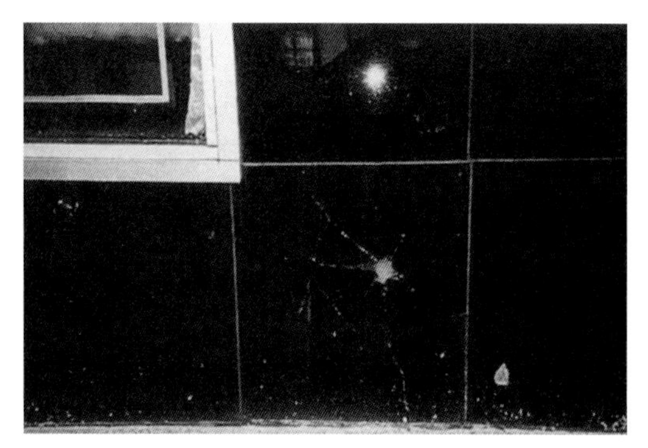

**Figure 25.5.** The most common causes of structural glass breakage include accidental impact and vandalism. This street-level panel requires replacement.

**Figure 25.6.** In reinstalling or replacing structural glass, asphaltic mastic is daubed on the back of the panels, which are eased into position against the substrate.

glass was set in position by rocking the panel until the flattened mastic was forced into the backup surface, providing a keying action. When the cement was set, the joints were pointed with a pointing cement, which, like the mastic, was provided by the glass manufacturer. Panel edges could be protected with cork tape 1⁄16 inch thick, which was set back ⅛ inch from the front of the glass.[24] In locations where high moisture was expected, such as tub surrounds, and the backup substrate was masonry, the panels were sometimes attached with cement rather than mastic. For ceiling applications, small screws with felt washers were used in addition to the larger daubs of mastic.[25]

## CONSERVATION

Although structural glass is a fairly durable material it is vulnerable to breakage. Conservation of existing glass is important because it is no longer produced domestically.

### DETERIORATION

Structural glass does not warp, craze, fade, or stain easily and is not affected by most acids. Impact, however, causes the glass to break. Stresses caused by thermal expansion and contraction can lead to glass movement and breakage. Where structural alterations have been made, cracks, holes, and chips may be evident.

Another common form of deterioration is hardening of the mastic adhesive. When the mastic becomes brittle over time, movement can occur between the glass and substrate and can compromise the assembly's structural integrity.[26] Because of the narrow joints, such movement can cause stresses in adjacent panels and breakage if the slab detaches and falls. Water infiltration also damages the mastic, accelerating the hardening process.

Pointing cements also gradually deteriorate. Weathering and other stresses can damage joint cements. The darker shades of structural glass absorb a significant amount of heat, causing the panels and walls to be subjected to greater thermal stress. Although structural glass is often heat tempered, the joints on dark facades are exposed to more thermal expansion and contraction.

### DIAGNOSTICS AND CONDITION ASSESSMENT

Most failures in structural glass systems are readily obvious: panels are visibly cracked, damaged, missing, out of alignment, or delaminating; joints are deteriorated; or water intrusion is evident. Because many exterior and interior installations are less than 15 feet above the ground, the panels can be easily reached. Gently push on a panel to see if it is still securely adhered to the substrate. If a wall has been subjected to severe water damage, then removal of selected panels may be necessary to determine the stability of the mastic and the substrate.

### CONSERVATION TECHNIQUES

Because opaque structural glass is no longer produced in the United States, it should be repaired whenever possible. Maintenance is straightforward. The glass can be cleaned with a mixture of water and ammonia or detergent. Joints can be repaired with traditional joint cement (with an integrated watertight surface), latex caulking, or glazing compound.[27] Silicone sealants are harder to control with fine joints.[28] Traditionally joint cement was colored to match the glass. New materials should also be tinted with pigments compatible with the historic joint patching material.[29]

Minor hairline cracks can be filled with caulking tinted to match the glass. One method for repairing chips and holes is to fill the defect with polyester resin adhesive tinted to match the glass. The surface can then be polished with fine sandpaper and buffed with polish.[30]

Another method is to fill the hole with glazing compound and paint the area with paint color matched by computer.[31]

Removal and reinstallation of structural glass panels that are detaching or misaligned are difficult because of the gradual hardening of the mastics and the fine joints between panels. No method is immune to glass breakage. Solvents can be injected behind the glass to soften the mastic, and then piano wire can be slipped behind the panels to cut through the mastic.[32] Another method is to direct steam at the face of the panel for approximately 10 minutes to soften the mastic. This method limits breakage but is time-consuming. Edge joints should be cut to prevent "oystering" of the glass edge. The panels can then be pried off, cut off with piano wire, or sawn off. When prying off glass panels, a block of wood should be used to protect the face of the glass from the crowbar or nail puller.[33]

Once cleaned, slabs can be reattached with a traditional hot-cup asphaltic mastic in daubs after the substrate is coated with an asphalt-based primer. The integrity of any existing shelf angles should be evaluated and supplemented as necessary. Horizontal edges should be protected with recessed cork tape. The joints can be pointed with joint cement, caulk, or glazing.

#### REPLACEMENT

When pieces are broken, severely damaged, or missing, finding an appropriate replacement is a problem. Occasionally glass shops, glass repair specialists, and architectural salvagers have stock left in warehouses, but finding the correct color or number of slabs is difficult.

One kiln in Czechoslovakia still produces black, white, beige, and mint green structural glass. It is distributed in the United States by Floral Glass and Mirror of Hauppauge, New York, but sizes and finishes are limited. The panels are sized metrically and are approximately ¼ inch thick. Differences in panel thickness can be adjusted with the mastic and mechanical fasteners, and metric panels can be cut to fit the American dimensions. In Japan NEG Industries manufactures NeoClad, an opaque colored glass available in white, beige, and gray. Asahi Corporation produces an opalescent, nearly opaque structural glass in white and light gray. As with the Czechoslovakian glass, limited colors, metric sizes, and the cost of shipping to the United States make matching the size, finish, reflectivity, and color of domestic glass problematic.

Various materials have been used or proposed for use, including spandrel glass, back-painted flat glass, Lexan, laminated glass, and solid surfacing. None, however, perfectly replicates the qualities of the historic material.[34]

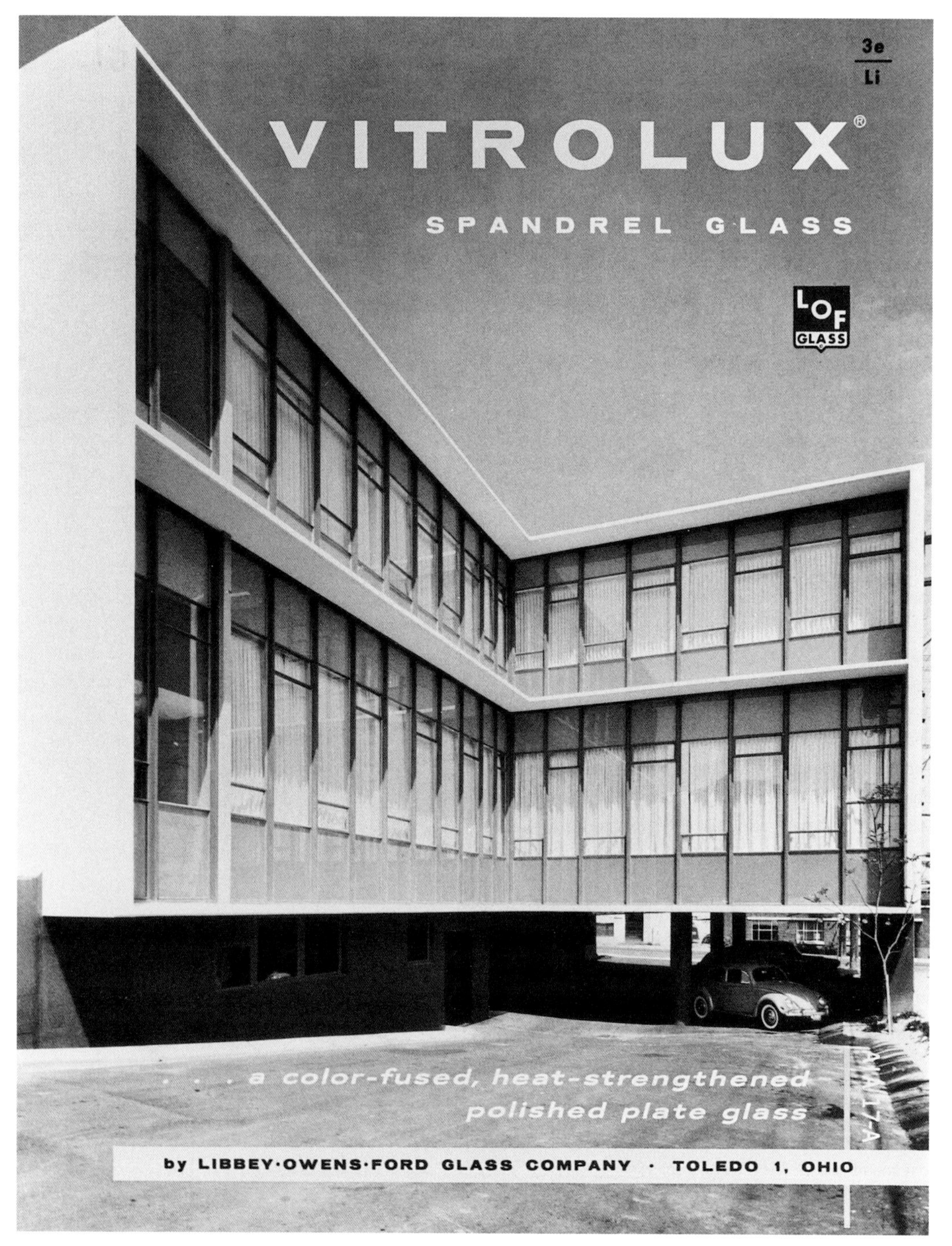


**Figure 26.1.** Vitrolux spandrel glass, Libbey-Owens-Ford's color-fused tempered plate glass, became popular for curtain wall construction in the late 1950s. It was used in the AAA Building (1960, John Graham and Company), Seattle, Wash. Vitrolux catalogue, Libbey-Owens-Ford Company, 1960.

ROBERT W. MCKINLEY

# 26 | SPANDREL GLASS

**DATES OF PRODUCTION**
1955 to the Present
**COMMON TRADE NAMES**
Spandrelite, Vitrolux

## HISTORY

Spandrel glass usually covers the space above and below horizontal strip windows where knee walls and spandrel beams are located. In the late 1950s *spandrel glass* referred specifically to ceramic-coated plate glass, but today the term is used broadly to include many types of transparent flat glasses used for spandrels.[1] Opacifiers of organic composition have also been applied to the transparent substrates since the early 1950s to create a remarkably varied palette for designers.

### ORIGINS AND DEVELOPMENT

Between 1880 and 1950 various translucent and opaque glasses were manufactured, mainly by the pot plate process, for glazing and facades. Because of their cost and limited color potential, these glasses were used only rarely in spandrels.[2] As architects searched for ever larger and flatter yet stronger, lighter, more durable, and less costly building materials, the potential of glass as a replacement for masonry and other materials became apparent.

One of the earliest glasses to be incorporated for wall spandrels was tempered structural glass. Even though structural glass was colored and opaque, its use for this purpose was limited.

The building boom following World War II encouraged innovation in construction methods, materials, and design. Manufacturers of ceramic enamel frits (fine powders) for bottle decoration and labels and for porcelain enamel panels were well established. Producing frits appropriate for spandrel glass required significant but relatively simple changes in ingredient mix and vehicle.

In 1955 Pittsburgh Plate Glass introduced Spandrelite, the first ceramic-coated glass.[3] Its popularity grew rapidly, and larger sizes and new colors were introduced in response to architectural demand. By 1958 spandrel glass was available in sizes as large as 5 by 7 feet, and manufacturers soon found it practical to provide an almost unlimited palette of custom colors and screen-applied patterns. National promotional campaigns were undertaken by Pittsburgh Plate Glass and Libbey-Owens-Ford, which called its ceramic-coated spandrel glass Vitrolux.[4]

The market for spandrel glass grew along with that for tempered transparent glass. Many small manufacturers invested in heat-strengthening and tempering furnaces to satisfy the growing demand for tempered glass patio doors. As a result, they found it convenient to offer ceramic-coated spandrel glass and simply purchased substrate glass from Pittsburgh Plate Glass and Libbey-Owens-Ford, among others. Since substrates and frits came from a limited number of sources, competing spandrel glass products differed primarily as a result of furnace design, performance, and firing practices.

### MANUFACTURING PROCESS

Spandrel glass manufacture begins with the production of plate glass or float glass. Although plain tempered glass could be used as spandrel glass, spandrel glasses are often ceramic coated.

Ceramic frits are made from glass and various colorants. The glass is quenched (cooled quickly) to break it into small fragments for grinding with the pigments necessary to produce a particular color and adhere to the substrate light when fired. The frit is mixed with a liquid vehicle so that a very thin coating can be applied to the substrate glass by brush, roller, spray, or silk screen after the substrate light is cut to the desired job size, ready for heat strengthening or tempering.

Once coated on one side with frit, the glass is fired in a special furnace. In vertical furnaces, sharp metal tongs are used to suspend each light. In horizontal furnaces, the glass is supported on horizontal rollers. Furnace temperatures of 1,200 to 1,500 degrees Fahrenheit are required to melt and bind the thin frit coating to the glass substrate.[5] To temper the glass, the surfaces of each light are quickly cooled by a blast of air.[6]

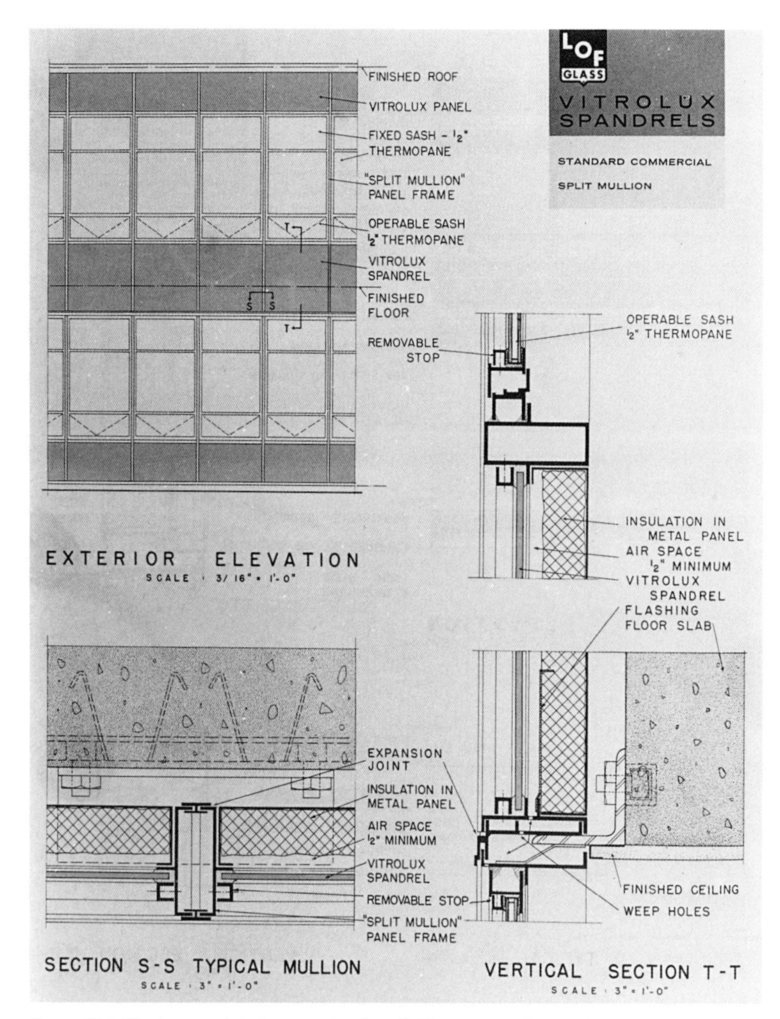


**Figure 26.2.** Vitrolux spandrel glass was often installed in curtain wall assemblies. Typical vertical and horizontal sections show how spandrel glass was set between mullions in glazing rabbets with a building sealant or gasket. Adequate clearances were needed to accommodate expansion and forces.

## USES AND METHODS OF INSTALLATION

Spandrel glasses with transparent substrates, which were popular components of postwar design, helped many commercial buildings become known as glass boxes. Lever House (1952, Skidmore, Owings and Merrill) on Fifth Avenue in New York City is one of the best known. The back surface of its original wired glass spandrels had an opaque coating with a green tone. For the Museum of Modern Art annex (1951, Philip Johnson), the Pittsburgh Plate Glass Company produced ceramic-coated gray spandrels.[7] Based on this installation, the company further developed its ceramic-coated glass and introduced Spandrelite commercially in 1955.

Design flexibility was and continues to be one of the most important characteristics of spandrel glass. It can be made to match the adjacent window and indoor shading, lighting, and color, or to contrast strongly with these elements. Silk screen and other frit pattern and texture application techniques were adapted for the production line, thus making the product more economical. While these characteristics made spandrel glass enormously popular for large office buildings, it was also used for storefronts, shopping centers, schools, motels, and hotels.

Like adjacent transparent, tinted, and reflective glass windows, ¼-inch spandrel glass usually was glazed in glazing rabbets ½ to 1 inch wide and ¾ inch deep.[8] Setting blocks at sills and centering blocks indoors and at heads and jambs provided structural support and helped sealants remain weatherproof.

## CONSERVATION

Properly designed and maintained, spandrel glass is a reliable building material. However, it is subject to various types of deterioration. Broken glass usually requires replacement to match unbroken lights.

### DETERIORATION

Solar ultraviolet radiation is the cause of long-term but subtle change in spandrel glass—the color of its transparent substrate sheet, plate, or float glass. This change occurs so slowly over the years that it may not be evident unless a replacement for other reasons becomes necessary.

Dirt typically accumulates in relation to the magnitude and direction of prevailing wind. Near-neutral pH dirt will not damage spandrel glass. However, alkaline solutions and hydrofluoric acid can permanently etch or frost exposed surfaces.[9] The most common source of both is nearby concrete or mortar.

Some tapes, paints, soaps, crayons, and chalks, often applied to spandrel glass to mark glazed openings during construction, may protect the surface and cause visible variations when removed. Ceramic enameled spandrel glass is not subject to biological deterioration.

Since spandrel glass is heat strengthened when the frit is fired, its size and shape may not be altered in any way after firing. Such operations as drilling holes or cutting, nipping, filing, sandblasting, or grinding surfaces and edges may weaken a light sufficiently to allow breakage by a relatively small mechanical or thermal stress.

Breakage, one of the most common forms of deterioration, can be caused by bullets, vandalism, or hot metal splattered by welders. Spandrel glass can be abraded by sandblasting or grinding adjacent materials during or after construction. It may be pitted by wind-driven sand or broken by roof gravel. Insensitive repairs can also damage spandrel glass.[10]

Like all materials, spandrel glass absorbs and releases heat continuously by convection, conduction, and radiation.[11] Resultant temperature differences within spandrel glass and at interfaces can cause thermal stresses in excess of 5,000 pounds per square inch that may cause breakage of annealed glass.[12] Extremely severe edge damage that penetrates the central tension core of either a heat-strengthened or a tempered light of spandrel glass may cause it to break.[13]

A rare form of glass failure is cracking caused by nickel sulfide stones. Manufacturing specifications for all glass permit the inclusion of many types of stones.[14] Unfortunately, nickel sulfide changes its state and increases in size when exposed long enough to sufficiently high tem-

**Figure 26.3.** For the Rainey Rogers Building (1951, Philip Johnson), an annex to the Museum of Modern Art, New York City, the Pittsburgh Plate Glass Company developed an opaque, ceramic-coated spandrel glass that led to the commercial production of Spandrelite.

**Figure 26.4.** A two-color scheme, featuring red and charcoal spandrel glass, was used for the curtain wall of the Electricians Union Building (1956, Joseph McCarthy), San Francisco. The spandrel glass for this modern office building was Spandrelite, produced by the Pittsburgh Plate Glass Company.

peratures.[15] When nickel sulfide stones cook only partially during the tempering process, thermal changes after the glass has been installed can complete the cooking process and, in rare instances, cause a tempered panel to disintegrate.[16] Fortunately, heat-strengthened spandrels are not subject to this type of breakage.

### DIAGNOSTICS AND CONDITION ASSESSMENT

Careful visual examination of broken glass may reveal areas requiring maintenance and the cause of breakage, particularly if broken pieces remain in the opening or can be reassembled and taped to permit examination. When inspecting spandrel glass, a survey system must first be developed. Each light must be examined, opening by opening and floor by floor.[17] Detailed information may be obtained with the aid of binoculars or a higher-powered spotting scope after cleaning. Inspecting glass from the ground or from adjacent buildings or rooftops may be appropriate and is less expensive than working from swing stages or scaffolding.

Observe whether weep holes are functioning and determine the condition of rabbet sealants and gaskets. Also, any cracks should be carefully assessed.

A crack pattern delineating innumerable small pieces (measuring from ½ by ½ inch to 2 by 2 inches) indicates that the light has been tempered. Cracks that appear to originate at a cat's-eye with little displacement, in or out, could be caused by a change of state of a nickel sulfide stone. If a black speck can be found in the crack between the two sides of the cat's-eye, save it for laboratory analysis. Using a mass spectrometer, the nickel sulfide composition of the stone can be confirmed.

A similar break pattern might originate at a bullet hole or at the point of impact of a hard-point missile. Displacement or fallout may occur in either case.

If the crack pattern delineates only five to fifteen large, irregular pieces of glass and most are retained in the opening without displacement, the light has been heat strengthened. Under the proper lighting conditions, smaller lines can be seen on the cracked edge surface of each piece. They appear as ripples emanating from the break origin.[18]

### CONSERVATION TECHNIQUES

Spandrel glass systems require regular inspection and maintenance. Glazing, framing, flashing, and weep and vent systems will require regular maintenance and cleaning. More meticulous inspections should be made every five years. In urban and industrial settings, more frequent cleaning will be required.

**Figure 26.5.** Finding spandrel glass to match historic colors is difficult. When an inappropriate replacement is installed (as in the left lower panel), the visual flaws are immediately apparent and compromise the building's character.

Because the costs of site access and rehabilitation work are high, careful planning, design, and execution are important. To develop an appropriate repair program, the original shop drawings and maintenance records should be studied. Detailed repair proposals, following a condition survey, may involve repairing identified deterioration, covering up identified deterioration with the addition of a new layer over the existing wall, and removing and replacing severely deteriorated glass with similar new glass that matches the historic glass as closely as possible. When adding a new layer, it may be possible to attach new spandrel glass glazed off-site in structural gaskets over the old with screw-applied retaining clips if the old wall can carry an additional 4 pounds per square foot.

Spandrel glass can be cleaned with high-pressure water spray to remove particulates. Surface residues, such as fingerprints, interleaving paper scum, and glass-cutting oils, may be removed with a number of detergents. Adhesives from tapes and labels, heavy grease, and tar deposits can usually be removed with mineral spirits or isopropyl alcohol. Because reaction contaminants such as hard water salts adhere very tightly if permitted to dry, it may be necessary to remove them by undercutting with an acidic chemical. Manual cleaning can also be undertaken with a low-foaming detergent, rinsing the glass thoroughly with water and removing all remaining contaminants with a lint-free cloth.[19]

Surface corrosion may look somewhat like deposited dirt. If it is very slight and hard to see, as in its initial stages, it might be removed with an extremely careful application of fine cerium oxide abrasive in demineralized water.

Leaking walls can be resealed after rabbet surfaces are carefully cleaned. New sealants and gaskets should be selected not only for their durability but also to match as closely as possible the visual appearance of the original sealant system. Structural glazing gaskets are easier to manage than sealants and require less sealing surface cleanliness for good performance. When properly applied to adequately cleaned and primed surfaces, silicone sealants can last longer than other types. Dark gray or black sealants are usually more physically stable.

## REPLACEMENT

Obtaining identical replacements for spandrel glass is difficult and expensive. Manufacturers should be provided with professional photographs of the original glass and a representative sample, which they can examine, test, and analyze. Often manufacturers are asked to supply and glaze a proposed production sample in the old wall for direct comparison and color photography.[20] Each opening must be measured because settling and the adjustment of expansion joints may mean that a different size and shape of light will be required for each opening. Before the glass is reglazed, glazing rabbets must be cleaned thoroughly, weep and vent holes must be cleared, and flashing joints must be resealed.

# PART VI
# FLOORING

LINOLEUM

RUBBER TILE

CORK TILE

TERRAZZO

VINYL TILE

**Figure 27.1.** Advertisements for linoleum claimed that it was sanitary and easy to maintain. Advertising felt-based products as linoleum, a practice of some floor product manufacturers during this period, was illegal. *Ladies' Home Journal,* March 1924.

BONNIE WEHLE PARKS SNYDER

# 27 | LINOLEUM

**DATES OF PRODUCTION**
1864 to c. 1974

## HISTORY

Linoleum is a resilient sheet flooring, historically available in several thicknesses and countless patterns, which was acclaimed in its heyday for its water-resistant, sanitary, and heat- and sound-insulating qualities. Frederick Walton invented linoleum in 1863, naming it for the Latin terms for its chief components, flax (*linum*) and oil (*oleum*). Made of oxidized linseed oil and ground cork or wood flour calendered (pressed with heated rollers) into sheets with a burlap (originally linen) backing, it was the floor covering of choice for numerous residential and commercial uses from the 1860s through the 1940s.

### ORIGINS AND DEVELOPMENT

Walton noticed the leathery skin that developed on paint left in the can too long and realized that this oxidized linseed oil had the potential to contribute to the manufacture of waterproof products. He patented a process for manufacturing linoxyn (oxidized linseed oil) in 1860. In 1863 he received a patent for "Improvement of the Manufacture of a Wax Cloth for Floors," the product that eventually led to the development of linoleum.[1] The ingredients of the imported linoleum products today are still similar to those specified by Walton in 1863.

Walton began the manufacture of linoleum under the name Walton, Taylor and Company in a factory at Staines, England, in 1864. Within the year, the company's name was changed to the Linoleum Manufacturing Company. In 1872 Walton and the firm of Joseph Wild and Company began the American Linoleum Manufacturing Company on Staten Island.[2] In 1886 Michael Nairn and Company, of Kirkcaldy, Scotland, which had begun operation as a floorcloth factory, established an American branch of its linoleum business in Kearny, New Jersey. The company subsequently became associated with the Dominion Oilcloth and Linoleum Company in Montreal and the Congoleum Company in the United States.[3] Also in 1886 the George W. Blabon Company of Philadelphia installed the first American-built linoleum calender in its Nicetown, Pennsylvania, plant.[4] Armstrong began linoleum manufacture in 1908 as an offshoot of its cork stopper business.

During its early years, the linoleum industry was larger in Europe, especially in Great Britain, and the product's acceptance was higher there than in the United States. At about the time Armstrong entered the market, there was more linoleum capacity than demand in this country. Seven companies were producing linoleum in the United States, and English and German manufacturers were actively competing in the U.S. market.[5]

The real market success was due primarily to Armstrong's advertising and marketing techniques. Starting in 1913 with advertisements in trade papers and sample books and explanatory literature for suppliers, Armstrong began an aggressive program for marketing linoleum. In 1917 the company committed to an advertising and marketing campaign unprecedented in the industry. In addition to the advertising literature describing the material's many advantages, this campaign included training a sales force and developing new designs. Whereas linoleum's first uses were commercial and utilitarian—in residences, primarily in bathrooms and kitchens, where sanitation was important—the new designs were to make it appropriate for use in living rooms and dining rooms as well. Armstrong hired Frank Parsons, a leading authority on home decoration, to produce a decorating book showing how linoleum could be used in all rooms of a house.[6]

Linoleum's sanitary qualities, its resiliency, and its sound- and heat-insulating qualities became its selling points. However, disadvantages were inherent in its advantages. Its resiliency and insulating qualities were due to ground cork (or cork dust) and wood flour. But while cork was the better ingredient for these purposes, it was dark and necessitated the addition of large amounts of whiting or white pigment to counteract its darkness when lighter shades were desired. This diminished the effect of the cork. Wood flour was lighter in color but was not as resilient as cork. Thus natural- and brown-hued linoleum,

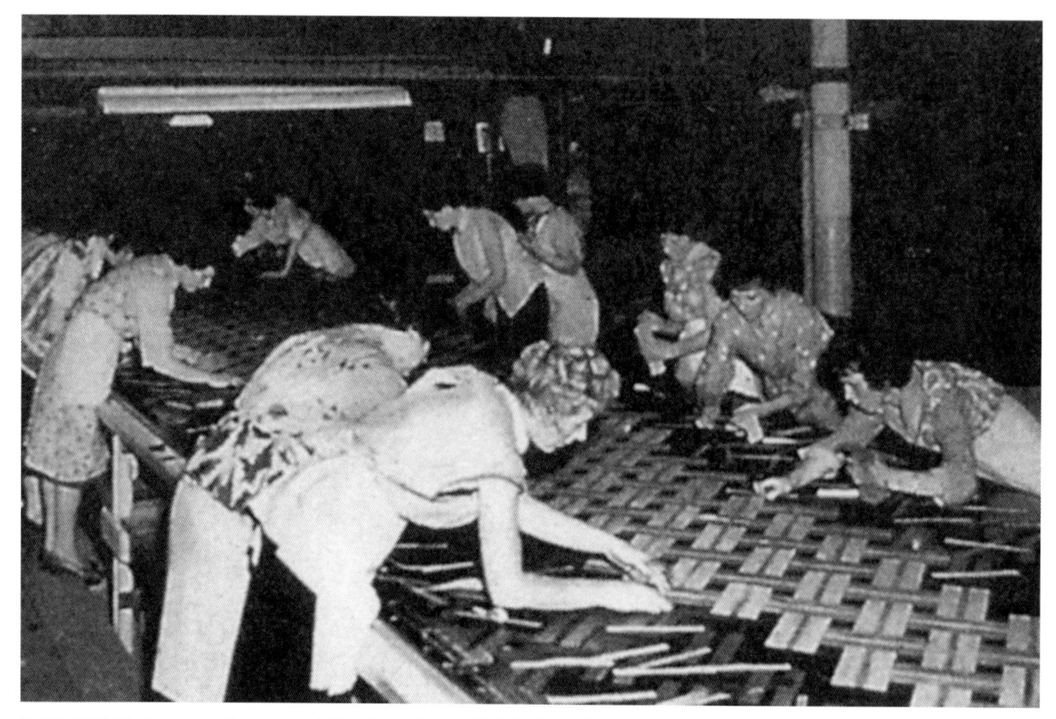

**Figure 27.2.** Workers piece (hand assemble) a design for straight-line inlaid linoleum. After assembly, the pieces—cut from solid linoleum sheets—were calendered onto a burlap backing.

which required the least pigment, was the most resilient.[7] Likewise, linseed oil, which made linoleum water resistant and therefore washable and sanitary, was also yellowish or brownish and had a tendency to darken further with age. For these reasons, it was impossible to produce pure white, rich blue, and purple linoleum.[8]

During the industry's infancy, manufacturers did not realize the product's potential for integrating design colors with the body of the product; instead, the early designs were painted on the surface of the solid-color body, as they had been with floorcloths. This early fault was soon corrected by technological advances that allowed for the development of several types of inlaid linoleum. These new types of linoleum were characterized by designs integrated with the resilient material during the manufacturing process.

## MANUFACTURING PROCESS

The process for making linoleum began with repeatedly spraying boiled linseed oil over sheets of scrim (thin cotton cloth) and allowing these to dry in a heated environment until the oil buildup reached the thickness of an inch or more. The boiling and the spraying and drying processes caused the linseed oil to absorb oxygen and polymerize, making linoxyn. The sheets, called skins, were then taken down, ground up (including the cotton scrim), put into pans, and heated until the granules softened. Pine rosin and kauri gum were added, and the mixture was cooled and ground again to create linoleum cement, the binding medium into which the remaining ingredients were added.[9]

This cement, with the addition of ground cork or wood flour (or both), whiting, and pigment, was pressed into sheet form by the heated rollers of a calender. Passing through more rollers, the sheet was pressed onto a canvas (burlap) backing, which was coated with a layer of paint, usually red oxide. The product was then seasoned in linoleum stoves (steam-heated rooms). During the seasoning, which took at least three weeks, the oxidation and polymerization, begun in the cement manufacturing process, continued until the product was hard and leathery. The hardened linoleum was given a surface treatment of polish, wax, or lacquer to increase soil resistance and provide a better finish.[10]

The earliest improvement was the integration of design into the body of the material through the manufacturing process. The first of these products were the marble, granite, and jaspé linoleums. Mixing together linoleum cement granules of various colors before passing them through the calender created the flecked effect of the granites, developed in 1879. Partially mixing the colored granules before calendering them produced marble linoleum, which resembled the rock for which it was named, and jaspé, a striated effect in which different colors ran in blended streaks parallel to the length of the linoleum.[11] However, before the turn of the century Walton developed a method for manufacturing inlaid linoleum with defined geometric designs.

The first such product was molded or stenciled inlaid linoleum.[12] To produce it, workers standing at long, sectioned metal tables covered with perforated trays divided into stencil shapes spread linoleum cement granules of various colors into the stencil sections to form the design. When the trays were removed and the linoleum cement was run through the calender, the granules fused together and onto the canvas backing, forming a sheet of linoleum with the design and the resilient material fully integrated.[13] A slight blending of the linoleum grains where sections of colors met was characteristic of this product. Typical of the process were designs in imitation of encaustic tile.

In 1898 Walton opened a factory to produce straight-line, or Walton, inlaid linoleum. To manufacture straight-line inlaid linoleum, various colors of green (unseasoned) linoleum were cut with a special machine that pressed blades, set into the pattern shapes, onto the sheet goods.[14] Workers then assembled the variously colored shapes into the design in a process called piecing. The product's name is derived from the sharp lines separating the colored shapes, a quality that characterizes this type of linoleum and distinguishes it from molded inlaid linoleum.[15]

Another design-influencing manufacturing innovation was embossing. Embossed molded inlaid linoleum possessed a texture created by subjecting calendered molded inlaid sheets to embossing presses consisting of plates fitted to elements of the design, such as mortar lines. The plates pressed these elements below the surface of the material, so that the remaining portions, such as tiles, stood out in relief.[16] Embossed molded inlaid linoleum was introduced by Armstrong in 1926 and soon became its most popular linoleum line.[17]

Although many combinations and variations of plain and inlaid linoleum were introduced, and sometimes later discontinued, those mentioned here remained the main types. One particularly successful variation was spatter linoleum, a unique and recognizable product produced by Armstrong just after World War II. Originally it was produced in only black and white with multicolored spatters, but Armstrong marketed thirty-seven color combinations of the product over the next ten years.[18]

**Figure 27.3a–c. a.** Applying felt under linoleum with paste was a common installation practice by the late 1920s. **b.** When the linoleum was installed, edges of the adjoining linoleum overlapped and seams were subsequently cut with a special tool. **c.** Unpasted areas near the seams were sealed with waterproof cement and the linoleum was then rolled and seams were weighted with sandbags until the floor was dry.

**Figure 27.4.** Clean, attractive, functional floors were the hallmark of a modern kitchen in the early twentieth century. White, cream, and light gray were desirable colors, as well as pale greens and blues.

From the earliest integrated designs resembling marble and granite to the inlaids' tile look, linoleum designs essentially imitated other materials, especially natural materials. In addition to tiles and slate, linoleum imitated wood planks and parquet; 3-foot-wide strips of wood-grain linoleum were sold as borders to surround area rugs. Even extremely complex carpet designs were replicated in both inlaid and printed linoleum. Many of the designs remained in production for long periods. However, while basically imitative, linoleum design also followed current design trends. The simple lines of the Art Deco style, for instance, fostered production of some outstanding and distinctive linoleum designs.

During the period between the two world wars, linoleum became an important economic player in the floor coverings industry and a major component of modern interiors. However, after World War II, vinyl tiles eclipsed linoleum's share of the floor-covering market.

Much of the resilient flooring found in old houses today is not actually linoleum but rather a felt-based product first marketed in America around 1910. This was a bitumen-impregnated felted paper bearing a printed design. A popular version was the room-size "rug" printed with a border and center design.[19]

### USES AND METHODS OF INSTALLATION

The linoleum industry distinguished the various products by type—the two major divisions being plain and inlaid—and by gauge (thickness). Typical gauges included 0.250-inch battleship, 6-millimeter (.235-inch) battleship (heavy), 0.187-inch battleship (medium), 0.142-inch battleship (light), and A, B, C, D, and E gauge plain. The battleship gauges were the thickest and took the longest to mature in the linoleum stoves (six weeks on average). They differed from the A–E gauges in that the burlap to which they were applied was not backed (painted). The 0.250-inch battleship gauge was recommended for use in offices, stores, hospitals, banks, lodge rooms, elevators, and, of course, battleship decks. The thinner battleship gauges were recommended for similar uses where the traffic was less heavy or where cost was a factor. The colors were limited to brown, dark gray, and green. In A-gauge plain linoleum, light gray, black, blue, and tan were added to the color choices. Recommended for use in apartments or offices where traffic was heavy but not severe, this gauge took four weeks to mature. B, C, and D gauges were recommended for most residential applications. E gauge, manufactured primarily for the automobile industry, matured in only ten days.[20] Later the gauges were reduced to five: 6 millimeters (0.235 inch) and 0.187 inch for plain and jaspé; and heavy (0.125 inch), medium (.095 inch), and standard (.079 inch) for plain and inlaid linoleum.

Linoleum was originally installed directly over subflooring. However, because of seasonal contracting and expanding of wood subfloors, buckling and cracking were common problems. The solution was to glue a felted product to the floorboards before installing the linoleum.[21] For the thinner gauges, even the felt did not always keep irregularities in the subflooring from showing through. The manufacturers recommended starting by smoothing the floor as much as possible.

Pattern books and trade journals carried detailed directions for laying a linoleum floor. Installers applied linoleum paste to the felt to within 4 or 5 inches of the seams and then laid the sheets with lapped seams so as to cut through both layers in one stroke. They then lifted the unpasted edges, applied some paste to them, and rolled the floor thoroughly. They laid the linoleum on top of the felt by the same method.[22]

Because linoleum was often an interior design feature of a house or office, the manufacturers produced, in addition to the various gauges manufactured for different applications, some patterns with particular rooms or uses in mind. Straight-line linoleum was available in designs created especially for children's playrooms, for instance, with game boards inlaid in gaily colored pieces. Custom designs were also available. Many commercial installations featuring the company logo became the centerpiece for the office decorating scheme. Custom inlaid designs, while sometimes assembled at the factory, could also be created on the job by skilled installers.[23]

**Figure 27.5.** Linotile, a ¼-inch-thick linoleum tile, was produced in "Ford" blue for the Ford Rotunda (1935, Albert Kahn), Dearborn, Mich.

## CONSERVATION

Because of its utilitarian nature, linoleum, whether through overuse or neglect, is usually found in some state of deterioration. When conserving linoleum, it is important to analyze the deterioration and its cause, intervene as little as possible, test all intervention first, and where more or special expertise is required, consult a trained conservator.

### DETERIORATION

The deterioration of linoleum has three primary causes: wear, water, and chemical changes within the product itself. Wear results largely from abrasion from dirt and grit in high-traffic areas. Heavy furniture and high-heeled shoes cause indentations that can also seriously damage the surface. Where worn areas are severe and have exposed the burlap backing, water damage to the backing is also likely.

Even when the worn areas have not penetrated the surface, water may cause individual pieces of straight-line inlaid linoleum to separate and warp, allowing moisture to penetrate to the backing. Once exposed to water, the backing material can separate from the top layer.[24] Moisture and damp conditions will degrade the wood flour in linoleum and can also cause it to swell and lift or curl at the edges.

Linoleum is also susceptible to damage from exposure to alkali. Cleaning solutions containing alkalis will soften the linseed oil, destroy cork filler material, and attack paints on printed linoleum and felt-based floor coverings.[25] Alkali-damaged surfaces appear pitted or severely abraded. Contact with concrete subflooring will cause similar deterioration.

Changes in the chemical makeup of aged linseed oil can also cause deterioration. Linoleum becomes brittle as the linseed oil continues to oxidize over time. This is especially true of linoleum with a high proportion of filler (wood flour or cork) to linseed oil. Linseed oil also darkens over time, especially when kept in a dark environment. Thus light and medium-value colors can appear

**Figure 27.6.** Durability and quietness made linoleum suitable for use in schools, as in the kindergarten room at the Arlington School, Poughkeepsie, N.Y.

**Figure 27.7.** At the Congoleum-Nairn showroom in New York City in the late 1920s, Sealex linoleum was displayed on specially constructed racks.

substantially darker than they were originally. The colors may also vary throughout a room depending on their exposure to ultraviolet rays (daylight).

## DIAGNOSTICS AND CONDITION ASSESSMENT

In developing a conservation approach, it is important first to understand the specific setting in which the linoleum is found and evaluate its contribution to the design of the space. Next, areas of deterioration and damage caused by wear, neglect, or mistreatment should be identified and the causes of the distress analyzed. In addition, an analysis of the light and light sources may help determine which areas of the floor exhibit color closest to the original.[26]

The first rule for treating a material such as linoleum should be to intervene as little as possible. When intervention is necessary, it is important to understand the type of linoleum requiring conservation, research any considered treatment carefully, perform the treatment on a test area first, and in some instances consult a trained conservator.

## CONSERVATION TECHNIQUES

Redirecting foot traffic through a room, moving furniture to hide and protect the worn areas, and putting plastic or rubber cups under the legs of heavy furniture are the simplest conservation techniques. Protective covers, such as mats, can also be placed over areas subject to heavy foot traffic.

When cleaning the floor, the goal is to minimize the stress on the system while improving the floor's appearance. Anything used should be tested on a small patch before being applied to a large area. The mild soaps recommended by the linoleum manufacturer may have a pH that is too high for use on aged flooring.[27] Mild non-ionic soap with a neutral pH is recommended, but even this soap should be tested.[28]

Avoid prolonged contact with water and cleaning agents, especially those containing alkaline materials. Liquid bleaches will lighten the colors, and powdered cleansers will abrade the surface. These substances should not be used.

Producers often coated printed linoleum and bitumen-impregnated felt-based floorings with varnish or shellac. When working with these products, a small spot should be tested to identify the coating. Turpentine will remove varnish but may also soften the surface paint. Shellac responds to denatured alcohol.

If stripping layers of old wax is necessary, it is important to find a stripper that is strong enough to dissolve the wax without dulling the finish. Alkaline-based strippers should be assiduously avoided. Solvent-based strippers may work with less stress on the linoleum. Newer aqueous-based (often limonene-based) products are much safer for both the user and the environment and, if used carefully, may be appropriate.[29] Testing any stripping solution on an inconspicuous patch is essential.

When the backing material has separated from the top layer but the old adhesive and backing are not deteriorated, it may be possible to inject adhesives into an opening and then roll the floor with a heavy roller. Pieces that have come loose but are still in an acceptable condition can also be reinstalled with a compatible adhesive. When selecting an adhesive, consult the manufacturer about the product's compatibility with linoleum and take into account the subflooring material and atmospheric conditions.

Old adhesive should be removed from both linoleum and subflooring by mechanical, rather than chemical, means. Chemical removers may drive the adhesive material into the linoleum or floor substrate. When stripping adhesive from the subfloor, it is important to go back to the bare substrate. Scarifying the subfloor will give it a mechanical tooth and aid the adhesive bonding in the new application, although any deep gouges and irregularities will require filling. Mechanically scraping the adhesive off the linoleum may damage the backing or abrade the back surface, but this should not be a problem if the floor is to be relaid anyway.[30]

Linseed-oil darkening can be reversed by exposing the linoleum to daylight or fluorescent lighting. Under fluorescent light, reversal to the original (or close to the original) shade can be accomplished in just a few days. Continued exposure to daylight or fluorescent light will maintain the lighter hues in a nearly original value.[31]

Once a linoleum surface has been cleaned or repaired, a protective coating, such as wax or aliphatic polyurethane, should be applied. Aromatic polyurethane should be avoided.[32] Routine maintenance should consist of sweeping or dry mopping to remove the surface dust and dirt. When necessary the floor should be mopped with a damp (not wet) mop and a small amount of non-ionic, pH-neutral detergent and then rinsed the same way with clear water.

## REPLACEMENT

Because it is difficult to find replacement linoleum, it is important to conserve as much historic linoleum as possible. When replacement is necessary, any linoleum being removed should be documented and saved.

Some flooring distributors may still have a stock of linoleum, although it may be brittle and darkened. If such linoleum is used for patching, inserted pieces that are darker than the adjacent flooring can be corrected by exposing them to light sources containing ultraviolet rays.

New linoleum still available from European suppliers can be used to replace historic linoleum and replicate some historic patterns.[33] For simple geometric designs, skilled cutters may be able to produce custom, straight-line inlaid linoleum on-site. In some historic houses, artisans have replicated complex patterns by painting or silk-screening them on solid-color linoleum using acrylic paints. The characteristic blended line of molded inlaid linoleum is more difficult to reproduce. Using a stencil to apply the design and then blotting it or slightly shifting the template while the paint is wet may work. The granular appearance may be achieved by stippling two or more colors over each other.

When an acceptable replacement cannot be found, epoxy filler with very fine inert filler can be used for small patches. The proportion of inert filler to organic or polymeric binder in the patching material should be as close as possible to that of the linoleum.[34] When thoroughly dry, the patch can be painted with acrylic paint and sealed. Worn spots on printed linoleum can also be in-painted with acrylics.

**Figure 27.8.** This worn linoleum shows the effects of abrasion and indentations. Some edges are beginning to lift up, usually a sign of water damage.

Vinyl flooring is the successor to linoleum, and some vinyl manufacturers still produce several patterns that originated with linoleum. One of the most common is the random-size square and rectangular tile pattern originally known as Armstrong 5352.[35] Granite patterns are also available from several companies.[36]

**Figure 28.1.** Standard color (solid and marbleized) rubber tiles could be combined to create elaborate floor patterns, such as these advertised by the Goodyear Tire and Rubber Company. *Sweet's Architectural Catalogues,* 1931.

SHARON C. PARK

# 28 | RUBBER TILE

**DATES OF PRODUCTION**

c. 1894 to the Present

## HISTORY

Rubber is a polymer that has or appears to have elastic properties; it can have an organic source, such as rubber trees, or be synthetically produced from hydrocarbons.[1] Rubber of both types was used to produce rubber flooring—interlocking tile, rubber tile, and sheet rubber. Noted for its sound-deadening properties and resiliency, rubber flooring also provided a hygienic and water-resistant flooring surface.

### ORIGINS AND DEVELOPMENT

In 1839 Charles Goodyear, an American, developed a technique for curing rubber with heat and sulfur known as vulcanization.[2] This process prevented deterioration from heat or cold and was the key to the successful use of rubber in manufactured products.[3] In the late nineteenth century the demand for tires propelled the rubber industry into a global manufacturing concern resulting in a wide array of new products including rubber flooring.

Before the advent of industrial manufacture, there had been attempts to produce a waterproof or resilient flooring. An early reference to rubber flooring came from Goodyear's 1853 description of a type of painted floorcloth made from canvas, wallpaper, and a rubber varnish. Another early rubber flooring was a compound made of India rubber, ground cedar, and fibrous material that was masticated, rolled into sheets, vulcanized, and painted.[4] As early as the 1870s, rubber tiles 1 to 2 inches thick were used for courtyards of English hotels to deaden the sound of horses and carriages.[5]

Interlocking rubber tiles, the first use of manufactured rubber flooring for interiors, were invented and patented by the Philadelphia architect Frank Furness in 1894.[6] The New York Belting and Packing Company, which had offices in Philadelphia, claimed that it was the original manufacturer of rubber interlocking tiles.[7] These tiles, approximately 2 inches square and ⅜ inch thick, did not require mastic and stayed in place with the friction fit of the interlocking tabs.[8] By 1901 the Goodyear Tire and Rubber Company was producing a similar interlocking tile.

During World War I companies such as Goodyear suspended production of interlocking floor tiles. In 1921 Goodyear's president visited England and became interested in the North British Rubber Company's mottled block rubber flooring.[9] Goodyear introduced a hard, sheet rubber flooring in the early 1920s, and by 1924 numerous U.S. companies were producing sheet rubber flooring, which could be glued directly to subflooring as a result of improvements in adhesives.[10]

Most sheet flooring was ¼ inch thick and was marketed as extremely durable. However, it was initially expensive compared to other flooring materials.[11] Until about 1940 sheet sizes were limited to 4½ feet in width, so sheet rubber flooring was widely used for corridors and borders.

In the early 1920s rubber tiles, cut from sheets or cast in molds, became popular. These tiles could be used in large, expansive spaces, and the designs were compatible with various interiors from the revival styles of the 1920s and 1930s.[12] Patterns for flooring could be solid colors, marbleized, or paisley designs.[13]

### MANUFACTURING PROCESS

Natural rubber flooring was a pliable mixture of rubber, clay, chalk, barites, fibrous materials such as asbestos or ground wood, processing oils and waxes, sulfur (for curing), and organic pigments.[14] The amount of rubber within the compounded mix was generally less than 25 percent, which reduced the cost of the flooring while increasing durability.[15]

Synthetic rubber, in use since the 1930s, was more resistant than natural rubber flooring to deterioration from oxygen, ozone, solvent attack, and oily stains.[16] Rubber flooring technology also changed in the 1940s with a variety of poured resilient floorings.[17]

The manufacturing process involved preparing a rubber "dough," compounding it with additives (to impart strength and create colors and patterns), and finally

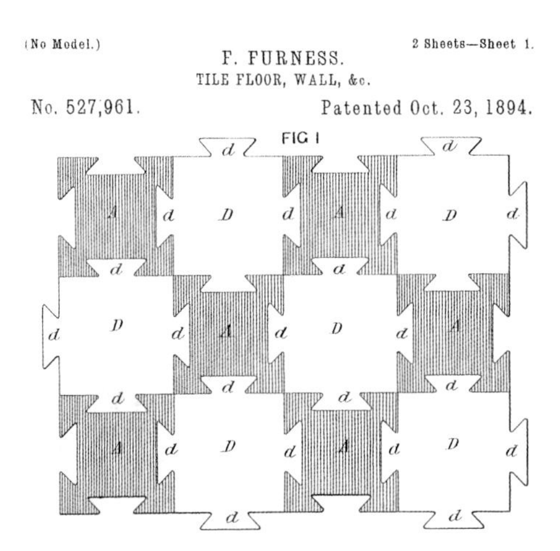


**Figure 28.2.** Philadelphia architect Frank Furness invented and patented the first interlocking rubber tile in 1894 and later a triangular version of these square tiles.

rolling it out into thin sheets that could then be cured or vulcanized. The dough was the result of a coagulation process in which latex from the natural rubber or the synthetic rubber was curdled.

Stored dry until needed, the dough was then sent to the compounding room, where the fillers, pigments, and vulcanizing agents were measured and added.[18] The compounding room was considered the most integral part of the rubber factory, and its operations were kept secret to protect the company's formulas.[19] Once compounded, the dough and additives were mixed and masticated in a mill.[20] The thick, rolled sections of rubber were dusted with talc powder and stored in preparation for calendering, die cutting, or molding.

The steam-driven calender machine consisted of two or more rollers that heated, rolled, and pressed the compounded mixed rubber into thin sheets.[21] Calender rolls had hollow chambers so that heat or steam could pass through and control the temperature of the rubber. The distance between the adjustable rollers established the thickness of the flooring.

Once the proper thickness was achieved, vulcanization (curing) was completed on either a calender roller with vulcanizing drums or on a hydraulic draw-out table where heat and pressure were applied.[22] Early interlocking tiles were generally die cut from sheets of rubber that had been cured under heat and pressure. Curing could also occur in molds and set into steam pans. The sheets or molded pieces could then be cut into squares or special sizes. Rubber tiles, specialty treads, coordinated countertops, and wainscoting could also be die cut or molded.[23]

## USES AND METHODS OF INSTALLATION

Interlocking tiles were popular for their quietness underfoot, flexibility, and waterproof qualities, particularly on boats and in kitchens and work areas where moisture was present. As the larger square tiles became noted for their resiliency, ease of maintenance, design potential, and durability, and as adhesives improved, they gained acceptance. Rubber tiles were used extensively in automobile showrooms, banks, athletic facilities, churches, elevators, hotels, libraries, residences, restaurants, and theaters. Rubber is also an excellent insulator, and its use was promoted in areas where electrical equipment was located, such as elevators, mechanical equipment rooms, and operating rooms. Hospitals used rubber flooring extensively for its hygienic properties.

Interlocking tiles were approximately 3/8 inch thick and approximately 2 inches in overall dimension and could be assembled as a mosaic of color and design. The tiles could be interlocked with a dovetail connector similar to a carpentry joint or with tabs from one tile (male) interlocking with an inverted joint of a receiving tile (female).[24] Tiles could be square, triangular, or rectangular. The predominant colors were black and white; red, green, blue, buff, gray, chocolate, and yellow were also popular. Interlocking tiles, for instance, are still in use in the basement rooms of the Heurich House (1894, John Granville Meyers) and the butler's pantry of the Anderson House, headquarters of the Society of the Cincinnati (1905, Arthur Little and Herbert W. C. Browne), both in Washington, D.C.

Rubber tiles were cut from rubber sheeting. In some respects rubber tiles were the most successful of the rubber flooring products because of the versatility of their designs. The marbleized, square checkerboard style with color accent borders is the most recognizable of the floor patterns, but tiles could also be set in basket weave, herringbone, brick bond, and simulated slate patterns.[25] Tiles were used in combination with strip borders and larger fields of rubber sheeting and could be adapted to a range of architectural styles.

By 1929 the Goodyear Company offered tiles in thirty standard colors and in thicknesses of 3/16, 1/4, and 3/8 inch with some specialty flooring being offered at 1/2 inch.[26] The thicker the tile, the higher the cost. Standard sizes for square tiles were 4, 6, 9, 12, 18, 24, and 36 inches, and standard sizes for rectangle tiles were 6 by 12, 9 by 18, and 12 by 24 inches. Custom orders for any size were available. It appears that 1/4-inch tiles were ideal for most flooring; 3/8-inch tiles were difficult to cut for tiles of more than 9 inches and required a great deal of rolling to obtain a well-adhered, smooth floor. Stedman Naturalized Flooring offered a similar range of products, but its standard sizes were different; for example, its 6-inch-square tile was really 6 13/16 inches square, and its largest square tile was 20½ inches square.[27]

Early interlocking floor tiles could be laid without any mastic or with a thin vegetable or animal-hide glue. Laying the later rubber sheets and tiles was often described as similar to laying linoleum. Rubber flooring could be laid directly on concrete as well as on wood subfloors. The subfloor had to be smooth, clean, and dry. By the 1930s it became standard to tack a floor lining of 1½-pound felt or strong cotton fabric onto the wood subfloor to absorb slight differential movements in the various materials.[28]

Tiles or sheets were set on a thin coat of specially prepared rubber or adhesive cement and were then rolled.[29] The rolling was done with a heavy steel or iron roller (100 pounds) about 12 inches long with a 7-inch-diameter drum. Rolling, which began about an hour after the floor was laid and lasted for at least 15 minutes per section, forced the flooring into the cement and removed all air bubbles, leaving the floor smooth and in full contact with the subfloor. The floor was to be left untouched for at least five days to allow the cement to dry thoroughly. It was then gently washed and either polished or left with a rubbed finish, depending on the manufacturer's instructions.[30]

**Figure 28.3.** More than 31,000 square feet of Goodyear rubber tiles were installed in the Richfield Oil Building (1929, Morgan, Walls and Clements), Los Angeles.

## CONSERVATION

Rubber flooring that has been maintained in good condition can be refreshed with cleaning and rewaxing. However, once the rubber or other compounds break down and begin to disintegrate, it is difficult to consolidate or conserve rubber flooring.

### DETERIORATION

All chemically unsaturated rubbers—natural and synthetic—are susceptible to degradation caused by oxygen, ozone, heat, light, pro-oxidant metals, abrasion, bacteria, and chemical reactions.[31] Because most rubbers involve vulcanization with sulfur, rubber continues to age and dry out long after the vulcanization process.[32] Chemicals in the rubber can migrate to the surface, creating a "bloom."

Oxidation, which is caused by oxygen and can be catalyzed by light, heat, and metals, is manifested by increasing brittleness and surface crazing, conditions known as shelf aging. Rubber surfaces may appear dull from exposure of inorganic fillers as the rubber deteriorates, a condition known as chalking. Degradation caused by ozone is described as atmospheric aging.[33]

Some rubber flooring and tiles have been damaged by contact with certain substances (copper, manganese, and fatty acids) or improper cleaning that has resulted in a softening of the rubber into a sticky or tacky surface.[34] Rubber is sensitive to some oils, solvents, and petroleum-based products, which can soften floors, attract dirt and debris, and cause staining.

### DIAGNOSTICS AND CONDITION ASSESSMENT

Before conservation treatments are considered, the polymeric composition and curing system of a rubber must be understood. Laboratory tests on samples of the rubber can identify various compounds.[35] The type of rubber and binder can be identified using infrared spectroscopy. Unfortunately, these tests are destructive. The scanning electron microscope can x-ray a sample scraped from a rubber surface below any protective coatings. The presence of sulfur indicates a cured product, and the presence of aluminum and silica generally confirms that an earth or clay filler was used. Samples for testing should be taken from inconspicuous areas or from already deteriorated tiles.

These tests can help establish whether a cleaning solution would have an adverse chemical reaction. If testing is not done, only mild cleaning, emulsion wax, and light buffing treatments should be used. Small, inconspicuous areas of tile can be used for test patches to evaluate

**Figure 28.4.** For the telegraph room of this Los Angeles, Calif., office building (late 1920s), rubber tiles in varying shapes, colors, and geometric patterns were used.

cleaning and waxing products. If paint, tar, or other debris is to be removed from a rubber floor, a small area should be tested before chemical or solvent products are applied. If solvents are used, they should be wiped off quickly to avoid damaging the rubber around and under the substance being removed.

## CONSERVATION TECHNIQUES

Rubber floors should be swept frequently and damp mopped on occasion. A small amount of ammonia can be added to cold water for mopping with a clear water rinse afterwards, but do not flood the floor with water as it might penetrate the seam and cause the cement or mastic to deteriorate. Using a sponge mop rather than a cotton string or rag mop gives better control of the water. Interlocking tiles, particularly red and green tiles, tend to bleed their color when in contact with water.

Extensively soiled floors can be washed with a mild vegetable soap, light oil soap, or Ivory soap, using cool to warm water. Abrasive cleaners such as scouring powders should be avoided; these could eat the cotton fibers, affect the mottling of certain tiles, and dull the finish. Naphtha, turpentine, and pine oil cleaners with solvents can make the rubber base sticky and should be avoided.[36]

Conservation options include protective agents and protective barrier coatings. Protective agents (antidegradents) include a variety of proprietary amines and phenols but are not practical for architectural applications. Application of a protective coating—either a wax or polish—creates a barrier that can help reduce oxygen and ozone diffusion in rubber flooring and limit contact with harmful oils, acids, and petroleum distillates.[37] If a rubber floor is dry, is dull in finish, or shows cracks, a wax or polish should be applied.[38] Such coatings will improve the appearance of dull rubber by reducing light scattering.

When coating a floor with wax or polish, care should be taken to protect the rubber from solvent attack.[39] If waxing is desired, use a water-based emulsion floor polish.[40] These coatings can generally be applied without buffing. Using a hard paste wax on rubber floors with a light buffing has proven successful, even though these waxes must be applied on top of a water-based emulsion polish for a durable finish. Because of the petroleum distillates in paste wax, a test area should be tried and only a light coating applied. If polish is to be stripped for a thorough washing, 2 ounces of ammonia per gallon of warm water is recommended; again, avoid leaving water on the floor for long periods. Hot water should not be used. Modern acrylic floor finishes generally should be avoided as they are difficult to remove and provide too high a gloss finish.

Because historic rubber is often brittle, it is best to preserve it in place undisturbed. In a surprising number of cases, rubber tiles approaching a hundred years of age are still in good condition.

### REPLACEMENT

It may be possible to replace severely deteriorated rubber tiles with new ones that match in color and size. Both solid and marbleized tiles are still available from a number of producers. If suitable tile is found but the thickness does not match the historic tile, additional subflooring can be installed to create a level floor. For interlocking tiles, which generally had no adhesive, it may be possible to relocate tiles in good condition from a closet or from under a counter to a more prominent area where tiles are broken.

When a good match is not available, custom tiles can be considered. Rubber companies could produce custom interlocking and square tile made from either a rubber or synthetic compound.[41] Another alternative for a custom interlocking tile would be to die cut one from a square tile. Rubber replacement tiles that approximate the color and appearance of the original can also be considered.

**Figure 28.5.** Interlocking rubber tiles at the Anderson House (1905, Arthur Little and Herbert W. C. Browne), Washington, D.C., show drying and chalking. Natural, vulcanized interlocking rubber tiles are particularly prone to long-term aging and drying, which causes them to become brittle.

**Figure 29.1.** Cork tile was marketed specifically for its sound-deadening properties, which made it suitable for buildings such as libraries and churches. Ease of maintenance also made square- and beveled-edged cork tile an attractive choice as a floor covering. Cork tile catalogue, Armstrong Cork Company, 1936.

ANNE E. GRIMMER

# 29 | CORK TILE

**DATES OF PRODUCTION**
c. 1899 to 1975
**COMMON TRADE NAME**
Kencork

## HISTORY

Cork is a natural wood product from the outer bark of the cork oak, a species of tree native to the Mediterranean region.[1] Ground cork bark is used to manufacture cork floor and wall tiles, as well as a variety of cork insulation and composition products. A porous but inert material, cork has a cellular structure that imparts resiliency. Other physical properties include light weight, compressive strength, and low thermal conductivity, desirable for insulation products.[2]

### ORIGINS AND DEVELOPMENT

Historically, cork was used primarily to make bottle stoppers. Manufacturers soon learned that the resulting waste could be turned into other products.[3] Most early uses of cork waste did not require additional processing. Gradually, however, other types of cork products made from the ground-up chips and shavings began to appear.[4]

One important product was corkboard, which used large quantities of cork waste and was valuable as insulation. Initially, asphalt was chosen to fuse together the shavings. However, the strong smell made it unsuitable for certain uses. John T. Smith, an American inventor, discovered a process, patented in 1892, by which intense heat melted the resin in the cork, thereby providing a natural binder to hold the cork particles together.[5] This breakthrough was the foundation for the production of cork tile flooring.

Cork floor tiles were probably first manufactured in the United States about 1899 by the David E. Kennedy Company in New York City, although it is likely that they were produced earlier in Germany, where corkboard was produced in the early 1890s.[6] The Armstrong Cork Company, which became one of the largest manufacturers of cork tile, began production in 1904 and assigned distribution rights to the Kennedy Company in 1910.[7]

Before 1920 production of cork tile was limited. In the mid-1920s, when the United States was enjoying a period of prosperity and an accompanying building boom, cork tile gradually gained acceptance as a floor covering for new construction. The increased popularity of cork flooring materials was also due in part to the use of additives that made the tiles more impervious to moisture, dirt, food, grease, ink, and mild acids. In 1927, the peak year of production before World War II, manufacturers produced approximately 2,983,500 square feet of cork tile valued at $600,000.[8]

### MANUFACTURING PROCESS

To make the tiles, cork shavings were ground up and placed in large cast-iron molds. The molds were compressed under high pressure, sometimes up to 5,000 pounds per square inch, and baked in a kiln at a temperature of approximately 450 to 600 degrees Fahrenheit for 7 to 10 hours.[9] A belt conveyed the molds through the length of the kiln, about 120 feet, in approximately 45 minutes. The intense heat melted the natural resin in the cork and, combined with the pressure, caused the cork particles to fuse together without the addition of any binder or glue.

By the late 1920s cork flooring was often given a partially penetrating factory finish of pyroxylin or nitrocellulose lacquers that formed a surface impervious to moisture, dirt, and many stains. Although cork flooring finished in this way did not require additional treatment, it was usually waxed for extra protection.[10]

After World War II manufacturers strengthened cork tile by adding phenolic resin to the cork particles as a binder; later, urea resins were added. Resin-reinforced cork tiles are baked for longer, 10 to 12 hours, but at a lower temperature (about 250 degrees Fahrenheit). These changes in the manufacturing process created more resilient tiles, slightly lower in density than traditional cork tiles, with greater uniformity of color.[11]

The colors of traditional cork flooring tiles were natural shades of brown. The darker shades were the result solely of hotter temperatures and longer baking times. Basically there were, and still are, three shades of cork tile—light,

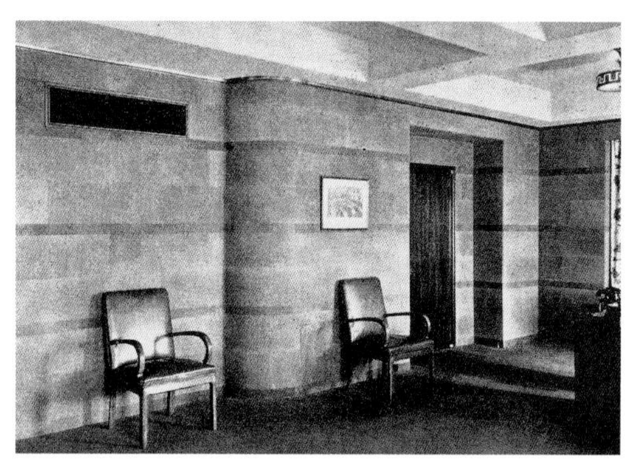

**Figure 29.2.** In cork tile manufacture, molds loaded with cork shavings were placed under enormous hydraulic presses. The level of compression, which determines the tiles' density, was maintained by inserting steel rods in slots in the mold. Molds were typically baked from 7 to 10 hours (top).

**Figure 29.3.** Cork tile could be used to line interior walls as well as floors. Bands of cork tile in varying shades gave this office a modern appearance (bottom).

medium, and dark. Beveled tiles were introduced in 1935, and tongue-and-groove tiles also were marketed.[12]

Early cork tiles were sold unfinished, exactly as they came out of the molds or presses. Application of a finish was left to the consumer after installation. Product literature recommended sanding the cork tile floor after installation to achieve a smooth, level finish and then waxing it. An alternative was to finish the sanded tiles or carpet with a lacquer-based sealer or two coats of a tung oil–based penetrating sealer and then apply one or two coats of wax. After laying untreated cork carpet or cork tile, some flooring specialists suggested first sealing it with a pliable (nonbrittle) oil sealer containing a phenolic resin, such as Bakelite, and then buffing and waxing for further protection and ease of maintenance.[13]

## USES AND METHODS OF INSTALLATION

Cork tile was notable as one of the most resilient hard-surface flooring materials.[14] For this reason, although moderately expensive, cork tiles were often used in libraries and schools.[15] Cork tiles also were promoted as natural products distinctive for their comfort and sound-deadening qualities and for the physical as well as visual warmth imparted by their natural colors. The material was popular for lobbies, churches, auditoriums, libraries, hospitals, and museums.[16] In 1941, for example, more than 25,000 square feet of cork tile flooring was installed in the National Gallery of Art in Washington, D.C.[17] Cork floors were also advertised for their sanitary qualities and their ease of cleaning and maintenance.[18] Probably for these reasons, cork tiles were also used extensively in many buildings at the Mayo Clinic (1914, Ellerbe and Round; 1928, Ellerbe, Inc.) in Rochester, Minnesota. All these features were significant in cork being used as both a residential and a commercial flooring material.

Although cork tile had been used widely during the 1920s, architects still viewed it as a new, modern material as late as the 1930s. Richard Neutra, for example, who researched new materials for his own house near Los Angeles, known as the Van der Leeuw House (1932), selected cork tile.[19] Walter Gropius also used cork tile for his own residence (1938) in Lincoln, Massachusetts, as did Frank Lloyd Wright for the bathroom floors and walls in Fallingwater (1935) in Mill Run, Pennsylvania.

Cork tiles, which were generally either 3⁄16 or ½ inch thick, were available in several shapes, including square, oblong, and rectangular as well as border strips, and in a wide variety of sizes traditionally based on multiples of 3, up to 18 by 36 inches. During the 1930s tiles as small as 2 by 2 inches and as large as 18 by 48 inches were avail-

**Figure 29.4.** The warm colors of the elaborate hand-laid cork design installed in the Bowery Savings Bank (1893, McKim, Mead and White), New York City, "lent dignity" to the room, according to the Armstrong Cork Company.

able.[20] In the 1940s stock sizes became more standardized, the most common being 9 by 9 or 12 by 12 inches; rectangular or oblong tiles were available in 9 by 18 or 12 by 24 inches; and border strips ranged from 3 by 18 to 12 by 36 inches.[21] Tiles could also be specially ordered to create custom designs.[22]

Cork is a natural product and expands and contracts according to changes in temperature and moisture. To prevent buckling, the cork must reach its normal point of expansion before it is installed. Therefore, it was and still is necessary to deliver the cork tiles to the site several days before laying a cork floor, remove the wrapping, and spread them out over the floor in the room where they are to be laid.

One of the earliest methods of installing cork tiles over a new concrete floor involved the use of brads. While the underfloor was still green (not completely set), 1 inch of asbestos concrete was troweled on a true surface with a sidewalk finish to ½ inch below the finished floor level. (This composition made a fibrous bed that would hold a nail.) Once dry, a special waterproof mastic was applied in which the cork tiles were bedded. The tiles were laid so that all joints were closed and cemented, and the surface was then weighted. Finally, the tiles were bradded into place with headless brads that were driven beneath the surface of the tiles to hold them in place while the mastic set up, a process that took about 12 hours.[23]

By 1925 some companies still recommended the use of brads when installing cork tiles, but this method was superseded by a simpler one in which mastic or an elastic waterproof cement was used to glue the tiles to the subfloor and then the tiles were weighted down to secure them to the floor.[24] When installing cork tiles over a concrete subfloor, it was first necessary to fill any cracks or seams. Plaster of paris was originally recommended for this purpose, but later a portland cement mix or a latex filler was generally preferred. After this dried it was important to make sure that the overall surface was smooth, clean, and dry. When the concrete surface was fully prepared, the cork tiles could either be laid directly in mastic applied to the concrete, or glued with linoleum paste to a layer of heavy felt (1½ pounds per square yard) or building paper that had been glued to the concrete with waterproof linoleum paste or cement.

## CONSERVATION

Cork tiles can be damaged by excessive moisture, sunlight, and ultraviolet light, abrasion, grease, and oil staining. Thin scratches may sometimes be removed by machine buffing, small gouges may be filled with a mixture of finely ground cork particles, and individual cork tiles that are too deteriorated to repair may be replaced selectively with new tiles.

### DETERIORATION

All cork floors, particularly those in high-traffic areas, are susceptible to damage and abrasion from dirt and grit once their protective coating has worn off. Traditional and factory-waxed cork tiles can be damaged by water and excessive dampness, which solubilize the binder, loosen the bound cork shavings, and thus cause the tiles to shrink and buckle. Resin-reinforced cork tiles are less sensitive to moisture but are not water resistant.

As a wood product, traditional unfinished, factory-waxed, and resin-reinforced cork tiles are also subject to deterioration caused by sunlight and ultraviolet radiation. Both traditional and resin-reinforced cork tiles will fade over time, and they can become brittle, dry, and crumbly.

Cork is also very porous, and traditional cork tiles are very susceptible to grease and oil staining. Alkaline cleaners, such as those based on sodium hydroxide or ammonia, will degrade cork, as will many organic solvents, abrasive scouring powders, and caustic cleaners.

### DIAGNOSTICS AND CONDITION ASSESSMENT

Before undertaking conservation work on cork tiles, it is important to identify whether they are unfinished or resin reinforced. The date of installation and original construction documents may provide clues about the type of tile. It is also important to identify the finish on the tiles, which is frequently wax or polyurethane.

Cork tile floors should be inspected for worn areas, surface loss, evidence of the original finish, loose tiles, disfiguring stains, and sources of water penetration.

### CONSERVATION TECHNIQUES

Maintenance of traditional cork tiles and factory-waxed cork flooring should include periodic vacuuming or sweeping with a soft broom and occasional damp (but not wet) mopping. To avoid excessive abrasion, mats can be placed on heavily used areas.

Less frequently, cleaning and buffing with a paste or liquid wax is recommended to remove ingrained dirt. Although the use of water and detergents is generally not advised for this kind of cork flooring, careful damp mopping using a diluted solution of linseed oil–based liquid cleaner or a few drops of mild phosphate detergent, rinsing with clean water, and drying with a clean mop to remove all traces of water may sometimes be necessary to remove ground-in dirt and mud.

Resin-reinforced cork tiles are more impervious to water and can be cleaned with a damp mop. Washing with water and a neutral pH detergent is acceptable from time to time, as long as any excess water or detergent is quickly removed. After it is washed, the floor should be waxed and buffed.[25]

Loose cork floor tiles may be reglued or reattached to the floor substrate using a water- or solvent-based adhesive. Reglued tiles should be weighted for approximately 24 hours after gluing.

Light or thin scratches on the surface of traditional cork floor tiles can sometimes be buffed out using a machine buffer with a lamb's wool pad. Small gouges or holes in a cork floor tile can be patched using a putty made up of a mixture of cork shavings and clear shellac.[26] The area to be patched should be surrounded with masking tape to protect the cork surface; the putty is then pressed into the hole and the top of the patch smoothed with a putty knife. After the patch has dried for approximately 30 minutes, it should be sanded with 00 grade steel wool. If the patch appears too noticeable and is obviously duller than the rest of the cork floor surface, clear varnish or lacquer may help it blend in.

Although sanding, an abrasive technique, has traditionally been used to maintain and rejuvenate discolored and uneven cork tile floors, it will remove too much cork and should generally be avoided.

### REPLACEMENT

When historic cork flooring is so damaged or deteriorated that it cannot be repaired, it is possible to replace individual tiles by removing only the most deteriorated, damaged, or worn tiles and replacing them with new cork tiles that match the historic color.[27] The fact that cork is a traditional material with subtle variations in color may make it easier to find a suitably close color match. However, it may be necessary to match the faded color of the original tile by staining unfinished tile. Because a floor surface may have worn unevenly over the years, it may be necessary to carefully sand down replacement tiles so that they are flush with adjacent tiles.

Replacement cork flooring materials are still manufactured, but these are available only as import products sold by a number of suppliers in the United States. Contemporary cork floor tiles are generally available in the same three basic shades of light, medium, and dark. Cork tiles come in a variety of finishes that include polyurethane (matte or glossy) and the more traditional waxed finish. Unfinished traditional cork tiles are also available. Most cork tiles today are the standard 12 by 12 inches and are commercially available in thicknesses of 3/16 and 5/16 inch. Other sizes and tile thicknesses—1/8, 1/4, and 1/2 inch—can be custom ordered to match historic tiles.

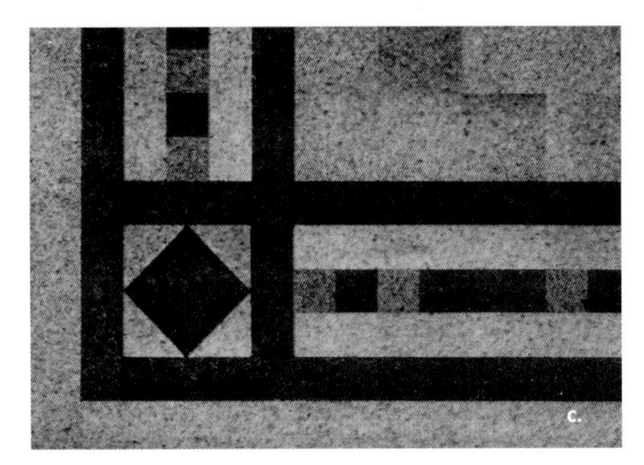

**Figure 29.5a–c.** Armstrong's suggested cork tile designs (**a.** no. 700, **b.** no. 720, and **c.** no. 740) could be created using dark, medium, and light shades of cork. Precision molding created tiles with uniform densities and smooth surfaces, and the unique cellular structure of compressed cork contributed to its "restful resilience" and "rich beauty," according to the company.

# Terrazzo

THE FLOORING OF CENTURIES

SHAWSHEEN PHARMACY, Lawrence, Mass. Adden & Parker, Architects
A GALASSI INSTALLATION OF TERRAZZO FLOORING

*—known to architects since the days of* MICHAEL ANGELO

*—equal in beauty and durability to any flooring*

*—low cost*

*—sanitary*

*—endless variety of architectural treatment*

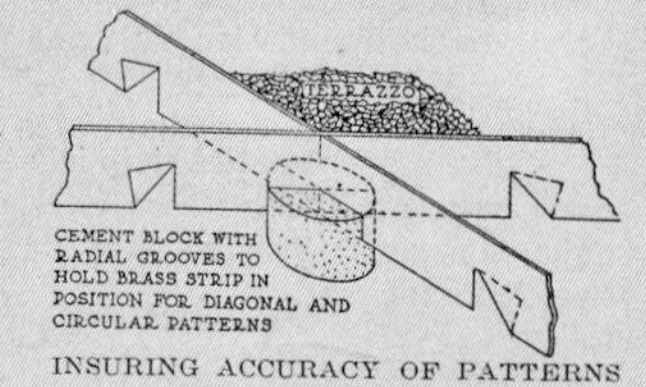


INSURING ACCURACY OF PATTERNS

## Preventing Cracks in Terrazzo Floors

THE cutting up of the surface into small squares or panels, to take care of any expansions or contraction, and the locating of the joints strategically so as to control the cracking directly above the strain of cross steel beams, will eliminate ragged and unsightly cracking.

The most effective means thus far devised is the provision of brass or metal strips, which adds very little to the cost of installation. The strips provide straight line openings for eventual cracks and enable the architect to carry out any decorative scheme of floor in keeping with surroundings, impossible heretofore with terrazzo work on account of prohibitive cost in doing it by the old method.

Under our "Terrazzolay" method, the metal strips are deep enough to cut both the wearing surface and the underbed supporting it, thus dividing the pavement clear to the foundation slab, and rendering it independent of settlement movements. Rigidity is insured by special forms of anchorage as shown in details.

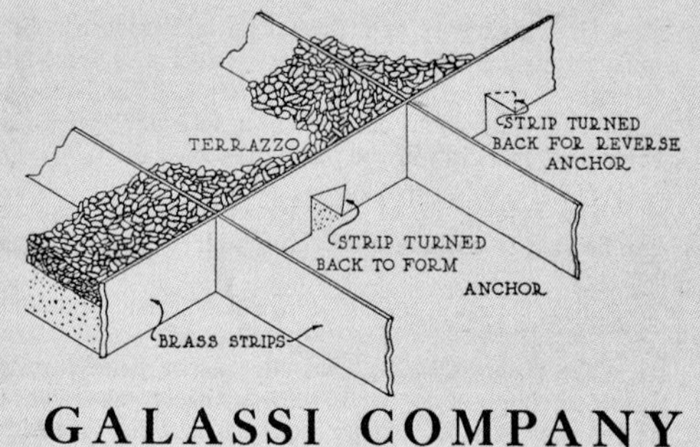


GALASSI COMPANY

TERRAZZO AND TILEWORK OF EVERY DESCRIPTION

345 Lexington Avenue
NEW YORK

11 Bennett Street
BOSTON

**Figure 30.1.** Brass strips were installed in terrazzo floors to prevent cracking and to separate sections of the design. To ensure rigidity, strips were sometimes set on small cement blocks with grooves for diagonal and circular patterns. *Suggesting a Standard Specification for Terrazzo Work,* Pasquale Galassi, 1923.

WALKER C. JOHNSON

# 30 | TERRAZZO

**DATES OF PRODUCTION**
c. 1895 to the Present
**COMMON TRADE NAMES**
Punte, Seminato

## HISTORY

Terrazzo is a nonresilient floor surface made of small pieces of marble or other hard stones embedded in a portland cement matrix. When it has cured, it is ground and then polished to a smooth finish or left in a rusticated state. Terrazzo, known for its durability and ease of maintenance, can be formed in attractive and colorful patterns with the aid of divider strips. Used as early as the 1890s, terrazzo did not gain widespread acceptance until shortly before 1920. Advances in technology, such as the development of epoxy cements, have helped terrazzo remain competitive as a floor material in recent years.

### ORIGINS AND DEVELOPMENT

Terrazzo is descended from the art of mosaic, which originally was applied as a decorative treatment to walls and later to floors.[1] The technique was developed in Alexandria, Egypt, when the country was under the artistic influence of the Greek empire.[2] Mosaics have traditionally been formed by handsetting small pieces of stone, ceramic, or glass in a decorative pattern onto a mortar base.

The Romans initiated further developments, distinguishing two varieties of mosaic. *Opus musium* was a glass mosaic used for wall decoration, and *opus lithostrotum* was a mosaic used for floors. The latter version was divided further into three subgroups: *opus tesselatum, opus vermiculatum,* and *opus sectile.*[3] *Opus tesselatum* and *opus vermiculatum* involved different methods of setting small pieces of tile or broken marble, porphyry, or travertine fragments into a cement base, often to form pictorial scenes, or setting larger marble tile to create linear borders of repeating patterns. *Opus sectile* was comparable to marquetry, in which thin sections of cut marble were set within a larger field of marble to create decorative patterns and figures. Eventually, fields of decorative mosaic work were laid between larger slabs of marble. The designs were almost always geometric, with triangles, squares, lozenges, hexagons, and octagons most prevalent.[4]

Over time these various methods for mosaic flooring lost their unique definitions and were blended together by technique, design, and construction. After the fall of Rome, the mosaic tradition was continued in the early Christian and Byzantine churches. In Venice the art of mosaic has been practiced continuously to the present day.

By the eighteenth century cruder forms of mosaic flooring had developed in and around Venice. Terrazzo is directly descended from these forms. Twentieth-century terrazzo floors are derived from a technique known as *pavimento alla Veneziana* (Venetian pavement), in which marble fragments ranging from ½ inch to 2 inches long are laid closely together in a cement base.

A less expensive technique that evolved was *seminato,* in which irregular marble chips were sprinkled onto a cement base coat. Larger pieces, measuring approximately 3 to 4 inches long, were sprinkled onto the base first. Medium-size chips (½ inch long) and then smaller chips (less than ¼ inch long) were added to fill the gaps. Finally, the slab was rolled until the chips settled into the cement base. After the floor had set, it was ground and polished.

Both these methods became known generically as *terrazzo,* a term applied to any floor in which pieces of stone are bonded in a cement bed.[5] The technique was introduced in the United States in the late 1890s.[6] Concrete mosaic, as terrazzo was also known in the early years of the industry, did not gain acceptance for several decades, partly because early installations were prone to cracking but also because the material received only scant mention in trade journals.[7]

### MANUFACTURING PROCESS

Initially, designers formulate designs and select colors from precast terrazzo samples. Specifications outline the type of marble chips, matrix mix, and method to be used.[8] Sample blocks are usually ordered from the installer and used to guide the installation. The installer, in turn, prepares shop drawings and samples.

**Figure 30.2.** Before electricity terrazzo installers used galleras (long-handled tools) to polish floors.

**Figure 30.3.** Terrazzo's durability made it desirable for floors in high-traffic areas, such as bus stations.

Traditional terrazzo floor surfaces are manufactured from a mixture of 70 percent or more marble chips and 30 percent or less portland cement matrix over a concrete base. Coloring pigments are sometimes added to the white portland cement to obtain a desired shade. Resinous terrazzo toppings are obtained by mixing the same proportion of marble chips with a synthetic binder.[9]

Decorative chips, typically marble, are chosen for their color and strength.[10] Other stones, including onyx, travertine, and serpentine, can also be used. Alundum and Carborundum are used to vary the texture and provide a somewhat slip-resistant surface.[11] Chips are graded by size, varying from No. 1 (between ⅛ and ¼ inch) to No. 8 (1 to 1⅛ inches). A proportional mixture of sizes is most common.[12]

Divider strips are used to localize cracking caused by shrinkage, subdivide areas for different colored terrazzo, and provide guides for construction. They also ensure uniformity in the topping thickness and create easily workable panels. With portland cement toppings, divider strips should be located at column lines or over beams; they should be no more than 6 to 10 feet apart for large field areas.[13]

### USES AND METHODS OF INSTALLATION

Most stone and concrete floors in the United States were mosaic until the mid-1920s, when American architects became aware of terrazzo's design potential. Terrazzo was well suited for the smooth, curvilinear designs of the Art Deco and Moderne styles, prevalent from the late 1920s through the 1940s. Unbonded methods offered stability, and marble and pigmented cements provided architects with new opportunities for floor designs.

In addition to design considerations, technological advances in the mosaic and terrazzo industry made the material more economical and durable. Before about 1919 terrazzo floors were laid in large monolithic slabs. This method presented problems in twentieth-century office tower applications, where terrazzo toppings were prone to cracking, particularly over structural elements. In 1919 the L. Del Turco and Brothers Company of Harrison, New Jersey, introduced a method to subdivide terrazzo surfaces with brass divider strips.[14] That year these brass strips were first installed in a building constructed in Philadelphia.[15] Before the development of metal strips, single rows of mosaic tesserae were used to subdivide larger panels of terrazzo.[16] Eventually, strips of brass, copper, nickel silver, or zinc were used to prevent or at least localize cracking. The strips were placed at regular intervals in the underbed to provide continuous joints throughout the entire length and width of the room. In this way either curved or geometric patterns could be formed and performance could be improved.[17]

Another factor that influenced terrazzo's popularity was the invention of the electric grinding machine. Before about the late 1910s the mosaic or terrazzo floor surface had to be ground down manually by workers using a *gallera,* a long pole onto which a pumicelike stone was clamped.[18] With the introduction of electric grinders for stone and terrazzo finishing, the terrazzo trade gained speed and accuracy, and overall costs were reduced.

Terrazzo soon overtook mosaic in popularity for public buildings and apartment buildings. Its potential for industrial buildings was also explored.[19] The trade was referred to as mosaic-terrazzo throughout the 1920s, but by the 1930s terrazzo became the predominant practice. As the industry began to expand, so did the need for an organization to support the growing number of installers. The

National Terrazzo and Mosaic Association was incorporated in 1931 to develop recommended terrazzo practices and report on completed installations.

Terrazzo is not limited to use as a flooring material. The terrazzo material can be precast into a number of forms for architectural use, including murals.[20] Most common are cove moldings, wainscoting, and either precast or cast-in-place terrazzo stairs.[21] By the late 1930s terrazzo had become a nearly ubiquitous flooring material for hospitals, airports, and schools, among other building types.

A number of installation methods have been used historically and continue to be used. These installation systems are divided into two classes: bonded and unbonded. Bonded systems are in turn subdivided into four classes: bonded underbed, monolithic, chemically bonded, and thin-set, a relatively new method using a resinous topping.[22] For each system, once the topping is applied and has cured, surfaces are ground smooth using mechanical means.

In the bonded underbed method, which has a minimum thickness of 1¾ inches, the structural slab is soaked with water, and then a slurry coat of portland cement and water is applied. A mortar underbed ⅞ inch to 1½ inches thick is poured over the slab. Divider strips are inserted in the underbed, and a layer of portland cement ½ to ⅝ inch thick is then poured over the top. Once the topping has cured, it is ground and finished.

Monolithic bonded systems are installed by pouring a portland cement terrazzo topping ½ to ⅝ inch thick directly on the structural slab without an underbed. Divider strips are used in this application to separate terrazzo colors and provide dividers at control joints that have been saw cut into the concrete slab. This system is applied where less than 2½ inches for topping is available but it is more susceptible to cracking.[23] The chemically bonded system is nearly identical to the monolithic system, except that a bonding agent is applied to the structural concrete before the topping.[24]

The unbonded installation method, also known as the sand cushion method, separates the mortar underbed and terrazzo topping from the structural slab with a bond-breaking film over a bed of sand. The mortar underbed in this system is typically 1⅞ inches thick, with metal wire reinforcement. The portland cement terrazzo topping is ½ to ⅝ inch thick. Movement in the slab is absorbed by the sand bed, typically ¼ inch thick, so that any expansion in the underbed and topping occurs at the divider strips. Divider strips are used most effectively in this system.

**Figure 30.4.** The Manhattan Terrazzo Brass Strip Company's 1931 floor designs featured various combinations of field and border units.

## CONSERVATION

Terrazzo is an extremely durable flooring material. Failure almost always is a result of forces acting on structural systems that support the terrazzo.[25] Wearing properties are also influenced by maintenance practices, the quality of materials, and the quality of installation.

### DETERIORATION

Cracking, the most common form of failure, is the result of movement and differential settlement or expansion and contraction of the structural system supporting the terrazzo topping. Cracking can occur regardless of the installation method. Once cracks appear, the terrazzo topping can spall along a crack, particularly next to control joints.[26]

In the design of sand-bed systems, metal divider strips are installed as part of the design in part to provide a fault line so that cracks are not so noticeable. The sand cushion provides a fault zone so the structure can move below and not telegraph this movement through the terrazzo topping. When movement does occur, it is controlled. Cracking in bonded or monolithic installations is telegraphed directly through the topping, regardless of the location of the divider strips.[27] In such situations, the faults are obvious and more frequent, and repair is more difficult.

Terrazzo floor surfaces can be etched or pitted by acids or caustic cleaning and by floor-stripping materials, causing the cementitious bonding agent and some aggregates to disintegrate.[28] Contact with engine oils and automobile transmission fluids, which will soak through and create a ghosting pattern, should be avoided. Other materials that can stain and discolor terrazzo surfaces include ink, tobacco, ketchup, mustard, and waxes.

Some blue and red marbles and elastic polymer setting beds are affected by ultraviolet light and are inappropriate for exterior installations because they fade. In general, however, terrazzo aggregates are quite durable and have a high abrasion resistance and low porosity.[29] Over time abrasion will wear down terrazzo surfaces, particularly in high-traffic areas such as entrances to doors or on steps along railings.

The structural system may not be properly designed to support the terrazzo installation specified. Cracks will appear if the floor slab's modulus of elasticity is significantly different from that of the terrazzo. This occurs most frequently with bonded installations.

Improper installation is another cause of failure and is evident almost immediately after the terrazzo topping is polished. If a floor is loaded before it is completely cured, failure can occur. The use of accelerators and curing agents in a bonded system can lead to delamination of the terrazzo topping.

## DIAGNOSTICS AND CONDITION ASSESSMENT

Assessment of terrazzo floor deterioration should begin with a visual inspection to determine the type of failure and its causes. Look for evidence of structural movement or uneven cooling and heating. Past maintenance practices should be discussed with the building owner, and it is often useful to have an experienced terrazzo installer investigate problems.

Destructive testing may be required to ascertain whether bonded systems have delaminated. If necessary, structural engineers can establish the modulus of elasticity of the subfloor structure to determine if it is compatible with the terrazzo, as well as conduct impact and loading tests. In some cases destructive testing with assistance from an installer is necessary to ascertain why a terrazzo floor is failing. Core samples can also be taken for analysis by a testing laboratory to evaluate the topping, mortar bed, and subfloor.

## CONSERVATION TECHNIQUES

General maintenance for portland cement–based terrazzo floors should include sweeping with compounds that do not contain oil or sand, which can penetrate, discolor, stain, and abrade the floor.[30] Occasional mopping is desirable to remove ordinary dirt, soil, and dust. Only non-ionic neutral detergents (with a pH ranging from 7 to 10) should be used to clean terrazzo.[31] All water, mops, and buckets must be clean, and the floor should not be permitted to dry before rinsing so dirt will not be reabsorbed. All-purpose or harsh cleaners, sealers, waxes, and soaps containing water solubles, inorganic or crystallizing salts, and harmful acids or alkalis should be avoided.[32]

Portland cement terrazzo typically has a 70 percent density of marble-chip exposure at the surface. The remaining 30 percent of portland cement binder is porous enough to require the application of a penetrating solvent-type sealer immediately after final polishing of the new installation. Periodic application of water-based acrylic sealer is generally recommended to prolong hydration of the cementitious binder and improve durability.[33] Resinous epoxy or polyester systems do not require a penetrating sealer. Instead, surface sealers are recommended.

Because terrazzo is used in many public buildings with heavy traffic, stains are a common occurrence. Before undertaking any work to remove a stain, the staining agent must be identified. Depending on the type of stain, removal may be accomplished by either dissolving or absorbing the agent with a poultice.[34] Without proper knowledge of the potential effect of a stain remover on the terrazzo and the substance being removed, stains can be set rather than removed, and pigmented portland cements can be damaged. Therefore, chemicals should be used only if less aggressive methods are unsuccessful.

Water-based stains should be removed with water. Grease can usually be removed by blotting with a cotton cloth and soap. Another common stain is tobacco, which can be removed with a poultice of trisodium phosphate, chlorinated lime, and talc. This method can also be used for urine. Inks vary in content, so removal methods are likewise varied. A method that works for common writing ink is a solution of sodium perborate in hot water, which can be made into a poultice with whiting.[35]

To prevent failure from recurring, structural systems supporting terrazzo floors must be adequately stabilized before making any repairs. Small areas of damaged terrazzo can be repaired by removing damaged areas to the depth of the terrazzo topping and placing a matrix that matches the original in the void. In this process marble pieces are pressed into the matrix and ground after the patch has set. It is advisable to work with a terrazzo installer first to obtain an acceptable marble-and-binder combination to match adjacent areas.

## REPLACEMENT

When large areas of terrazzo require replacement, severely deteriorated terrazzo must be cut out. The structural subsurface is then examined for compatibility with replacement terrazzo. It is usually best to replace entire areas within divider strips. If a whole floor is being replaced, the location of metal divider strips must be documented in order to recreate the original floor design.

The original marble chips and binder should be visually examined and tested if necessary to develop specifications for new areas. Physical data can also be compared with the original specifications if they are available. As with the patching process, obtaining marble chips to match the color and grading of the original mix is one of the most challenging aspects of replacement. When a terrazzo formula matching the original is not available, an acceptable match can be developed by "bracketing" mixes and chips to satisfactorily approximate the original mix. Once this is achieved, the components of the topping mix are recorded for duplication in the final installation.

For floors where cracking has been a problem, it may be necessary to integrate expansion joints to control cracking, particularly over beams. Selection of an installation type depends on the structural subfloor and depth available for the floor. An unbonded floor 2½ inches deep offers the best results, but if that depth is not possible, a bonded installation must be used. It may be possible to mix bonded and unbonded systems in a repair program if they are separated by a control joint, the subflooring is adequate, and different colors are used in the mixes. Seldom will new terrazzo surfaces match existing terrazzo unless the existing surface is resurfaced and the new terrazzo is formulated to match.[36] Grinding, buffing, and sealing, according to practices recommended by the National Terrazzo and Mosaic Association, complete the replacement process.

**Figure 30.5.** Differential settlement in the entrance of this Washington, D.C., apartment building has caused the terrazzo to crack. The durability of most terrazzo installations is largely a function of the structural stability of the system supporting the terrazzo.

Photo features Kenflex flooring made of Bakelite Vinyl Resins by **Kentile, Inc.,** Brooklyn, N. Y.

**There's always room for this kind of beauty**

Flooring made of Bakelite Brand Vinyl Resins gives you a greater selection of colors and patterns that satisfy clients' desires. You not only have more to please them with, you can assure them that the bright lustrous beauty of flooring made of Bakelite Vinyl Resins will last years longer.

Remember: because of Bakelite Vinyl Resins, the impervious nature of this flooring shuts out soil and wear, resists chemicals, cleansers, scuffs and scars. Therefore, you can assure excellent economy and very low maintenance.

These are attractive advantages for any building. You can provide them easily by making "flooring made of Bakelite Vinyl Resins" your standard specification.

**BAKELITE COMPANY,** *A Division of Union Carbide and Carbon Corporation* UCC 30 East 42nd Street, New York 17, N. Y.

The term Bakelite and the Trefoil Symbol are registered trade-marks of UCC

**Figure 31.1.** Kentile's Kenflex vinyl tile flooring used vinyl resins supplied by the Bakelite Company, a division of the Union Carbide and Chemical Company. As resins needed for vinyl tile became more affordable in the 1950s, production increased dramatically. *Architectural Record,* October 1955.

KIMBERLY A. KONRAD; PAUL D. KOFOED

# 31 | VINYL TILE

**DATES OF PRODUCTION**

c. 1933 to the Present

**COMMON TRADE NAMES**

Azphlex, Corlon, Cortina, Flexachrome, Flor-Ever, Floron, Kentile, Lonseal, Plastile, Terraflex, Themetile, Vina-Lux, Vincor, Vinylflex, Vinylite

## HISTORY

Vinyl floor tiles are generally made from polyvinyl chloride-acetate or pure polyvinyl chloride, various fillers, and pigments. They are offered in 9-by-9 and 12-by-12-inch squares with thicknesses ranging from 1/16 to 1/8 inch. Physical properties include stain, indentation, and abrasion resistance; dimensional stability; and ability to resist moisture. Vinyl tiles are thermoplastic and subject to shrinkage and expansion caused by excessive heat.[1]

### ORIGINS AND DEVELOPMENT

Several years before the chemist Leo Baekeland introduced Bakelite, the first completely synthetic plastic, in 1909, he experimented with its potential commercial applications. Bakelite floor tiles were just one product he made by impregnating paper with Bakelite and exposing it to high temperatures.[2]

Although experiments with plastics were being conducted in the first quarter of the twentieth century, plastics did not enter the resilient flooring industry for practical use until the late 1920s, when they were used as a substitute binder in asphalt tiles. Asphaltic binder produced only dark-colored tiles (black, brown, green, and red), but the synthetic resin coumarone-indene permitted lighter-colored tiles.[3] By the late 1940s asphalt tiles consisted of 35 percent coumarone-indene resin and contained little or no asphalt.

The first vinyl floor covering was a semiflexible vinyl asbestos tile called Vinylite, made in 1931 by Carbide and Carbon Chemicals Corporation. This tile was exhibited in 1933 as part of the Vinylite House at the Century of Progress Exposition in Chicago.[4] However, appreciable amounts of vinyl tile were not produced until after World War II.[5] During the war, military applications, such as floor matting for bombers, required a major portion of vinyl resins, but the military also stimulated development in asphalt and vinyl tile technology.[6]

As the price of vinyl resin became more affordable in the postwar years, vinyl flooring made tremendous inroads in the hard-surface flooring market.[7] In roll or tile form, flooring made of polyvinyl chloride-acetate competed with rubber, asphalt, and linoleum.[8] By 1947 at least twenty-two firms were producing or had announced plans to produce various types of plastic floor coverings. This number had increased to thirty-four firms by 1952.[9]

Many manufacturers of other flooring products began producing vinyl tile because existing equipment used to manufacture asphalt and rubber tile could be readily adapted to produce vinyl asbestos tiles. The ability to convert to new materials with existing machinery and the decreasing cost of vinyl resins are two reasons the market for vinyl tile expanded so rapidly.[10]

### MANUFACTURING PROCESS

Polyvinyl chloride-acetate, the result of the copolymerization of vinyl chloride and vinyl acetate, is the most common plastic resin used in the manufacture of floor tiles. The tiles can be plasticized in a wide range of hardnesses. Like most vinyls, it is a naturally clear material, allowing an unlimited color range to be produced.

All vinyl tiles are manufactured by the same basic process, which involves mixing the materials under heat and pressure to produce a soft, moldable dough that is then passed through a series of calender rollers to reduce the thickness of the sheet in steps down to the required gauge. Color pigments are added during the mixing stage to create various design effects, such as variegating, marbling, and flashing, thus ensuring that in the finished product the design effect is present through the full thickness.[11]

The squares are cut out of the hot material either with a square knife and ejector plate or by punching the tiles with a male and female punch. The process of punching tiles produces a large quantity of waste; however, since the material is thermoplastic, waste can be reheated and used again.

On cooling, the tiles harden into their finished form and require no curing. Thermoplastic resin-bonded tiles

**Figure 31.2.** Materials are mixed under heat and pressure, and the resulting dough passes through a series of calenders to attain the proper gauge.

are ideally suited to large-scale production using continuous methods of manufacture. Provided the raw materials are at hand, vinyl tiles can be manufactured and packaged in as little as 6 minutes.

### USES AND METHODS OF INSTALLATION

Twentieth-century applications for vinyl tile range from domestic, commercial, and military to industrial. Beginning with the first installation in the three-room apartment of the Vinylite House in 1933, vinyl tile use grew dramatically during the 1950s. Vinyl flooring tiles were and continue to be available in three basic types: vinyl asbestos tile, vinyl composition tile, and backed vinyl tile. Vinyl asbestos tiles were offered in the standard 9-by-9 and 12-by-12-inch sizes and in 1⁄16-, 3⁄32-, and 1⁄8- inch thicknesses. Solid vinyl tiles were made in two thicknesses, 3⁄32 and 1⁄8 inch, and came in 9-, 12-, 18-, and 36-inch squares.[12]

Vinyl asbestos tile is composed of vinyl resin binder, usually vinyl chloride–vinyl acetate copolymer (30 percent); asbestos fiber fillers and crushed limestone aggregate (6 percent); and plasticizers, stabilizers, and color pigments.[13] The tiles owe much of their wearing strength to the 7R-grade asbestos fibers; 7R is a waste product of the asbestos mining industry and one of the shortest gradings with a length of less than 1 millimeter.[14] The gray tint caused by the use of asbestos fibers is the only color limitation of vinyl asbestos tile.

Backed vinyl tile consists of a wearing layer of vinyl resins, plasticizers, pigments, and fillers overlaid on a backing that may be alkali resistant or waterproof, or provide extra comfort and resilience. Backed vinyl, also referred to as cushioned vinyl, is formed by applying a vinyl laminate to a highly filled and relatively cheap backing. Backing tiles include asphalt-impregnated felt, linoleum, cork, or even solid vinyl made from reused scrap.[15]

Introduced in 1945, solid vinyl tile, which is more expensive than vinyl asbestos tile, is composed through the full thickness of straight polyvinyl chloride resin, plasticizers, fillers, and pigments formed under pressure while hot.[16] Solid vinyl, also referred to as homogeneous tile, produces exceedingly tough, flexible tiles. They can contain from 35 to 50 percent binder to create an opaque

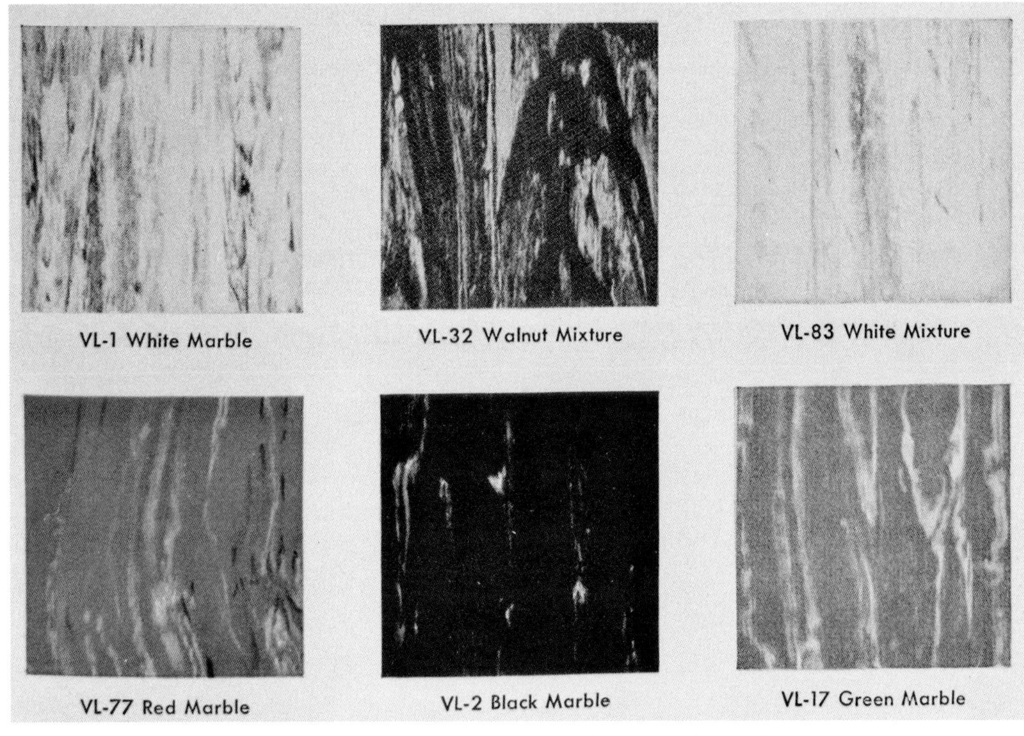


**Figure 31.3.** Pigments added in the mixing stage are used to create colors and patterns, such as marbleizing.

tile and up to 100 percent binder for the higher-priced translucent or transparent vinyl tiles.[17]

Vinyl tiles were usually applied to a rigid base, such as smooth concrete.[18] For wood bases, an asphalt underlay was recommended, either hot or cold, at least ½ inch thick.

Vinyl asbestos tile is recommended for application below, above, and on grade. Tiles are typically fixed to the floor by means of a soft asphaltic adhesive that is plastic and water resistant; it forms an impervious layer between the subfloor and the tile backing. Bitumen was used extensively in the early 1950s because it was the most successful in sealing the subfloor surface and was compatible with the binders in the tile.

## CONSERVATION

Vinyl tile typically falls into one of three major categories: vinyl asbestos tile, vinyl composition tile, and solid vinyl.[19] Each type has distinct chemical and physical properties that make it more or less suitable for a given existing substrate and environment. Failure is usually caused by poor design and improper specifications, moist environments, poor workmanship, improper substrate conditions, or a combination of these factors. Common failures include bubbling, cracking, oozing or bleeding of the mastic or adhesive through joints, and localized or widespread tile delamination.[20]

## DETERIORATION

Polyvinyl chloride, the binder component in vinyl tile, degrades by a process known as dehydrochlorination, in which hydrogen chloride is eliminated from the polymer backbone and forms conjugated double bonds.[21] This accounts for the gradual but pronounced yellowing of the vinyl tile surface when exposed to direct sunlight (ultraviolet light). Direct ultraviolet exposure can cause cleaving of the polymer backbone—a breakdown of the carbon–carbon bonds—that may result in superficial fading and chalking of the tile surface. This type of degradation usually occurs even when vinyl tile contains an ultraviolet-stabilizer additive.

High-moisture conditions can cause tiles to swell and generate volumetric stresses at adhesive bonding planes, chemically or physically break material, and cause gypsum-containing leveling-course substrate materials to deteriorate. Moisture may also induce migration of oily plasticizers from within the vinyl tile, and introduce bond-inhibiting compounds at the interfacial bonding planes between the adhesive and the tile and between the adhesive and the concrete.[22]

**Figure 31.4.** Vinylite floor tile was among the vinyl products used for the Pierce Foundation's demonstration apartment at the 1933 World's Fair in Chicago.

If water vapor can permeate the underlying concrete slab (or other substrate material) at a greater rate than it can permeate the existing vinyl tile flooring system, failure can occur. High moisture can promote swelling and generate stresses at the bonding planes between the tile and adhesive or the substrate and adhesive, which can lead to localized bubbling and delamination. Specifying an ineffective vapor retarder instead of a true vapor barrier beneath concrete slabs can also lead to moisture-related failure.

Moisture can also cause failure if an alkali-sensitive adhesive or unsuitable leveling-course material was specified. A cementitious leveling-course material is often used to level and smooth out an existing cementitious substrate. If it contains an additional gypsum component (other than the very small amount of gypsum additive already found in the portland cement component) or if it is composed primarily of gypsum, this material may become volumetrically unstable. Available moisture reacts with the alkali aluminate compounds within the portland cement and sulfate from the gypsum component to form ettringite, a very large molecule that causes stresses within the hardened cementitious system that can lead to a gradual disintegration of such material.[23] Moisture can also result in cyclic dissolution and recrystallization of the gypsum component. Both alkali reactions and gypsum recrystallization weaken the leveling course and affect the vinyl tile.

## DIAGNOSTICS AND CONDITION ASSESSMENT

Before undertaking any repairs, the type and cause of failure should be identified. Once existing conditions have been evaluated, remedial actions can then be developed, implemented, and maintained.

Whenever possible the original specifications, project correspondence, construction documents, records of generic or proprietary materials used, technical literature, and material safety data sheets should be reviewed to complement site and laboratory studies. An on-site condition survey is necessary to determine the type or mode of failure and may involve some destructive exploration.

Samples and photographic documentation may be taken for laboratory studies. Sampling typically consists of taking a substrate core with the associated leveling course, adhesive, and tile. Often subjective physical testing is undertaken to determine relative adhesion characteristics of the existing tile at localized areas representing good and bad performance characteristics.

Moisture emission rate characteristics of the flooring substrate should be evaluated during the field studies. A number of tests that are easily performed may be inadequate or too imprecise to definitively state whether moisture conditions are suitable. However, these tests, including the plastic sheet test (Standard Test Method for Indicating Moisture in Concrete by the Plastic Sheet Method, ASTM D4263) and use of an electronic moisture meter, can provide data that will indicate whether excessive moisture is present.[24]

For a more accurate evaluation, moisture-emission test kits can be used to gauge whether moisture conditions are suitable for installation of vinyl tile. The Moisture Emission Test Unit, developed by the Rubber Manufacturers Association, measures the rate at which water vapor permeates the concrete slab and its top surface into the atmosphere.[25] Emission rates are calculated in terms of pounds of water per 1,000 square feet of surface area per 24 hours. Values above 3 pounds of water per 1,000 square feet per 24 hours represent an unsuitable environment for organic-based adhesives and vinyl tile flooring.

In addition to testing on-site, petrographic and microscopic laboratory studies can be performed to confirm the type of failure and condition of installed vinyl tile flooring. These procedures also provide practical information as to whether failures should be classified as adhesive (a clean separation of adhesive from either the tile backside or the substrate bonding planes) or cohesive (an integral failure that occurs within the body of the adhesive material or within other adjacent materials), or a combination of adhesive and cohesive failures.[26]

Another useful laboratory technique for identifying organic compounds and polymeric materials used in tile and adhesives is Fourier transform infrared (FTIR) spectrophotometry. This technique helps identify the chemical type of adhesive present; whether the adhesive has been durable during its service life or whether it has chemically degraded in its service environment; potential bond-inhibiting compounds (for example, substrate sealers and curing compounds or oils); whether and to what extent the existing adhesive had been improperly mixed, misproportioned, or contaminated or improperly stored before use; and chemical characteristics of any other vinyl tile system component that may have played a role in the failure mechanism.[27]

## CONSERVATION TECHNIQUES

Vinyl floors typically have low maintenance requirements. Periodic sweeping and damp mopping remove dirt that can abrade and stain tiles. Sweeping compounds containing oils, sands, or abrasive materials should be avoided. Abrasive cleaners and highly alkaline cleaners, which can attack binders, fillers, and pigments, should not be used, especially on vinyl asbestos tile and vinyl composition tile.[28] Synthetic detergents in dilute concentrations are usually effective.[29]

Once tiles have been cleaned, thoroughly rinsed, and are dry, floor polish should be applied and maintained to seal pores and prevent absorption of liquids. Extremely worn vinyl asbestos and vinyl composition tiles can be sealed by this process to extend service life.

Whether tile installation failure is localized or widespread, its cause and the extent of damage should be ascertained before selecting repair methods and materials.[30] Establish on-site mock-ups of the chosen repair scheme to evaluate future durability of repairs. The American Society for Testing and Materials (ASTM) has developed standards pertaining to resilient flooring. Manufacturers of replacement tiles and adhesives should be informed of any previous studies and advised of potential problems. It is imperative to evaluate the current moisture emission rates of the vinyl flooring substrate to determine whether and to what extent moisture conditions will affect efforts to conserve a vinyl tile floor.

Vinyl tile flooring materials (tile and adhesives) that may contain asbestos should be submitted to a testing firm approved by the U.S. Environmental Protection Agency to confirm whether the materials contain asbestos, some of which are potential carcinogens, and should be considered health hazards during construction.[31] If vinyl flooring contains a form of asbestos classified as a hazard and that flooring is to be removed, an asbestos-abatement program must be followed in accordance with all applicable federal, state, and local regulations regarding the handling and abatement of the asbestos-containing material.[32] If a vinyl asbestos tile floor is in good condition, it may be best to leave the tile intact and protect it with a sealer.[33] When whole tiles are completely loose and do not require any mechanical or chemical methods of removal, replacement tiles can be installed and total abatement may not be necessary.

Conservation of an existing vinyl tile installation may require removal and replacement of tiles at localized areas. Finding replacement tiles to match existing tiles can be accomplished by contacting manufacturers. Depending on the project size, a custom order may be possible.

Selection of a suitable tile adhesive is paramount when selectively replacing vinyl tiles. Important durability characteristics include resistance to oxidation and sensitivity to moist alkaline environments. Adhesives on the market today include asphaltic mastic, latex emulsions, such as styrene-butadiene and acrylic-based materials, and synthetic hydrocarbon resins and wood resin blends.[34] These various adhesive types are not limited to any specific type of resilient flooring. Adhesives containing a polyvinyl acetate binder component are susceptible to chemical deterioration, particularly in a moist, alkaline environment such as concrete substrate.

Although all parties are responsible for a successful tile installation, the contractor has primary responsibility for good workmanship. The repairs should maximize the tiles' adhesion to their underlying substrate.[35] Before installation the substrate's moisture emission rate should be confirmed to be within acceptable threshold levels, the existing adhesive and coatings must be removed, and the substrate should be mechanically scarified where necessary.

If a remedial vinyl installation is at risk, then all parties involved must understand and acknowledge the inherent risk of proceeding with this type of approach, especially if a less breathable material such as vinyl composition tile or solid vinyl tile is being considered.

After replacement tiles have been installed, a more aesthetically uniform look can be accomplished by applying a finish wax or polish to limit any perception of surface imperfections such as fine cracks, scratches, and surface abrasion.[36] Cleaning procedures recommended by the manufacturer should be strictly followed after the tiles have been installed.

In some cases it may be possible to reuse existing tiles that are debonded but intact, particularly tiles other than vinyl asbestos tiles, which are brittle. The tiles designated for reuse may have curled or cracked edges and may be noticeably bowed, but they could still be used if they are reconditioned or restored to their original flat state. This can be done by subjecting the tiles to sustained loads on a flat surface at optimum temperature and humidity conditions.[37]

## REPLACEMENT

When a vinyl tile floor is beyond repair and conditions do not permit the installation of vinyl tile in a traditional manner, several alternatives can be considered. Under certain conditions, especially when a concrete substrate has a high moisture-emission rate, a remedial vinyl asbestos tile or vinyl composition tile floor may be considered too prone to long-term failure. This can happen if a slab-on-grade concrete substrate does not contain a true vapor barrier beneath the slab or if water infiltrates the substrate.

If a concrete or other type of substrate can be subjected to prolonged, accelerated, forced drying conditions until moisture conditions are acceptable, and if an exterior moisture source will not elevate moisture levels, then a typical remedial vinyl tile installation can be implemented. Likewise, a French drainage system can be installed beneath the existing slab to channel water away from the slab and foundation. Although this engineering procedure is effective, it often does not lower moisture levels enough for reapplication of a remedial vinyl tile floor. Also, if a true vapor barrier system could be installed over the existing substrate but beneath a high-performance, built-up overlay (such as a polymer-modified, cementitious topping compound), then a conventional vinyl tile floor could be reinstalled.

# PART VII
# ROOFING, SIDING, AND WALLS

ASPHALT SHINGLES

PORCELAIN ENAMEL

ACOUSTICAL MATERIALS

GYPSUM BOARD

BUILDING SEALANTS

**Figure 32.1.** Important factors in the selection of roof shingles were beauty, protection, and style. Through its dealers, Bird and Son sold a wide variety of hexagonal, square, and three-tab shingles. Catalogue, Bird and Son, 1935.

MIKE JACKSON

# 32 | ASPHALT SHINGLES

**DATES OF PRODUCTION**
1903 to the Present
**COMMON TRADE NAMES**
Fiberglas, Hex-A-Tite, Neponset Twin, Setab Shingles

## HISTORY

Asphalt, a bitumen commonly made from refined petroleum, is used to produce asphalt shingles.[1] These typically consist of individual units of roofing felt coated with asphalt and colored mineral or ceramic granules. Asphalt shingles are classified by their size, shape, weight, color, exposure, life expectancy, and fire and wind resistance.[2] Since the 1970s glass fiber–reinforced felts have been used to produce shingles that are otherwise similar to asphalt shingles.

### ORIGINS AND DEVELOPMENT

The earliest composition roofing products were felted or woven fabrics covered with a bituminous substance, such as pine tar and sometimes sand. This type of roofing was used in Newark, New Jersey, and Boston during the 1840s. These products were later improved by saturating and coating fabrics with asphalt and "talc, sand, powdered limestone or gravel" to impart color and improve durability.[3]

The S. M. and C. M. Warren Company, based in Cincinnati, is credited with developing the first true composition roofing product. In 1847 the company experimented with coal tar, a readily available by-product of the gas lighting industry.[4] Rolls of felt were saturated with coal tar and covered with fine gravel. Prepared, or roll, roofing followed in the last quarter of the nineteenth century. By the early twentieth century, ornamental surfaces on such roofing were commonplace.

The transition from roll roofing to asphalt shingles occurred in 1903. Herbert M. Reynolds, a roofing contractor and manufacturer of prepared asphalt roofing in Grand Rapids, Michigan, invented the asphalt shingle, which he hand cut from a roll of "stone-surfaced" roll roofing.[5] However, Reynolds never patented his method of cutting prepared roofing sheets into shingles. Even though most roofing companies focused on improving roll roofing, manufacturers slowly began to add shingles to their product lines.

The earliest asphalt shingles were cut into small shapes (8 by 12½ inches) and imitated wood shingles. In the industry's early years, surfacing materials were typically natural materials such as black, green, or red slate. A considerable range of patterns was produced, and many patterns and attachment systems were patented.[6] Rectangular and hexagonal shapes were the most common.

As the industry developed between 1900 and 1920, shingle sizes became more varied. Individual shingles were produced in sizes ranging from 8 by 12½ inches to 10 by 16 inches. The first important change was the evolution toward the multi-tab strip shingle. As early as 1906, Bird and Son marketed its Neponset Twin, a 12½-by-20-inch shingle with a slot tab dividing it to look like two shingles.[7] This concept caught on quickly within the industry; larger pieces reduced the installation costs, while the tab lines simulated the scale and effect of wood shingles. In 1911 the Prepared Roofing Manufacturers Association was formed to promote asphalt products and improve product quality.[8]

Another turning point occurred in 1916 when the National Board of Fire Underwriters published a booklet on shingles that advocated the elimination of wood roofing shingles as a fire hazard.[9] At the same time the board developed a standardized classification system for fire resistance of roofing materials.[10] Asphalt shingles were available with higher flame-resistance ratings than ordinary wood shingles, and manufacturers trumpeted this fact to promote sales.

The number of patterns and sizes reached its zenith in the 1920s.[11] Angled-cut patterns included diamond, hexagonal, and octagonal shapes. Shingles with random-cut ends to create a more rustic or thatched effect were available, as were curved and scalloped shingles. Thicker felts and additional granule coatings helped create shadow effects. Variations in tabs and interlocking shingles made it easier to keep installations aligned and offered extra protection against uplifting winds. The proliferation of shapes and patterns ended in the early

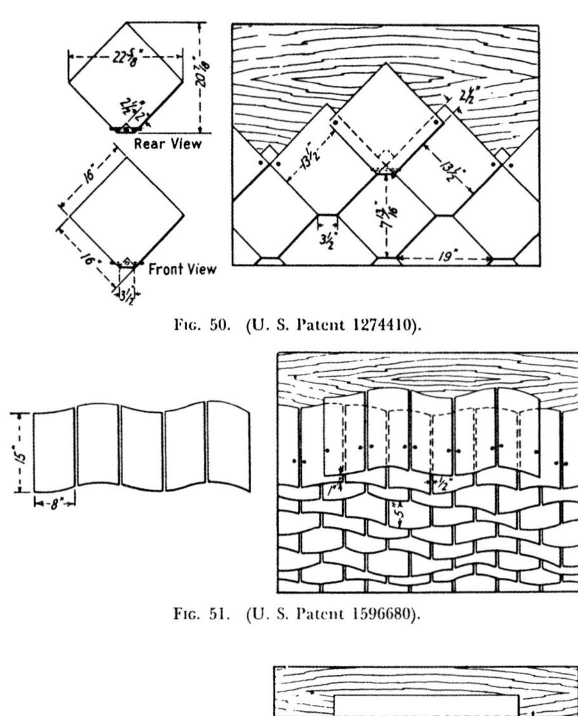


FIG. 50. (U. S. Patent 1274410).

FIG. 51. (U. S. Patent 1596680).

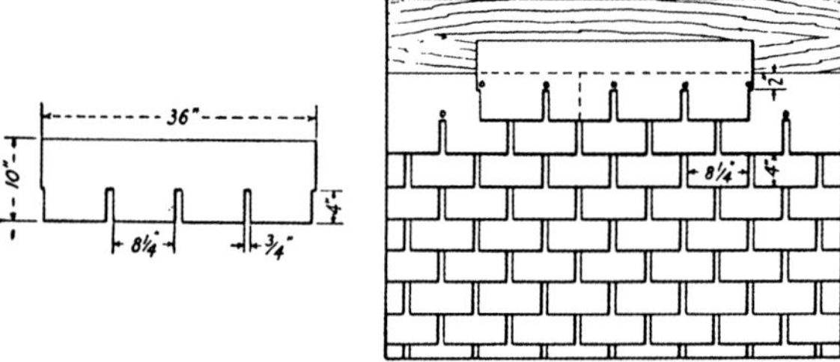


FIG. 52. (U. S. Patent 1150298).

**Figure 32.2.** Countless asphalt shingle patterns—hexagonal, thatched roof effects, conventional rectangles for multi-tab shingles—and methods of attachment were patented (above).

**Figure 32.3.** The Flintkote Company's 1931 promotional material illustrated its multi-tab and square-cut shingles and offered a siding pattern to imitate bricks, a relatively new development in the early 1930s (right).

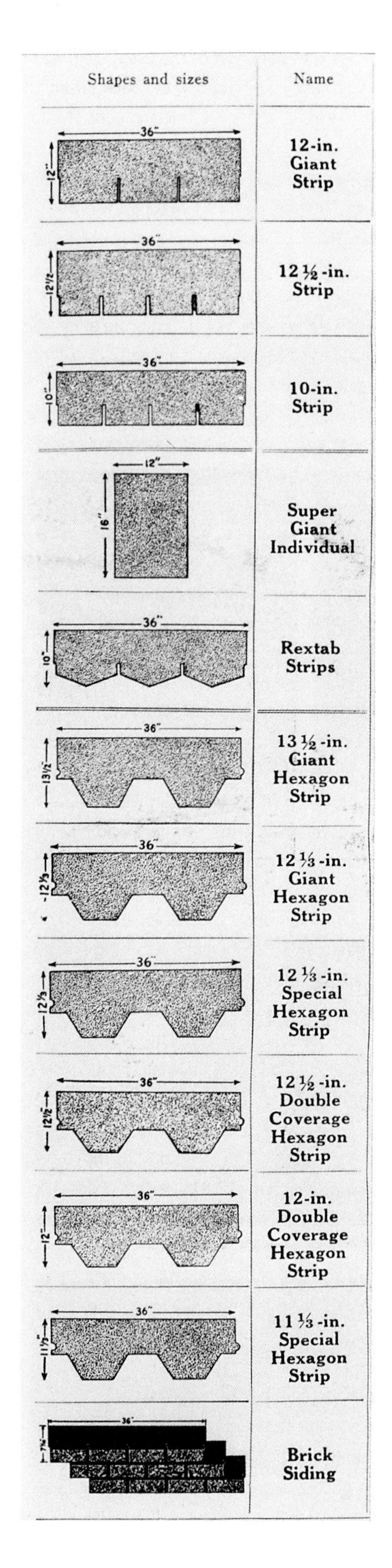

| Shapes and sizes | Name |
|---|---|
| 36" × 12" | 12-in. Giant Strip |
| 36" × 12½" | 12½-in. Strip |
| 36" × 10" | 10-in. Strip |
| 12" × 16" | Super Giant Individual |
| 36" × 10" | Rextab Strips |
| 36" × 13½" | 13½-in. Giant Hexagon Strip |
| 36" × 12⅓" | 12⅓-in. Giant Hexagon Strip |
| 36" × 12⅓" | 12⅓-in. Special Hexagon Strip |
| 36" × 12½" | 12½-in. Double Coverage Hexagon Strip |
| 36" × 12" | 12-in. Double Coverage Hexagon Strip |
| 36" × 11⅓" | 11⅓-in. Special Hexagon Strip |
| 36" | Brick Siding |

1930s with the onset of the Great Depression, when manufacturers were forced to cut back production because of the overall decrease in building activity. As the industry contracted, shingle dimensions and shapes became more standardized. By 1935 the 12-by-36-inch, multi-tab shingle, which is standard today, was available from all the major companies.[12]

While size and pattern became standardized, the range of available colors grew dramatically. By the early 1930s colors had evolved from the early slate colors to blended colors, in which a variety of color granules were used. Decorative, mottled effects could be achieved within a production run of shingles, or a variety of colors could be mixed together. The development of special rollers in the mid-1930s made it possible to imprint textures into the granular surface. In 1939 the Ruberoid Company began to offer asphalt shingles with a wood-grain texture. Textured shingles enjoyed great popularity in the 1940s and 1950s. In the 1950s the 12-by-36-inch multi-tab shingle was the predominant format, and bright blue shingles were particularly popular.

The most recent major change in the asphalt shingle industry occurred in the 1970s with the development of glass fiber–reinforced felts. The increased tensile strength of the glass-reinforced felts meant that thinner and lighter-weight shingles could be produced. By laminating several layers of these felts, more dramatic shading effects were possible.

### MANUFACTURING PROCESS

The manufacture of asphalt roofing shingles begins with the production of dry felts, the primary ingredients of which were originally wool rags, cotton, and paper. Later, wood fiber became the main ingredient.[13] The felts are produced in rolls on floating loopers and then are saturated with an asphalt flux, a specifically formulated distillation of crude oil.[14]

The saturated felts are passed through a pan that applies the coating to the asphalt, which historically may have included mineral fillers (also known as stabilizers), such as powdered stone, slate flour, or powdered oyster shells.[15] Next the sheets pass under a shower of mineral or ceramic granules, which provide the final wearing surface, and then are sprayed with water to lower the temperature.[16] Patterned rollers can be used to imprint textures or patterns in the surface.

The final step is cutting the material into roll roofing or individual shingles with a rotary knife. The process may also include dusting with talc or mica particles to keep the rolled or stacked materials from sticking to each other.

Color is created from fillers and mineral or ceramic granules, which are available in a wide range of colors and are usually blended to create a particular shade. The colors varied with the fashions of the day or with regional preferences and technology. Ceramic-fired granules, which were first used around 1930, produced colors different from those of natural crushed stones. Such granules were hard stones that could be coated with a ceramic slurry (frit) and fired.[17] The addition of asbestos increased fire resistance.

### USES AND METHODS OF INSTALLATION

Asphalt shingles quickly replaced wood shingles as an affordable roofing material. Since they were first developed, asphalt and fiberglass shingles have dominated the residential roofing market.

Shingles are installed in an overlapping method to produce a surface that sheds water even though it is made up of many small pieces with open joints between them.[18] The amount of exposed shingle varied from shingle to shingle. On low-slope roofs, two layers of felt underlayment were recommended.[19]

Methods of fastening the shingles also varied. Many of the earliest shingles were nailed to roof decks with nails driven through tern-plate disks.[20] Disks made from thick roofing felts and cork have also been used. Ultimately, large-headed, galvanized roofing nails became the predominant method of fastening shingles. Once shingles were fastened, a spot cement was often applied under the center of each tab with a putty knife or caulking gun.[21]

## CONSERVATION

The durability of asphalt shingles depends on aging, maintenance practices, and product quality. All asphalts change over time and are subject to physical and chemical weathering. Despite these effects, a typical asphalt shingle roof will last twenty years with little obvious deterioration.

### DETERIORATION

Constant exposure to moisture and drying will break down the asphalt-and-felt-saturated base. Mineral and ceramic granules are highly resistant to exposure, but asphalt and fiberglass elements degrade. If the soluble constituents of the asphalt are leached away, shingles can warp, buckle, blister, and lose their granular covering.[22] Thinner shingles may curl if the felt was inadequately saturated.[23]

Exposure to oxygen causes the asphalt to oxidize, leading to hardening and deterioration of the shingle.

**Figure 32.4.** Properly installed asphalt shingles are applied over roofing felt, but shingles were sometimes installed directly over existing roofing systems (top).

**Figure 32.5.** Taking the house and its roof on the road in a veritable mobile showroom was the way this company sold its asphalt shingles (middle).

**Figure 32.6.** The National Park Service hand cut tabless shingles using a custom jig to create the historic pattern for the Harry Truman Home, Independence, Mo. (bottom).

Light acts as a catalyst for this form of deterioration, and this chemical change is often irreversible.[24] Other factors that may affect durability include exposure to corrosive chemicals and organic growth.

Failure typically occurs not with the shingles but with the system components, particularly connections, which are frequently sources of water infiltration. When the asphalt shingles connect to vertical surfaces adjacent to the roof, a special flashing material is necessary to ensure a continuous barrier against water intrusion. Surface application of a bituminous emulsion is only a temporary expedient. As these materials age, they lose their flexibility and ultimately crack at the point of connection. The best roofing systems use separate flashing materials, such as step flashings, that extend under and above the shingles and up along the wall of the adjacent element.

Valleys where two roof slopes come together will generally be subjected to twice the volume of rain water as typical shingles and are consequently the first areas to fail. Valleys constructed of lapped shingles are particularly vulnerable to the additional weather exposure. Separate valley linings, which are thicker and more durable than regular shingles, are recommended as a longer lasting valley detail. At the intersection of a low-slope, asphalt shingle roof with a steeper one, the differential of thermal expansion between the two surfaces and the tendency for snow and ice to build up frequently result in leaks.

Damage can result from walking on a roof on a hot day or from a falling branch. Shingles can be dented and the mineral granules pushed away, exposing the felt core to advanced deterioration.[25] On a cold day, the entire shingle can be cracked and broken if walked on or hit.

## DIAGNOSTICS AND CONDITION ASSESSMENT

Roof leaks are the most obvious sign of failure. Aside from assessing the origin of potential leaks, it is important to evaluate the condition of individual shingles. Scraping shingles can determine how well the surface granules are adhering to the asphalt. A visual inspection can also identify serious hardening and cracks. The accelerated erosion of the surface granules or buckling and warping of the shingles are the most critical signs of impending failure. Although external conditions have the most immediate effect, other factors should be investigated. Inadequate attic ventilation can contribute to a shorter life span; it will cause consistently high temperatures and can result in moisture condensation below the felt underlayment, thus causing deterioration of the roof decking and shingles.

## CONSERVATION TECHNIQUES

Little information about the conservation and repair of individual shingles is available; most relates to preventing and stopping leaks. The use of paints, coatings, or water repellents that can extend the life of asphalt shingles is currently not an industry practice. One new technique being marketed is adding a metal (zinc) strip along the upper roof so that during rains minute quantities of the metal are washed down the surface of the shingle; the metal acts as a fungicide to prevent organic growth. Another new installation technique is adding a layer of roll roofing below the areas of the roof subject to ice dams. The effect is that of installing a low-slope roofing material so that water forced under the shingles is kept off the roof deck. Similar materials are used for valley flashings.

Regular maintenance practices such as cleaning gutters and downspouts can prevent many problems. Trees and branches that can abrade the roof should be trimmed, and adequate surface ventilation is needed to avoid prolonged dampness. Downspouts from upper roofs, such as dormers, should not discharge onto lower roofs but should be extended to the gutter system. Physically damaged shingles usually require replacement. It is good practice to provide owners of a new roof with several extra bundles of shingles so that matching pieces are available for repairs.

A major conservation issue related to asphalt shingle roofs is algae discoloration. Dark streaks are an unsightly result of the growth and by-products of several airborne types of algae, the most common being *Gloeocapsa*.[26] This condition is found most frequently in the southern and eastern coastal areas. The discoloration can be lightened by washing the roof with a solution of household bleach, trisodium phosphate, and water. Scrubbing or high-pressure water treatments are not recommended because these techniques can dislodge the surface granules; moreover, they will not permanently kill the algae, and staining may recur.

## REPLACEMENT

Asphalt shingles and their modern fiberglass-reinforced varieties are still in regular production and available in the same 12-by-36-inch dimensions that have been standard since the 1930s. Red, green, and blue-black shingles are all still in production, and some hexagonal patterns are available. For most roofing projects, finding a new asphalt shingle similar to the existing one should not be difficult.

Replacing an asphalt shingle of a particular shape or color blend not in regular production today is more difficult. The basic technological apparatus used to make shingles can be modified with custom mixes of granules to duplicate virtually any historic shingle, and it is even possible to change the cutting dies to produce a special shape. Unfortunately, the minimum quantity of shingles that most manufacturers are willing to produce is far beyond the quantity needed for a typical roofing project.

To create a shingle that accurately duplicates a historic pattern, new shingles can be cut from roll roofing. Many manufacturers still make granular roll roofing in a variety of colors. Dark green, red, black, and white are readily available and are typical of the colors of early asphalt shingles. Asphalt shingles can easily be cut (from the back side, with a utility knife) to duplicate the shape of an earlier shingle. Cutting a small quantity of shingles from rolled roofing—a typical house would require about ten squares—would not be difficult. These shingles will be slightly lighter than typical shingles, but they can be sealed down with roofing mastic. Standard shingles without tabs can also be purchased, and tabs can be cut to match a historic pattern.[27]

**Figure 33.1.** Republic Steel was one of several companies that manufactured and supplied iron specifically engineered for enameling. In addition to being flat and free of waves, enameling iron maximized enamel adherence. *Architectural Porcelain Enamel,* 1938.

THOMAS C. JESTER

# 33 | PORCELAIN ENAMEL

**DATES OF PRODUCTION**
c. 1924 to the Present
**COMMON TRADE NAMES**
Glasiron Macotta, Mirawall, Porcelite, Porcelok, V-Corr, Veos, Zourite

## HISTORY

Porcelain enamel is a thin coating of glass fused to metal at temperatures above 800 degrees Fahrenheit.[1] Also known as vitreous enamel, porcelain enamel is a nonporous composite material that can be manufactured with a range of physical and chemical properties.[2] Iron, steel, aluminum, and stainless steel are the most common substrates for architectural enamels.

### ORIGINS AND DEVELOPMENT

Porcelain enamel has been used for centuries for jewelry, vases, and other art objects, but the application of porcelain enamel to sheet steel first occurred in Austria and Germany in the mid-nineteenth century.[3] The material's potential in building construction was explored around 1890 by Theodor Bergmann, owner of a German enameling plant, who applied enameled steel sheets resembling shingles to his house in Gaggenau, near Stuttgart.[4]

Advanced enameling technology was introduced in the United States during the last quarter of the century, and many early American enameling facilities were operated by German immigrants or relied on German supplies.[5] The industry expanded rapidly in the early twentieth century as enameling became more firmly rooted in engineering and inorganic chemistry.[6] In 1909 the American Rolling Mill Company, based in Middletown, Ohio, developed an iron low in carbon and manganese content expressly for enameling.[7] By the 1920s porcelain enamel was widely used for kitchenware, appliances, and bathroom fixtures. To promote new products, organize research, and develop new markets, including construction, manufacturers formed the Porcelain Enamel Institute in 1930.

### MANUFACTURING PROCESS

Porcelain enamel is made by fusing a glassy composition to a metal substrate, often cast iron or steel. The enamel, or frit, is a mixture of minerals such as silica, feldspar, and quartz; fluxes such as borax, soda ash, and iron oxides; and opacifiers. The raw materials are combined, smelted at high temperatures, quenched (cooled), and ground in a mill.

Historically, porcelain enamel was applied to metal substrates using either a dry or a wet method. Most enamels were applied to cast iron using the dry method, where the frit was applied to the heated substrate through a sieve.[8] In the wet process enamel was suspended in a slip (emulsion) of water, clay, and frit. The mix was applied by dipping or spraying. Ground coats, which were typically dipped, formed the bond with the metal substrate; cover coats, which were sprayed on, provided opacity, color, and acid resistance. Thickening agents were added to control the flow of the enamel and ensure proper thickness. Porcelain enamel components made of iron and steel were subsequently dried and then fired at temperatures between 1,400 and 1,600 degrees Fahrenheit.

After World War II new enameling steels were developed, including titanium steel, which Inland Steel produced in 1947 and sold as Ti-Namel.[9] By 1950 enamel formulations had also been developed for aluminum.[10] Aluminum enamels were lighter, stronger, more corrosion resistant, and more workable than steel.

A variety of finishes and textures have been used for architectural enamels. During the 1930s glossy finishes, also known as lustrous or glazed finishes, predominated. Textured finishes like the white-and-tan "oatmeal" mottle and stipples were popular in the late 1940s and were used for many commercial chains, including Kresge.[11] Manufacturers also used corrugated, crimped, and embossed metal substrates.[12] Matte and semimatte finishes were not used extensively until the 1960s.

### USES AND METHODS OF INSTALLATION

Porcelain enamel sheets, panels, tiles, and shingles all originated in the mid-1920s, but true commercial markets did not develop until the early 1930s.[13] The White Castle restaurant chain probably pioneered the use of porcelain enamel sheets in 1925, when it designed and constructed

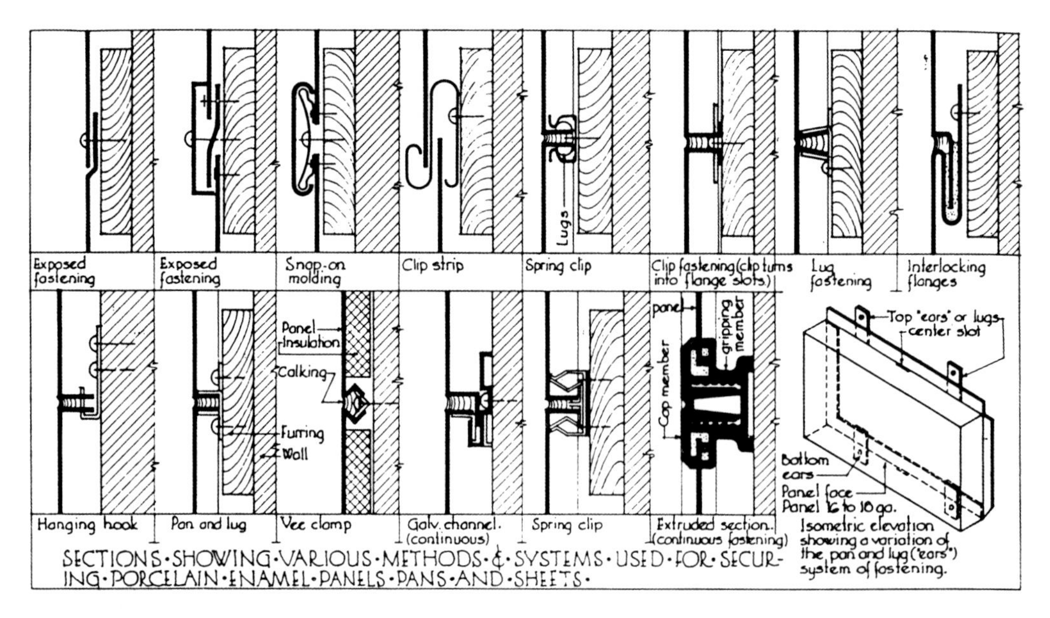


**Figure 33.2.** In the late 1930s porcelain enamel sheets and panels could be attached to wood, masonry, and concrete backup with a dizzying array of systems. Some were simply sheets applied with screws, while others required sequential installation. Joints could be covered with moldings or battens or filled with caulking.

a St. Louis restaurant with a porcelain enamel interior.[14] In 1929 in Wichita, Kansas, White Castle opened the first building featuring porcelain enamel panels on both the interior and the exterior. The company's superintendent of construction, L. W. Ray, invented and patented a key-lock device for its exterior panels. Success with the panels prompted the company to market other porcelain enamel building types, notably gas stations.[15]

Porcelain enamel shingles had been exhibited by the German industrialist Theodor Bergmann at the 1893 World's Columbian Exposition in Chicago.[16] However, they were not produced in the United States until 1924, when the Columbian Enameling and Stamping Company built a house roofed with shingles in Terre Haute, Indiana.[17] As early as 1924 the Detroit-based Vitrified Metallic Roofing Company also began producing shingles.[18] Most shingles were shallow, flanged pans with a convex appearance, not larger than 12 square inches.

Porcelain enamel tiles were also developed in the mid-1920s. In 1925 the Chicago Foundry Company produced experimental tiles that attached to walls with interlocking lugs and screws.[19] The same year the Porcelain Tile Company of Chicago introduced a product called Porstelain.[20] These flanged tiles, made of 20-gauge stamped steel, were cemented in grooved insulation boards measuring 4 by 5 feet. Originally, these were available in two standard sizes (4½ by 4½ inches and 3 by 6 inches), but larger sizes (6 by 6 inches and 6 by 12 inches) were introduced after the Youngstown Pressed Steel Company purchased the company in the early 1930s. The trade name was changed to Veos (an acronym for vitreous enamel on steel).[21] Other manufacturers soon appeared.[22]

Porcelain enamel shingles remained popular with restaurant chains. Howard Johnson's adopted its signature bright orange shingles, manufactured by Wolverine Porcelain Enamel, in the late 1930s.[23] Other chains that used shingles included White Castle, White Tower, Stuckey's, and A&W. However, porcelain enamel shingles were expensive to make and, like porcelain enamel tiles, struggled to compete with other materials.[24]

The largest area of growth—in architectural panels—coincided with the increasing popularity of the Moderne style.[25] Many stores, theaters, restaurants, and particularly gas stations—Esso, Mobil, Texaco, Shell—used porcelain enamel to establish images of modernity. Architects became attracted to porcelain enamel because of its purported color permanence, flatness, weathering characteristics, and resistance to abrasion.

In 1932 the Ferro Enamel Corporation built the "world's first porcelain enameled house" with a porcelain enamel roof, chimneys, siding, wall, and light fixtures.[26] National expositions in the 1930s also provided opportunities to demonstrate how porcelain enamel could be used for construction. Two porcelain enamel houses were erected for the Century of Progress Exposition in Chicago in 1933: the Armco-Ferro House, designed with 13-inch-wide porcelain enamel sheets, and the Stran-Steel House, constructed of panels 2 feet wide and 8 to 10 feet long. Manufacturers also exhibited porcelain enamel products at the Great Lakes Exposition in 1936.[27] By the late 1930s a large number of companies were selling architectural panels.[28] Architects began using porcelain enamel panels for storefronts, tollbooths, schools, and offices.

While tiles, shingles, sheet stock, and corrugated roofing were usually prefabricated, virtually all flanged panels and spandrel panels were custom sized and installed sequentially on-site.[29] The simplest porcelain enamel components were flat sheets used for exteriors and wainscoting held in place with screws and battens.[30] Beginning in the late 1930s, flanged veneer panels (pans) became the most common panel type. Panel cores varied from insulation board to paper honeycomb to foam core; some were not laminated at all.[31]

Most manufacturers developed their own attachment and fastening methods. Panels could be fastened together and to masonry, wood, and steel backup with screws, springs, hooks, clips, lugs, or clamps.[32] Some techniques relied on moldings or battens to conceal joints and hold panels in place. Others had joints filled with caulking.[33] By the early 1940s architects could choose from at least seventeen methods.[34] Because some building codes either did not permit or limited the use of porcelain enamel, the Porcelain Enamel Institute adopted its first standard specification in 1946.[35]

Following World War II new formulations enabled thinner coatings of enamel to be applied to the base metal. One example was Mirawall, introduced in 1947 and touted as "enamel wallpaper."[36] Mirawall was a thin sheet of porcelain enamel backed by Masonite with an adhesive. It was used for walls in diners, restaurants, hospitals, supermarkets, and bathrooms. In 1949 the Kawneer Company of Niles, Michigan, developed the first enamel for aluminum, which quickly gained acceptance for spandrel panels, storefronts, and signs.[37] However, steel has remained the preeminent metal substrate for enameling.

New markets also emerged. Before the war porcelain enamel in houses had been limited to bathroom tiles, sinks, and appliances.[38] Between 1947 and 1950 the Lustron Corporation, a short-lived but notable housing venture, developed and sold approximately 2,500 prefabricated houses made almost entirely of porcelain enamel components.[39] Panels measuring 2 by 2 feet were used for exterior walls, and panels 2 by 8 feet were used for interior walls.

**Figure 33.3.** Porcelain enamel panels are installed sequentially to a wood-framed structure in the late 1930s. The panels were fastened by hand.

In the early 1950s architects began to use porcelain enamel for spandrel panels in curtain wall systems.[40] Eero Saarinen designed one of the first true porcelain enamel curtain wall systems for the General Motors Technical Center (1950) in Warren, Michigan.[41] In 1954 the Porcelain Enamel Institute retained William Lescaze to develop a porcelain enamel curtain wall system.[42] Lescaze stressed the importance of creating a "complete exterior wall system" to "go beyond the skin," a reference to the use of porcelain enamel as a veneer facing.[43]

The distinction between veneer panels and composite panels was not always great. In fact, many manufacturers produced panels that could be used as facing and in curtain wall systems. Composite panels for curtain wall systems, however, generally required thicker insulation to meet fire-rating requirements in building codes.[44] Throughout the 1960s composite panels continued to be used, although today facing panels are more prevalent.

## CONSERVATION

As a composite material of glass fused to metal, porcelain enamel's chemical and physical properties vary widely.[45] Resistance to corrosion, stresses and strains between the glass and metal bond, manufacturing quality, and maintenance practices all affect an enamel's

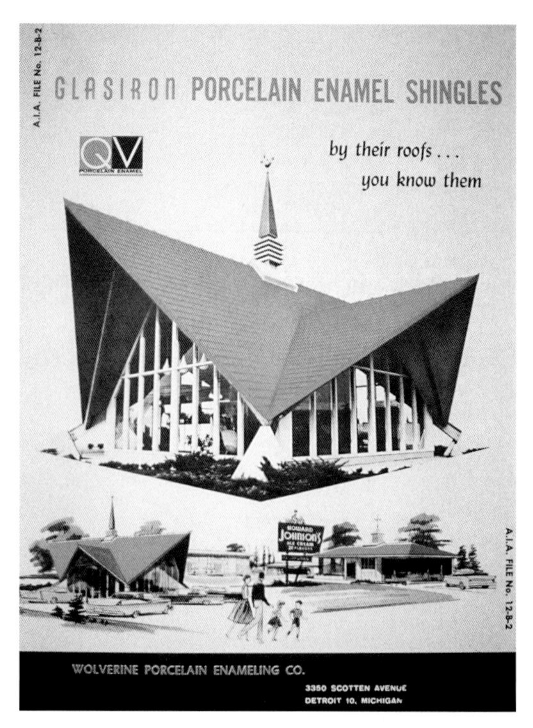


**Figure 33.4.** White Castle's Wichita, Kans., store, which opened in 1929, probably exemplified the first use of flanged porcelain enamel panels (above).

**Figure 33.5.** The Wolverine Porcelain Enameling Company supplied Howard Johnson's orange Glasiron porcelain enamel shingles (right).

**Figure 33.6.** Standard Oil was one of many oil companies to build service stations using porcelain enamel beginning in the mid-1930s.

**Figure 33.7.** In this Lustron house and garage (1950), Chesterton, Ind., the only elements not made of porcelain enamel were the windows and structural frame.

durability and weathering characteristics. Generally quite hard and abrasion resistant, porcelain enamel is also resistant to common organic solvents and oils.

## DETERIORATION

Corrosion of the glass coating can be caused by water, acids, alkalis, or atmospheric pollution.[46] When porcelain enamel corrodes, physical and chemical changes occur on the enamel surface.[47] Temperature, humidity, and accumulation of water all influence the extent of weathering and may cause pitting of the enamel surface.[48] Changes to the enamel surface also appear as hazing (fading) and can be detected by reduced gloss and color loss.

The most common form of deterioration is corrosion of the base metal, frequently steel, which can be caused by direct impact and manufacturing defects. Research indicates that base-metal corrosion probably causes spalling of the porcelain enamel coating.[49]

Direct impact can fracture the chemical bond between the glass and metal, exposing the base metal. Thicker, less flexible porcelain coatings, such as the mottled finishes, are more prone to chipping. Vandalism, accidents, and careless installation can all cause chipping.

The most common manufacturing defect is inadequate enamel coverage, which can cause corrosion if the porcelain enamel does not cover the panel edges or welded clips. Once corrosion begins, it can spread to the panel face. Deteriorated or missing joints between panels allow water infiltration and may lead to this type of deterioration. Manufacturing defects may cause crazing and warping.[50]

## DIAGNOSTICS AND CONDITION ASSESSMENT

The condition of historic porcelain enamel panels should be assessed before making any repairs. Architectural drawings, specifications, and shop drawings may reveal the metal substrate, attachment method, and original caulking material. To evaluate the panels' structural stability, physically press on each panel to detect loose or detached panels. Because the attachment fasteners are usually concealed, it may be necessary to remove selected panels by cutting the fasteners between the panels with a reciprocating saw. Tape can be applied to the panel surfaces and panel flanges to prevent chipping when the attachment clips in the joints are cut.

The condition of the panel joints should be evaluated to detect any avenues of water infiltration. Oil-based knife caulks, extruded plastics, and synthetic rubbers, such as Thiokol, were all used historically. Today a variety of improved caulking materials and sealants are available to supplement or replace deteriorated caulking materials, which can usually be matched with the original joint color.

**Figure 33.8.** Eero Saarinen and the Wolverine Porcelain Enameling Company developed a porcelain enamel curtain wall panel with a honeycomb core for the General Motors Technical Center (1950), Warren, Mich.

## CONSERVATION TECHNIQUES

Although there are no accepted techniques for conserving architectural porcelain enamel, a number of experimental conservation measures can be considered.[51] These techniques are adapted from conservation techniques for museum objects, processes for refinishing bathroom fixtures, and field experiments and practices. The reversibility of untested treatments should be evaluated and weighed against other options.[52]

Cleaning may be desirable to reveal the original finish, especially if the panels are sound and show little or no corrosion, or as preparation for a coating on corroded areas. Test panels should be used to evaluate the effectiveness of various cleaning solutions. Cleaning should begin with warm water and a clean cloth. Large installations may require low-pressure water or steam. Stubborn dirt can be removed with a 1 percent solution of trisodium phosphate, but acid-based cleaners should be avoided.[53]

Grease and oil can be removed with alcohol-based solvents. Paints can be removed with proprietary strippers, but caustic ones, which can etch the porcelain enamel, should be avoided.

Exposure of the base metal is a common form of deterioration. When in otherwise good condition, panels should be repaired by removing corrosion products and applying a protective coating on the affected areas. Epoxy, urethane, and lacquer coatings have been used on porcelain enamel.[54] Irreversible repairs that require etching the panel surfaces as preparation for a coating should be avoided. Limited application of coatings to arrest corrosion in specific areas is cautiously recommended.

Pigmented lacquers are commonly used by conservators for retouching porcelain objects and were also used historically by porcelain enamel sign manufacturers to repair minor manufacturing defects.[55] Urethanes have been used widely to repair cast-iron bathroom fixtures. The ability of a lacquer or urethane to retain the desired color, adhere to the metal substrate, and withstand ultraviolet light is an important consideration.[56]

Surface-tolerant epoxies may offer longer protection than lacquers and urethanes. Corrosion must first be removed carefully with a wire brush or Dremel tool. When epoxy is applied near panel edges, proper surface cleaning and preparation are required to avoid reactions with sealant residues (often oils).[57] A durable paint, such as asphaltic urethane, can be applied to match the porcelain enamel color.

Polyester fillers are used to repair porcelain enamel bathroom fixtures and fill in small areas of chipped enamel where the base metal is not exposed; they have been used experimentally for architectural panels.[58] However, it may be difficult to achieve a suitable color match and sand the filled-in areas without scratching the surrounding glass surface. Such repairs are made principally for aesthetic reasons.

## REPLACEMENT

Selective replacement of severely deteriorated porcelain enamel components is possible. In curtain wall systems, replacement may be necessary if corrosion compromises the adjacent curtain wall stops that hold panels in place.

Each type of porcelain enamel must be approached differently. Flanged panels, concrete-backed panels, and spandrel panels can be replaced with new ones that match the historic color and finish. Solid colors are relatively easy to match, but stippled and mottled panels with more than one color are more difficult and expensive to match.

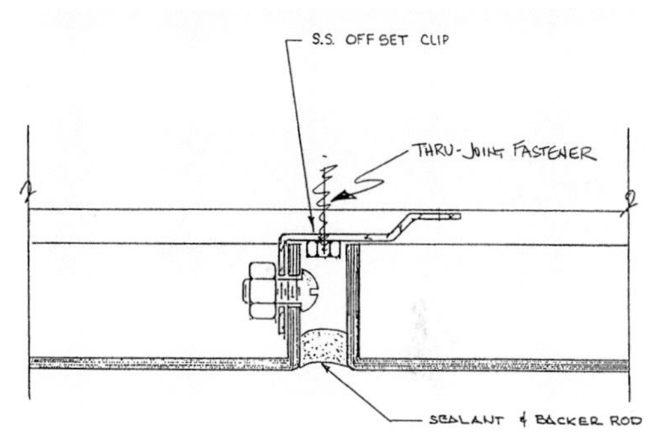


**Figure 33.9.** Dented and chipped panels on a former Texaco gas station in Austin, Tex., reveal (top) that direct impact has fractured the bond between the enamel coating and metal substrate. When chipping occurs, the exposed metal substrate corrodes.

**Figure 33.10.** When replacement panels are installed next to existing panels, an offset clip and through-joint fastener can be used to attach new panels (bottom). Panel joints are then sealed.

When a flanged panel is replaced, the surrounding caulking is typically removed and a reciprocating saw is used to cut the clips holding the panel in place. A custom-designed panel is then inserted with shortened stainless steel clips and through-joint fasteners.[59] The final step is to reseal the perimeter joints around the new panel.

**Figure 34.1.** The perforations in acoustical tiles absorbed unwanted noise, making them suitable for schools, offices, and restaurants. Tile thicknesses, ranging from ½ inch to 1 inch, determined the sound-absorption coefficient. Catalogue, Owens-Corning Company, 1954.

ANNE E. WEBER

# 34 | ACOUSTICAL MATERIALS

**DATES OF PRODUCTION**
c. 1892 to the Present

**COMMON TRADE NAMES**
Acoustex, Acousti-Celotex, Acoustifibroblock, Acoustone, Akoustolith, Cabot's Quilt, Corkoustic, Cushocel, Hushkote, Kalite, Kilnoise, Macoustic, Nashkote, Perfatile, Quietile, Rumford Tile, Sabinite, Sanacoustic, Silen-Stone, Softone, Temcoustic

## HISTORY

Acoustical materials are used in buildings to absorb sound, reduce reverberation, or isolate sound by limiting its movement from one room to another. Acoustical materials can generally be classified as felt-and-membrane systems, acoustical plasters, and board-and-tile materials.

### ORIGINS AND DEVELOPMENT

One of the earliest materials with purported acoustical properties was Cabot's Quilt. Made of cured eelgrass enveloped in sheets of paper, Cabot's Quilt was patented in 1892 and marketed for its insulation and "sound deadening" properties; it remained in production through the early 1930s.[1] Other varieties appeared, including a fireproof form laminated with mica.

The scientific study of architectural acoustics in the United States began in earnest in 1895, when Wallace Clement Sabine, a Harvard professor, was hired to investigate and correct the acoustical quality of the university's recently completed lecture room at the Fogg Museum. From this pioneering study and subsequent experimentation, Sabine developed a theory of architectural acoustics based on a room's volume, the construction material, and the reverberation time.[2] He concluded that the most common problem in architectural acoustics was excessive reverberation.[3]

In 1911 Sabine, in collaboration with Raphael Guastavino, developed Rumford Tile, a mixture of clay, peat, and feldspar.[4] When the tiles were fired, the peat burned off, creating a porous tile that absorbed sound. St. Thomas's Church (1914, Cram, Goodhue and Ferguson) in New York City was the first building to use Rumford Tile.[5] Another tile produced by Sabine and Guastavino was Akoustolith, made of pumice and portland cement and patented in 1916. Rumford and Akoustolith tiles were used extensively in churches.

### USES AND METHODS OF INSTALLATION

The flood of materials and systems developed in the early 1920s to combat excessive reverberation included felt-and-membrane systems, board-and-tile materials, and acoustical plasters. All aimed to reduce reverberation by increasing absorption and were widely used in offices, schools, theaters, and apartment buildings.

The felt-and-membrane systems developed first and consisted of felt, an absorber, covered by a membrane, usually fabric. Ideally the membrane was completely permeable by sound waves; they would pass through it and be absorbed by the felt. The membrane protected the felt from dirt and wear and provided a more attractive finish.

In the 1920s the product line of Keasbey and Mattison, asbestos miners and manufacturers, included acoustical treatments of the felt-and-membrane method and Sabinite sound-absorbing plaster.[6] Felt and membranes were recommended for acoustical correction in existing structures. An air space was left between the felt, which was applied directly to the plaster substrate, and the canvas membrane.[7] Acoustically this system functioned as long as the membrane remained permeable, which generally meant that it was not painted.[8]

The main problems with the felt-and-membrane system were the time-consuming method of applying furring, felt, and membrane as three distinct elements, the inability to paint the membrane without impairing the system's acoustical properties, and its vulnerability to wear and damage. Several companies developed products to deal with these problems.

One early leader in the acoustical materials field was Johns-Manville, the large producer of many asbestos-based materials. In fact, asbestos fiber was the basis of many of its absorptive acoustical products. Absorption occurs when sound waves can reflect themselves out of existence in small spaces. Fibrous materials provide these conditions, and the fireproof nature of the asbestos fiber made it an attractive component for commercial acoustical materials.

The basic composition of the Johns-Manville felt-and-membrane system, known as Nashkote, was canvas laminated to ¾-inch Akoustikos Asbestos Felt. Generally

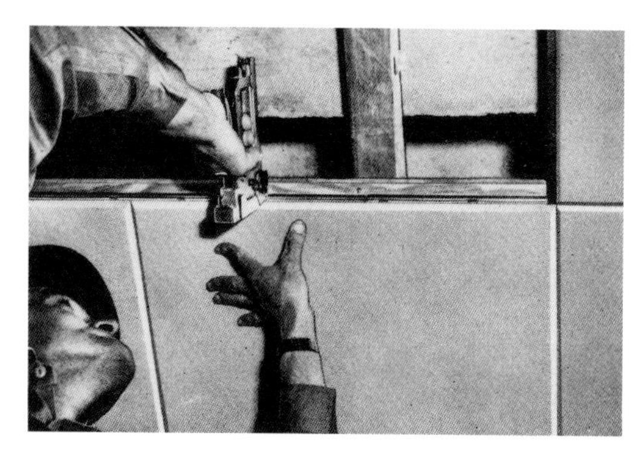

**Figure 34.2.** Acoustical tiles are stapled to furring strips with aluminum splines that support the tile in the kerfed edges. Tiles could also be attached with adhesives, nails, screws, or suspended T-bar systems.

it would be glued or nailed to a solid substrate, thus eliminating the application of three separate elements and increasing construction efficiency. Nashkote came in several varieties, with perforated or solid canvas and oilcloth as optional finishes. Some were paintable, such as Nashkote AIS, which the company recommended finishing with textured paint to achieve the appearance of a sand-finish plaster, and the oilcloth was washable.[9]

A nearly identical product, Cushocel, was manufactured by the Acoustical Corporation of America.[10] In 1931 texture-painted Cushocel was installed as a wall and ceiling finish in four courtrooms in the New Jersey State House annex (1931, J. Osborne Hunt and Hugh A. Kelly). It was applied to plaster wall and ceiling substrates and does look like sand-finished plaster.[11]

While Nashkote and Cushocel solved the problem of applying three different elements, felt-and-membrane systems existed only in rolls or large floppy sheets and therefore were less convenient than boards or plasters.[12] The membrane was upgraded to make it more durable, paintable, and removable. Johns-Manville accomplished this end with Sanacoustic tile, a perforated, baked enamel metal sheet with a special rock wool lining. The metal could be scrubbed and painted without impairing either its acoustical qualities or its appearance. The 12-by-24-inch tiles were made to snap into a T-bar system and could be removed for renovations, cleaning, or ceiling changes. The Sanacoustic tile was intended for ceiling use, and the companion product, Sanacoustic Holorib, a V-crimp style perforated sheet over the rock wool pad, was intended for use on walls.[13]

Because of their obvious practical limitations, the felt-and-membrane products fell into disuse.[14] Throughout the industry, board-and-tile products became the standard, along with acoustical plasters.

In the 1920s acoustical plaster, which was less expensive than other systems, was recommended for new work. The early acoustical plasters were generally a standard gypsum plaster with the addition of a fibrous or porous aggregate material, which made the plaster more permeable and better able to absorb sound. The fibers could be wood, mineral wool, cork, or asbestos.

Keaseby and Mattison produced one of the earliest plasters, Sabinite.[15] In the 1930s Sabinite, which was made from inorganic aggregates and gypsum plaster, became a product of United States Gypsum.[16] United States Gypsum manufactured Sabinite in three grades: A, for general use; B, for wainscots or other areas subject to abuse or where a hard, troweled surface was required; and 38, a moisture-resisting formulation.

In the 1920s the R. Guastavino Company offered both plaster and artificial stone under the name Akoustolith. The plaster was processed with a lightweight aggregate and could be tinted with the addition of pigments.[17]

Another fiber acoustical plaster was Macoustic, manufactured eventually by the National Gypsum Company as part of its Gold Bond line. It was advertised as fireproof, so the fiber was likely either mineral wool or asbestos.[18]

The Kelley Island Lime and Transfer Company manufactured Kilnoise Acoustical Plaster, in which "asbestos and rock wool fibers are woven together horizontally to form a mat, bonded by foamed lime."[19]

Certain-Teed Products Corporation's acoustical plaster, a blend of pumice with gypsum or portland cement, was known as Kalite. It was used in two prominent New York City buildings in the 1930s—the U.S. Courthouse (1936, Cass Gilbert) and Radio City Music Hall (1932, Raymond Hood et al.). In both buildings the plaster was used flat as well as cast into elaborate patterns for ceilings.[20]

An alternative to the fibrated or modified aggregate plaster was one with an activating ingredient that would give off gas when mixed with water, thus creating holes into which sound waves could be absorbed. The Cleveland Gypsum Company manufactured such a product, Hushkote Sound Absorbent Plaster, which had a suedelike texture as a result of an activating ingredient that reacted like yeast.[21] After World War II spray-applied plasters also appeared.[22]

Manufacturers generally recommended that the plasters be applied with a rough texture. Some manufacturers even recommended nail stippling, in which a grid of nails was pressed into the surface of wet plaster to create a perforated surface.[23] This perforation served the same function as the holes in the perforated tiles, letting sound

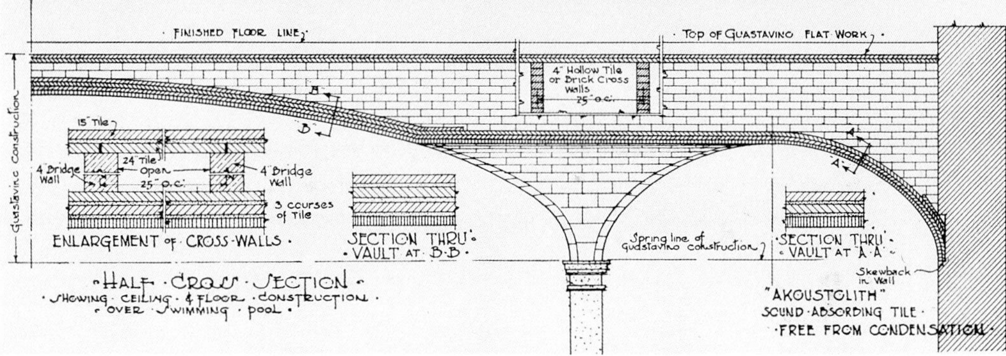


**Figure 34.3.** Akoustolith sound-absorbing tile, produced by the R. Guastavino Company, formed the vaulted ceiling of the swimming pool at Smith College (J. W. Ames and E. S. Dodge), Northampton, Mass.

waves pass into the porous material without obstruction from finish treatments.

Recommendations for finishing and decorating plasters and tiles were similar. Standard paint was generally not acceptable. Most manufacturers recommended spray painting with thinned paints or water-based paints. Staining was considered acceptable by some manufacturers. Many products were available in a range of standard colors, and the plasters could be tinted in the field.

The sound-absorbing board-and-tile products developed in the 1920s were made from a surprising variety of materials. While wood and asbestos fibers and mineral wool were the most common, sugarcane, flax, cork, gypsum, and portland cement were also used.

Because building insulation is based on providing trapped air spaces within components, it is not surprising that the earliest products to claim to be acoustical were actually developed as insulators. Among the earliest was Universal Insulite, a wood-fiber board, which could be inserted within wall or floor construction to provide deadening or used as a finish material.[24]

In 1924 the Union Fibre Company was manufacturing Acoustifibroblock, a molded unit made of flax fiber and rock wool. This economical product had a finished surface and so did not require a membrane. However, it could not be painted and was difficult to clean.[25]

By 1926 the Boston Acoustical Engineering Company was in the market with two sound-absorbing board materials, Silen-Stone and Acoustex.[26] Silen-Stone was "a decorative artificial stone tile" made from sand and portland cement whose absorptive properties were obtained from "globular cells of various sizes" induced by "a unique process." It was installed over furring to resemble ashlar or any other stone pattern.[27]

Acoustex, made of wood fiber and cement, had "a beautiful and interesting but unobtrusive texture." It was originally intended to achieve a monolithic appearance by butting rough-sawn joints that would not be visible in the surface's rough texture.[28] By 1931 it was made only with beveled edges, putting it firmly in the tile aesthetic.[29]

In 1935 Atlantic Gypsum, the manufacturer of Acoustex, offered a new product, A.A.T., "a decorative acoustical finish especially designed for redecoration of old acoustical plasters and tiles." It could be applied to a surface with no loss in absorption. By 1935 acoustical materials installed some ten years earlier were apparently showing some wear.[30]

The Calicel Company also made both acoustical tile and an acoustical imitation stone. The base material for both products was Calicel, an expanded stone aggregate said to resemble a petrified sponge. For the tile, this material was mixed with a mineral bonding agent, molded in hydraulic presses, and kiln dried to create a ceramic acoustical tile. The tile could be installed either by gluing or nailing, or could be suspended on a concealed spline system.

The Calicel imitation-stone material, known as Calistone, was the same product as the tile but without the beveled edges. The colors were adjusted with the addition of colored granules to its surface. It could also be molded, and molded Calicel aggregate, as well as Calistone and Calicel tile, was installed in the Pennsylvania Railroad Station (1935, McKim, Mead and White) in Newark, New Jersey.[31] The tile on the ceiling was originally decorated with gold-leaf lines and blue water-based paint.[32] Three companies that began making these products in the 1920s—Armstrong, Celotex, and United States Gypsum—are still major forces in the acoustical product field.

By 1925 the Celotex Corporation had developed the idea of taking standard Celotex (a felted, sugarcane fiberboard), perforating it, and calling it Acousti-Celotex.[33] Because the sound waves passed into the sound-absorbing medium through the perforations, it could be painted without impairing its acoustical efficiency or left in its "pleasing, natural tan color."[34] By 1935 Celotex was making two versions of Acousti-Celotex, one of cane fiber and one of mineral fiber.[35]

United States Gypsum approached the acoustical market from two angles, manufacturing both Sabinite acoustical plaster and acoustical tiles.[36] Its 1935 product line included Acoustone, a mineral wool tile; Quietile, a less expensive wood fiber tile; and Perfatile, a perforated metal and rock wool tile. Acoustone was the premier product.[37] Although it came in only one color, it could be decorated in the field. The literature notes that "interiors in the modern trend may be delightfully handled in tile with a metallic luster in many shades and colors." The metallic luster was to be achieved by applying bronze-powder lacquer.[38] Acoustone continues today in the United States Gypsum product line as a ceiling tile.

By 1930 the Armstrong Cork and Insulation Company was manufacturing Corkoustic, a pure cork ceiling tile. It was available only in "its natural rich brown" but could be painted with water-based paint and even stenciled.[39] Corkoustic could also be applied to walls in an ashlar pattern to imitate stone. By 1937 the company had added Temcoustic, a wood fiber tile, to its products and had begun offering a line of accessory products for acoustical installations, including several types of cements for various applications.[40]

Most tiles were mounted with adhesives or wood screws to wood furring, plaster, gypsum board, or concrete substrates. However, in 1929, the first suspended acoustical ceiling was installed in the Philadelphia Savings Fund Society Building (1932, Howe and Lescaze) in Philadelphia. This system incorporated cast gypsum panels with a concealed suspension system.[41] However, even though a number of spline systems were available during the 1930s, tiles applied with adhesive continued to be the most prevalent type.

After World War II tiles and plaster took over the market, virtually eliminating the board materials. An important development in the application of acoustical tiles in commercial and institutional interiors was the lay-in ceiling panel and exposed grid in the early 1950s.

In November 1953 Owens-Corning, a newcomer in the acoustical materials field, advertised Textured Ceiling Board, which was installed on an exposed T suspension system.[42] In a parallel development the Edwin F. Guth Company advertised Gratelite ceilings, a luminous ceiling system with the diffusing panels supported on an exposed T system.[43] The 1954 Owens-Corning catalogue shows both the ceiling board and Stria tile installed in the exposed T system. Stria was also shown installed in a concealed spline system.[44]

Exposed T suspension systems gained favor because of cost, accessibility, and industry standards for ceiling board, light fixtures, and mechanical equipment diffusers. Ceiling board became the universal standard for commercial and institutional construction. As in other sectors of the building industry, the trend in twentieth-century acoustical materials has been away from labor-intensive systems and toward standardized products. The wide range of tile-and-board sizes and complicated setting systems developed into today's grids of 2 feet by 2 feet and 2 feet by 4 feet, and hand-applied plaster has yielded to sprayed-on coatings.

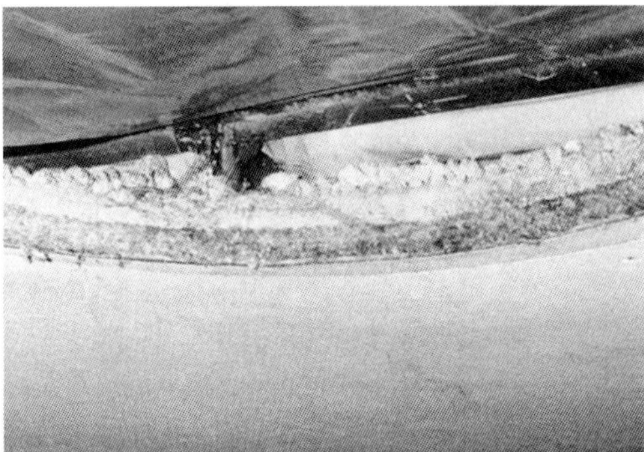

**Figure 34.4.** Acoustone tile was specified for the ceiling of the remodeled employee dining room (1949) of the Northern Trust Company (1905, Frost and Granger), Chicago. The striated tiles were applied flush to one another, creating the effect of a flat plane (top).

**Figure 34.5.** The ballroom ceiling of the American Red Cross Building (1917, Trowbridge and Livingston), Washington, D.C.—consisting of plaster on metal lath, an asbestos-based felt, and perforated, painted canvas—shows deterioration and was replaced (bottom).

# SEE YOUR DEALER
# IN LUMBER AND BUILDING MATERIALS

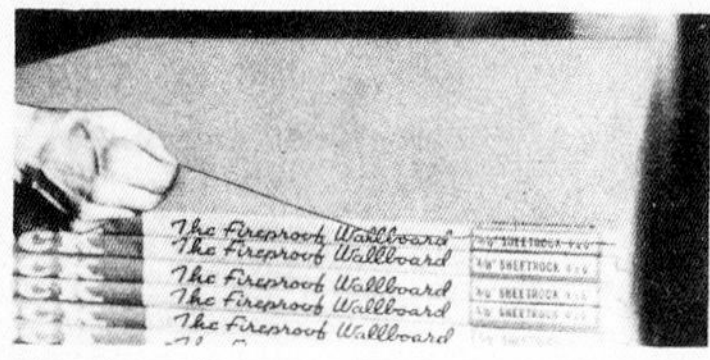


THE SHEETROCK ZIP-CORD PACKAGE

When you think about remodeling or building with wallboard remember there is a dealer in your town who can help you. He can answer your questions; he can recommend materials; he can get you a reliable contractor; he can point out the advantages of different materials so you will have a basis for making a choice. He can supply you with wallboard and any other construction materials which may be needed. The chances are he handles Sheetrock and recognizes it as a superior product for wallboard construction. Always you can order Sheetrock and be sure of getting it—because it is stocked in almost every community and it is always carefully marked with the name "Sheetrock" for your protection.

# UNITED STATES GYPSUM COMPANY

**Figure 35.1.** By the mid-1930s gypsum board was becoming a staple building material available in lumber yards. Sheetrock, the United States Gypsum Company's fireproof wallboard, was even sold in "zip-cord" packages for easy transportation to and use on the job site. Catalogue, United States Gypsum Company, 1937.

KIMBERLY A. KONRAD AND MICHAEL A. TOMLAN

# 35 | GYPSUM BOARD

**DATES OF PRODUCTION**

c. 1894 to the Present

**COMMON TRADE NAMES**

Adamant, Bestwall Firestop, Gold Bond, Gyp-Lap, Sackett Board, Samson Board, Sheetrock

## HISTORY

Gypsum board is a factory-produced panel composed of a set, noncombustible gypsum core typically encased in paper. This face paper is folded around the long edges to reinforce and protect the core, while the ends are cut square and finished with tape. Long edges can be either squared, rounded, beveled, or tapered to allow joints to be reinforced and concealed with tape and gypsum cement.[1] Gypsum board is lightweight and fire resistant and conducts little heat or sound. It is also vermin proof and noted for its ease of installation.

### ORIGINS AND DEVELOPMENT

The first gypsum board was patented in 1894 by the inventor Augustine Sackett.[2] It measured 32 by 36 inches with six layers of soft, unsized felt paper separating five thin layers of gypsum. This multi-ply board varied in thickness from ¼ inch to ½ inch and generally weighed about 1½ pounds per square inch. Sackett continued to refine his product at his Pamrapoe, New Jersey, plant, and in 1898 the Sackett Wall Board Company introduced Sackett Board, consisting of only four sheets of paper separating three layers of gypsum.[3] Although this board was marketed as quickly providing a hard finish surface and thus a substitute for lath and plaster, in practice it was most commonly used as a plasterboard, the backing surface to which plaster was applied.[4]

In 1907 Stephen Kelley founded the Samson Plaster Board Company and shortly thereafter patented a process for the manufacture of 32-by-36-inch, two-ply boards using chip paper instead of felt paper.[5] The United States Gypsum Company entered the gypsum board industry shortly after acquiring the Sackett Wall Board Company plants and the Samson Plaster Board Company at Passaic, New Jersey, in 1909. After the United States Gypsum Company took over the Lockwood Paper Company and began producing its own paper, Clarence Utzman, a company engineer, discovered a method of making a plaster wallboard with folded paper edges.[6]

In 1916 John and Joseph Schumacher invented a method of producing long boards faced with a paper that was sized to retard permeation. The boards were successively manufactured in stacks and trimmed uniformly.[7] These boards were ⅜ inch thick and generally 48 inches wide.[8]

Standardization of gypsum board and use of a production line helped increase output before World War I. One of its first applications during the war was to replace flammable fiberboards in military barracks.[9] Because of its cost advantage over wood and particularly its fire resistance, gypsum board began to displace wood and metal lath for plaster backing by World War I.[10]

Between 1920 and 1925 United States Gypsum introduced a number of gypsum building boards. Sheetrock wallboard, Sheetrock tile board, rock lath, and Gyp-Lap sheathing all had a large impact on the cost and speed of new construction. Made in widths of 32 to 48 inches with a thickness of ⅜ inch, Sheetrock gypsum board provided a finished wall surface when attached to the interior support members.[11] United States Gypsum issued patent license agreements to several manufacturers, including the National Gypsum Company, Certain-Teed Products Corporation, Celotex Corporation, Ebsary Gypsum Company, Newark Plaster Company, and Texas Cement Plaster Company.[12]

Demand for gypsum board increased during World War II as drywall construction became preferable to plastering because of its use in low-cost housing, and because conservation orders limited consumption of such strategic materials as metal and lumber. The extensive use of gypsum as a substitute building material in war housing and other temporary structures revealed its potential and influenced its widespread use in the postwar era when the demand for housing skyrocketed.[13]

Although the basic manufacturing concept has changed very little, a variety of improvements have increased production volume and board strength and have reduced board weights, simplifying installation and lowering manufacturing costs.

In the late 1920s, following the lead of the concrete industry, experimentation with various fillers, aggregates,

**Figure 35.2.** In the forming end of gypsum board machinery, foam and gypsum were spread between the paper layers (top). The introduction of foam filler into gypsum boards created lighter products.

**Figure 35.3.** Workers install gypsum board in 1946 (bottom). Joint cement was applied over the joint and nail heads, a perforated metal strip was centered over the joints, and a finish coat of cement was applied.

and air entrainment (microscopic bubbles) led to the manufacture of a lighter wallboard.[14] In 1927 the National Gypsum Company registered the trademark Gold Bond for this new gypsum board.[15]

Improvements in the surface finish of gypsum board were numerous. The heavy, unsized felt paper used to face early products was replaced with a less permeable, multi-ply paper produced from recycled newspaper.[16] To compete with the decorative and performance features of fiber wallboards, varied surfaces were introduced. Wood veneer finishes were launched in 1937, followed by decorative pastel surfaces and functional vapor-retarding, water-resistant, and foil-insulating layers in the 1940s. Vinyl-covered boards, introduced in the late 1950s, provided a wall system that required no additional decoration.

From its early years gypsum board was recognized for its fire-retarding qualities.[17] Increased protection could be obtained by using multiple layers of regular wallboard. Research in the 1940s led to products with special additives, including asbestos fibers, mineral wood, glass wool, and fiberglass strands, which enabled builders to install fire-retarding wall systems with fewer layers. In 1946 Certain-Teed Products introduced its first ½-inch fire-rated gypsum board.[18] Further developments led to its ⅝-inch Bestwall Firestop, consisting of small amounts of glass fibers and unexpanded vermiculite, which helped retain the shape and strength of the gypsum core after the chemically bound water was driven off by extreme heat.[19] By the mid-1950s a number of companies were producing fire-rated gypsum board.

## MANUFACTURING PROCESS

Gypsum is a calcium sulfate combined with water and is found in its natural state as a grayish white rock existing in extensive deposits all over the world.[20] Its use in the building industry accounts for more than 90 percent of total mined volume.[21] In its manufacture into gypsum board, the raw material is quarried and crushed into a fine powder.

At the mill the ground gypsum is placed in a calcining kettle or rotating cylinder and heated to 340 to 350 degrees Fahrenheit, causing more than three-fourths of its chemically bound water to evaporate.[22] During the heating process, a rotating shaft in the center of the kettle agitates the material.[23] The calcined plaster is combined with starch and retarder in a mixing chamber and fed to a 4-foot-wide cupped conveyor belt that runs through reservoirs of water, where the mixture absorbs enough water to form a slurry—a thin, watery mixture of fine gypsum particles. The mixture is fed into a continuous wet mixer, conical in shape with a propeller-like mixing device.[24]

From the mixer the plaster is poured onto an extended piece of paper on a forming table, covered by a top paper, and drawn through a series of rollers. Heaters are used along the line to set the plaster edges before the paper edges are scored and folded over.

The calcined gypsum begins to set as it is moved to the automatic cutter. Once cut and partially set, the boards are conveyed to the dryer for final curing. Today when packed, wallboard is generally twin-mounted—two boards are placed face-to-face with the edges taped together.[25]

## USES AND METHODS OF INSTALLATION

Gypsum board became attractive for drywall construction because its use eliminated the need to wait several weeks for numerous coats of plaster to dry. Instead, the interior walls could be installed and finished in a matter of hours. This product was also appealing to house builders since it was fireproof and inexpensive, made from an inexhaustible and readily available raw material.[26]

Since World War II gypsum board has been used in countless commercial, industrial, and residential structures. Sheetrock tile board was designed to be used as wainscoting or an entire wall finish, while Sheetrock wallboard has found widespread use for ceilings, walls, and partitions.[27]

Two basic methods have been and continue to be used for single-layer application—horizontal and vertical. In both cases the boards are attached directly to the interior studs. The framing, whether wood or metal, should not be less than 16 inches on center with header strips inserted where necessary to provide a nailing support. Ceilings should be erected first at right angles to the joists, with joints staggered. Boards used for walls should be hung from the ceiling first.

Nails should generally be spaced not less than ⅜ inch from the edges and ends of the boards, nailing first to intermediate supports and second to edges. Nails should be driven straight and with the heads recessed slightly below the surface.[28]

Boards may be attached to wood or metal using different types of drywall screws. For wood framing, the screws should penetrate the supports by ⅝ inch, while screws used for metal framing should penetrate at least ⅜ inch.[29]

Joints are concealed with reinforcing tape and joint adhesive or cement. Despite the many types of tapes and adhesives, the method is the same. The cement is spread over the joint and nail heads, and tape is centered over the joint and pressed into the cement. A second, thin coat is applied to hide the tape with the surface leveled and the edges feathered. Nail holes should be filled flush and all the cement allowed to dry completely. Once dry, the covered joints and nail holes can be lightly sanded.[30]

Corners require special treatment. For inside corners, cement is spread on each side and the reinforcing tape is folded along the center crease and pressed into the angle. After drying, additional coats are applied. For outside corners it is recommended that a metal corner bead be nailed in place and several coats of joint cement applied to each side.[31]

Installation methods have changed over the years to create smoother, more even wall surfaces. In the early 1900s the boards were produced with a round, folded edge to provide a reinforced nailing edge. This was replaced by the square, folded edge, which allowed the boards to fit together snugly; later, tongue-and-groove edges were introduced to eliminate ridging. In 1937 a process was developed to produce wallboards with recessed edges. This created a valley at the joint where two boards were fitted together, allowing for a flat, smooth joint treatment. In the 1950s a wallboard with a tapered edge was introduced and is still used today.

The evolution from special nails to drywall screws with special points and threads has helped prevent the problem of popping nails. Even joint tape has changed from perforated paper tape to metal strips to self-adhering glass fiber tape.[32]

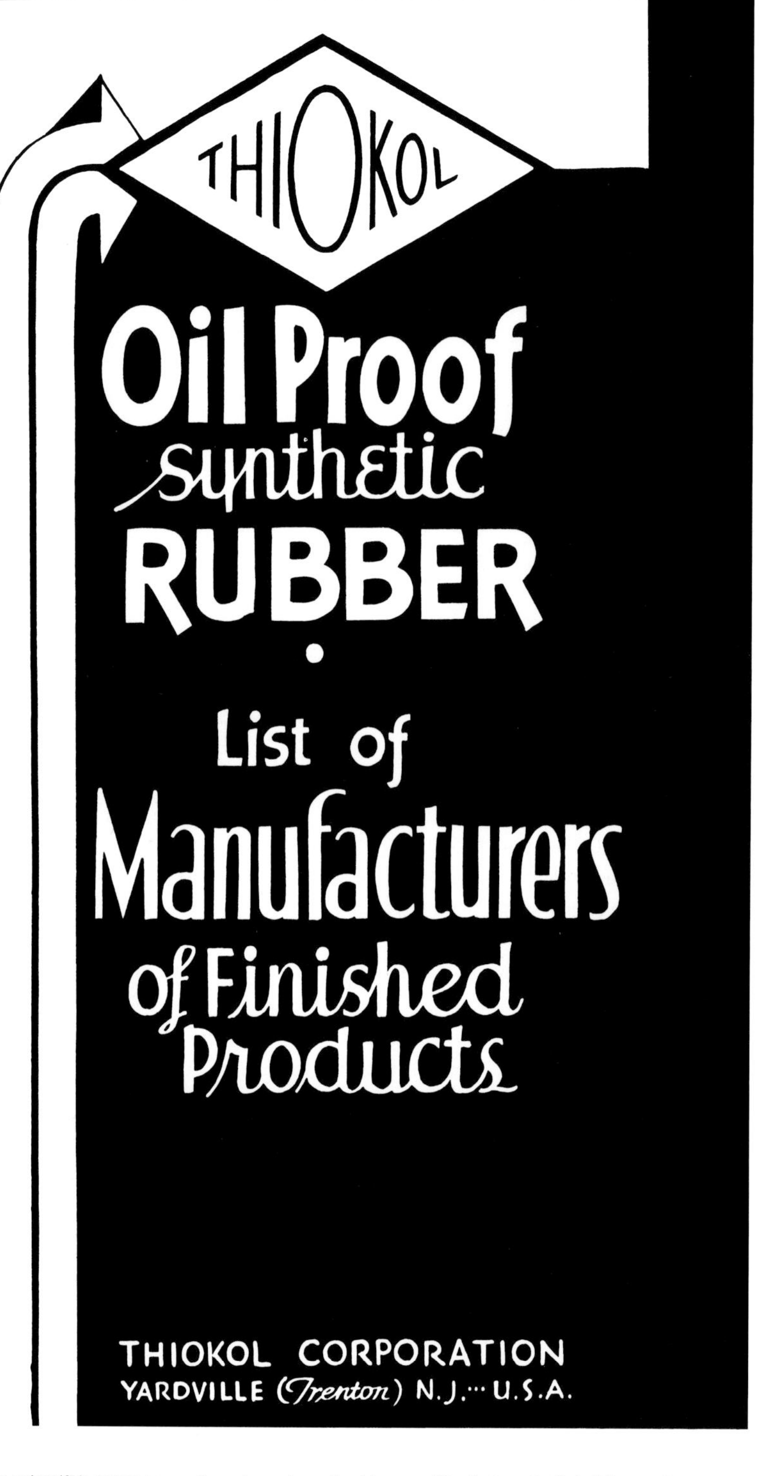


**Figure 36.1.**

Thiokol, a rubber based on a polysulfide polymer, was the primary ingredient in one of the first synthetic building sealants available. Catalogue, Thiokol Corporation, 1935.

MICHAEL J. SCHEFFLER AND JAMES D. CONNOLLY

# 36 | BUILDING SEALANTS

**DATES OF PRODUCTION**
c. 1950 to the Present
**COMMON TRADE NAME**
Thiokol

## HISTORY

Elastomeric sealants in building facades provide the first defense against water penetration and air infiltration through building joints. Modern joint sealants are typically synthetic elastomeric materials that adhere to joint walls and accommodate joint movement. Sealants can be either single-component sealants, which do not require mixing, or multicomponent sealants, which require mixing before application. The four principal characteristics of sealants are adhesion, cohesion, elasticity, and weathering ability.[1] The most common types of elastomeric sealants used to seal joints between building elements are silicone, polyurethane, polysulfide, acrylic, latex, and butyl-based products.

### ORIGINS AND DEVELOPMENT

Oil- and resin-based caulks were the only building joint sealants before about 1950. Oil-based caulk, used as early as 1900, was often termed *putty.* Oil-based caulk typically was made primarily with calcium carbonate and drying oils such as linseed, fish, soybean, tung, and castor.[2] Some oil-based caulks and glazing compounds also contained lead or asbestos.[3]

Construction requiring sealed joints before the 1950s typically was bearing masonry having joints that experienced little movement. Oil-based caulks, including those manufactured before the 1950s, met this need, despite their limited movement capability. These caulks are not considered elastomeric sealants. To enhance performance, some oil-based caulk manufacturers recommended adding oakum or equivalent filler to joints to reduce the amount of caulk required.

As the popularity of curtain wall construction, which experiences greater movement than masonry, increased in the 1950s, the demand for more elastic sealants also increased. Polysulfide sealant, the first widely used elastomeric sealant, is a high-performance sealant based on a synthetic polysulfide polymer or rubber.[4] The first polysulfide polymer was developed by the Thiokol Chemical Corporation in 1929, and the polysulfide sealants available in the 1950s were all based on Thiokol products. Until 1961 these products consisted of two-component products in separate containers.[5] Among the early manufacturers of a one-part polysulfide sealant were the Dicks-Armstrong-Pontis and A. C. Horn Companies.[6]

Silicone polymers, used in silicone sealants, were discovered in the early 1800s. In the United States, research on silicone polymers for sealants began in the 1930s at Dow Corning and General Electric. The first silicone sealants were two-component products developed in the early 1950s.[7]

The first silicone building sealant was manufactured by Dow Corning around 1960.[8] General Electric introduced one in 1963. Both were one-component products. Silicone sealants typically consist of a high percentage of the silicone polymer and may contain fumed silica, calcium carbonate, ground quartz, carbon blacks, talcs, a plasticizer, and other compounds.[9]

Butyl sealants are manufactured from gaseous hydrocarbons (butyl rubber, a synthetic rubber, also contains a small percentage of isoprene). They became available to the construction industry in the mid-1950s, although butyl rubber was being manufactured earlier for other uses.[10] Butyl rubber was developed to replace natural rubber, which was in great demand just before and during World War II. Butyl rubber–based sealants are one-component sealants that may include talc powder, calcium carbonate filler, polybutene, mineral spirits, and adhesion promoters.

The first production of acrylic sealants, based on acrylic polymer technology that dates to 1843, took place in the 1920s.[11] Acrylic polymers are based on thermoplastic resins formed by polymerization of the esters of acrylic acid. In 1958 Tremco Manufacturing Company developed the first acrylic sealant for buildings. Acrylic polymer–based sealants typically are one-component solvent curing products that include fillers such as

**Figure 36.2.** These curtain wall lights were sealed with preformed butyl rubber gaskets in the late 1950s. Such gaskets could be installed easily, provided good resistance to heat, and were weathertight.

calcium carbonate with silica and solvents such as xylol, catalysts, thixotropic agents, and plasticizers.

Later sealants included urethanes and latex. Urethane sealants, based on polyurethane synthetic rubber, were first used in joints where abrasion resistance was required. The first urethane sealants available were multicomponent products.[12] Latex caulks were developed in the 1950s and typically are based on acrylic, styrene-acrylic, vinyl acrylic, polyvinyl acetate, and styrene butylene. These caulks are based on a colloidal suspension of rubber resin (synthetic or natural) in water. Latex caulks were not available until the mid-1960s. Typical components of a latex caulk include water, ground calcium carbonate, plasticizers, small amounts of mineral spirits, ethylene glycol, surfactants, and pigments.

### USES AND METHODS OF INSTALLATION

Typical applications requiring sealants include joints between stone veneer, concrete, or metal cladding panels; between stone panels and flashing; masonry expansion and coping joints; joints at window and door perimeters; and joints in horizontal surfaces, such as plazas and terraces. Sealants are usually applied as a viscous liquid that is either pourable for horizontal applications or nonsagging for vertical surface joints. While simple butt joints are the most prevalent, fillet joints, lap joints, glazing beads, and glazing heel beads have also been used historically.

Joint backup materials have been recommended by sealant manufacturers since the late 1950s. The depth of the sealant in joints has been controlled with foam backer-rod fillers made of polyethylene, polyurethane, or polyethylene bond-breaker tape. Some sealants formed bubbles while curing because of air or gases released from closed-cell backup materials cut during sealant installation. To help eliminate this problem, a backer rod of open-cell foam with a continuous skin was developed.

Silicone-based sealants are most frequently used for nonporous surfaces in high-movement joints. They are commonly used for metal and glass cladding substrates. Silicone sealants were also used for two-sided structural glazing joints in storefronts during the 1960s.[13] In such joints, sealant is placed in vertical joints between glazing elements in place of mullion supports; the glass elements are held in place along the top and bottom edges with mechanical supports.

Urethane-based sealants are typically used for porous surfaces in high-movement joints. These sealants are commonly used for cladding joints. Besides urethane, polymer-urethane sealants include fillers, colorants, plasticizers, thixotropic agents, adhesion-promoting additives, and solvents.

Acrylic sealants are widely used in small-scale construction because they have limited movement capability. Latex caulks, which also have limited movement capability, are used in light building construction and residential construction. Isobutylene-based sealants are typically used for glazing joints and splice seals in window units.

## CONSERVATION

Before conserving building sealants, it is important to understand common types and causes of deterioration, as well as basic diagnostic and assessment techniques. Conservation techniques for sealants are generally limited to removal or selected removal and replacement, but they require a thorough understanding of previous causes of failure and materials to achieve a lasting repair.

### DETERIORATION

Water, exposure to ultraviolet light, and freeze–thaw cycles cause weathering, which is indicated by chalking, discoloration, cracking, or softening of the surface area.

Loss of adhesion results in separation of the sealant from the substrate. This separation is usually visible but may not be obvious in compressed joints. Coatings or contaminants on the substrate surface may limit or prevent proper adhesion. Contaminants may include previous sealants, sealers, form-release agents, or bond breakers.

Cohesive failure occurs when the internal integrity of the sealant deteriorates. This is evidenced by cracking parallel to the joints' interface. Loss of elasticity, which is distinguished by hardening of the sealant, can give rise to cohesive and adhesive failure.

Important factors that contribute to these forms of deterioration include inappropriate sealant selection and improper joint design. Either condition can result in a sealant that does not have the extension or compression capability to accommodate substrate movement. Inherent problems in the sealant and poor substrate conditions also can contribute to deterioration.

Sealants must be properly configured in the joint to accommodate movement. Sealants are typically installed with polyethylene bond-breaker tape, compressible-foam backer rods, or other means to prevent the sealant's adherence to the back as well as the sides of the joint. Sealant that adheres to the back as well as sides tends to tear before its potential movement capability is attained. Sealant installed when the substrate or ambient temperature is too cold (below 45 degrees Fahrenheit) may not accommodate expansion of the substrate and compression of the joint when heated. Conversely, if the sealant is installed when temperatures are too high (above 90 degrees Fahrenheit), the joint may be at its narrowest dimension, and the sealant may be unable to accommodate expansion of the joint when cooled.[14]

A common problem with sealants, particularly silicones, is staining. Staining can be evidenced by dirt on the sealant itself or by migration of the sealant plasticizer into the adjacent substrate. The amount and viscosity of the silicone fluid in a silicone sealant plays an important role in staining.[15]

Silicone sealants can fail if the substrate is improperly primed or cleaned or if water gets into the joint, behind the sealant, and into absorbent backup material. As with other types of sealants, many silicones do not have good long-term adhesion if continuously wet or if in standing water for extended periods.

Problems with multicomponent urethanes are usually related to incomplete or improper mixing. If mixing is incomplete, portions of the sealant will be fluid, or uncured. Incomplete curing can also result from using materials beyond their shelf life, using components stored in inadequately sealed containers, and improper formulation. Curing at temperatures above 120 degrees Fahrenheit can make some urethanes vulnerable to premature degradation in ultraviolet light, leading to severe reversion. Under ambient air temperatures of 80 to 90 degrees Fahrenheit, dark-colored substrates in direct sunlight can reach temperatures exceeding 160 degrees Fahrenheit. High temperatures affect both the cure and durability of sealants. Some urethane sealants cured at high temperatures may later become fluid; they may develop bubbles to a depth of ⅛ inch and be gummy at lower levels.

Single-component urethanes, in contrast to multicomponent urethanes, cure by reacting with moisture in the air. Under high-temperature and high-humidity curing conditions some urethanes, particularly single-component ones, will develop excessive bubbling and blistering. Some are made with moisture scavengers such as calcium oxide, which are used to retard setting until placement occurs. If particles of the scavenger are too large and are exposed on the surface, they will react with rainwater and can cause stains on windows.

## DIAGNOSTICS AND CONDITION ASSESSMENT

It is important to understand the causes of sealant failure and the factors that may contribute to it. It is also important to understand the wall or paving system of the building being conserved and to look for adjacent construction conditions that could lead to sealant failure. Characteristic evidence of sealant deterioration or failure includes cracking, wrinkling, sagging, surface bubbling, splitting, and staining of the sealant itself or of adjacent surfaces.

Original construction drawings and repair documents should be reviewed to determine which sealants (and sealant primers) were specified, whether backer-rod and bond-breaker tape were used, and whether joints were sized and spaced properly.

The extent and location of the failure should be determined through visual examination, either directly or with a field microscope. The sealants and substrates should be identified and existing conditions evaluated through visual and tactile examination. The sealant profile and joint configuration should be noted from removed samples.[16] When diagnosing sealants in the field, the principal sealant characteristics—adhesion, cohesion, elasticity, and weathering—should be evaluated.

Other techniques for field diagnostics include a field adhesion test performed by cutting across the width of the sealant joints and on both sides of the joint along the edge of the substrate. The partially detached sealant is then pulled away from the substrate at a 90-degree angle, and the elongation is measured when adhesive or cohesive failure occurs.

In-depth analysis of sealants is performed in a laboratory. Samples are first evaluated for conditions such as excessive softness, tackiness, fluidity, and cracking. Microscopic examination is used to check for uniformity, voids, air pockets, and pockets of unreacted sealant.

Chemical tests, such as the Fourier transform infrared (FTIR) test and solvent extraction, can be performed to identify polymeric binders and plasticizing agents. Filler contents and the proportions of components can also be determined through laboratory tests. Color can be an indication of sealant type, improper mixing or curing, or other conditions.

**Figure 36.3.** In-place field adhesion pull tests provide information about a sealant's adhesive, cohesive, and elastic properties.

### CONSERVATION TECHNIQUES

After assessing the condition of sealants used in a building, it is possible to determine whether the sealant can be selectively replaced or will require complete replacement. In many cases deteriorated sealants will require replacement to prevent water penetration. It may be possible, however, to retain existing sealant that is performing adequately at specific locations.

### REPLACEMENT

When replacing a sealant is necessary, it is important to select an appropriate replacement. Currently, silicone- or polyurethane-based sealants are most often selected for joints that will experience significant movement. However, selection of a specific sealant type depends on the joint width, substrate, movement requirements, aesthetics, intended service life, previously used sealant, and cost.

Proper selection of sealant for compatibility with the substrate should also be evaluated. If the substrate is not the same on both sides of the joint, a different preparation may be required for each substrate. Some sealants cannot bond to certain other sealants. When selecting a sealant, test the adhesive in both dry and wet conditions. Sealants other than silicone can rarely provide strong, durable bonds to existing silicone. Also, silicones will not cure properly when installed over isobutylene sealants. Avoid installing a new sealant over an existing sealant, as this may create a poor bond. Sometimes sealant installed on glazing is susceptible to poor adhesion because of contaminants on the glass surface.

Some sealants cannot bond to fluoropolymer architectural finishes without a primer and special preparation. In some instances solvent-based sealant placed over thermoplastic coatings can partially dissolve the coating. As the sealant shrinks, the bond between paint and substrate will fail. Many sealants require a primer coating on the substrate before application of the sealant.

# NOTES

**Editor's note:** *Sweet's Catalogues,* referred to frequently in the text, were published under three different titles throughout their history: *Sweet's Catalogue of Building Construction* (1906–15); *Sweet's Architectural Catalogues* (1916–32); and *Sweet's Catalogue File* (1933 to the present). *Sweet's Catalogues* were published by the F. W. Dodge Corporation until 1961. Since 1962 McGraw-Hill has been the publisher. For simplicity, only titles and years are given here.

### INTRODUCTION

1. Barclay G. Jones, "Developments of a Methodology for Making Indirect Estimates of the Built Physical Environment," September 13, 1994, 7. This unpublished paper is based on an extensive bibliography and several exhaustive field studies.
2. Olivier Zunz, *Making America Corporate, 1870–1920* (Chicago: University of Chicago Press), 68ff.
3. George S. Brady, *Materials Handbook* (New York: McGraw-Hill, 1929). Boron, carbon, chromium, cobalt, manganese, nickel, silicon, tungsten, and vanadium were the most notable.
4. Albert L. Batik, *A Guide to Standards* (Parker, Colo.: Serendel Research Institute, 1989), 11.
5. W. Bernard Carlson, "Academic Entrepreneurship and Engineering Education: Dugand C. Jackson and the MIT-GE Cooperative Engineering Course, 1907–1932," in *The Engineer in America: A Historical Anthology from Technology and Culture*, ed. Terry S. Reynolds (Chicago: University of Chicago Press, 1988), 367–98.
6. Bernard M. Baruch, chairman, *American Industry in the War: A Report of the War Industries Board, March 1921* (New York: Prentice-Hall, 1941), 17–19.
7. Ibid., 224–26.
8. Ibid., 235.
9. Ibid., 235. The broader importance of the War Industries Board's efforts was that it offered a ready-made model for government intervention during World War II. Baruch wrote in the foreword to this 1941 volume, "At times the reader will have to raise his eyes from these pages to wonder whether he is reading what took place a quarter of a century ago or what is unfolding before us daily." Ibid., vi.
10. "Non-Shatterable Glass," *Automotive Industries* 41 (October 9, 1919): 731.
11. David O. Woodbury, *Battlefronts of Industry* (New York: John Wiley and Sons, 1948), 151.
12. R. W. Stone, "Historical Sketch," in *Gypsum Deposits of the United States* (Washington, D.C.: Government Printing Office, 1920), 33–34.
13. Adelbert Philo Mills, *Materials of Construction: Their Manufacture and Properties* (New York: John Wiley and Sons, 1915; reprinted 1922, 1926, 1939, 1955). A comparative examination of this standard source and others bears out this statement.
14. "Slag Wool," *Scientific American* 1 (November 1885), 10, 21; (February 1886), 114.
15. See, for example, "Silver Lining Model Home," *American Builder* 57 (October 1935): 70–71, for a demonstration house by the Reynolds Aluminum Corporation.
16. Emily Ann Thompson, "'Mysteries of the Acoustic': Architectural Acoustics in America, 1800–1932" (Ann Arbor, Mich.: University Microfilms International, 1992), 259–324.
17. David E. Nye, *Electrifying America: Social Meanings of a New Technology* (Cambridge, Mass.: MIT Press), 261.
18. Harry L. Derby, "The Future of Industrial Research: The View of Industrial Management," in *The Future of Industrial Research* (New York: Standard Oil Development Company, 1945), 59. The growth in the next decade was equally substantial, although nowhere near as dramatic, for by 1940 more than 2,350 establishments employed more than 70,000 research personnel.
19. U.S. Department of Commerce, *Report of the Panel on Engineering and Commodity Standards of the Department of Commerce Technical Advisory Board,* February 2, 1965 (Washington, D.C.: U.S. Department of Commerce, 1965), cited in David Hemenway, *Industrywide Voluntary Standards* (Cambridge, Mass.: Ballinger, 1975), 81.
20. Hemenway, 81.
21. "Building Nearly 50 Percent More Homes," *Business Week,* January 23, 1937, 26–29.
22. Home Owners Loan Corporation, Appraisal and Reconditioning Division, *Master Specifications for Reconditioning* (Washington, D.C.: Home Owners Loan Corporation, Appraisal and Reconditioning Division, 1940), 1.
23. "Research Program in Building Materials Launched," *Architectural Record* 82 (October 1937): 37.
24. Dean Richmond, *The Story of National Gypsum Company: In Commemoration of Its 35th Anniversary* (Buffalo, N.Y.: Baker, Jones, Hausauer, 1960), 25–26, 40.

25. R. G. Wallace, "600 Fighting Retailers," *Printers Ink* 157 (December 31, 1931): 3–4; "Celotex—Houses from Sugar Cane: Mr. Bror Dahlberg and His Company," *Fortune* 19 (February 1939): 80–84.

26. "Synthetic Houses," *Scientific American* 149 (October 1933): 149.

27. Harry Parker, Charles Merrick Gay, and John W. MacGuire, *Materials and Methods of Architectural Construction* (New York: John Wiley and Sons, 1962), 310; Sheldon Hochheiser, *Rohm and Haas: History of a Chemical Company* (Philadelphia: University of Pennsylvania Press, 1986), 57.

28. Vinyl tile is yet another example. First introduced in combination with asphalt, it did not become widely accepted until after World War II.

29. J. V. Sherman, "Research Can Re-House America," *Barron's* 20 (August 26, 1940): 20; "Albert Farwell Bemis Foundation, 1934–1954," Manuscript Collection MC 66, Institute Archives and Special Collections, Massachusetts Institute of Technology, Cambridge.

30. Donald M. Nelson, *Arsenal of Democracy: The Story of American War Production* (New York: Harcourt, Brace, 1946), 35–36.

31. Neoprene, formerly called Duprene, is a polymerized chloroprene then produced from acetylene. Butyl rubber is a variation of Buna rubber made from petroleum. George S. Brady, *Materials Handbook: An Encyclopedia for Purchasing Agents, Engineers, Executives, and Foremen* (New York: McGraw-Hill, 1940), 411–12.

32. Glenn D. Babcock, *History of the United States Rubber Company: A Case Study in Corporation Management* (Bloomington: Bureau of Business Research, Graduate School of Business, Indiana University, 1966), 381.

33. Russell P. Wibbens, "Structural Glued Laminated Timber," in *Composite Engineering Laminates,* ed. Albert G. H. Dietz (Cambridge, Mass.: MIT Press, 1969), 73.

34. "SPAB Prohibits Non-Essential Construction to Conserve Raw Materials," *Commercial and Financial Chronicle* 154 (October 16, 1941): 602; "Amounts of Critical Metals Necessary for Home Construction," *American Building* 63 (October 1941): 61; "Use of Copper Banned for Building Work," *Engineering News* 127 (October 23, 1941): 127.

35. U.S. Bureau of the Budget, *The United States at War: Development and Administration of the War Program by the Federal Government* (Washington, D.C.: Committee of Records of the War Administration, 1947), 109–14. The board's success was somewhat limited. The flood of war orders requesting everything possible produced a number of competing claims for the same materials. Additional problems arose because of the different rates of production in various industries.

36. Van Renssalaer Sill, *American Miracle: The Story of War Construction around the Globe* (New York: Odyssey, 1947), 36.

37. Ibid., 36.

38. "Shortage of Major Building Materials," *American Building* 69 (January 1947): 64–65; "Brick Famine?" *Economist* 151 (October 26, 1946): 676–78; "Building Materials Stay Short," *Business Week,* November 8, 1947, 25; R. M. Bleiberg, "Building Materials Feel Sharp Slump," *Barron's* 9 (May 2, 1949): 9; "Material Shortages Appear," *Engineering News* 144 (June 1, 1950): 23.

39. Richmond, 43; "Dry Wall Construction Popular," *American Builder* 69 (October 1947): 124–25.

40. Charles Z. Carroll-Porczynski, *Asbestos: From Rock to Fabric* (Manchester, England: Textile Institute, 1956), 362–92.

41. William Edward Sinclair, *Asbestos: Its Origin, Production, and Utilization* (London: Mining Publications, 1959), 328–31.

42. Robert H. Brown, "Aluminum Alloy Laminates: Alcad and Clad Aluminum Products," in Dietz, 227.

43. "Sandwich Wall Construction Developed for Aircraft Is under Study as Boon to Low Cost Housing," *Heating and Ventilating* 44 (December 1947): 110; "Metal-Faced Sandwich Material, Developed for Aircraft Construction, Is Strong, Lightweight Structural Material," *Architectural Forum* 84 (June 1946): 169.

44. "Aluminum Door Frames," *Modern Metals* 5 (November 1949): 22–23; "Aluminum for Building," *Architectural Forum* 97 (September 1952): 152–53; E. H. Laithwaite and E. W. Skerry, "Aluminum Cladding of Buildings," *Journal of Applied Chemistry* 7 (May 1957): 216–31; Jurgen Joedicke, *A History of Modern Architecture* (New York: Praeger, 1959): 134.

45. The qualification is necessary because of the temporary construction lull in 1960–61.

46. "Commerce Department Launches Plan to Aid Research for Building Industry," *Engineering News* 138 (March 27, 1947): 473; Julius J. Harwood, "Emergence of the Field and Early Hopes," in *Materials Science and Engineering in the United States,* ed. Rustum Roy (University Park: Pennsylvania State University, 1970), 1–10.

47. Mark Erhlich, *With Our Hands: The Story of Carpenters in Massachusetts* (Philadelphia: Temple University Press, 1986), 147.

48. Peter Cook, *Experimental Architecture* (New York: Universe Books, 1970), 41–67; S. M. Parkhill, "Bulbous Buildings," *Compressed Air Magazine* 63 (July 1950): 18–21; Fred N. Severud and Raniero G. Corbelletti, "Hung Roofs," *Progressive Architecture* 37 (March 1956): 99–107.

49. Albert G. H. Dietz, "The Potentialities of Plastics in Building," *Architectural Record* 107 (March 1950): 132–37; Dietz, "Today's Newest Building Materials," *Factory Management* 115 (June 1957): 114–18.

50. Michael F. X. Gigliotti, "Plastics for Interior Walls," *Progressive Architecture* 37 (February 1956): 138–41; *Advanced Materials. Policies and Technological Challenges* (Paris: Organization for Economic Co-Operation and Development, 1990), 21.

51. "Defacto Ban against Plastics," *Progressive Architecture* 55 (September 1974): 23.

52. A. G. Ward and E. L. E. Westbrook, "Rheology of Building Mastic: Caulking Compounds for Gun Application," *Society of Chemical Industries Journal* 67 (October 1948): 389–98.

## PART I. METALS

### 1. ALUMINUM

1. In 1821 the ore in which aluminum is contained was discovered in the village of Les Baux in southern France. The term *bauxite* is derived from the name of the village. E. I. Brimelow, *Aluminum in Building* (London: Macdonald, 1957), 25.

2. In 1845, having refined this process, the German scientist Friedrich Woehler produced small amounts of aluminum in sufficient quantity to determine its physical characteristics, such as its low density and ductility. Ibid., 25.
3. During this period aluminum was more precious than gold or silver. Napoleon III's most honored guests were served dinner with place settings of aluminum, while others received place settings of silver and gold. Paul Weidlinger, *Aluminum in Modern Architecture*. Vol. 2, *Engineering Design and Details* (Louisville, Ky.: Reynolds Metals Company, 1956), 14.
4. Ibid., 14.
5. "Setting the Capstone," *Harper's Illustrated Weekly,* December 20, 1884, 844.
6. Neither Hall nor Héroult knew of each other's work until Héroult applied for a patent in the United States. Hall had already received a patent on April 2, 1889. Hiram Brown, *Aluminum and Its Applications* (New York: Pitman, 1948), 11–12.
7. Weidlinger, 15.
8. In 1907 the Pittsburgh Reduction Company changed its name to the Aluminum Company of America. By 1955 the Aluminum Company of America (Alcoa) would become the largest producer of aluminum in the world with an annual output of 600,000 metric tons. Brimelow, 28.
9. "Aluminum in Architecture," *Architectural Forum* 53 (August 1930): 255–59. Of these three early Chicago skyscrapers, only the Monadnock is extant.
10. Wrought aluminum includes sheet and extruded aluminum.
11. Kent R. Van Horn, ed., *Aluminum*. Vol. 3, *Fabrication and Finishing* (Metals Park, Ohio : American Society for Metals, 1967): 81.
12. Monoplane designs built in this way continued in wide use after the war. Trevor I. Williams, *A History of Technology* (Oxford: Clarendon Press, 1878), 88.
13. D. J. Bleifuss and B. J. Fletcher, "Aluminum Alloys for Engineering Structures," *Civil Engineering* 9 (August 1939): 481.
14. Van Horn, 81.
15. Margot Gayle, David W. Look, and John G. Waite, *Metals in America's Historic Buildings,* rev. edn. (Washington, D.C.: U.S. Department of the Interior, National Park Service, 1992), 149.
16. Van Horn, 660.
17. Weidlinger, 72–73.
18. P. J. Carlisle, A. D. Deyrup, and A. O. Short, "Vitreous Enamels for Aluminum," *Finish* 6 (January 1949): 42; "Aluminum Siding for Residential Construction Has Baked Paint Coat," *Architectural Forum* 92 (March 1950): 210.
19. J. C. Fistere, "Use of White Metals," *Architectural Forum* 55 (August 1931): 232–40.
20. Gayle, Look, and Waite, 86.
21. H. R. Dowswell, "Walls, Floors, and Partitions in the Tall Building," *Engineering News-Record* 106 (February 19, 1931): 319–21.
22. Harold W. Vader, "Aluminum in Architecture," *Architectural Record* 70 (December 1931): 461.
23. Brown, 7.
24. Van Horn, 81.
25. Weidlinger, 9.
26. Talbot Hamlin, *Forms and Functions of Twentieth-Century Architecture.* Vol. 4, *Building Types* (New York: Columbia University Press, 1952), 165.
27. Brimelow, 30.
28. James S. Russell, "Icons of Modernism or Machine-Age Dinosaurs?" *Architectural Record* 177 (June 1989): 145.
29. Burton H. Holmes, "Alcoa Building: Lightweight Construction," in *Structural Engineering Techniques for Architects and Engineers,* 3d edn. (New York: Reinhold Publishing, n.d.), 48–51.
30. Henry-Russell Hitchcock, *Architecture: Nineteenth and Twentieth Centuries* (New York: Penguin, 1977), 559–61.
31. Frederick S. Merritt, *Building Construction Handbook,* 2d edn. (New York: McGraw-Hill, 1965), 52–57.
32. Chris Adams and David Hallam, "Finishes on Aluminum—A Conservation Perspective," in *Saving the Twentieth Century: The Conservation of Modern Materials,* ed. David W. Grattan (Ottawa: Canadian Conservation Institute, 1993), 283.
33. In general, aluminum should not come in contact with damp building materials of any kind, particularly when free access of oxygen is prevented and galvanic cells can be established.
34. Merritt, 52–57.
35. Brimelow, 314, 323.
36. This type of corrosion is electrolytic in nature; the shielded areas of the aluminum behave anodically to the exposed aluminum.
37. Van Horn, 757–59.
38. Ibid., 758–59.
39. Ibid., 759–61.
40. Ibid., 764–66.
41. Adams and Hallam, 283.
42. Van Horn, 768.
43. Ibid., 773–74.
44. Ibid., 771.
45. Ibid., 772.
46. Heat-treated sections that are welded lose strength because the weld metal is essentially a cast aluminum material. Merritt, 52–58.

**2. MONEL**

1. The nominal composition of Monel is 67 percent nickel, 30 percent copper, 1.40 percent iron, 1 percent manganese, 0.15 percent silicon, and 0.15 percent carbon.
2. In fact, Monel was found to have the optimum corrosion-resisting combination of nickel and copper. H. S. Rawdon and E. C. Groesbeck, *Bureau of Standards, Technical Bulletin No. 367* (Washington, D.C.: U.S. Government Printing Office, 1928), 409–46; and F. L. LaQue, *Journal of the American Society of Naval Engineers* 53 (1941): 29–64.
3. Hugh R. Williams, "Monel Metal: Points of Superiority of This New Natural Alloy in All Fields for Non-Corroding Steel," *Scientific American* Supplement 88 (August 16, 1919): 99.

4. For the content of original ores, see A. Stansfield, "Monel Metal," *Canadian Society of Civil Engineering, Transactions* 23 (1909): 303. For current manufacturing processes, see Walter Betteridge, *Nickel and Its Alloys* (New York: Halsted Press, 1984), 82.
5. John F. Thompson and Norman Beasley, *For the Years to Come: A Story of International Nickel of Canada* (New York: G. P. Putnam's Sons, 1960), 21–22. Much of the early history of the nickel industry described here was extracted from this history of the International Nickel Company (Inco).
6. Ibid., 27–29.
7. Ibid., 23–24. Once the metallurgists at Orford realized that the ore contained both copper and nickel, they started to research methods of separating the metals. These efforts eventually led Thompson to discover and patent a process for separating nickel and copper, later known in the industry as the Orford Process. The success of extracting useful metals from the Sudbury ores provided the essential precursor for the discovery of Monel. For the Orford Process and experimentation at Orford, see Thompson and Beasley, 89–92.
8. Before joining Inco, Monell had been the chief metallurgical engineer and assistant to the president of Carnegie Steel Company. Ibid., 140–45.
9. Ibid., 158.
10. Ibid., 159.
11. Ibid., 160–61. See also Margot Gayle, David Look, and John Waite, *Metals in America's Historic Buildings*, rev. ed. (Washington, D.C.: U.S. Department of the Interior, 1992), 39, in which the following anecdote is related: "The first bar was stamped 'Monell Metal' in honor of Ambrose Monell, then president of International Nickel. When the name was registered as a trademark one 'l' was dropped because the law prohibited the use of a family name as a trademark."
12. Williams, 98. This job, which required 265,000 square feet of material, was undertaken only after the Penn Central Railroad was satisfied that the alloy would perform adequately. For Penn Central testing, see "Monel Metal Sheets and Their Characteristics," *Metal Worker Plumber and Steam Fitter* 92 (July 25, 1919): 94.
13. Monel had been specified for a number of architectural applications in the 1910s, but sales were far below the level that could justify continued production. In fact, company executives had on more than one occasion considered dropping the alloy altogether, but one obstacle stood in the way: the trademark for the alloy had been derived from the name of the company's president.
14. Thompson and Beasley, 162–65.
15. Ibid., 180.
16. The first advertisement for Monel appeared in the October 4, 1919, issue of *The Saturday Evening Post*, a rather misguided effort, given that the company had no products for public consumption. Ibid., 189–90.
17. The company touted Monel's low weight-to-strength ratio as an advantage for many architectural applications—roofing, handrails, and decorative metalwork.
18. Ibid., 187–88.
19. Monel sinks were available from the company in three standard types and in a range of sizes. *Model Kitchens Help Make Model Homes* (New York: International Nickel Company, n.d. [c. 1930]). For other domestic applications, see *Modernizing the Home* (New York: International Nickel Company, n.d. [c. 1930]).
20. Thompson and Beasley, 105. In the late 1890s Ludwig Mond, a magnate in the British chemical industry, stumbled on a method to produce nickel tetracarbonyl. This process later became known as the Mond process and was central to Mond's new nickel business during the first two decades of the twentieth century.
21. When Monel was first introduced, its high-yield strength made it impossible to extrude. Because of this limitation, many architectural forms, such as moldings, handrails, and storefronts, could not be as readily fabricated in Monel as single pieces. Advances in metalworking technology, particularly improvements in the lubrication of dies that have virtually eliminated problems of galling (metal sticking to the die), have made it possible to extrude Monel, but the metal's relatively high cost and the ready availability of extruded aluminum forms have kept a market for extruded Monel from developing. *Fabricating, No. IAI-121* (Huntington, W.Va.: Inco Alloys International, 1987), 15. In early Monel applications where it was imperative to imitate an extruded section, the radius of sharp corners could be reduced to less than the thickness of the metal by scoring the sheet under the bend. *Practical Design in Monel Metal for Architectural and Decorative Purposes* (New York: International Nickel Company, 1931), 21.
22. *Welding, Brazing and Soldering Monel Metal and Pure Nickel: No. 1 Working Instructions* (New York: International Nickel Company, 1931), 7.
23. On uses for cold-rolled sheets, see *Polishing Monel Metal and Pure Nickel* (New York: International Nickel Company, n.d. [c. 1930]), 1. For full-finished sheets and No. 5 finish, see *Modern Domestic Service Equipment: A Manual on the Uses of Monel Metal in the Kitchen, Pantries, and Laundries* (New York: International Nickel Company, 1933), 17.
24. Shot and blocks were available for casting; the shot had a minimum nickel content of 65 percent, a minimum copper content of 25 percent, a maximum iron content of 2.50 percent, a maximum manganese content of 1 percent, a maximum carbon content of 0.25 percent, and a maximum content of 2 percent silicon to improve the flow of the metal. The block had similar contents of nickel, copper, and iron but only 0.50 percent manganese, 0.20 percent silicon, and 0.20 percent carbon. Inco's literature suggests that the silicon in the shot, while obviating the need for the foundry to add material (a feature that made it more popular than block), may oxidize and cause porosity in the casting. For improved quality the company recommended that block be used and silicon added to the molten Monel. See *Hand Forging Monel Metal: No. 2 Working Instructions* (New York: International Nickel Company, n.d. [c. 1930]), n.p.
25. *Polishing Monel Metal and Pure Nickel,* 2.
26. *Hand Forging Monel Metal: No. 2 Working Instructions,* n.p.
27. "Nickel Copper Alloy Roof Protects Museum's Treasures," *American*

*Artisan,* February 1958. Reprinted by Inco as "How the Metropolitan Museum Cuts Roofing Maintenance. . . ."

**28.** For casting uses, see *Architectural and Ornamental Monel Metals* (New York: International Nickel Company, n.d. [c. 1930]); for uses for forgings, see Williams, 99.

**29.** *Monel-Plymetyl* (New York: International Nickel Company, n.d. [c. 1930]), 1–2; *Monel Notes* 3 (London: Henry Wiggin and Company, Ltd., c. 1930), 35.

**30.** Gaseous chlorine, bromine, and ammonia in combination with excess water will cause severe corrosion of Monel; these gases are infrequently encountered under normal atmospheric conditions, however. Inco has found that steam condensate at 160°F with a vapor phase containing 70 percent carbon dioxide and 30 percent air as noncondensable gases was extremely corrosive. The corrosive properties of steam have been noted in light of the practice of steam cleaning buildings. *Monel Alloys: 400, 401, 404, R-405, X-500,* 5th edn. (Huntington, W.Va.: Inco Alloys International, Inc., 1984), 16–17.

**31.** *Monel Alloys,* 20, and Betteridge, *Nickel and Its Alloys,* 83.

**32.** Acids containing chromates, dichromates, nitrates, peroxides, and other oxidizing compounds will also affect Monel. *Monel Alloys,* 21.

**33.** *Practical Design in Monel Metal for Architectural and Decorative Purposes,* 11.

**34.** The material presented here has been extracted from Inco literature of the 1930s, collected from a few telephone interviews with architects and manufacturers who have had the occasion to work with Monel, and compiled from the research of Laura J. Culberson, "The Deterioration of 'Non-Corroding' Nickel Alloys in Architectural Applications: A Study of Nickel Silver, Monel Metal, and Chromium-Nickel Stainless Steels" (master's thesis, Columbia University, May 1994).

**35.** In harsh environments the corrosion products may indicate the presence of an element peculiar to one of the white metals; greenish residues may be indicative of copper-bearing metals (nickel silver and Monel), and "rust" stains may be present on some stainless steels. In the absence of these visual clues, it is often difficult (or impossible) to eliminate one of the several white metals.

**36.** A Monel surface can have a white or silver color and a polished finish, a pewterlike color and finish, or a black color and a matte finish. Where it has been exposed to the atmosphere, it may have a slate gray or grayish green to mottled green to brown or black color and a matte finish. When regularly exposed to sunlight and "the cleansing effect of the free sweep of rain," Monel will not develop a patina but will acquire a translucent film similar in color to the virgin metal. *Practical Design in Monel Metal for Architectural and Decorative Purposes,* 15.

**37.** Monel loses its magnetic properties when heated to 212°F (100°C). For more on magnetism, see Stansfield, 306; for variation in magnetic properties, see *Monel Alloys,* 3; for the loss of magnetic properties with elevated temperature, see *Monel Notes,* 1.

**38.** To identify Monel, the following tests may be used: (1) Place a drop of 10 percent hydrochloric acid on a sample of the unknown metal and allow it to react for 1 minute. (2) Add a drop of 10 percent potassium ferricyanide and observe for 30 seconds. A red, green, or yellow-brown drop or no color may indicate Monel, 20 percent nickel silver, or 18 percent nickel silver. (3) Add a drop of nitric acid (concentrated) to the sample and observe after 1 minute. A cloudy, blue-green color indicates Monel or 20 percent nickel silver. (4) Add a drop of nitric acid and observe after 5 minutes. A yellow-green color indicates Monel; confirm by adding a drop of 1:1 nitric acid and observing after 8 minutes. A bright green color identifies Monel. *Rapid Identification (Spot Testing) of Some Metals and Alloys* (New York: International Nickel Company, 1948). For other spot-testing methods, see Fritz Feigl, *Spot Tests in Inorganic Analysis,* 6th edn. (New York: Elsevier Publishers, 1972).

**39.** Of the more sophisticated tests, the following are the most commonly available: spark tests, spectroscopes, mobile spectrometers, x ray fluorescent spectrometers, portable x rays, and thermoelectric and eddy current devices. Unlike the simpler tests, many of these more complex devices can provide a quantitative analysis of the metal. For more on these and other testing methods, see R. Newell, R. E. Brown, D. M. Soboroff, and H. V. Makar, *A Review of Methods for Identifying Scrap Metals,* Information Circular No. 8902 (Washington, D.C.: U.S. Department of the Interior, Bureau of Mines, 1980).

**40.** Because acrylic lacquers are easily scratched and deteriorate when exposed to ultraviolet rays, the wax treatments for outdoor bronze sculpture may be a course to pursue. For more on bronze conservation, see Phoebe D. Weil, "Maintenance Manual for Outdoor Bronze Sculpture: Particularly Intended for Sculpture That Has Received Conservation Treatment, Including a Protective Coating of Incralac," in *Maintenance Manual for Outdoor Bronze Sculpture* (St. Louis, Mo.: Washington University, Technology Associates, 1983).

**3. NICKEL SILVER**

**1.** William Campbell, "A List of Alloys," Appendix III, Report of Committee B-2 on Non-Ferrous Metals and Alloys, *American Society for Testing Materials Proceedings* 22 (1922): 224–27. This list of more than 150 nickel silver alloys provides the percentage of copper, nickel, zinc, iron, tin, lead, and other elements included in each variant.

**2.** I. I. Kornilov, *Nickel and Its Alloys* 1, trans. A. Aladjem (Jerusalem: Israel Program for Scientific Translations, 1963), 5.

**3.** Alfred Bonnin, *Tutenag and Paktong* (Oxford University Press, 1924), 22.

**4.** Kornilov, 5.

**5.** F. B. Howard-White, *Nickel, An Historical Review* (Princeton, N.J.: Van Nostrand, 1963), 23–24.

**6.** Ibid., 457.

**7.** Frederick Bradbury, *History of Old Sheffield Plate* (Sheffield, England: J. W. Northend, 1968), 137.

**8.** Samuel J. Rosenberg, *Nickel and Its Alloys,* National Bureau of Standards Monograph No. 106 (Washington, D.C.: U.S. Government Printing Office, 1968), 86.

**9.** Robert C. Stanley, *Nickel, Past and Present* (Toronto: International Nickel Company of Canada, 1934), 48.

10. "A Plant Devoted to Making German Silver," *Iron Age* 93 (April 9, 1914): 895. The Seymour Manufacturing Company, Seymour, Conn., the American Brass Company, Waterbury, Conn., and the Riverside Metal Company, Burlington County, N.J., were among the first companies to produce nickel silver in the United States.

11. Bradbury, 140; Earl Chapin May, *Century of Silver, 1847–1947: Connecticut Yankees and a Noble Metal* (New York: Robert M. McBride and Company, 1947), 48.

12. R. V. Hutchinson, "Casting Nickel-Silver, A Copper Nickel-Zinc Alloy," *The Metal Industry* 17 (January 1919): 13–14; C. P. Karr, "The Difficult Art of Casting German Silver," *The Foundry* 40 (November 1912): 486–87; Charles Vickers, *Metals and Their Alloys* (New York: Henry Carey Baird and Co., 1923; R. A. Wood, "German Silver," *The Metal Industry* 13 (June 1915): 229–32 (part 1), and 13 (August 1915): 322–23 (part 2).

13. Gas pores, or blowholes, appear as spherical cavities in nickel silver, and their presence makes the metal unsound. These pits are due either to improper melting or incomplete deoxidation of the metal. Among other problems, gas pores cause hairline cracking, spills, and blisters in rolled sheets and introduce serious complications with finishing treatments, such as electroplating.

14. Karr, 486.

15. W. Betteridge, *Nickel and Its Alloys* (New York: Halsted Press, 1984), 139.

16. E. G. West, *Copper and Its Alloys* (Chichester, England: Ellis Horwood, 1982), 121.

17. "Nickel Silver," *Journal of the Franklin Institute* 183 (February 1917): 167.

18. "A Plant Devoted to Making German Silver," 895.

19. "White Metal Inside and Outside: City Bank Farmers Trust Company, New York City," *Metalcraft* 6 (April 1931): 165.

20. Anne Lee, "The Chicago Daily News Building," *Architectural Forum* 52 (January 1930): 31.

21. "Nickel Silver Lends Its Beauty to the National Bank of Commerce, Houston, Texas," *The Metal Arts* 2 (November 1929): 503–4.

22. Margot Gayle, David W. Look, and John G. Waite, *Metals in America's Historic Buildings*, rev. edn. (Washington, D.C.: U.S. Department of the Interior, 1992), 35.

23. Ibid., 36.

24. "White Metal Inside and Outside," 164.

25. Caleb Hornbostel, *Materials for Architecture* (New York: Reinhold Publishing, 1978), 334.

26. When cleaning metals, the condition of the water will determine the efficiency and performance of the chosen cleaner.

27. For more information on the types and application of coatings, see Gayle, Look, and Waite, 127.

28. Martin Weaver, *Conserving Buildings* (New York: John Wiley and Sons, 1993), 210.

#### 4. STAINLESS STEEL

1. Chromium is found in chromite ore, which is 30 to 35 percent chromium and 15 to 20 percent iron.

2. Martensitic, ferritic, and precipitation-hardened stainless steels are magnetic; austenitic stainless steel is nonmagnetic or only slightly magnetic. American Iron and Steel Institute, *Stainless and Heat-Resisting Steels* (Washington, D.C.: American Iron and Steel Institute, 1974), 29–30.

3. Robert A. Hadfield, "Alloys of Iron and Chromium, Including a Report by F. Osmond," *Journal of the Iron and Steel Institute* 2 (1892): 49.

4. Carl A. Zapffe, *Stainless Steels* (Cleveland, Ohio: The American Society for Metals, 1949), 10–15. In 1904 Léon B. Guillet of France developed methods of producing iron-chromium alloys with a low carbon content.

5. American Iron and Steel Institute types are mentioned without explanation in the U.S. Steel section in the 1940 *Sweet's* catalogue but not again until 1943. In 1942 composition ranges for AISI stainless steels were published. See "Composition Ranges of Standard Stainless Steel Type Numbers," *Iron Age* 49 (February 26, 1942): 68.

6. Elwood Haynes, president of Haynes Stellite Company, claims in his article "Stellite and Stainless Steel" (*Iron Age* [June 17, 1920]: 1,723–24) to have discovered in 1911–12 the same type of stainless steel as Brearley. Both Brearley and Haynes applied for patents, and eventually both were accepted. The American Stainless Steel Company was formed in 1920 and was granted use of both these patents for the manufacture of stainless steel.

7. Zapffe, 19–20. Turbine blades are still a major product of ferritic stainless steel.

8. Other uses include spinning wire and applications requiring maximum carbonization and oxidation resistance, such as fertilizer storage tanks. Zapffe, 21–24.

9. Zapffe, 216.

10. Several variations on the traditional method have been developed to process scrap metal produced in manufacturing. Oxygen-blowing processes that can use scrap metal to charge the furnace were developed by the late 1940s but were replaced by the argon-oxygen process and the vacuum decarburization process in the early 1970s. Induction furnaces have been used for specialized steels because they can be closely controlled, but since no refining or purifying is possible, this method is not widely used. A. G. E. Robiette, "Steelmaking Trends and the Role of Ferro-Chrome," in *Stainless Steel Special, A World Survey* (London: Metal Bulletin, 1972), 63–71. See also Zapffe, 243–48; and L. Colombier and J. Hochmann, *Stainless and Heat Resisting Steels*, trans. Edward Arnold, 2d French edn. (Norwich, England: Fletcher and Son, 1967), 480.

11. Zapffe, 239–66.

12. These patents had large composition ranges, unlike the stricter composition ranges of the AISI standard types. Some of the more important patents listed are Brearley's (no. 1,197,256) and several that include the 18–8 range of the Type 302 stainless steels. "Composition of Patented Stainless Steels Vary," *Iron Trade Review* 84 (September 8, 1927): 593.

13. Curtis C. Snyder, "Polishing Chrome-Nickel-Alloy Steel Requires Care and Special Compounds," *Iron Age* 126 (August 7, 1930): 35–36; F. W. Stuyvesant, "Diversity of Advice Clouds Problem of Polishing Stainless

Steel," *Steel* 87 (December 25, 1930): 48–49. There may have been a No. 5 finish at one time, but it does not appear in advertisements or articles. No. 1 finish is hot rolled, annealed, and descaled; it is used in industrial applications where smoothness or finish is not important. No. 2D finish is generally an intermediate finish for forming difficult, deep-drawn articles that will be polished after forming. No. 2B finish is produced in the same manner as No. 2D finish, except that the sheet receives a light, cold-rolled pass on polished rolls; this finish is used for almost all deep-drawn applications. No. 4 finish is a general-purpose polished finish used commonly for restaurant and kitchen equipment and is produced with an abrasive of approximately 120–150 mesh. No. 6 finish is a dull satin finish having a lower reflectance than No. 4; it is commonly used for architectural applications where a high luster is not desired or where a contrast with brighter finishes is needed. No. 7 finish, produced by grinding to a 200 mesh and buffing, is an almost mirrorlike finish with the grit lines still visible and is used for architectural and ornamental applications. No. 8 finish is a mirror finish free of grit lines commonly used for press plates, small mirrors, and reflectors. This finish is produced by buffing with fine abrasives and buffing rouge. Colombier and Hochmann, 105–6. Explanations of electropolishing development and processes can be found in Zapffe, 275–77, and Colombier and Hochmann, 530–31. Rustless Iron and Steel Corporation's process is promoted in "Electropolishing of Stainless Steel," *Iron Age* 156 (July 12, 1945): 64–64B, 130.

**15.** *Sweet's Catalogue File*, 1950.

**16.** Ibid.

**17.** "Stainless Steel and Iron," *American Architect* 120 (October 26, 1921): 359.

**18.** For a more complete listing of trade names and manufacturers for chrome steels and chrome-nickel steels, see John Cushman Fistere, "The Use of White Metals," *Architectural Forum* 55 (August 1931): 232–40.

**19.** "New Decorative Medium for Architects, Stainless Steel for Interior and Exterior Facades, An Interview with Oscar B. Bach," *Architect and Engineer* 126 (August 1936): 23–24, 47. Bach's work is also featured in Ernest Eberhard, "Architectural Uses of Chrome Nickel Steel, the Metal Used for the Chrysler Tower and the Empire State Building," *American Architect* 138 (August 1930): 42–47.

**20.** *The Architectural and Domestic Uses of Stainless Steel* (New York: Electro Metallurgical Company, 1937).

**21.** Fistere, 233–40; George S. Herrick, "White Metal Used in Architecture," *Iron Age* 123 (November 21, 1929): 1,361–63.

**22.** William Van Alen, "The Structure and Metal Work of the Chrysler Building, Part 2," *Architectural Forum* 53 (October 1930): 493–98; R. H. Shreve, "The Empire State Building: The Window-Spandrel-Wall Detail and Its Relation to Building Progress," *Architectural Forum* 53 (July 1930): 99–104.

**23.** Because alloy steels must be poured at temperatures exceeding 3,000°F, they have to be melted in electric furnaces and poured quickly. H. A. Cooper, "Cast Stainless Steel . . . for Architectural Ornamentation," *Metalcraft* 6 (May 1931): 218–20.

**24.** J. M. Quinn, "Making Cast Stainless Steel," *Iron Trade Review* 77 (July 23, 1925): 185–88, 225.

**25.** "Stainless Steel Used Widely in Modern Bar Designs," *Steel* 92 (April 3, 1933): 24.

**26.** "Stainless Steel Used as Basic Material in Service Station Design," *Iron Age* 56 (November 1, 1945): 74. The service station featured in the article was a prototype design, and it is unclear whether it was built.

**27.** American Iron and Steel Institute, *Curtain Walls of Stainless Steel* (Princeton, N.J.: American Iron and Steel Institute, 1955), 20–23. Standard Robertson Q-Panel square-ribbed panels were used for the General Electric Turbine Building.

**28.** "New Finish for Sheet Metal," *Sheet Metal Worker* 36 (July 1945): 39, 72; "Rigid Sheets Made by Roll Embossing," *Iron Age* 155 (April 5, 1945): 64. Other provisions for preventing waviness include limiting panel sizes, allowing movement, and using either continuous backings or stiffening members. See also American Iron and Steel Institute, 99–128.

**29.** *Architectural and Domestic Uses of Stainless Steel*, 10–17.

**30.** American Iron and Steel Institute, *Carpenter Stainless Steels Working Data* (Washington, D.C.: American Iron and Steel Institute, n.d.), 141–42.

**31.** For more information on industrial cleaning techniques for stainless steel, see American Iron and Steel Institute, *Cleaning and Descaling Stainless Steels* (Washington D.C.: American Iron and Steel Institute, 1983), 27–45.

**32.** Margot Gayle, David W. Look, and John G. Waite, *Metals in America's Historic Buildings*, rev. edn. (Washington D.C.: U.S. Department of the Interior, 1992), 99.

**33.** Horst Peppa, Chicago Ornamental Iron Company, to Robert Score, April 7, 1993. We are also indebted to Mr. Peppa for the information related to the repair of scratched, dented, and warped stainless steel.

**34.** Gayle, Look, and Waite, 148.

**5. WEATHERING STEEL**

**1.** *Steel* is the general term for iron alloys containing less than 2 percent (by weight) of carbon. Most structural steels are plain low-carbon steels, containing less than 0.25 percent weight carbon. Resistance to atmospheric corrosion is enhanced and strength increased by alloying with small amounts of copper, phosphorus, silicon, manganese, molybdenum, nickel, chromium, titanium, or zirconium. Toughness and yield strength can be further improved with small additions of vanadium or niobium and by "controlled rolling" to relatively low temperatures in manufacturing, a process that reduces ferritic grain size. *Metals Handbook*, 10th edn., vol. 1 (Metals Park, Ohio: ASM International, 1990–91), 399–400.

**2.** American Society for Testing and Materials (ASTM) specifications A242 (for the original grade of architectural or structural shapes), A325 (for nuts, washers, bolts, rivets, and so forth), A374 (cold rolled), A375 (hot rolled), A588 (for a heavier grade of structural shapes), A606, Type 4, and A87I (for tubular structures and poles) establish acceptable ranges of alloy composition, yield strength and tensile strength, and corrosion resistance for weathering steel. Bethlehem Steel's Mayari R steel and U.S. Steel's USS Cor-Ten A have been produced under the ASTM A242 Type I specification. Mayari R-50 was manufactured under ASTM A588, Grade B. USS Cor-Ten B was produced as ASTM A588, Grade A. Kawasaki Steel's River-Ten products also fit these specifications.

3. A. Parrish, ed., *Mechanical Engineers Handbook*, 11th edn. (London: Butterworth, 1973), 5–45.
4. R. B. Madison, "Unpainted Steel for Permanent Structures," *Civil Engineering* 36 (February 1966): 68–72; "Whither Weathering Steel?" *Modern Steel Construction* 22 (2d quarter, 1982): 8–10.
5. J. J. Carey, "Weathered Steel—A New Concept in Corrosion Control," *Engineering Journal*, November 1964, 23–28.
6. John Dinkaloo, "The Steel Will Weather Naturally," *Architectural Record* 131 (August 1962): 148–50; "John Deere Sticks of Steel," *Architectural Forum* 121 (1964): 76–85; "Architecture: The Plowmen's Palace," *Time*, August 7, 1964, 64–65.
7. "Chicago Civic Center," *Architectural Forum* 116 (May 1962): 110, 129.
8. Pedro Albrecht and A. H. Naeemi, "Performance of Weathering Steel in Bridges," National Cooperative Highway Research Program Report No. 272 (Washington, D.C.: Transportation Research Board, National Research Council, 1984).
9. M. Pourbaix and L. de Miranda, "On the Nature of the Rust Layers Formed on Steels in Atmospheric Corrosion as a Function of Alloy Composition, Environmental Composition, Temperature and Electrode Potential," in *Passivity and Its Breakdown on Iron and Iron Base Alloys*, ed. R. W. Staehle and H. Okada (Houston: NACE, 1976), 47–48; H. Okada et al., "The Protective Rust Layer Formed on Low-Alloy Steels in Atmospheric Corrosion," *Fourth International Congress on Metallic Corrosion*, Amsterdam, September 7–14, 1969 (Houston: NACE, 1969); L. A. Wurth, "The Role of Chromium in Weathering Steel Passivation," *Materials Performance* 3 (January 1991): 62–63.
10. S. E. Thomasen and C. S. Ewart, "Detailing of Weathering Steel Facades," in IABSE-ECCS Symposium, *Steel in Buildings* (Zurich: International Association for Bridge and Structural Engineering, 1985), 174.
11. "Whither Weathering Steel?" 9.
12. Thomasen and Ewart, 174.
13. "Cor-Ten Steel, Corrosion Resistant Steel," *Architectural and Engineering News*, February 1965, 49.
14. Complete recladding of a nine-story building in Austin, Texas, was required fifteen years after construction because of inside corrosion of insulated weathering steel panels. The panels as well as the welds attaching the panels to the structure were corroding. The panel lap joints were poorly designed and did not exclude wind-driven rain; they corroded severely. Thomas J. Rowe and Karl A. Strand, "Evaluation of Weathering Steel Facade Panels at the Jesse Jones Communications Complex for the University of Texas at Austin," unpublished report, Wiss, Janney, Elstner Associates, WJE No. 880254, May 26, 1989.
15. John A. Frauenhoffer, "Weathering Steel Cladding Failure," *ASCE Journal of Performance of Constructed Facilities* 1 (May 1987): 96.
16. "Whither Weathering Steel?" 9.
17. Before being sealed at all edges, faying surfaces prone to corrosion must be cleared as completely as possible of loose or impacted matter (high-pressure hydroblasting gives a reasonably thorough effect) and then dried well by compressed air or heat. Joints and gaps can often be sealed effectively using weather-resistant, flexible, long-lasting caulks, such as RTV silicones, applied to surfaces properly primed with appropriate paint. Deteriorated structures may be taken apart. The faying surfaces are then cleared thoroughly by blasting, and coated with paint. Structures are then reassembled with "formed in place gaskets" of caulk sealing the joints.
18. Rowe and Strand.
19. Frauenhoffer, 96.
20. In many municipalities, local law prohibits outdoor silica (sand) or similar abrasive blasting for reasons of public safety, sanitation, and disposal. Alternatives are blasting with high-pressure or ultra-high-pressure water, with small additions of fine, nonsilica abrasive. In some instances comprehensive blasting is not necessary, since visually distracting cleaned or repaired areas are small in proportion to the structure as a whole or may affect only some views of the building.

## PART II. CONCRETE

### 6. CONCRETE BLOCK

1. The blocks are hardened by chemical reactions between the material components rather than by firing. Blocks can be moisture controlled (type 1) or non–moisture controlled (type 2). The purpose of moisture control is to limit shrinkage and cracking as the concrete cures. In addition to concrete blocks, other types of units formed by this process include concrete bricks, sand-lime bricks, and cast stone. By the 1920s the terms *concrete masonry unit* and *concrete building unit* were sometimes used to mean not only concrete blocks but also other concrete products, such as concrete bricks or tiles, that were often made by the same companies.
2. The oven-dry weight of a single block is 125 pounds per cubic foot or more for a normal-weight block, 105 to 125 pounds per cubic foot for a medium-weight block, and less than 105 pounds per cubic foot for a lightweight block. Normal-weight blocks are manufactured with aggregate of sand and gravel or of sand and crushed stone. Medium-weight blocks contain aggregate such as air-cooled slag. Lightweight blocks may contain coal cinders, scoria, or pumice, among others aggregates. National Concrete Masonry Association (NCMA), "ASTM Specifications for Concrete Masonry Units," TEK 36A (Herndon, Va.: NCMA, 1986). Special lightweight aggregates are used to reduce the weight of the blocks without reducing their strength. Christine Beall, *Masonry Design and Detailing for Architects, Engineers, and Contractors*, 3d edn. (New York: McGraw-Hill, 1993), 70.
3. Ibid., 67.
4. Ibid., 67.
5. U.S. Patent no. 674,874. H. Kempton Dyson, "Concrete Block Making

in Great Britain," *Concrete and Constructional Engineering* 3, Nos. 3–6, (1908–9): 224–30, 291–98, 383–90, 463–66. William Torrance, "Types of Hollow Concrete Blocks Used in the United States and Their Patents," *Concrete and Constructional Engineering* 1 (July 1906): 206–14. For further information on the history of concrete block, see Joseph Bell, *From Carriage Age to Space Age: The Birth and Growth of the Concrete Masonry Industry* (Herndon, Va.: National Concrete Masonry Association, 1969).

**6.** A few of the better known companies were the Miracle Company, Minneapolis; the Dykema Company, Grand Rapids, Mich.; the Winget Company and the Blakeslee Company, Columbus, Ohio; the Cement Machinery Company, Jackson, Mich.; the Besser Company, Alpena, Mich.; the Century Cement Machine Company, Rochester, N.Y.; and the Ideal Concrete Machinery Company, South Bend, Ind.

**7.** S. B. Newberry, "Hollow Concrete Block Building Construction in the United States," *Concrete and Constructional Engineering* 1 (May 1906): 118.

**8.** Ibid., 118.

**9.** Harvey Whipple, *Concrete Stone Manufacture*, 2d edn. (Detroit: Concrete-Cement Age Publishing Company, 1918), 230.

**10.** The National Concrete Masonry Association established itself as a separate organization from the Portland Cement Association in 1942 and is the primary professional group of concrete product manufacturers.

**11.** Bell, 19–32.

**12.** Straub was granted U.S. Patent no. 1,212,840.

**13.** *Better Buildings with Straub's Patented Cinder Concrete Building Block* (Philadelphia: Philadelphia Partition and Building Block Company, 1927).

**14.** Bell, 96–98.

**15.** While the formula for cement to sand to aggregate varied in the period literature, the general recipe was 1 part cement to 2 to 3 parts sand to 4 to 6 parts aggregate. For the aggregate, it was recommended that the gravel or stone be sifted through a wire mesh to screen stones larger than ½ inch. While the proportion of water to cement varied with the type of mix, the general recommendations were that the concrete should be as wet as possible as long as the resulting block did not stick to the metal or sag when removed from the mold box. Most of the prescriptive literature gave formulas for the proportional mix, but the recommendations varied. For examples, see Maurice M. Sloan, *The Concrete House and Its Construction* (Philadelphia: Association of American Portland Cement Manufacturers, 1912), 206; William A. Radford, *Cement Houses and How to Build Them* (Chicago: Radford Architectural Company, 1909), 14, 34; and *Dykema Cement Stone Molds, Cement Stone Machines, Cement Appliances, Cement Building Plans* (Grand Rapids, Mich.: K. Dykema and Son, 1905), 5–7.

**16.** For a detailed discussion of the history of the curing process, see Bell, 103–25.

**17.** For more information on rockfaced concrete block, see Pamela H. Simpson, "Cheap, Quick and Easy: The Early History of Rockfaced Concrete Block Building," in *Perspectives in Vernacular Architecture* 3, ed. Thomas Carter and Bernard L. Herman (Columbia, Mo.: University of Missouri Press, 1989), 108–18.

**18.** Mary M. Stiritz, "Oakherst Place Concrete Block District," National Register of Historic Places nomination form, May 5, 1987. "Concrete Cottages for Workingmen," *The Building Age* 1 (April 1910): 169.

**19.** Gustav Stickley, "Concrete Bungalow, Showing Economy of Construction," *The Craftsman* 24 (March 1912): 665–75.

**20.** William Grant, *Manufacture of Concrete Masonry Units*, 2d edn. (Chicago: Concrete Publishing Company, 1959), 230–34.

**21.** In drying from saturated to ordinary service conditions, concrete blocks typically shrink about 0.035 percent. National Concrete Masonry Association, "Control of Wall Movement with Concrete Masonry," TEK No. 44 (Herndon, Va.: NCMA, 1972), 2.

**22.** A table of appropriate cleaners for various types of stains is provided in National Concrete Masonry Association, "Removal of Stains from Concrete Masonry Walls," TEK No. 45 (Herndon, Va.: NCMA, 1972), 2–3.

**23.** In reinforced concrete masonry, the steel incorporated in the wall generally provides crack resistance. In unreinforced concrete masonry, bond beams or horizontal reinforcement may be used. Bond beams are structural elements composed of specially shaped masonry units filled with concrete or grout that is reinforced with steel; they are used, for example, as lintel beams over wall openings. Horizontal joint reinforcement consists of metal reinforcement embedded in the joints at intervals of 8, 16, or 24 inches, depending on the wall configuration. Guidelines for placement of joint reinforcement are provided in National Concrete Masonry Association, "Maintenance of Concrete Masonry Walls," TEK No. 3 (Herndon, Va.: NCMA, 1972).

**24.** "Maintenance of Concrete Masonry Walls," 4. A mortar mix of 1 part portland cement to 2 parts dry masonry sand is recommended for repair of cracks.

**25.** Sealant repair of cracks tends to be much more visually intrusive than mortar repair. The use of sand spread across the sealant surface may help disguise the repair.

**26.** Mortar joints with a concave profile are most effective in shedding water. Joints in a V-shaped, weathered, or beaded profile are also effective. Other profiles should be avoided.

**27.** ASTM C270 provides guidelines for mortar for nonreinforced masonry and ASTM C476 for mortar for reinforced masonry. Guidelines for selection of mortar types are provided in National Concrete Masonry Association, "Building Weathertight Concrete Masonry Walls," TEK No. 8 (Herndon, Va.: NCMA, 1977), 3.

**28.** Through-wall flashings should be installed below copings and at the base of parapets. Parapets are especially vulnerable to water-related distress because they are exposed to thermal and moisture stresses on both sides and at the top of the wall. If possible, through-wall flashings should also be installed at the base of the parapet.

**29.** On rough-surface block, a first-coat application of a fill material may be required to fill surface pores and provide a smooth, dense surface.

**30.** Fill coats may be latex, styrene-butadiene, portland cement, or portland cement with an acrylic latex or polyvinyl acetate binder.

31. For any coating application, the surface must be free of dirt, dust, grease, oil, and efflorescence and must be properly prepared. The coating must be mixed and applied according to manufacturer's recommendations. Clear coatings and clear penetrating sealers are generally used to improve water resistance and may be used for masonry that does not require a fill coat. Clear penetrating sealers such as silane-based coatings render the surface hydrophobic yet are permeable so that water vapor can escape from the block. Test samples are required to evaluate whether the treatment will change the masonry's appearance and how effective the coating or sealer will be in protecting the concrete block. If water is penetrating the wall or if efflorescence is occurring, the sources of water must be located before any type of remedial coating is applied.
32. When replacement block is needed for a historic building, various standards and specifications are available to guide the manufacture and use of concrete block: for aggregates, "Standard Specification for Concrete Aggregates" (ASTM C33) and "Standard Specification for Lightweight Aggregates for Concrete Masonry Units" (ASTM C331); for concrete blocks, "Standard Specification for Loadbearing Concrete Masonry Units" (ASTM C90) and "Standard Specification for Non-Loadbearing Concrete Masonry Units" (ASTM C129). Other ASTM standards provide guidelines for solid concrete masonry units and concrete bricks. American Society for Testing and Materials, *ASTM Standards in Building Codes* 2 (B211–C960), 30th edn. (Philadelphia: ASTM, 1993). 33.
33. In addition to unit weight and type, ASTM defines blocks as grade N or S. Grade N is for use above or below grade, in moderate or severe exposures, and where blocks are exposed to weather. Grade S is for use only above grade and for less severe exposures and requires a protective coating if exposed to the weather. Freeze–thaw durability of concrete blocks is based on absorption characteristics.

**7. CAST STONE**

1. In February 1929 the American Concrete Institute (ACI) tentatively adopted specification P3-A-29T, which defined *cast stone* as "a building stone manufactured from portland cement concrete, precast and set in place as trim or facing on or in buildings and other structures." This definition was later reaffirmed by the ACI in 1942. C. G. Walker, "Proposed Specification for Cast Stone," *Journal of the American Concrete Institute* 13 (February 1942): 313.
2. The term *artificial stone* was commonly used in the nineteenth century, but by the early twentieth century cast stone was also referred to as *art stone*, *béton*, *cut art stone*, and *composite stone*, among other names. The perception of cast stone as an imitative substitute for quarried stone rather than a material with its own characteristics also contributed to its many synonyms. The majority of early products were therefore identified either as a type of stone or by the manufacturer's or inventor's name. This practice continued even after 1929, when the ACI clearly defined the term *cast stone* in its specification P3-A-29T. See L. A. Falco, "The Development and Use of Cast Stone," *Proceedings of the American Concrete Institute* 25 (1929): 497; and Walker, 313.
3. The compressive strength of cast stone is measured in pounds per square inch (psi).
4. Michael A. Tomlan, "Cast Stone: Its History and Use in the United States to 1914," unpublished manuscript (1974), 4.
5. Carl W. Condit, *American Building Art: The Nineteenth Century* (New York: Oxford University Press, 1960), 226–27.
6. "Frear Artificial Stone, Patented 1868," *Journal of the Society of Architectural Historians* 13 (March 1954): 27. *Frear's Patent Artificial Stone, Stucco, Mastic Cement, Etc. and Pressing Machine* describes the product as "THE BEST BUILDING STONE EVER YET USED" and boasts that all types of architectural elements resembling the stone of Nova Scotia or the brownstones of New York and Philadelphia could be produced for much less than natural stone.
7. Q. A. Gillmore, *A Practical Treatise on Coignet-Béton and Other Artificial Stone* (New York: Van Nostrand, 1871), 85–86.
8. Tomlan, 6.
9. Gillmore, 73.
10. Theodore H. M. Prudon, "Simulating Stone, 1860–1940: Artificial Marble, Artificial Stone, and Cast Stone," *Association for Preservation Technology Bulletin* 21 (1989): 83. Between 1869 and 1870 M. François Coignet purchased the rights to produce artificial stone in America under his own French patents. In 1869 Coignet's agent, Louis Mangeon, formed the Coignet Agglomérée Company. Later that year Mangeon sold the patent rights to John C. Goodridge, Jr., who formed the New York and Long Island Coignet Stone Company (later the New York Stone Contracting Company) in Brooklyn, N.Y. The firm was one of the largest in the country at the time and provided artificial stone for a wide variety of projects. "Beton Structures," *Manufacturer and Builder* 9 (October 1877): 234–35. The Union Stone Company of Boston, Mass., also began to manufacture artificial stone in the 1870s after purchasing the American rights to the patents of M. Sorel, a prominent French chemist who had discovered that the oxychloride of magnesium produced an extremely durable hydraulic cement. This cement formed the basis of Union Stone, which the company used to manufacture building elements such as window sills and lintels. See Gillmore, 94–95; and "Artificial Stone," *Manufacturer and Builder* 3 (November 1871): 246.
11. For a list of American firms that produced cast stone, see Adrienne B. Cowden, "The Art of Cast Stone: The History of the Onadaga Litholite Company and the American Cast Stone Industry" (master's thesis, Cornell University, August 1995).
12. Joseph B. Mason, "Architectural Use of Cast Stone," *Architectural Forum* 49 (1928): 255–57. By the early twentieth century, manufacturers had formulated an infinite number of colors by substituting or combining different aggregates and mineral coloring agents. Manufacturers eventually developed semistandard formulas to match particular natural stones. For examples of these formulas, see H. L. Childe, *Manufacture and Uses of Concrete Products and Cast Stone* (London: Concrete Publications Limited, 1930), 16; and "Reconstructed Stone," *Concrete and Constructional Engineering* 30 (May 1935): 317–19.

13. C. G. Walker, "Recent Developments in the Manufacture and Use of Cast Stone," *Journal of the American Concrete Institute* 7 (March–April 1936): 474.

14. After 1942 the Cast Stone Institute's influence began to diminish. It published its last membership directory in 1953 and formally disbanded in the 1960s. The association was reincorporated in 1986. "This Institute Has a Mission," *Concrete* 45 (January 1937): 73.

15. Fred Wiegel, "Testing Cast Stone," *Journal of the American Concrete Institute* 3 (September 1931): 33–36.

16. The Portland Cement Association and the American Society for Testing and Materials began to develop standards in the 1920s as well. Prudon, 87.

17. C. G. Walker, Chairman, ACI Committee 704, "Proposed Specification for Cast Stone," 313.

18. Standard SS-S-721 required a minimum compressive strength of 4,000 psi and allowed for an absorption of 10 percent. "Federal Specifications for Cast Stone Adopted," *Concrete* 40 (March 1932): 20.

19. "Cast Stone Samples for Color and Finish," *Concrete* 43 (August 1935): 8.

20. Walker, "Recent Developments in the Manufacture and Use of Cast Stone," 474.

21. *Slump* refers to the amount a given sample slips down vertically immediately after its removal from a slump cone. It is a relative measure of consistency and stiffness of a given mixture and thus its water content and workability. The higher the water content of a mixture, the higher the slump; similarly, the lower the water content, the less a mixture will "slump" down.

22. Prudon, 85.

23. Albert A. Houghton, *Molding and Curing Ornamental Concrete* (New York: Norman W. Henley Publishing, 1911), 44.

24. Marcus T. Reynolds, "The New Delaware and Hudson Office Building at Albany, N.Y.," *Concrete-Cement Age* 6 (June 1915): 289.

25. According to the *Biennial Census of Manufactures*, in 1925 there were 262 cast stone manufacturers, which produced 534,164 tons at a value of $12,843,418. By 1939 only 156 firms remained, producing 132,447 tons valued at $3,258,452. The industry had recovered from a low point of 30,475 tons sold for $847,365 in 1933. But census statistics indicate that the demand for cast stone as well as what consumers were willing to pay for the product had dropped dramatically over the years.

26. "Cast Stone Group Takes Up Manufacture of Substitute Precast Products," *Concrete* 50 (July 1942): 19. Bror Nordberg, "Find New Uses for Cast Stone," *Rock Products* 42 (October 1939): 59–62.

27. T. H. Merriam, "'Volumitis' and the Concrete Products Industry," *Concrete* 36 (February 1930): 23–24.

28. Sara Stauffer, "Cast Stone: History and Technology" (master's thesis, Columbia University, 1982), 63–67.

29. Philip H. Perkins, *Repair, Protection and Waterproofing of Concrete Structures* (New York: Elsevier Applied Science Publishing, 1986), 48–49.

30. Harvey Whipple, ed., *Concrete Stone Manufacture* (Detroit: Concrete-Cement Age Publishing Company, 1918), 230.

31. Nicola Ashhurst, *Cleaning Historic Buildings*. Vol. 1, *Substrates, Soiling and Investigation* (London: DonHead, 1994), 116.

**8. REINFORCED CONCRETE**

1. Jasper O. Draffin, "A Brief History of Lime, Cement, Concrete, and Reinforced Concrete," *Journal of the Western Society of Engineers* 48 (1948), reprinted in *A Selection of Historic American Papers on Concrete, 1876–1926*, ed. Howard Newlon, Jr. (Detroit: American Concrete Institute, 1976), 30–31; Henry J. Cowan, *The Master Builders*. Vol. 1, *Science and Building: Structural and Environmental Design in the Nineteenth and Twentieth Centuries* (New York: John Wiley and Sons, 1977), 36; Carl W. Condit, *American Building: Materials and Techniques from the First Colonial Settlements to the Present* (Chicago: University of Chicago Press, 1968), 171; Cecil D. Elliott, *Technics and Architecture* (Cambridge, Mass.: MIT Press, 1992), 172.

2. Cowan, 36; Draffin, 31; Condit, 172; Elliott, 176; Ernest L. Ransome, "Reminiscence," in *Reinforced Concrete Buildings* (New York: McGraw-Hill, 1912), reprinted in Newlon, 291–93.

3. Walter Mueller, "Reinforced Concrete Construction," *Cement Age* 3 (1906): 321. See also Reyner Banham, *A Concrete Atlantis* (Cambridge, Mass.: MIT Press, 1986), 23–107.

4. Air entrainment provides durability in a freezing and thawing environment and also lowers permeability. High-range water reducers enable a significantly lower water-cement ratio, which improves strength and reduces permeability while maintaining good workability. Silica fume used in conjunction with high-range water reducers can dramatically improve strength and further reduce permeability. Polymer modifiers improve bond strength and tensile strength and reduce absorption and permeability. By reducing the permeability of the concrete, these materials curtail the penetration of aggressive salts such as chlorides and sulfates into the concrete. Thus, they help protect the concrete from aggressive materials and steel from corrosion-accelerating salts.

5. Peter Collins, *Concrete, The Vision of a New Architecture* (New York: Horizon Press, 1959), 63. See also Condit, 241, and Cowan, 79.

6. Condit, 174.

7. Cowan, 55.

8. Elliott, 196.

9. Condit, *American Building Art: The Twentieth Century* (New York: Oxford University Press, 1961), 172–74.

10. Elliott, 193–97.

11. The technical support, research and development, and educational activities of the Portland Cement Association (PCA) did much to expand the use of concrete and improve the quality concrete construction in the twentieth century. Many excellent publications are still available from PCA, including Steven H. Kozsmatka and William C. Panarese's primer *Design and Control of Concrete Mixtures*, 13th edn. (Skokie, Ill.: Portland Cement Association, 1988).

    PCA's research and educational contributions to the industry from about 1940 to 1980 were probably unsurpassed anywhere in the world. Other organizations such as ACI and PCI continue to provide guidance in the application of concrete technologies. ACI publishes an excellent reference work, *ACI Manual of Concrete Practice*, which is updated yearly.

ASTM Committees on Cement and Concrete provide standard specifications and test methods for cements, aggregates, admixtures, and concretes.

**12.** *Guide to Making a Condition Survey*, ACI 201.IR-92 (Detroit: American Concrete Institute, 1992); *Strength Evaluation of Existing Structures,* ACI 437R-91 (Detroit: American Concrete Institute, 1991); *Guide for Evaluation of Concrete Structures Prior to Rehabilitation,* ACI 364.IR-94 (Detroit: American Concrete Institute, 1994); *Guide for Repair of Completed Bridge Superstructures,* ACI 546.IR-80 (88) (Detroit: American Concrete Institute, 1980).

**13.** *Practice for Petrographic Examination of Hardened Concrete,* ASTM C856 (Philadelphia: American Society for Testing and Materials, 1983 [ed. note 1988]).

**14.** For guidelines on the repair of concrete with embedded reinforcing steel, see *Maintenance and Rehabilitation Considerations for Corrosion of Steel-Reinforced Concrete Structures,* NACE-RP-03090-90 (Houston: NACE, 1990).

**15.** In the 1980s a new class of penetrating sealers based on chemically reactive silanes was introduced to the concrete field. This generic class of sealers was found to be effective for protecting concrete and embedded steel from aggressive solutions, such as chloride-containing corrosion-accelerators, while having little effect on vapor permeability.

**16.** William T. Scannell and Ali Sohanghpurwala, "Cathodic Protection as a Corrosion Control Alternative," *Concrete Repair Bulletin* 6 (July–August 1994): 6–7.

**9. SHOTCRETE**

**1.** ACI Committee 506.2-77, "Specification for Materials, Proportioning and Application of Shotcrete," *ACI Manual of Concrete Practice, Part 5* (Detroit: American Concrete Institute, 1985), 7. Definitions for *dry-mix* and *wet-mix shotcrete* have gradually developed to an accepted standard over the past forty years. Several versions of these standards, developed by the ACI Committee 506 (formerly 805) can be consulted for additional information on shotcrete standards. In 1951 ACI Committee 805 published "Recommended Practice for the Application of Mortar by Pneumatic Pressure" (805-51). In this standard the term *shotcrete* referred only to the dry-mix shotcrete procedure; the wet-mix procedure was not mentioned. Fifteen years later a revised "Recommended Practice for Shotcreting" (506-66) was published by the ACI Committee 506 in the *ACI Journal.* This time both the dry- and wet-mix processes were described and comparisons made between them.

**2.** In the wet-mix process cement, sand, aggregate, admixtures, and water are combined before application. The final mixture is transferred from the mixing chamber of the delivery equipment through the delivery hose to the nozzle and projected onto a surface by means of compressed air introduced at the nozzle. Theodore R. Crom, "Dry-Mix Shotcrete Practice," in *Shotcreting* SP-14 (Detroit: American Concrete Institute, 1965), 15.

**3.** In both forms of application, the force created as the mortar-concrete materials hit the prepared surface compacts the material, eliminating the need to vibrate the mixture to ensure compaction. Typically, compaction is greater in dry-mix than in wet-mix shotcrete. The blowing action, actually a pressurized discharge using compressed air, is also known as shooting or gunning. The gunning procedure differentiates shotcrete from other concrete application practices such as poured-in-place or hand-placing procedures.

**4.** A wet-mix grout compound was used for repair early in this century. In 1906 the exposed-concrete, arched surfaces of the Paris–Lyons–Mediterranean railway were coated with a grout material similar to wet-mix shotcrete. The grout was not blown onto the surface but was pumped through holes in the structure and allowed to flow over the outer surfaces. A similar system was reportedly used for railway tunnel repairs for the Paris Metropolitan railway in 1909. G. L. Prentiss, "The Use of Compressed Air in Handling Mortar and Concretes," *Journal of the National Association of Cement Users* 7 (1911): 505.

**5.** Prentiss, 506. Akeley's initial intent was to form plaster animal casts. The first machine sprayed a wet "plaster mortar" (a "plastic mixture of cement, sand and water"). After some experimentation, in 1911 he developed and patented a machine consisting of a single-chamber pressure tank with a vertical flywheel. However, the wet plaster became lodged in the delivery hose, preventing successful distribution. For the second patent Akeley added a second chamber to the pressure tank and positioned the feed wheel horizontally, rather than vertically. This, according to Leon Glassgold, is the same basic configuration of modern-day shotcrete machinery. Leon Glassgold, "Evaluation and Design of Concrete Structures and Innovations in Design," in *Proceedings, ACI International Conference, Hong Kong*, 1991, SP-128, vol. 1 (Detroit: American Concrete Institute, 1991), 294–95.

**6.** Prentiss, 509; Crom, 15. Another source indicates that the cement show was held in Madison Square Garden two years earlier, in 1910.

**7.** An interesting variation on wet-mix shotcrete was the "concrete atomizer," described by its developer, Harold P. Brown, in 1915. In this process premixed mortar materials were subjected to high-pressure water vapor. The resulting mixture was then blown onto the surface using the combined force of compressed air and steam. This system was used for concrete repairs by the Lackawanna Railway for its Hoboken, N.J., terminal. Harold P. Brown, "Mixing, Curing and Placing Concrete with High Pressure Steam," *Journal of the American Concrete Institute* 2 (1915): 367.

**8.** Glassgold, "Evaluation," 295. Interest in the wet-mix process was not rekindled until the early 1950s, when the True-Gun, a pneumatic device for gunning a premixed mortar material, was introduced.

**9.** Ibid. Wet-mix shotcrete was formally recognized by the ACI in 1966.

**10.** *The Cement Gun*, Bulletin No. 1,200 (Allentown, Pa.: Cement Gun Company, 1937).

**11.** The Traylor Engineering and Manufacturing Company of Allentown, Pa., purchased the rights to the cement gun in 1916. Glassgold, "Evaluation," 293.

12. Ibid., 296.

13. The mix described in Carl Akeley's patent called for a fine-aggregate dry mix consisting of 1 part cement to 3 parts sand. Glassgold, "Evaluation," 301. Similar proportions were specified for a wet-mix formula in 1909 by J. W. Buzzell and William H. Larkin, Jr., who recommended 1 part cement, about 3 parts sand, and 5 parts of 2-inch broken stone mixed with water. The 1:3 mixture has been superseded, for the most part, with a leaner mixture closer to 1:4 or 1:5, which is less costly and reduces the likelihood of shrinkage cracks developing during drying. Prentiss, 506.

14. Glassgold, "Evaluation," 301. Latex-modified shotcrete has been in use in the United States for some time. Steel fibers and silica fume were introduced as admixtures in the 1970s and 1980s respectively. On accelerators, see T. A. Hoffmeyer, "Wet-Mix Shotcrete Practice," in *Shotcreting: ACI Publication SP-14,* 59. For field-mixed shotcrete, admixtures are added through the mixing water. For wet-mix shotcrete, liquid accelerators are often introduced to the mix at the nozzle. However, in the dry-mix method when using quick sets one has more control over the mixture when these are added to the base mixture. Glassgold, "Evaluation," 301.

15. As was the case for poured-in-place concrete, calcium chloride, other caustic accelerators, and quick-setting agents were added to dry-mix shotcrete from the early part of this century until about 1973, when nonchloride agents were introduced to the industry. Glassgold, "Evaluation," 301.

16. Ibid., 302.

17. Prentiss, 517.

18. Ibid., 513.

19. "Discussion on Cement Gun Mortar," *Journal of the National Association of Cement Users* 8 (1912): 550.

20. "'Gun-Stone' House at Watertown, Massachusetts," *Proceedings: Journal of the American Concrete Institute* 18 (1922): 131.

21. R. L. Bertin, "Construction Features of the Zeiss Dywidag Dome for the Hayden Planetarium Building," *Proceedings: Journal of the American Concrete Institute* 31 (May–June 1935): 450.

22. Hoffmeyer, 59.

23. Early reports on shotcrete describe the use of expanded metal lath or even metal fencing wire for reinforcing. Prentiss, 521.

24. J. Lamprecht, "Concrete Restoration in Water Impounding Structures," *Proceedings: Journal of the American Concrete Institute* 32 (1936): 555–56. An average depth of 1½ inches was recommended where reinforcing was to be covered with shotcrete. In such instances, the reinforcing would be placed approximately ½ inch from the substrate, providing for 1 inch of coverage over reinforcing.

25. Rebound sand pockets are most likely to build up on inside corners, at the base of walls, behind reinforcing bars, pipes, or other obstructions, and on horizontal surfaces. Crom, 22. For this reason, reinforcing should be spaced evenly and overlapped uniformly to assure proper coverage. T. J. Reading, "Shotcrete as a Construction Material," in *Shotcreting, ACI Publication SP-14* (Detroit: American Concrete Institute, 1965), 12.

26. Crom, 22.

27. "Recommended Practice for Shotcreting," 219.

28. As water passes through the cracks, it carries with it a small amount of leached cement from the shotcrete mixture. A reasonable accumulation of lime within the crack will fill the fissure, causing it to self-heal.

29. Carbonation is the reaction of portlandite (the alkaline phase in the cement paste) with carbonic acid (which is derived from atmospheric carbon dioxide). The resultant loss in alkalinity creates an environment in which steel corrosion occurs more rapidly than within noncarbonated concrete.

30. George W. Seegebrecht, Albert Litvin, and Steve H. Gebler, "Durability of Dry-Mix Shotcrete," *Concrete International* 11 (October 1989): 47–50.

31. Glassgold, "Evaluation," 302.

32. Glassgold, "Shotcrete," 589. ASTM Subcommittee C009.46, through the ASTM Committee C9, is presently reviewing, expanding, and upgrading the current ASTM standards for shotcrete.

33. The compressed air used for pneumatic application must be allowed to escape from the point of impact. Crom, 25.

34. The ACI has recommended the use of wire mesh where repairs are being made with shotcrete. T. J. Reading, "Shotcrete as a Construction Material," in *Shotcreting, ACI Publication SP-14* (Detroit: American Concrete Institute, 1965), 4.

**10. ARCHITECTURAL PRECAST CONCRETE**

1. *Architectural Precast Concrete*, 1st edn., MNL 122 (Chicago: Precast/Prestressed Concrete Institute, 1973), 104–17.

2. Anthony E. J. Morris, *Precast Concrete in Architecture* (London: George Goodwin, 1978), 32–48.

3. Walter W. M. Smith, "Ornamental Treatment of Concrete Surfaces," *American Architect* 101 (January 24, 1912): 33.

4. John J. Earley, "The Project of Ornamenting the Baha'i Temple Dome," *Proceedings: Journal of the American Concrete Institute* 29 (June 1933): 403–11; Earley, "Architectural Concrete of the Exposed Aggregate Type," *Proceedings: Journal of the American Concrete Institute* 30 (March–April 1934): 251–78; Frederick W. Cron, *The Man Who Made Concrete Beautiful* (Fort Collins, Colo: Centennial Publications, 1977), 38–48.

5. G. P. Lagergren, "The World's Largest Barn," *Architectural Concrete* (1938): 32–33.

6. Hugo C. Fischer, "The Navy's New Ship Model Testing Plant," *Journal of the American Concrete Institute* 35 (April 1939): 317–36.

7. Ibid., 327–28.

8. U.S. Patent no. 1,376,748 was issued on May 3, 1921.

9. Simms Embler, "Design Freedom with Exposed Aggregate Architectural Concrete Slabs," *Concrete* 57 (September 1949): 15–19.

10. Ralph Ironman and Richard S. Huhta, "Shocked Concrete Comes to the States," *Concrete Products* 63 (December 1960): 29–32.

11. Depending on the cement content, the shocking technique produces compressive strengths of 3,000 to 4,000 pounds per square inch (psi) after 24 hours and up to 10,000 psi at 28 days.

12. J. L. Peterson, "History and Development of Precast Concrete in the United States," *Journal of the American Concrete Institute* 50 (February 1954): 477–96.

13. *Architectural Precast Concrete,* 3–6.

14. "Guide for Evaluation of Concrete Structures Prior to Rehabilitation," ACI Committee 364, Report 364.1R, in *ACI Materials Journal* 90 (September–October 1993): 479–98.

15. Deborah Slaton, Stephen J. Kelley, and Harry H. Hunderman, "Cleaning Historic Facades," *Construction Specifier* 47 (July 1994): 55–61; David Hadden, "Cleaning Restoration of the Baha'i House of Worship," *Concrete International* 14 (September 1992): 44–47, 51.

16. If at all possible, cleaning of concrete should be done when the temperature and humidity allow rapid drying. Slow drying increases the possibility of recurring efflorescence and discoloration.

17. Whenever a cleaning compound or acid solution is used, the areas to be cleaned should be thoroughly saturated with clean water before the cleaning solution is applied. This prevents the solution from being absorbed deeply into the surface of the concrete. The cleaned area should then be thoroughly rinsed with clean water. pH paper should be applied to a dampened concrete surface to ensure that the surface is neutral. Cleaning solutions should not be allowed to dry on the panel faces. Residual salts can flake or spall the surface or leave difficult stains. During acid washing, all adjacent corrodible materials, glass, or exposed parts of the building should be protected. A spray-on plastic that can be stripped off can be used to protect glass and aluminum frames. Use dilute solutions of acid to prevent surface etching that may reveal the aggregate and slightly change the surface color and texture of a panel and, thus, affect the appearance of the finish. The cleaning method should not bleach or darken the surface. The entire panel or facade should be treated to avoid a mottled effect. Also, long-term side effects should be determined from previous projects.

18. *Removing Stains and Cleaning Concrete Surface*, IS124T (Skokie, Ill.: Portland Cement Association, 1988), 1–16.

19. For shallow repairs it may be necessary to remove large sand particles that may hinder working the repair mix into the repair area. For smooth surfaces concrete sand may be too coarse for use in the mortar. Coarse sand will impart a coarse grain texture to the concrete, and this will be very conspicuous against the surrounding smooth concrete; therefore, limestone or marble dust should be added to the sand, or a fine silica sand substituted. These fine aggregates also increase the water requirements of the mortar and produce a whiter surface. In large areas requiring repair, the coarse aggregate would normally be in the repair mix. If the surface texture is sandblasted, then coarse aggregate larger than ¼ inch should be blasted before use in the repair mix.

20. *Manual for Quality Control for Plants and Production of Architectural Precast Concrete*, MNL 117, 3d edn. (Chicago: Precast/Prestressed Concrete Institute, 1996), 25–26, 148–55.

21. Epoxies should conform to ASTM C881. Cracks wider than 0.25 inch should use a gel-type resin method incorporating a mineral filler or material with a long pot life. Some cracks extending downward from nearly horizontal surfaces may be filled by gravity; the minimum width of a crack that can be filled by gravity is a function of the viscosity of the material.

22. "Use of Epoxy Compounds with Concrete," ACI Committee 503, Report 503R-93, in ACI *Manual of Concrete Practice, Part 5* (Detroit: American Concrete Institute, 1994).

23. Chemical anchors (resin capsule or epoxy anchors) could be also considered for this application or for heavy loads. Because of full-depth bonding, resin capsule anchors are less likely to work loose under shocks or vibrations than are expansion inserts. However, chemical anchors may degrade because of creep (time-dependent deformation due to load) at temperatures in the range of 140 to 150°F. Such temperatures are typically experienced in warm climates, particularly in facade panels with dark aggregates. Also, chemical anchors may not be allowed in fire-rated connection assemblies. Manufacturers' recommendations for installation must be followed.

24. The type of solvent used in sealers, as well as the solids content, can affect the resulting color of the panel surface. Thus, neither the type nor the source of sealer should be changed during the project. Sealers may consist of the methyl methacrylate form of acrylic resins, which have a low viscosity and high solids content, or silane products, which are based on monomeric alkylaloxysilane (AS) or oligomeric alkylalkoxysiloxane (OAS). Generally, sealers having a high solids content tend to darken the panel surface. The amount of color change depends primarily on the type of material in the sealer and its concentration, as well as the porosity of the surface. The active ingredient of the sealer must be chemically resistant to the alkaline environment of the concrete. Also, sealers should be evaluated as to how well they penetrate panel surfaces that vary in absorption and texture. The new generation of penetrating sealers, generally silanes or siloxanes, develop their water-repellent ability by penetrating the surface to depths of ¼ inch, reacting with the cementitious materials in the concrete and making the concrete hydrophobic. The sealer's penetrating ability depends on the molecular size of the active ingredients, its viscosity, and the solvent-carrying material.

25. *Manual for Quality Control for Plants and Production of Architectural Precast Concrete.*

**11. PRESTRESSED CONCRETE**

1. American Concrete Institute, Committee 116, *American Concrete Institute of Concrete Practice 1993*, Part I (Detroit: American Concrete Institute). In ordinary reinforced concrete, the metal (usually steel) is positioned before placement of the concrete and thus is not stressed significantly until service loads are imposed. Interaction of the reinforcement and the concrete depends on the bond development between the two materials. This bond results from mechanical interaction between the deformed surfaces of the metal and concrete, or from chemical bonding of the two materials. Fortunately, the thermal coefficients of the concrete and steel are equal to three significant figures so that differential thermal

movements do not occur at normal temperatures. Prestressed concrete utilizes reinforcing steel of much higher tensile resistance; its primary function is to impose compressive stresses on the concrete that must be overcome by the tensile forces developed from service loads before tension can develop in the concrete.

2. Much less common is circular prestressing, used for tanks, in which the prestressing steel is wrapped around a concrete structure.
3. "U.S. Progress in Prestressed Concrete," in *Architectural Engineering* (New York: F. W. Dodge Corporation, 1955), 26.
4. Ibid., 27.
5. The first prestressed concrete bridge constructed with posttensioned concrete blocks was constructed in Madison County, Tenn., in 1950. Charles C. Zollman, "Dynamic American Engineers Sustain Magnel's Momentum," in *Reflections on the Beginnings of Prestressed Concrete in America*, ed. George Nasser (Chicago: Prestressed Concrete Institute, 1981), 51.
6. "U.S. Progress in Prestressed Concrete," 29.
7. Charles C. Zollman, "Magnel's Impact on the Advent of Prestressed Concrete," in *Reflections on the Beginnings of Prestressed Concrete in America*, 25.
8. The original bridge contract also provided for testing a prototype main span girder. This testing was conducted successfully on October 25–26, 1949, before an audience of more than three hundred engineers from seventeen states and five countries. Zollman, "Magnel's Impact," 25, 30.
9. *First United States Conference on Prestressed Concrete* (Cambridge: Massachusetts Institute of Technology, 1951). In 1991 the PCI became the Precast/Prestressed Concrete Institute.
10. The quality of American concrete production when the Walnut Lane Bridge was constructed concerned Magnel, and his specification requirements were relaxed to permit its completion. The major considerations that affect the design and performance of prestressed concrete are reflected in the various sections of the ACI Building Code as follows: anchorage seating loss, elastic shortening of concrete, creep of concrete, shrinkage of concrete, relaxation of tendon loss, and loss in posttensioning tendons.
11. Requirements for the design of prestressed concrete are contained in a number of codes, most notably ACI 318 "Building Code Requirements for Reinforced Concrete," which is updated periodically and is usually included by reference in other codes.
12. Zollman, "Dynamic American Engineers," 57–60.
13. ACI Committee 116 defines a double-tee beam as "A precast concrete member composed of two stems and a combined top flange, commonly used as a beam but also in exterior walls."
14. This beam also was tested to destruction before many officials and exceeded the requirements for heavy loads established by the American Association of State Highway Officials.
15. Ross H. Bryan, "Prestressed Concrete Innovations in Tennessee," in *Reflections on the Beginnings of Prestressed Concrete in America*, 128.
16. Zollman, "Dynamic American Engineers," 53.
17. Anderson was responsible for instrumentation and data gathering on the Walnut Lane Bridge and Pottstown beams, among other early structures. He later founded Concrete Technology Corporation, which was involved in design, production, and research for prestressed concrete and has been responsible for many innovations. It consistently produced the high-strength, zero-slump concrete that Magnel originally advocated for the Walnut Lane Bridge.
18. Arthur R. Anderson, "An Adventure in Prestressed Concrete," in *Reflections on the Beginnings of Prestressed Concrete in America*, 220.
19. Ibid., 243–44.

## PART III. WOOD AND PLASTICS

### 12. FIBERBOARD

1. Lyman's method charged the mass with hot water or steamed or compressed air in a cylinder and then projected the mass from the cylinder into the atmosphere, where the lower pressure caused a breakdown in the mass by the sudden expansion of the fluid within it. The corresponding patent for the processing of cane fibers into paper boards was granted to a Mr. Russell in 1867. "Wall Board Patent History," *Paper Trade Journal* 86 (May 31, 1928): 50–53.
2. U.S. Patent no. 21,077 was granted on August 3, 1858.
3. U.S. Patent no. 99,432 was granted on February 1, 1870.
4. U.S. Patent no. 111,611 was granted on February 7, 1871.
5. In 1878 Charles Hansen et al. secured a patent for improved paper panels to imitate wood. Paper panels of straw, wood, or rags were soaked in boiled linseed oil; when dried, they were varnished, lacquered, or painted for use as panels, ceiling, wainscoting, and other finishes. U.S. Patent no. 199,435 was granted on January 22, 1878. Other pre-1900 patents for decorative panels from fiberboard include U.S. Patent no. 331,469, granted on December 1, 1885, and U.S. Patent no. 361,686, granted on April 26, 1887.
6. "History of Laminated Wall Board Patents," *Paper Trade Journal* 86 (January 19, 1928): 45–47.
7. "Homasote Co. was formed in 1909 as the Agasote Millboard Co. to manufacture high-density wood fiber boards for use as the curved ceilings and sides of railroad passenger car and interior panels for ships. Later the boards went also into automobile sedan roofs, and for many years the company . . . was a leading producer in each of these three fields. In 1916 Agasote developed a cheaper product by repulping waste newspapers, mixing them with waxes, oils and other weatherproofing elements and pressing them into large gray panels." "Old Newspapers to New Houses," *Architectural Forum* 73 (December 1940): 531–32.
8. U.S. Patent no. 1,153,512 was granted on September 14, 1915.
9. Muench's "Outline of the Insulation Board Industry" (1947) explains that until 1914 the first commercially available insulation materials were quilt insulations such as the Cabot Company's Cabot's Quilt and the Union Fiber Company's Linofelt. Semirigid insulation felts such as the

Bohn Refrigerator Company's Flaxlinum and the Union Fiber Company's Fiberfelt were superseded by insulation board because the former could not be used as an interior finish or base for plaster. Carl G. Muench, "An Outline of the Insulation Board Industry," *Paper Trade Journal* 125 (July 31, 1947): 48–51.

**10.** These included U.S. Patents no. 1,063,941 (Lewis, 1913), no. 1,121,951 (Thickens, 1914), no. 1,214,803 (Lowary, 1917), no. 1,285,433 (Sidwell, 1918), no. 1,334,637, and no. 1,334,796 (Robinson, 1920).

**11.** Kimberly A. Konrad, "The Beaver Board Companies: Manufacturers of Beaver Board Wallboard" (unpublished paper prepared at Cornell University, February 1994).

**12.** Tempered Presdwood was used to make the forms for the Minneapolis Water Tower, a concrete structure. Masonite Corporate Advertising Files, c. 1930.

**13.** Some early trade names for insulation board include J-M Board (Johns-Manville); Feltex (Phillip Carey Company); Beaver Insulating Board (Beaver Products Company, later acquired by Certain-Teed); Universal Insulite (International Insulation Company); Building Board (Celotex Company); and Structural Insulation, Insulating Lath, and Canec Insulation (Masonite). Wallboards included Beaver Board and Beaver Tile (Beaver Board Company); Satincote, Smoothcote, Ins-Lite, and Graylite (Insulite); C-X TexBord, C-X Utility Board, and C-X Green and Blue Label Wallboard (Celotex Company); and Quartrboard (Masonite). Hardboards included Presdwood, Tempered Presdwood, and Presdwood Temprtile (Masonite); Bronzelite (Insulite); and Hardboard, Tempered Hard Board, and Hard-Board Tile (Celotex).

**14.** F. P. Kollman, Edward W. Kuenzi, and Alfred J. Stamm, *Principles of Wood Science Technology*. Vol. 2, *Wood Based Materials* (New York: Springer Verlag, 1975), 552.

**15.** By 1967 the United States led the industrialized nations in consumption of wood-based sheet materials, including plywood and particleboard. Plywood was in greatest demand, followed by particleboard, and then fiberboard. Kollman, Kuenzi, and Stamm, 317.

**16.** Logs were reduced to chips by means of a rotating disk chipper.

**17.** The machinery typically used in this stage included grinders, shredders, defibrators, beaters, Claflins, Jordans, and disk refiners, which were used to break down fibers into a size appropriate for felting into a board.

**18.** Otto Suchsland and George E. Woodson, *Fibreboard Manufacturing Practices in the United States*, Handbook No. 640, Forest Service, U.S. Department of Agriculture (Washington, D.C.: Government Printing Office, 1987).

**19.** The board-forming machinery generally allowed for matting of long fibers into continuous sheets of board; these were pressed on one side and drained through a screen on the other.

**20.** Howard F. Weiss, "Man-Made Lumber: Revising the Structural Arrangements of Saw-Mill Waste to Make Possible Its Utilization," *Scientific American* 130 (April 1924): 251.

**21.** In the 1920s very little finishing was done; however, by the 1940s the largest amount of labor in insulation board plants was expended in finishing large boards as they came out of the dryer. Muench, 50.

**22.** In 1937 it was estimated that "80 percent of [the pulp for] wood fiber boards [was] produced by the mechanical process, about 20 percent by the sulphate or a combination process [Kraft process] and a very small percent by the sulphite process (less than 2 percent)." U.S. Department of Commerce, Bureau of Foreign and Domestic Commerce, Forest Products Division, *Wall Boards and Insulating Materials* (Washington, D.C.: U.S. Department of Commerce, 1937), 12. By 1979 the U.S. Department of Agriculture noted that all pulping processes in the manufacture of fiberboard were mechanical and that chemicals were no longer used for dissolving the natural bonding agent of wood. By this time chemical treatments were generally related to the fireproofing of hardboard. Suchsland and Woodson, 13.

**23.** "Processed Wood Products," *Architectural Forum* 69 (September 1937): 225.

**24.** "History of Laminated Wall Board Patents," 45.

**25.** In 1918 the use of phosphate of soda in the development of fireproof wallboards was predicted because it had been used to "render paper somewhat immune to fire" for some time. Joseph Poesi, "Some Interesting Notes on the Manufacture of Wall Board," *Building Age* 40 (January 1918): 36–38.

**26.** During this period the Masonite Corporation, because it held these patents, was the sole distributor to other hardboard suppliers—Celotex, Johns-Manville, Armstrong Cork, National Gypsum, Flintkote, Insulite, Certain-Teed, Hawaiian Cane, and others.

**27.** "Small Plant Set-Up for Insulation and Hardboard Manufacture," *Paper Trade Journal* 131 (October 19, 1950): 42.

**28.** Muench, 50.

**29.** *Celotex Manual for Architects* (Chicago: The Celotex Corporation, 1937), 10. The Celotex Corporation's 1938 annual report indicated that more than 600,000 homes in the United States were insulated with Celotex products. "Contestants in Olympic Winter Games Will Gather Here," *Celotex News* 5 (February 1932): 1.

**31.** Frank R. Walker, *The Building Estimator's Reference Book* (Chicago: Frank R. Walker Company, 1940), 998–1,003.

**32.** "Insulation Board for Home Building," *American Building* 61 (December 1939): 57.

**33.** *Interiors Beautiful* (Lockport, N.Y.: The Upson Company, Fiber Board Authorities, n.d.), 16.

**34.** Boards intended for application over flat roofing were generally 22 by 47 inches and were applied over roofing felt or resin-sized building paper. Where high levels of humidity were expected, a vapor cutoff was generally applied over the roof deck before the insulation was installed. This type of insulation could also be applied to concrete, gypsum, unit tile, and steel roof decks by embedding the insulation units in a mopping of hot bitumen. "Insulation Board for Home Building," 63.

**35.** "Details of Wall Board Paneling," *Building* Age 38 (February 1916): 63.

**36.** In modern decorative treatments, metal moldings of plain or polished chromium, copper, brass, and bronze are used to conceal joints, corners, and angles. This type of decoration is used in restaurants, bars, cafés, stores, soda fountains, beauty parlors, store displays, bathrooms, and kitchens. Walker, 990.

**37.** *Masonite* (Chicago: Masonite Corporation, 1935), 20.

### 13. DECORATIVE PLASTIC LAMINATES

**1.** D. J. Duffin, *Laminated Plastics* (New York: Reinhold Publishing, 1958), 10. In translucent laminates, cellulose and rag paper are used as fillers.

**2.** The term *decorative laminate* was adopted in 1927 with the commercial introduction of the material. Herbert Chase, "Laminated Plastics Make Rapid Advances in U.S.A.," *British Plastics and Moulded Products Trader* 8 (November 1936): 277.

**3.** Leo Baekeland became interested in photography and phenolics after studying chemistry at Ghent University. He invented Velox photography paper, which he sold to Kodak, and then came to the United States in 1902 to generate interest in phenolic resins. His February 7, 1907, patent involves the combination of phenol and formaldehyde into a synthetic resin applied as a varnish that would harden to an infusible, insoluble condition by the completion of a chemical reaction aided by heat and pressure. Bakelite, his trademark, was introduced to the public on February 5, 1909, before the American Chemical Society in New York. The first commercially produced phenolic laminate is credited to J. P. Wright, founder of the Continental Fiber Company, in 1911. Robert Freidel, *Pioneer Plastics* (Madison: University of Wisconsin Press, 1983), 107.

**4.** The General Bakelite Company licensed manufacturers to use Baekeland's patented heat-and-pressure method with the proviso that the phenolic resin varnishes had to be purchased from Bakelite. The Baekeland license policy also dictated the materials that the resin could be used to produce. Duffin, 220–21.

**5.** O'Conor had perfected the process for making rigid laminated sheet from kraft paper and liquid Bakelite at Westinghouse by winding and coating the paper on a mandrel, slitting the resulting uncured tube, and flattening it with a press. The patent for this process was applied for in 1913 and issued to Westinghouse on November 12, 1918. Steven Holt, "The Formica History: It Isn't What You Think," in *Formica and Design*, ed. Susan Grant Lewin (New York: Rizzoli, 1991), 17–39; and Jeffrey Meikle, "Plastics," in *Formica and Design*, 39–59.

**6.** Other companies entering the laminating business at this time included Monsanto in 1922, National Vulcanized Fiber in 1922, Sythane Corporation in 1928, and Spaulding Fiber Company in 1923. Holt, 27.

**7.** The one property in which phenolics suffer in comparison with other thermosetting resins is their lack of color stability. Almost all discolor in time as a result of exposure to ultraviolet rays. *Plastics Catalog: 1944 Encyclopedia of Plastics* (New York: Plastics Catalogue Corporation, 1943), 58.

**8.** Stanley Newcombe, lecture to the Plastics Historical Society, January 13, 1933, Institute of Materials, Hobart Place, London.

**9.** U.S. Patent no. 1,863,239 was granted on July 14, 1932, and no. 1,904,718 was granted on April 18, 1933. Other technical improvements quickly followed as continuous designs were developed on gravure cylinders. Holt, 27.

**10.** The gases from which the raw materials are derived are ammonia, carbon dioxide, hydrogen, and carbon monoxide. Two urea-formaldehyde resins introduced by 1931 were (1) Beetle, produced by Beetle Products Company of Oldbury, England, and sold through the Beetleware Division of American Cyanamid Company in New York (Beetle urea-formaldehyde molding powder was first used as a wall paneling system for the British Broadcasting Corporation Studios in 1931 as a result of the corporation's ban on the use of the German material Trolit); and (2) Plaskon. *1944 Encyclopedia of Plastics*, 121.

**11.** The high cost of processing urea-formaldehyde prevented its use as a binder for the full thickness of the laminate. Meikle, 49.

**12.** Melamine had been known for more than a century; the compound of carbon, nitrogen, and hydrogen was first synthesized in 1834 by Justus von Liebig. Research on the production of chemicals from cyanamide in the 1930s resulted in the commercial development of melamine. A filler melamine material is considerably harder, more heat resistant, more resistant to acids and alkalis, and less affected by boiling water than are the ureas. *Modern Plastics Encyclopedia: 1947*, 3 vols. (New York: Plastics Catalogue Corporation, 1947), 1:93.

**13.** Polyester resins were typically used to impregnate a glass fiber cloth. Early applications include airplane doors and wing tips and automobile bodies. C. A. Redfarn, *A Guide to Plastics* (London: Iliffe and Sons, 1958), 52.

**14.** Ibid., 54.

**15.** During World War II, when the use of metals was limited, plastics were used as a substitute material. Laminates were used in various applications including airplanes and automobiles. J. Harry DuBois, *Plastics History, U.S.A.* (Boston: Cahners Publishing, 1972), 198.

**16.** They were targeted at the existing and expanding markets for schools, public buildings, and especially houses as part of the greater desire for more efficient and easier maintained residences and the housewife's liberation. Holt, 28.

**17.** ColorCore, introduced by Formica in 1982, is a solid laminate that eliminates the problem of a dark core material. Holt, 33.

**18.** The resin solution is in the form of a varnish, the resin (phenol, urea, or melamine) being combined in solvents such as water or alcohol. George H. Clark, "How Laminates Are Made," *Modern Plastics* 15 (October 1938): 260. Mechanization allowed the paper to be saturated in a continuous process. There were two methods for impregnating the filler: (1) passing it through a bath of resin and then between squeeze rolls to remove the excess resin; and (2) transferring the resin varnish to a surface of the filler by a rotating roller. Premix methods were also used when the bonding resin in powdered form was added to the fiber before or during its formation into sheet form. *1944 Encyclopedia of Plastics*, 741.

**19.** In building a laminate, a translucent sheet is placed against the finish plate of the press followed by a printed sheet whose pattern is on the front (if a plain color is desired, the printed sheet is omitted). Next come the colored background sheet, filler stock, and the backing sheet. Core sheets can be impregnated with less expensive resins (phenolic) than those used for the liner (urea or melamine). Clark, 261.

20. *Modern Plastics Encyclopedia 1956* (New York: Plastics Catalogue Corporation, 1956), 93–94, 134.

21. Duffin, 59.

22. Clark, 261.

23. The liner sheets can contain any pattern or design, although the inks used must be non-reactive with and unaffected by the resin through all stages of the laminating process. *Modern Plastics Encyclopedia: 1953* (New York: Reinhold Publishing, 1953), 308.

24. Clark, 261.

25. Metals used for increased structural strength include stainless steel, aluminum, and magnesium. *Modern Plastics Encyclopedia 1941* (New York: Plastics Catalogue Corporation, 1941), 628.

26. *Formica: Plastics Finishing Material* (Cincinnati: Formica Insulation Company, 1935), 2.

27. *A Handsome and Durable Storefront Bulkhead* (Cincinnati: Formica Insulation Company, n.d. [c. 1931]), 2.

28. Frederick S. Snow, A. Caress, and Robert J. Schaffer, "The Practical Use of Plastics in Building and Construction Work," *Chemistry and Industry* 21 (July 31, 1943): 288–89.

29. "Plastics for Interior Design," *Modern Plastics* 17 (March 1940): 38, 74.

30. George J. Crosman, Jr., "Laminated Ureas for the Building Industry," *Modern Plastics* 12 (June 1935): 13–15, 61–63.

31. Robert L. Davidson, "Illuminated Sign Display," *Modern Plastics* 17 (March 1940): 25–27, 70, 72.

32. Formica Company advertisement, *Modern Plastics* 15 (October 1937): 281.

33. "The Queen Mary," *Modern Plastics* 13 (May 1936): insert.

34. W. G. Steiner, "A Year's Progress in Decorative Laminates," *Modern Plastics* 14 (October 1937): 292.

35. The kitchen of this house was reconstructed in 1989 at the Queen's Museum as part of the exhibition "Remembering the Future: The New York World's Fairs from 1939 to 1964." Holt, 31.

36. The laminate's characteristics are determined by the bond formed during polymerization. Degradation occurs if the polymer chain is broken in any way. J. Delmonte, "Permanence of the Physical Properties of Laminates," *Modern Plastics* 17 (May 1940): 78.

37. Ibid., 65.

38. The severity of both time and temperature determines the amount and type of change that takes place. In general, phenolic resins resist heat to temperatures of 300°F, ureas to 270°F, and melamines to 400°F. Herbert R. Simonds, *A Concise Guide to Plastics* (New York: Reinhold Publishing, 1957), 50.

39. High temperatures cause the rapid carbonization of both the resin and the cellulosic base and the eventual disintegration of the combination. Slow carbonization can take place at lower temperature depending on the length of the exposure and the temperature. S. W. Place, "Effect of Heat on Phenolic Laminates," *Modern Plastics* 17 (September 1940): 59.

40. High temperatures may not cause deterioration in all cases but may result in additional polymerization of the thermosetting materials and, thus, improvement of the material's physical properties. "Standard Practice for Determining the Effect of Heat on Plastics, Designation D794-82," *Annual Book of American Society for Testing and Materials Standards,* vol. 08.01 (Philadelphia: American Society for Testing and Materials, 1982), 297.

41. John Morgan, *Conservation of Plastics* (London: Museums and Galleries Commission, Conservation Unit, and the Plastic Historical Society, 1991), 26.

42. Phenolic-based laminates have a slower rate of water permeation than ureas or melamines. Delmonte, 86.

43. John Morgan, "Cleaning and Care of Plastic Artifacts," *Polymer Preprints* 33 (1992): 644.

44. John Morgan, "A Joint Project on the Conservation of Plastics by the Conservation Unit and the Plastics Historical Society," in *Saving the Twentieth Century*, 46–47.

45. Patrick Cook and Catherine Slessor, *An Illustrated Guide to Collectible Bakelite Objects* (Secaucus, N.J.: Chartwell Books, 1992), 123.

46. Deformation under load is highly dependent on time and temperature, because plastics are more susceptible at high temperatures. An increase in temperature results in a decrease in strength. Delmonte, 50.

47. Laminates are most susceptible to damage under stress from loads to the edge of the laminate (parallel to the plies) or flatwise (at right angles to the plies). L. E. Caldwell, "Properties of Laminated Phenolics," *Modern Plastics* 20 (August 1943): 83.

48. Delmonte, 50.

49. G. M. Kline, R. C. Rinker, and H. F. Meindl, "Resistance of Plastics to Chemical Reagents," *Modern Plastics* 18 (December 1941): 64, 88.

50. Ibid., 88.

51. "Flexural Strength of Adhesive Bonded Laminated Assemblies, Designation D1184-93," in *Annual Book of American Society for Testing and Materials Standards*, vol. 15.06 (Philadelphia: American Society for Testing and Materials, 1993), 71.

52. Lucius Gilman and Stanley Lowell, "Adhesives for Phenol-Formaldehyde Laminates," *Modern Plastics* 26 (March 1949): 95–98.

53. Susan Mossman, "Plastics in the Science Museum, London: A Curator's View," in *Saving the Twentieth Century*, 31.

54. Morgan, *Conservation of Plastics*, 14–15.

55. "Standard Test Method for Liquid Penetrant Examination, Designation E165-92," in *Annual Book of American Society for Testing and Materials Standards*, vol. 03.03 (Philadelphia: American Society for Testing and Materials, 1992), 54.

56. *Specifications Bulletin No. 106A* (Cincinnati: Formica Corporation, 1959).

57. *Specifications Bulletin No. 107A* (Cincinnati: Formica Corporation, 1959).

58. Exposing the laminate to the ultraviolet part of the light spectrum is extremely damaging. Morgan, "A Joint Project," 48.

59. Ibid., 48.

60. A resinous spray can also be used to restore the protective surface covering of the laminate and prevent further degradation. The spray produces a film with properties similar to those of the surface coating that has worn away. Application of this spray may require the abrasion of the laminate surface, which raises certain ethical questions.

61. *The Popular Science Illustrated Do-It-Yourself Encyclopedia* (New York: Arlrich Publishing Company, 1956), 3:595.
62. Before the laminate is replaced, all the materials—the laminate, the adhesive, and the surface—should be stabilized in temperature and humidity for at least 48 hours. *Specification Bulletin No. 106A* (Cincinnati: Formica Corporation, 1959).
63. *The Popular Science Illustrated Do-It-Yourself Encyclopedia*, 595–97.

**14. PLYWOOD**

1. Although plies are usually bonded at right angles, other angles have been used historically. Center sheets are known as cores, and outside plies are known as faces and backs. The terms *plywood* and *laminated wood* are sometimes used interchangeably. However, *laminated wood* is usually used to describe wood layers bonded with the grains running parallel. Albert G. H. Dietz, *Materials of Construction: Wood, Plastics, Fabrics* (New York: Van Nostrand, 1949): 164.
2. Thomas D. Perry, "Plywood Is Engineered Wood," *Mechanical Engineering* 74 (October 1952): 794–95.
3. Thomas D. Perry, "Seventy-Five Years of Plywood," *Southern Lumberman* 193 (December 15, 1956): 280. Evidence suggests that the term *plywood* was already in use in Europe at this time. Charles M. Haines, "The Industrialization of Wood: The Transformation of a Material" (Ph.D. diss., University of Delaware, 1990), 231.
4. Perry, *Modern Plywood* (New York: Pitman, 1942), 26–27.
5. Perry, "Seventy-Five Years of Plywood," 280–84.
6. Ibid., 244.
7. *Plywood in Retrospect*, monograph no. 1, Portland Manufacturing Company (Tacoma, Wash.: Plywood Pioneers Association, 1967), 3.
8. Haines, 261.
9. Armin Elmendorf, "Plywood and Its Use in Automobile Construction," *Journal of the Society of Automotive Engineers* 6 (May 1920): 299–306.
10. Haines, 331.
11. "Plywoods for Present and Post-War Products," *Product Engineering* 15 (August 1944): 545–46.
12. Perry, *Modern Wood Adhesives* (New York: Pitman, 1944), 49. Soybean glue could be used for hardwood veneers with high moisture contents. However, because soybean glue stained thin veneers, it was more commonly used for Douglas fir plywood.
13. Eric Schatzberg, "Ideology and Technical Choice: The Decline of the Wooden Airplane in the United States, 1920–1945," *Technology and Culture* 35 (January 1994): 64.
14. Perry, *Modern Wood Adhesives,* 58–59.
15. Haines, 332.
16. Alexander Mora, *Plywood: Its Production, Use, and Properties* (London: Timber and Plywood, 1932), 361–66.
17. Andrew D. Wood and Thomas G. Linn, *Plywoods: Their Development, Manufacture, and Application* (London: W. and A. K. Johnston, 1950), 232.
18. "How Plywood Is Made," *Engineering News-Record* 144 (May 28, 1948): 82–85.
19. Perry, *Modern Plywood,* 151–76.
20. Perry, "Electricity for Plywood," *The Technology Review* 46 (December 1943): 80–82.
21. "Wood Goes to War," Report D1426 (Madison, Wisc.: U.S. Forest Products Laboratory, December 1942). By applying fluid pressure through bags of rubber, plywood was molded for aircraft and boat hulls. The Vidal and Duramold processes of bag molding are described in Herbert Simonds and Carleton Ellis, *Handbook of Plastics* (New York: Van Nostrand, 1943), 434–36. After the war molded plywood was used extensively for furniture.
22. Before the late 1920s plywood use was limited to doors. During the late 1920s plywood was introduced for paneling. "Building Products That Build Profits," *American Lumberman,* September 11, 1948, 148.
23. Haines, 337–38.
24. "Plywood in Retrospect," monograph no. 14, Harbor Plywood Corporation (Tacoma, Wash.: Plywood Pioneers Association, 1974).
25. George W. Trayer, "Data on Structural Use of Plywood from Two New Test Series," *Engineering News-Record* 113 (August 9, 1934): 172–74.
26. For builders' interest in plywood, see, for example, "Durbin Pioneers Plywood Houses," *American Builder* 59 (August 1937): 69–71, 119–20. Interlocking wall units are described in "Novel Plywood Wall Units Cut Costs in Seattle House," *American Builder* 59 (July 1937): 80–82. The John B. Pierce Foundation's contribution to prefabrication with plywood was made in 1939. See "Research in Low-Cost Housing Yields a Solution in Plywood," *Architectural Record* 86 (September 1939): 41–45.
27. Perry, "Seventy-Five Years of Plywood," 288–89.
28. Jacques Willis, "How to Build and Sell the Dri-Bilt House," *Timberman* 39 [annual plywood supplement] (December 1937): 23.
29. Thomas S. Hines, "Neutra's All-Plywood House," *Fine Homebuilding,* February–March 1984, 28–33. Kocher's house is described in "Plywood House," *American Builder* 61 (June 1939): 68.
30. Perry, "Seventy-Five Years of Plywood," 288.
31. Haines, 335–36.
32. Perry, *Modern Plywood,* 239; Frederick F. Wangaard, "Interior Applications of Plywood," in *Architectural Engineering* (New York: F. W. Dodge Corporation, 1951), 401–4.
33. Louis H. Meyer, *Plywood: What It Is, What It Does* (New York: McGraw-Hill, 1947): 71.
34. U.S. Plywood's first installation of prefinished plywood paneling (16 by 96 inches) was in its New York office, built following the war. "Advanced Plywood Techniques in New Weldwood Building," *American Builder* 68 (December 1946): 136.
35. P. C. Bardin, "Graining and Finishing Plywood Wall Panels," *Industrial Finishing* 35 (June 1959): 49-60; John E. Hyler, "Prefinishing Fine Hardwood Panels," *Industrial Finishing* 36 (March 1960): 74–76.

**15. GLUED LAMINATED TIMBER**

1. This material has also been referred to historically as laminated wood, glulam, built-up arches, and shop-grown timbers.
2. Albert G. H. Dietz, "Glued Timbers Tailored to Fit," in *Architectural Engineering* (New York: F. W. Dodge, 1955), 393.

3. M. L. Selbo and A. C. Knauss, "Glued Laminated Wood Construction in Europe," *Journal of the Structural Division: Proceedings of the American Society of Civil Engineers 7* (November 1958): 1,840–44.
4. Selbo and Knauss, 1,840–41.
5. T. R. C. Wilson, *The Glued Laminated Wooden Arch,* Technical Bulletin No. 691 (Madison, Wisc.: U.S. Department of Agriculture, 1939), 87.
6. Charles A. Nelson, *History of the Forest Products Laboratory: 1910–1963* (Madison, Wisc.: U.S. Department of Agriculture, Forest Products Laboratory, n.d), 75.
7. According to Steinhaus, McKeown Brothers Company was one of the earliest laminators to introduce gluing as a routine plant operation. Max Steinhaus, "Historic Notes on Wood-Laminating in the United States," *Forest Products Journal* 14 (January 1964): 49.
8. Max Hanisch, Jr., to D. Dague, July 10, 1978. Courtesy of Andreas Jordahl Rhude of Sentinel Structures.
9. Built with three-hinge arches that provide a clear span of 64 feet, the building still stands today, although it was converted into a library in 1983. Biographical information on Hanisch is found in his obituaries. See *Peshtigo Times,* June 29, 1950; *Marinette Eagle Star,* June 24, 1950. I am grateful to Andreas Jordahl Rhude of Sentinel Structures for providing this information.
10. A. D. Freas and M. L. Selbo, *Fabrication and Design of Glued Laminated Wood Structural Members,* Technical Bulletin No. 1069 (Washington, D.C.: U.S. Department of Agriculture, 1954), 11.
11. Typical finishes included stains, paints, varnishes, wax, and shellac. R. J. Waddington, "Laminated Wooden Arches," *Progressive Architecture* 31 (April 1950): 87.
12. Verne Ketchum, "Fabrication of Laminated Timber Members," *Civil Engineering* 13 (February 1943): 76–79. See also Russell P. Wibbens, "Structural Glued Laminated Timber," in *Composite Engineering Laminates,* ed. Albert G. H. Dietz (Cambridge, Mass.: MIT Press, 1969), 73–90.
13. Adhesives vary in their curing properties; some, such as casein, cure at room temperature while others require higher temperatures. See Freas and Selbo, 16–26.
14. Ignorance about glued laminated timber among engineers and architects is seen in a number of articles from 1939 into the 1950s. See, for instance, Alan D. Freas, "Laminated Timber Permits Flexibility of Design," *Civil Engineering* 22 (September 1952): 757; C. Pantke, "Modern Timber Construction," *Mechanical Engineering* 61 (November 1939): 798; I. D. S. Kelly, "Timber Research and Timber Structures," *Civil Engineering* 9 (December 1939): 727–30; "Use of Wood and Aluminum as Structural Materials Discussed by Division," *Civil Engineering* 17 (August 1947): 452.
15. "Glued-Laminated Wood Construction," *Modern Plastics* 19 (February 1942): 44.
16. "Glued Arches for Building Construction Stand Rigorous Tests," *Scientific American* 156 (May 1937): 311. This building received extensive coverage in the press. See, for example, L. W. Teesdale, "New Building Demonstrates Modern Wood Construction," *Engineering News-Record* 14 (April 11, 1935): 525–27; T. R. C. Wilson, "Latest Developments in Glued-Arch Construction," *American Lumberman,* December 21, 1935, 46–47; Gary M. Peterson, "FPL Classic Structure: Another First," in *Classic Wood Structures,* ed. ASCE Task Committee on Classic Wood Structures (New York: American Society of Civil Engineers, 1989), 61–67.
17. In 1937 Summerbell Roof Structures of Los Angeles added a gluing department under license from Unit Structures. Steinhaus, 40.
18. To introduce glued laminated timber to the military, Unit Structures had to suggest the substitution of glued laminated arches for wood trusses in their first military contract. Max C. Hanisch to Richard Caletti of Standard Structures, March 4, 1976; courtesy of Andreas Jordahl Rhude.
19. John S. Samelson, "Laminated Lumber Speeds Huge Expansion at Great Lakes Naval Training Station," *American Builder and Building Age* 64 (May 1942): 52.
20. Some estimates from the 1950s asserted that three-fourths of all new churches were built with laminated timber. For the use of glued laminated timber in churches, see "'Shop-Grown' Timber Is Finding Wide Use," *Midwest Engineer* 8 (June 1955): 19; "Laminated Arches for Dignified Economy," *Architectural Record* 100 (October 1946): 108–9; "Inside . . . or Out . . . Arches Form the Shape of This Church," *American Builder* 77 (February 1955): 122–23; and M. W. Jackson, "Why Not Use Timber?" *Civil Engineering* 22 (July 1952): 489.
21. Charter members of the American Institute of Timber Construction were Unit Structures, Peshtigo, Wisc.; Arch Rib Truss Corporation, Los Angeles; Associated Wood Products Company, Berkeley, Calif.; Attwell Construction, Everett, Wash.; Laing-Garrett Construction, Phoenix; Miller-Gardner Fabrication Company, Springfield, Ore.; Rilco Laminated Products, Saint Paul, Minn.; Standard Structures, Healdsburg, Calif.; Summerbell Roof Structures, Los Angeles; Timber Fabricators, Perry, Fla.; Timber Structures, Portland, Ore.; and Vo-Vec Roof Structures of Compton, Calif. This information was provided by Andreas Jordahl Rhude.
22. Mayer was general manager of Timber Structures. See "Laminated Timber Association Formed," *Engineering News-Record* 148 (November 13, 1952): 145.
23. Wibbens, 73.
24. U.S. Commercial Standard CS253-63.
25. *Designing Structural Glued Laminated Timber for Permanence,* Technical Note No. 12 (Englewood, Colo.: American Institute for Timber Construction, November 1986), 1.
26. *In-Service Inspection, Maintenance, and Repair of Glued Laminated Timbers Subject to Decay Conditions,* Technical Note No. 13 (Englewood, Colo.: American Institute for Timber Construction, December 1986), 1.
27. Ibid., 1.
28. *Use of Epoxies in Repair of Structural Glued Laminated Timber,* Technical Note No. 14 (Englewood, Colo.: American Institute for Timber Construction, January 1990), 1–4. Shrinkage and swelling of wood and

epoxies are particularly critical when large areas are being filled. The American Institute for Timber Construction does not recommend the use of epoxy without mechanical reinforcement for wood stressed in tension parallel to the grain, longitudinal checking or splits, or glue-line delamination. For a description of the gap-filling epoxy repair technique, see "Injecting New Life into a Veteran Structure," *Technology and Conservation* 7 (Summer 1982): 10–12, 14.

**29.** Wood *Handbook: Wood as an Engineering Manual,* rev. ed. (Madison, Wisc.: U.S. Department of Agriculture, Forest Products Laboratory, 1987), section 10, 6; Wibbens, 73.

#### 16. FIBER REINFORCED PLASTIC

**1.** H. Weisbart, "The Processing of Glass Fiber Reinforced Acrylics," in *Reinforced Plastics Conference* (London: The British Plastics Federation, 1964); G. L. Whicker, "Inorganic Fiber Reinforced Thermoplastics," in *Reinforced Plastics Conference* (London: The British Plastics Federation, 1964); Roger Jones, "Filled and Reinforced Polyolefins: Review and Preview," in *SPI 26th Conference Proceedings* (New York: SPI, 1968), 533; A. Bennett, "Crocodolite Asbestos as a Reinforcing Fiber in Synthetic Resins," in *Reinforced Plastics Conference* (London: The British Plastics Federation, 1964); "Carbon Fiber: A Progress Report from Courtaulds," *Reinforced Plastics* (April 1973): 108; J. E. Gordon, *The Science of Structures and Materials* (New York: Scientific American), 1988.

**2.** Dominick Rosato, *Plastics Encyclopedia and Dictionary* (New York: Hanser, 1993), lv. In 1954 more than 80 percent of fiber reinforced plastic (FRP) was glass-fiber-reinforced polyester. Glass-fiber reinforcement may be in the form of continuous threads for filament moldings, woven cloths, chopped strand mat, continuous mat ("rovings"), and to a lesser extent epoxy resins.

**3.** George Lubin, *Handbook of Fiberglass and Advanced Plastics Composites* (New York: Reinhold Publishing, 1969), 143.

**4.** Ibid., 12.

**5.** L. A. R. Waring, "Reinforcement," in *Glass Reinforced Plastics,* ed. Brian Parkyn (London: Iliffe, 1970), 124.

**6.** Brian Parkyn, ed., *Glass Reinforced Plastics,* 5.

**7.** Rosato, 15. The name is believed to come from the combination of dysfunctional alcohol (alk) and a dysfunctional acid (yd), which is also the basis of thermoset polyesters.

**8.** M. Kaufman, *The First Century of Plastics* (London: The Plastics and Rubber Institute, 1962), 85; R. H. Norman, M. H. Stone, and W. C. Wake, "Resin-Glass Interface," in *Glass Reinforced Plastics,* ed. Parkyn, 213.

**9.** "Glass Fabric Used in Laminating Phenolil Resin," *British Plastics and Moulded Products Trader* 10 (October 1938): 286. See also K. L Forsdyke, "Developments in the Applications Technology," *The British Plastics Federation 13th Reinforced Plastics Congress* (Brighton, England: British Plastics Federation, 1982).

**10.** Lubin, 12. In 1942 the first boat was produced by Bassons Industries; the need for a mold release agent had not been appreciated, however, and workers rolled the entire assembly into the Bronx! Other products included terminal boards using fiberglass melamine laminates for the U.S. Navy. In 1944 Bassons Industries produced the first vessel using a filament-winding technique. Rosato, liii.

**11.** In 1951 it became the Reinforced Plastics Division. It was renamed the Reinforced Plastics–Composites Division in 1967.

**12.** There are a number of machine-molding techniques ranging from matched die molds to wound filament, in which a resin-impregnated reinforcing strand is wound onto a mandrel, which forms the shape. Continuous sections can be produced by extrusion of a mix of resin and reinforcement under pressure or pultrusion, where the mixture is pulled through a die. Matched molds can be injected with a mixture of resin and reinforcement or make use of precut and sometimes preshaped reinforcement mats impregnated with resin (pre-pregs).

**13.** See Lome Welch, "Design of Structures," in *Glass Reinforced Plastics,* ed. Parkyn. Philip Morgan, *Glass Reinforced Plastics* (London: Iliffe, 1961), 193–204.

**14.** Libbey-Owens-Ford Glass Company, British Patent no. 760,530. Initially sheet materials were made by contact molding, but mechanized processes soon developed because of extensive demand. This was one of the first patents for a continuous manufacturing process for FRP sheets.

**15.** J. R. Crowder, "Cladding and Sheeting," in *Glass Reinforced Plastics,* ed. Parkyn, 86. For translucent products, a close match is required between the refractive index of the resin and the fiber if the sheet is to be transparent, and a high resin content (not less than 75 percent) is required if the material is to weather well.

**16.** Penny Sparke, *An Introduction to Design and Culture in the Twentieth Century* (London: Routledge, 1992), 134. Plaskon was established to produce large housings for scales manufactured by the Toledo Scale Company. Plaskon's work resulted in some of the first uses of pure white urea-formaldehyde.

**17.** Cotterell Butler, "War-Time Building Practice," *The Builder,* May 24, 1940, 611–14.

**18.** *SPI 12th Annual Meeting* (Chicago: Society of the Plastics Industry, 1956), section 10. In Britain 30 percent of the total reinforced plastics production went into roofing and sheeting, equaled only by military uses and far exceeding boats and cars. Patrick Moxey, "Miscellaneous Applications," in Morgan, *Glass Reinforced Plastics,* 295. In the United States FRP sheet for construction purposes represented 14 percent of the total use of glass-resin materials.

**19.** David Powell, "Reinforced Plastics in Building—The European Example," in *SPI Reinforced Plastics/Composites Division 26th Conference* (Chicago: Society of the Plastics Industry, February 1971).

**20.** John Berkson, "Light-Transmitting Panels," in *Plastics in Building,* ed. Charles Koehler (Washington: D.C.: Building Research Institute, 1955); Harold Hartman, "Translucent, Structural Curtain Wall Panels in Major Architectural Projects," in *21st Annual Meeting of the Reinforced Plastics Division* (Chicago: Society of the Plastics Industry, 1966), 1–5.

**21.** In 1948 Kalwall (then Robert R. Keller and Associates) began developing prototypes for the Kalwall panel. These were not like the current product, but after examination of architects' requirements a system similar to the current one was developed, although it had a staggered grid rather than the rectangular one now used. From 1956 the FRP sheets were manufactured in-house. The early panels had several forms of insulation. Because of problems with building codes, the marketing emphasized the aluminum framing. The panel was described as a "sandwich" panel and later a "composite" panel.

**22.** Robert Keller and James Ellet, "Reinforced Plastic Exterior Curtain Walls," in *SPI 12th Conference* (Chicago: Society of the Plastics Industry, 1957), 1–10.

**23.** NASA Outline Specification Part A Section 24A (May 1963), architects-engineers Urbahn-Roberts-Seelye-Moran. The panels provided light to interiors of buildings more than 90 feet high and nearly 100 feet wide.

**24.** In 1955 the first postwar all-plastic house was constructed by Goodrich Chemical for the Seventh Plastics Exhibition in New York; the following year the London Ideal Home, designed by Peter and Alison Smithson, was built. The Ionel Schein ("Snail") House, built in 1956 in Paris, was based on a central, hollow-steel column that also acted as a rainwater pipe, using eight prefabricated sections each with walls 30 millimeters thick (comprising an insulating foam core between two layers of reinforced polyester) and a floor of 80 millimeters made up of several layers of polyester and honeycomb plastic. E. K. H. Wulkan, "Experimental Plastic Houses," *Plastic* 16 (July 1963): 333.

**25.** A. T. Waidelich, "Plastics in Structural Panels," in *Plastics in Building,* ed. Koehler, 51–56; J. E. Gordon, *Structures* (London: Penguin, 1980), 297. The development of materials made of two layers separated by a lighter one gained considerable impetus during the war. According to Gordon, the idea of a honeycomb paper core was first proposed by a circus proprietor in 1943. However, Rosato states that Arthur Howald, a Plaskon engineer, made the first reinforced honeycomb core the same year (lii).

**26.** Martin Pawley, *Design Heroes: Buckminster Fuller* (London: Grafton, 1992), 120.

**27.** Albert G. H. Dietz, "Contemporary Shell Structures," *Modern Plastics* 49 (March 1972): 91–187. Dietz, noted for his work on the Monsanto house, was a professor in MIT's department of building engineering. He also directed the plastics research laboratory at MIT and has written widely on many building and engineering topics, particularly wood engineering and plastics in buildings.

**28.** Albert G. H. Dietz, *Plastics for Architects* (Cambridge, Mass.: MIT Press, 1969), 18.

**29.** Frank Heger and Richard Chambers, "Design, Analysis and Economics of Fiberglass Reinforced Plastics World's Fair Structures," in *SPI 21st Annual Meeting* (Chicago: Society of the Plastics Industry, 1966), 1–17.

**30.** "Monsanto-MIT Molded Module," *Progressive Architecture* 35 (December 1955): 71; "Monsanto Tests Bents for House of Future," *Progressive Architecture* 37 (February 1957): 101.

**31.** Albert G. H. Dietz, Frank J. Heger, and Frederick J. McGarry, "Engineering the Plastics House of the Future," in *SPI 12th Annual Meeting of the Reinforced Plastics Division* (Chicago: Society of the Plastics Industry, 1957), 1–10.

**32.** Albert G. H. Dietz, M. E. Goody, F. J. Heger, F. J. McGarry, and R. P. Whittier, "Engineering the Plastics House of the Future," *Modern Plastics* 34 (June 1957): 143. A hand lay-up, vacuum-bag manufacturing method using 55 percent glass was selected.

**33.** Irving Skeist, ed. *Plastics in Building* (New York: Reinhold Publishing, 1966), 78.

**34.** Julian Kestler and Stuart Wood, "Reinforced Plastics in Building Construction Abroad," in *SPI Reinforced Plastics/Composites Division 25th Meeting* (Washington, D.C.: Society of the Plastics Industry, 1970). Although the United States headed the overall development of the use of FRP, it was suggested that reinforced plastics were playing a more important role in building applications outside America. France led in the proportion of FRP used in building, with 44 percent of total FRP production being used for construction. See also Moxie, "Miscellaneous Applications," in Morgan, *Glass Reinforced Plastics,* 30.

**35.** David Powell, "The Designer's View of the Use of Plastics as Structural and Non-Load Bearing Units," in *The Use of Plastics for Load Bearing and Infill Panels Symposium* (University of Surrey, 1974), 68–69.

**36.** Ibid.

**37.** Welch, 193–204.

**38.** Albert G. H. Dietz, "Paper and Plastics in Building," *Technical Association of the Pulp and Paper Industry* 39 (October 1956): 38A; Frederick Rang, "Building Code Regulation of Plastic Building Materials," in Koehler, *Plastics in Building:* ". . . there was simply no performance [standard] for light-transmitting materials . . . industry had known only one light-transmitting material for hundreds of years and the building codes uniformly reflected this."

**39.** Ian McNeill, "Fundamental Aspects of Polymer Degradation," in *Polymers in Conservation* (Manchester, England: Royal Society of Chemistry, July 1991), 14, 28. There are several causes of atmospheric degradation, one of the commonest of which is oxygen, either alone or more commonly in combination with ultraviolet or thermal degradation. Nitrogen dioxide and sulfur dioxide in moist conditions will accelerate hydrolytic attack on some polymers, and nitrogen dioxide can also work as a photosensitizer, producing singlet oxygen, which will act as a free radical and initiate breakdown of the molecular chain. Polyamides and polyurethanes are particularly sensitive to attack by nitrogen dioxide, resulting in chain scission and cross-linking. Ozone acts on some polymers such as polyethylene, polystyrene, rubber, and polyamides to cause scission of the backbone. The products are shorter chains and can then undergo further degeneration. Ibid., 30. Most polymers are hydrophobic so hydrolysis, where it does occur, will proceed slowly, generally requiring humid conditions and a pH of less than 7. The results are breaks in the chains with a rapid loss of physical properties. The main groups affected are natural proteins, polysaccharides, and polymers containing ester, amide, urethane, and carbonate links. Biodegradation and microorganisms are specific in attack and may be able to adapt to new synthetic polymers. Those sensitive to attack are aliphatic polyesters, polyethers, polyurethanes, and polyamides. Fungi require oxygen, acidic conditions of pH 4–4.5, and a temperature of about 35°C. Actinomycetes and bacteria prefer less acidic conditions (pH 5–7) and will accept a wide temperature range, the optimum being 60°C. Biodegradation is influenced by chain length, and polymers that have been subject to photo-oxidation may be more susceptible to attack.

**40.** W. H. Moss and A. B. Lang, "The Use of Accelerated Weathering Methods in the Development of Reinforced Plastics Products for Outdoor Use," in *BPF Reinforced Plastics Conference* (London: British Plastics Federation, November 1964).

**41.** Data available in 1970 suggest that such erosion led to a considerable loss of mechanical strength. K. A. Scott and Jacob Matthan, "Weathering," in *Glass Reinforced Plastics,* ed. Parkyn, 226.

**42.** "Reinforced Plastics Cladding Panels," *Building Research Establishment Digest* 161 (January 1974). Tests have shown that commonly used fire-retardant sheets undergo surface deterioration at 2.5 to 3 times the rate of general purpose resins. Aurele Blaga, "GRP Composite Materials in Construction: Properties, Applications, and Durability," *Industrialization Forum* 9 (1978): 27–32; J. R. Crowder, "The Weathering Behavior of Reinforced Plastics Sheeting," in *BPF Reinforced Plastics Conference* (London: British Plastics Federation, 1964), 27-3; coatings of polyvinyl film or a weather-resistant lacquer were used to overcome this degradation. Skeist, 92.

**43.** Personal communication with Ruben Pleydell-Bouverie, November 17, 1992, regarding panels on a one-story factory. He had noticed ripples as the facade heated up and small areas of trapped air expanded, but because of the material's good fatigue resistance he did not consider these critical.

**44.** Bernard Noel, "Le Comportement des plaques de polyester fibres de verre en toiture et en bardage," in *Society of Plastics Engineers 22nd Annual Conference* (Montreal: Society of Plastics Engineers, 1966), 1–6.

**45.** F. R. Jones, J. W. Rock, A. R. Wheatley, and J. E. Bailey, "Environmental Stress Corrosion Cracking in GRP," in *13th Reinforced Plastics Congress* (Brighton, England, November 1982), 127–33.

**46.** Scott and Matthan recommend regular washing with water to loosen any accumulated dirt and refinishing every seven years. Scott and Matthan, 230.

**47.** The removal agents will contain various types of solvent and may be in liquid, gel, or paste form. Gels and pastes are particularly useful when longer evaporation time allows the agent to remain on the surface longer; they can also be worked into a weathered surface with a stiff brush. Those incorporating chlorinated hydrocarbons should not be used on polyurethane finishes, which are often used for the final coating on FRP panels. Transport authorities often have wide experience in the main types of graffiti in their locality and the most suitable locally available products for their removal. M. J. Whitford, *Getting Rid of Graffiti* (London: E&FN, 1992), 139; appendix B3 provides a list of worldwide organizations.

**48.** Samuel Oleesky, "Repairing Reinforced Plastics," *Modern Plastics* 29 (February 1952): 99–106.

**49.** Chemical strippers should not be used to remove coatings unless there is a specific recommendation to do so. Generally, hot air strippers should be used only by an expert. For larger areas grit blasting in conjunction with filling may be used instead of sanding. Discussion with Ruben Pleydell-Bouverie, plastics consultant, regarding remedial work to the FRP panels at Olivetti Training Centre by James Stirling, November 17, 1992. The remedial work included washing down with a domestic cleaner to remove organic growth, followed by rinsing with water. Cracks were repaired using a compatible resin with good dimensional stability so that it would not shrink away from the edges of the repair. This is followed by a heavy grit blasting using 600 grit, a fine spray filler with a very low viscosity applied by a specialist to fill any slight undulations, a further light grit blasting using 400 grit, and a final decorative coat of about 30 microns.

**50.** Kalwall Corporation, *Technical Bulletin* (October 1991).

**51.** Powell, "The Designer's View of the Use of Plastics."

**52.** "Reinforced Plastics" [manual], Owens-Corning Fiberglas Corporation, July 1972, 200.

## PART IV. MASONRY

### 17. STRUCTURAL CLAY TILE

**1.** Frederick C. Merritt, ed., *Building Construction Handbook,* 2d edn. (New York: McGraw-Hill, 1965), section 2, 20. Before about 1930, structural clay tile was designated by the American Society for Testing and Materials as hollow tile. See "We're on Our Way to Larger Markets," *Brick and Clay Record* 63 (October 30, 1923): 625; and E. Giles, "Call It Structural Clay Tile," *Brick and Clay Record* 75 (April 1928): 111.

**2.** Cecil Elliott, *Technics and Architecture* (Cambridge, Mass.: MIT Press, 1992), 45.

**3.** Sara Wermeil, "Structural Hollow Tile," *Building Renovation,* Spring 1994, 42.

**4.** Elliott, 46.

**5.** Peter B. Wight, "Origin and History of Hollow Tile Fire-Proof Floor Construction," *Brickbuilder* 6 (April 1897): 74, as cited in Elliott, 46.

**6.** Wermeil, 43.

**7.** Ibid.

**8.** F. E. Emery, "Using Burned Clay Masonry," *Civil Engineering* 1 (October 1931): 1,191.

**9.** Sources of information on classification include Charles E. White, Jr., *Hollow Tile Construction* (Philadelphia: David McKay Company, 1924); and C. E. Fritz, *Materials of Engineering, Part 2* (Scranton, Pa.: International Textbook Company, 1936). On the manufacturing process, see E. W. Knapp, "How They Licked Hard Problems in Hollow Tile Manufacturing," *Brick and Clay Record* 72 (March 27, 1928): 476–79; (April 10, 1928): 554–57; (April 24, 1928): 646–48, 650; and (May 8, 1928): 706–12.

**10.** For a detailed description of flat-arch construction, see Joseph Kendall Freitag, *Architectural Engineering,* 2d edn. (New York: John Wiley and Sons, 1901).

11. Freitag, 44.
12. Wermeil, 42.
13. In the composite reinforced concrete system, in which the clay tile blocks served as stay-in-place forms, structural calculations accounted for a portion of the clay tile in determining the strength of the floor system. See Harry C. Plummer and Edwin F. Warner, *Principles of Tile Engineering: Handbook of Design* (Washington, D.C.: Structural Clay Products Institute, 1947), 282.
14. Frank R. Walker, *The Building Estimator's Reference Book,* 9th edn. (Chicago: Frank R. Walker Company, 1940), 660; "Standard Sizes and Weights of Hollow Tile Adopted," *Brick and Clay Record* 63 (October 30, 1923): 625.
15. "Face Tile for Building," *Brick and Clay Record* 75 (July 1929): 28.
16. *Sweet's Architectural Catalogues,* 1932.
17. Ibid.
18. Ibid.
19. In situ test procedures traditionally used on brick walls that can be used for structural clay tile are discussed in *Proceedings: National Workshop on Unreinforced Hollow Clay Tile* (Oakridge, Tenn.: Center for Natural Phenomena Engineering, 1993).
20. For example, see *Building Code Requirements for Masonry Structures* (ACI-ASCE 530) (Detroit: American Concrete Institute, 1992). The masonry design provisions of Chapter 24 of the *Uniform Building Code* (Whittier, Calif.: International Conference of Building Officials, 1991) can also be consulted.
21. See, for example, the numerous tables in Plummer and Warner.

**18. TERRA COTTA**

1. One of the best-known early terra cotta plants in England was the factory of George and Eleanor Coade of London, which was in operation from the 1770s until the 1830s. This terra cotta, called Coade stone, gained popularity among architects such as Sir William Chambers and Sir John Soane for its economy and versatility. Both stock patterns and custom designs for architectural ornament were produced.
2. In the 1860s the firm of Hovey and Nichols, seed and flower dealers, purchased a plant in Indianapolis for the manufacture of garden urns and statuary. The high cost of shipping the completed ceramics caused them to move the pottery works to Chicago in 1868. In 1869 the firm was reorganized as the Chicago Terra Cotta Company, whose investors included Loring. He hired James Taylor as superintendent of the works; Taylor had worked for James Marriott Blashfield, an English architectural terra cotta firm, and brought the standards of English practices to the American firm.
3. The firm opened its Boston plant when it received a contract for terra cotta for the Boston English High and Latin School. A few years earlier the commission to prepare terra cotta for the Boston Museum of Fine Arts (1870, Sturgis and Brigham) went to the Blashfield factory in England because American firms could not supply the terra cotta.
4. Cecil D. Elliott, *Technics and Architecture: The Development of Materials and Systems for Building* (Cambridge, Mass.: MIT Press, 1992), 64. Post used terra cotta manufactured in Chicago as cladding for a residence on 36th Street, terra cotta manufactured in New Jersey for the Brooklyn Historical Society, and 2,000 tons of terra cotta on the Produce Exchange Building (1882) in New York City. Kimball used terra cotta creatively for ornament, as in the Montauk Club (1890) in Brooklyn. Elliott notes that Kimball used friezes of American Indian life for the Montauk Club.
5. A group of employees of the Chicago Terra Cotta Company left the firm to form True, Brunkhorst and Company in Chicago; it was reorganized as the Northwestern Terra Cotta Works after the failure of the Chicago Terra Cotta Company. Walter Geer, *The Story of Terra Cotta* (New York: Tobias A. Wright, 1920), 54–56.
6. Chicago's key role in the development of the American terra cotta industry is seen in the fact that four major terra cotta manufacturers—Chicago, American, Midland, and Northwestern—were headquartered in Chicago for periods between 1869 and the 1960s. The American Terra Cotta and Ceramic Tile Company, founded in 1888 by William Day Gates, manufactured terra cotta for more than seventy years. One of the firm's branches produced Teco art pottery, a favorite of the Mission and Prairie styles. Geer, 54–56; see also Timothy V. Barton, "Architectural Terra Cotta: Ornament Comes of Age in Chicago," in *The L.A. Modernism Show* (San Francisco: James H. Barry Company, 1992), 16; and Sharon S. Darling, *Chicago Ceramics and Glass* (Chicago: Chicago Historical Society, 1979).
7. The first appearance of extruded elements is variously attributed; most sources note that these were in use by the late 1920s or early 1930s.
8. William D. Hunt, Jr., *The Contemporary Curtain Wall* (New York: F. W. Dodge Corporation, 1958), 291–92.
9. Ibid.
10. The development of terra cotta for floor and wall construction and for fireproofing, commonly used around iron or steel structural members, parallels the history of architectural terra cotta.
11. *Architectural Terra Cotta: Standard Construction* (New York: National Terra Cotta Company, 1914); and *Terra Cotta: Standard Construction* (New York: National Terra Cotta Company, 1927).
12. Theodore H. M. Prudon, "Looking Back: Architectural Terra-Cotta," *Building Renovation,* May–June 1993, 60. Prudon notes, "By 1900 the industry published general standards, including details and specifications data. Despite these improvements, the 1920 edition of the AIA's *Handbook of Architectural Practice* cautioned architects to conduct a thorough review of shop drawings, particularly the size and jointing details and the setting drawings."
13. Barton, 16.
14. The American Terra Cotta Company, for example, created a terra cotta called Pulsichrome to imitate limestone. Ibid.
15. The design for the facades of the Reebie Brothers Warehouse (1923, George S. Kingsley) in Chicago took inspiration from the archaeological investigations at King Tut's tomb in Egypt in 1922. The polychrome terra cotta facade features various Egyptian motifs, statues of the firm's owners as Ramses, and hieroglyphics about the firm's services.

The terra cotta sculpture was designed by Fritz Albert, sculptor for the Northwestern Terra Cotta Company. In New York City the Fred French Building (1927, H. Douglas Ives and Sloan & Robertson) combined a complex series of setbacks to meet the city's zoning requirements with highly colored polychrome terra cotta ornament in an unusual facade. Designed by the company's own architect, the French Building incorporated Eastern and mythological motifs, including the rising sun and Mercury, the winged messenger.

**16.** In the McGraw-Hill Building in New York City (1931, Raymond Hood, Godley & Foulihoux), for example, the facades were clad with machine-made terra cotta components, 51 by 16 by 4 inches, with a blue-green glaze. The building name was incorporated in terra cotta at the top of the facade. Elliott, 57.

**17.** Although the principal terra cotta firms had been producing other fired clay products in addition to architectural terra cotta, the Midland Terra Cotta Company closed in 1939, the Northwestern Terra Cotta Company in 1956, and the American Terra Cotta Company in 1966.

**18.** Vertical stresses can result when, in response to building loads, the terra cotta expands while the concrete structural members shrink or the steel structural members deflect. Differential expansion can cause building loads normally carried by structural elements to be transferred to the terra cotta cladding.

**19.** The borescope, a fiber-optic viewing device with a light source, can be inserted through a mortar joint to view conditions in the terra cotta cladding or backup wall.

**20.** Severely damaged elements requiring replacement or loose elements requiring reinstallation are good choices for removal to provide inspection openings. Temporary covers should be installed over openings until the elements are replaced. In some instances the openings must be filled with brick masonry to provide support for the components above.

**21.** Unfortunately, there are no standardized tests for analyzing terra cotta. Therefore testing for similar materials, including brick, ceramic veneer, and stone, are adapted for use in analyzing terra cotta. These tests may include 24-hour water absorption and 5-hour boil absorption to determine saturation coefficient, permeability of the glaze, moisture expansion, thermal expansion, compressive strength, shear strength, and glaze adhesion. Specialized glaze tests based on American Society for Testing and Materials (ASTM) standards have been developed for discoloration, opacity and reflectance, fading, crazing, abrasion, and hardness. The tests listed here are used for terra cotta or ceramic veneer. The ASTM Subcommittee E06.24 on Building Preservation and Rehabilitation Technology, Task Group E06.24.07 on Terra Cotta, is currently developing a guide to the repair and conservation of terra cotta.

**22.** Recent testing of new terra cotta manufactured for replacement elements has revealed that compressive strengths vary from approximately 4,000 to 12,000 pounds per square inch.

**23.** Chemical cleaners have also been found to be effective in removing existing, inappropriate paint coatings from terra cotta. Care must be taken to ensure that the glaze is not damaged during paint removal.

**24.** Breathable coatings prevent moisture from entering the terra cotta while permitting any moisture vapor within to escape. Shallow spalls may require only painting to prevent moisture from penetrating the body of the terra cotta. Undamaged areas of glaze are left uncoated in this repair.

**25.** When time or construction constraints prohibit the use of new terra cotta, substitute materials can be considered. Materials that have been used as replacements for terra cotta include cast aluminum, fiber-reinforced polymers, glass fiber–reinforced concrete, precast concrete, and cast stone. Each material has advantages and disadvantages.

**19. GYPSUM BLOCK AND TILE**

**1.** U.S. Department of Commerce, National Bureau of Standards, *U.S. Government Master Specification for Calcined Gypsum,* Circular No. 248 (Washington, D.C.: U.S. Government Printing Office, 1925): 2–4. Calcined gypsum is also known as plaster of paris. It was produced for casting (class C), general use (class G), or finishing (class F), as designated by the *U.S. Government Master Specification for Calcined Gypsum* (Washington, D.C.: U.S. Government Printing Office, 1924). Calcined gypsum results from the partial dehydration of gypsum by heat. It is to contain not less than 60.5 percent by weight of $CaSO_4 \cdot \frac{1}{2} H2O$. Its tensile strength is to be not less than 200 pounds per square inch (psi), and its compressive strength not less than 1,000 psi. Class G calcined gypsum, used for producing block and tile, was to "all pass a No. 14 sieve, and not less than 40 per cent nor more than 75 per cent shall pass a No. 100 sieve. Unretarded calcined gypsum is to set in not less than 10 nor more than 40 minutes; retarded calcined gypsum is to set in not less than 40 minutes nor more than 6 hours." Standards for gypsum were adjusted over time; in 1939 the content of material designated as gypsum ($CaSO_4 \cdot \frac{1}{2} H2O$) was to be 64.5 percent by weight. Forrest T. Mayer, *U.S. Bureau of Mines Information,* Circular No. 7049 (Washington, D.C.: U.S. Government Printing Office, 1939).

**2.** ASTM Specification C52-41, first adopted in 1925 and revised in 1927, 1933, 1941, and 1954, stated that the weight of the combustible material should not exceed 15 percent of the dry tiles. Tiles were to be "rectangular with straight and square edges and true surfaces, or of special shapes. . . . The surfaces might be scored, but not so much as to reduce the thickness of the shell materially." They were to afford a suitable bond for plaster. They were to be of any convenient length and height but in general not greater than 30 inches in length and 12 inches in height. This became the standard size, although a number of other sizes were made. Tiles were also cast in custom shapes for use as fireproof casing for steel columns and girders.

**3.** David H. Newland, "The Gypsum Resources and Gypsum Industry of New York," *New York State Museum Bulletin* 283 (November 1929): 8.

**4.** George P. Grimsley, "The Gypsum of Michigan and the Plaster Industry," in *The Geological Survey of Michigan* (Lansing, Mich.: R. Smith Printing, 1904), 221–22.

**5.** The Detroit Fireproofing Tile Company of Pittsburgh, Pa., sold a product it called Gypsite, fashioned into "fireproof partitions"—specially shaped parts to cover steel columns for fireproofing and wall furring. Gypsite, found as raw material in the southwestern United States, contained calcium sulfate in addition to calcined gypsum and might contain up to 1 percent of phosphoric acid, which could cause efflorescence and attack the wire lath or reinforcing. Gypsite was marketed as "brown" plaster, for which it was said to be satisfactory. The Detroit Company's partitions were 1-inch-thick tiles bonded by staff columns cast in place between the inner faces of the tiles on 1-foot centers. The tile faces were corrugated, and the rear sides had indentations to receive columns. *Sweet's Catalogue of Building Construction,* 1906.

**6.** Newland, 10–11.

**7.** "Novel Building Method Utilizes Precast Gypsum Units," *American Builder and Building Age* 51 (August 1931): 62–64. The article describes the casting of the hollow gypsum tile and "a patented India rubber core, an advance on previous casting methods. The gypsum floor tiles are 4' to 6' long, with a smooth surface and with three hollow spaces running through their entire length."

**8.** "New Light Weight Gypsum Flooring System," *Rock Products* 35 (March 1932): 61.

**9.** "Architectural Acoustics," *Architectural Forum* 32 (June 1920): 249.

**10.** "Lightweight Fire-Protection for Steel," *Architectural Record* 100 (October 1946): 124–25.

**11.** U.S. Department of Commerce, National Bureau of Standards, *The Technology of the Manufacture of Gypsum Products,* Circular No. 281 (Washington, D.C.: U.S. Government Printing Office, 1926), 45–46.

**12.** Ibid., 45.

**13.** Other large mills active in the 1920s included those of the Universal Gypsum Company at Fort Dodge, Iowa, which could produce 4,000 square feet of gypsum tile per day, carrying out the process by hand and air drying the tiles, and the Ebsary Block Plant, in Garbutt, N.Y., which produced 2,500 blocks in 10 hours using aluminum molds on a 100-foot-long conveyor belt. The Rock Plaster Manufacturing Company in New York City made blocks by hand in iron molds set up on rubber mat bases. Four men each had charge of four tables, each with a gang of three molds. By the time a man made the circuit of all twelve molds, the first was ready to unmold. With this system the plant turned out 900 blocks in 10 hours. *Technology of the Manufacture of Gypsum Products,* 66.

**14.** *Sweet's Catalogue File,* 1936.

**15.** *Sweet's Catalogue of Building Construction,* 1,907–8.

**16.** H. H. Robertson Company, *Robertson Process Gypsum Roofs,* Bulletin No. 6 (1919).

**17.** *Sweet's Architectural Catalogues,* 1,924–25. The Gypsteel floor system continued to be advertised between 1928 and 1930 in *Architectural Forum, Architectural Record,* and *Pencil Points.*

**18.** "Reviews of Manufacturers' Publications," *Architectural Forum* 49 (August 1928): 202.

**19.** *Sweet's Architectural Catalogues,* 1,927–28.

**20.** A. M. Turner, "Effect of Aging on Calcined Plaster," *Rock Products* 36 (May 25, 1933): 20–23.

**21.** Virgil G. Marani, "The Use of Gypsum for Fire Protection," *Engineering Magazine* 48 (January 1915): 596–98.

**22.** *Sweet's Catalogue File,* 1936.

**23.** Charles Shaffer, Denver Service Center, National Park Service, to the author, March 29, 1994.

**24.** Since portland cement and lime plaster do not typically adhere to gypsum block, the United States Gypsum Company recommended applying metal lath to its gypsum tile before using portland cement plaster; the dictum that nongypsum plaster does not bond to gypsum block has exceptions. Charles Shaffer, National Park Service, has conducted tests on gypsum block produced in the Midwest around the turn of the century and has found that, because of impurities in the gypsum, portland cement does bond tightly to some samples. Tests would have to be conducted on the blocks in question before proceeding. Charles Shaffer, Denver Service Center, National Park Service, to the author, March 29, 1994.

**25.** Martin Weaver, "Nuts and Bolts: Fixing Plaster," *Canadian Heritage,* October 1981, 33–35.

**26.** Kathleen Randall, "Stouffer Vinoy Resort," *Building Renovation,* January–February 1993, 22–27.

**20. THIN STONE VENEER**

**1.** *Marble Engineering Handbook* (Mount Vernon, N.Y.: Marble Institute of America, 1962), 21.

**2.** *Standard Terminology Relating to Dimension Stone,* ASTM C119 (Philadelphia: American Society for Testing and Materials, 1993). *Building stone* is a natural rock of adequate quality to be quarried and cut as dimension stone for use in the construction industry. The following definitions of other stones used for thin stone veneers are useful: *Granite* (commercial definition) is a visibly granular, igneous rock generally ranging in color from pink to light or dark gray and consisting mostly of quartz and feldspars as well as one or more dark minerals. The texture is typically homogeneous but may be gneissic or porphyritic. Some dark granular igneous rocks, although not properly granite, are included in the definition. *Marble* (calcite, dolomite) is a carbonate rock that has acquired a distinctive crystalline texture by recrystallization, most commonly by heat and pressure during metamorphism, and is composed principally of the carbonate minerals calcite and dolomite, singly or in combination. Marble is stone capable of taking a polish. Stone in this category includes a variety of compositional and textural types, ranging from pure carbonate to rocks containing very little carbonate that are classed commercially as marble (for example, serpentine marble). Most marbles possess an interlocking texture and a range of grain size from cryptocrystalline to 5 millimeters. *Limestone marble* is a compact, dense limestone that will take a polish, is classified as marble in trade practice, and may be sold as limestone or as marble. *Commercial marble* is recrystallized limestone, compact microcrystalline limestone, and travertine that is capable of taking a polish and may be sold as either limestone or marble. *Travertine* is a variety of crystalline or microcrystalline limestone distinguished by a layered structure; pores and cavities commonly are concentrated in some of the layers, giving rise to an open texture. *Limestone* is a rock of sedimentary origin composed principally of calcium carbonate (the mineral calcite) or the double carbonate of calcium and magnesium (the mineral dolomite), or some combination of these forms. *Slate* is microcrystalline metamorphic rock most commonly derived from shale and composed mostly of micas, chlorite, and quartz; the micaceous minerals have a subparallel orientation and thus impart strong cleavage to the rock, allowing the latter to be split into thin but tough sheets.

3. Joseph Kendall Freitag, *Architectural Engineering,* 2d edn. (New York: John Wiley and Sons, 1901), 152–54.
4. Italian Institute for Foreign Trade, "Marble: Italian Culture, Technology and Design Presentation," New York, March 15–17, 1987.
5. Most quarries, however, promoted 1½-inch-thick veneer. National Building Granite Quarries Association, "Architectural Granite," in *Sweet's Architectural Catalogues,* 1932.
6. In the 1980s techniques for thermal finish and water-blast finishes were developed.
7. Diamond-studded cables were developed by the Italian marble industry and increased manufacturing efficiency. Traditional support techniques were combined with specially shaped and continuous steel support elements, which facilitated greater setting flexibility. Specialized anchors that engaged the back of the panels with stainless steel threaded fasteners were also developed, thus allowing veneer panels to be installed on a truss before shipping. Also, frame saws along with metal abrasives were being used for cutting stone as thin as ⅛ inch. As a result, composite panels that combined a very thin stone adhered to a honeycomb core or other backing material became available. The long-term durability of these relatively new products is not well known.
8. Today thin stone veneers are typically supported on shelf angles attached to the backup wall or building frame. The shelf angles often are constructed of stainless steel, aluminum, or mild steel with a non-corroding material separator between mild steel and stone. Sometimes metal tabs or dowels are fastened to the supporting leg of steel or stainless steel angles. Aluminum shelf angles are usually extrusions that include a tab. These tabs or dowels are dry set or set in adhesives in kerfs (slits or notches) or in oversized holes on the edge of the panels for lateral support. Anchors today are typically constructed of stainless steel or aluminum, rather than the painted or galvanized mild steel or copper used in the past. Usually, these anchors are placed intermittently along the edges of the panel but in some cases may be continuous along two opposite edges. Clip angles placed along the bottom of the panel may also provide gravity support for the panel. Sometimes a disk-shaped anchor is set into a kerf cut into the sides of two adjacent panels and fastened to the structure with a rod. Thin stone veneer may also be anchored to precast concrete backup panels or a steel frame with anchor components that provide both lateral and gravity support as described earlier.
9. *Sweet's Catalogue File,* 1944.
10. Antonio Consiglio, *A Technical Guide to the Rational Use of Marble* (Rome: Italian Marble Industry Association, 1970).
11. Christine Beall, *Masonry Design and Detailing,* 3d edn. (New York: McGraw-Hill, 1993), 93.
12. Stylolites are irregular surface features, generally parallel to a bedding plane, in which small toothlike projections on one side of the surface fit into cavities of complementary shape on the other surface; they are interpreted to result diagenetically by pressure solution.
13. See *Granite Dimension Stone* (ASTM C615), *Marble Dimension Stone* (ASTM C503), *Limestone Dimension Stone* (ASTM C568), and *Slate Dimension Stone* (ASTM C629) (Philadelphia: American Society for Testing and Materials, 1993).

**21. SIMULATED MASONRY**

1. Cast stone is generally considered a subclass of simulated masonry.
2. See, for example, Theodore H. M. Prudon, "Simulating Stone, 1860–1940: Artificial Marble, Artificial Stone, and Cast Stone," *Association for Preservation Technology Bulletin* 21 (1989): 81.
3. *Modernizing Magic* (Columbus, Ohio: Perma-Stone Company, 1954), 15.
4. Ibid.
5. U.S. Patent no. 2,095,641 was issued on October 12, 1937.
6. "Rostone—A New Industry," *American Builder* 55 (June 1933): 40–41. The company is still in business today, but production of Rostone ceased in 1954.
7. Philip Morgan, ed., *Glass Reinforced Plastic,* 3d edn. (New York: Interscience Publishers, 1961), 313.
8. U.S. Patent no. 1,852,672 was issued on April 5, 1932.
9. Ibid.
10. "Latest in Service Station Designs Use New Building Materials," *American Builder* 56 (March 1934): 70–71.
11. Rostone Company brochure (Lafayette, Ind.: Rostone Company, n.d.).
12. National Register of Historic Places nomination form, Beverly Shores-Century of Progress Architectural District, Wieboldt-Rostone House, Porter County, Ind., April 1986, part 7, continuation sheet, 1.
13. *Perma-Stone: For Beauty, Permanence, Strength* (Columbus, Ohio: Perma-Stone Company, n.d.).
14. James A. Burns, Paul's Insulation Company, to Clifton Utley, owner of a Rostone house in Beverly Shores, Ind., October 16, 1956.
15. Eldon Hambel of Cincinnati, Ohio, a former employee and part owner of the Perma-Stone Company, to the author, October 25, 1994.
16. *Perma-Stone: For Beauty, Permanence, Strength.* According to Hambel, Edward J. Miller of Youngstown, Ohio, received the first Perma-Stone patent around 1929. However, the author has been unable to locate these patents.
17. Aaron Chase of Louisville, Ky., a former dealer in the Louisville region, to the author, November 3, 1994. The Perma-Stone Company moved from Columbus, Ohio, to Louisville in 1983, where it still operates on a limited basis. It was purchased in 1973 by a group of investors who included a former licensed dealer. Perma-Stone is still being produced but not in large quantities.
18. U.S. Patent no. 2,095,641 was issued on October 12, 1937. This patent

refers to a previously patented process and explains how Formstone improves on that technique. While the previous process is not identified by name or patent number, it is likely Perma-Stone.

**19.** Ibid.

**20.** Ibid.

**21.** "Formstone's Glorious Past," *(Baltimore) Evening Sun,* February 24, 1992, B2.

## PART V. GLASS

### 22. PLATE GLASS

**1.** Plate glass may also be rough, ribbed, figured, sandblasted, and etched. It is also used to produce mirrors by coating one side of a sheet of plate with a film of silver. C. J. Phillips, *Glass: The Miracle Maker* (Chicago: Pitman, 1941), 255.

**2.** *Sweet's Architectural Catalogues,* 1933; *Sweet's Catalogue File,* 1936, 1948, 1952.

**3.** Plate glass differs from window or sheet glass, which is either hand- or mechanically blown into large cylinders and then split or drawn directly into sheets. Window glass does not require grinding or polishing and frequently possesses a visible waviness. Window glass is typically thinner, smaller, and of inferior clarity to plate glass. Frank W. Preston, "Evolution of the American Plate Glass Industry," *The Glass Industry* 11 (July 1930): 152; Ernst Lutz, "The Manufacture of Rolled Plate," *The Glass Industry* 11 (October 1930): 227.

**4.** In seventeenth-century France, Louis XIV's desire for large quantities of glass—especially mirrors—led to the establishment of a royal glass manufactory known as Saint-Gobain, located northeast of Paris. There the process for making large sheets of plate glass by casting was developed by Lucas de Nehou in 1688. Raymond McGrath and A. C. Frost, *Glass in Architecture and Decoration,* rev. edn. (London: The Architectural Press, 1961), 36. In the nineteenth century plate glass was also imported to the United States from England, Belgium, and Germany, as well as France.

**5.** One early American manufacturer was the Berkshire Glass Works, established in Berkshire, Mass., in 1847. In 1853 the company built another factory in Lanesboro; it was later purchased by Page, Harding and Company. On August 12, 1879, Thomas Gaffield, a window glass dealer, visited the Berkshire Glass Works and observed the making of rough plate glass. He stated that "the casting was done on an iron table 5 inches thick, 44 inches wide and 144 inches long" and noted that "the table with the roller and all the machinery for moving them cost about $1000 apiece." Thomas Gaffield, *Glass Journal* 3 (1879): 242. From 1847 to 1869 Gaffield was a partner in the firm Tuttle, Gaffield and Company, dealers in window glass. Gaffield had a wide-ranging interest in glass; he visited many glassworks and interviewed many glassworkers and manufacturers, conducted extensive experiments on the effect of sunlight on glass, and lectured on various aspects of glass. From 1896 to 1900 he was a member of MIT Corporation. From 1858 to 1881 he recorded his notes and observations about glass and glassmaking in two series of books: *Notes on Glass,* vols. 1–5 (1862–75), and *Glass Journal,* vols. 1–4 (1858–81). These, along with his extensive collection of books and other ephemera about glass and glassmaking, were presented by his widow to MIT in 1905.

**6.** Captain John B. Ford began his glassmaking activities with his sons Edward and Emory in 1865 by establishing a small glassworks in New Albany, Ind., which produced bottles, fruit jars, window glass, and some rolled rough plate glass. The enterprise failed, but they reacquired the glassworks, expanded it as the Star Glass Works about 1871, and began making polished plate glass. In 1871 and 1872 they were awarded two medals by the Indiana Board of Agriculture for their product. On November 19, 1874, a publication observed: "perfect polished plates have been turned out by this establishment measuring 7 × 17 feet." Gaffield, *Glass Journal,* 3:24.

**7.** Ford built one new plate glass works after another, managing each for a short time and then selling out to free himself for construction of larger and better plants. After two attempts to construct additional larger plants in the New Albany vicinity, Ford was successful in attracting New York City investors for the Creighton plant. In deference to them, the company changed its name to the New York Plate Glass Company. Production began in 1884, and the name was changed to Pittsburgh Plate Glass Company. A larger plant was built in Tarentum and a third in Ford City, Pa. Pearce Davis, *The Development of the American Glass Industry* (Cambridge, Mass.: Harvard University Press, 1949), 165–66.

**8.** During the 1870s one notable example was the American Plate Glass works in Jefferson County, Mo., where polishing halls measured 742 by 120 feet and production averaged 1,000 square feet per day. When the company came under new management in the late 1870s, this figure was reported to have doubled. *The Crockery and Glass Journal* 9 (April 10, 1879): 16.

**9.** In 1880 the Eleventh Census reported $4,172,482 in domestically produced polished plate and $781,550 in rough, cathedral, and skylight plate glass. In 1860 domestic production had been only 6 percent of total consumption, while in 1890 it constituted 82 percent of consumption of unsilvered polished plate and 90 percent of all types of nonpolished plate. Davis, 166.

**10.** The industry reaped the benefits of the increasing availability of better quality materials, while the construction of railroads lowered inland freight costs. Wood, coal, natural gas, and soda ash could be readily obtained from the vast forests, while supplies of suitable sand were found in New Jersey, Delaware, Pennsylvania, West Virginia, Illinois, and Missouri. Warren C. Scoville, *Revolution in Glassmaking* (Cambridge, Mass.: Harvard University Press, 1948), 39–40.

**11.** "Development of Plate Glass," *Commoner and Glassworker* 22 (March 2, 1901): 1, 13, 16.

**12.** J. W. Cruikshank, "Progress Made in Plate Glass Manufacture," *The Glass Industry* 2 (January 1921): 3.

**13.** Originally the annealing oven consisted of a chamber built above the melting chamber receiving waste heat. Later it was built as a separate furnace or oven that was loaded with ware during the day's run and then allowed to cool slowly during the night. S. R. Scholes, *Modern Glass Practice* (Chicago: Industrial Publications, 1935), 172.

**14.** Cruikshank, 5.

**15.** "Improvements in Flat Glass Manufacture," *American Glass Review* 47 (May 19, 1928): 15. The carrying operation before this crane took about seventeen men, most of whom were used as counterweights to balance the pot. Davis, 168.

**16.** The continuous annealing lehr was first used for plate glass by the Marsh Plate Glass Company of Walton, Pa., around 1897. United States Tariff Commission, *Flat Glass and Related Glass Products,* 2d series, no. 123 (Washington, D.C.: U.S. Tariff Commission, 1937), 82; "Sheet and Plate Glass Industry Makes Rapid Progress," *American Glass Review* 46 (May 28, 1927): 17. For a description of an annealing lehr, see Scholes, 172–73.

**17.** Davis, 254.

**18.** C. W. Avery, "Plate Glass Manufacture by the Continuous Process," *The Glass Industry* 11 (April 1930): 76.

**19.** The total number of workers needed in the continuous casting process was ten, compared to the fifty or more needed in the "discontinuous" process. Davis, 256.

**20.** The result of developments in both fields was that the gap between the two was partly bridged. No longer was window glass produced only in pieces thinner, smaller, and inferior to plate glass. While drawn glass was becoming thicker, larger, and of better quality, much of the plate glass manufacturer's effort was directed toward making plates thinner, smaller, and cheaper. To strengthen their position, plate glass manufacturers acquired window glass subsidiaries and vice versa. Frank W. Preston, "Evolution of the American Plate Glass Industry," *The Glass Industry* 11 (July 1930): 152.

**21.** In 1930 the Edward Ford Plate Glass Company and the Libbey-Owens Sheet Glass Company merged to form the Libbey-Owens-Ford Company. William Earl Aiken, *The Roots Grow Deep* (Cleveland, Ohio: The Lezius-Hiles Co., 1957), 12. This new twin-grinding process replaced the semicontinual grinding and polishing processes used before and involved mounting tables holding glass onto cars that traveled on rails under a series of revolving grinder heads. After passing under these grinding heads, which were fed progressively finer sand, the glass was washed and then ground smooth in the same manner. The surface was again washed, and the tables were passed under the polishers. At that point the plates were half finished. They were removed, inverted, reset, and sent through the system again. Scholes, 146.

**22.** This, however, was not the first invention involving the use of molten metal (tin) support in the manufacture of glass. The first patent of this system was granted to Henry Bessemer in 1848 (British Patent no. 12,101, September 22, 1848). Additional patent work was done by William Heal (1902) and H. K. Hitchcock (1905). It has been suggested that the introduction of other innovations, such as the continuous process and the Colburn, Fourcault, and Pittsburgh Plate Glass sheet-forming processes, took up much of the glass companies' available capital and personnel for research and development, delaying the development of the float process. David Boyd and John F. MacDowell, "Float Glass," in *Commercial Glasses: Advances in Ceramics* (Columbus, Ohio: The American Ceramics Society, 1984), 45.

**23.** In 1962 Saint-Gobain completed building a plant near Kingsport, Tenn., making it the third largest U.S. flat glass manufacturer. However, this $40 million plant used the twin-grinding process, not the float process, the newest innovation. At the opening ceremonies, David Hill, the president of Pittsburgh Plate Glass, was overheard to remark: "You are now looking at the most modern obsolete process in the world." That same year Pittsburgh Plate Glass had decided to become the first U.S. licensee of the float process. Richard Austin Smith, "At Saint-Gobain, the First 300 Years Were the Easiest," *Fortune,* August 1964, 148.

**24.** Orrick Johns, "The Coming of the Glass Skyscraper," *Engineering and Contracting* 65 (June 23, 1926): 258–59.

**25.** Insulated plate glass consisted of two panes of plate glass with a dead air space between them. Phillips, 252. Thermopane was introduced on April 26, 1944, through a press release to the Science Writer in Architectural League Headquarters. Libbey-Owens-Ford Board Records, Collection No. 066, Libbey-Owens-Ford Glass Company Records: 1851–1991, Ward M. Canaday Center, University of Toledo, Ohio.

**26.** Heat-absorbing plate glass is a product of a special chemical composition made to absorb a high percentage of the infrared heat rays of the sun. Its use provided a cooler atmosphere inside, and thus lower fuel costs, while also reducing sun glare. The Lever House (1952, Gordon Bunshaft, Skidmore, Owings and Merrill) in New York City was built in the early 1950s and used Solex heat-absorbing glass, first produced in quantity by Pittsburgh Plate Glass in 1951. *Sweet's Catalogue File,* 1948.

**27.** Tempered polished plate is made by reheating and sudden cooling; as a result the outer surfaces are in a state of high compression while the central portion remains in tension, producing a condition highly resistant to breakage. Tempered plate is typically three to five times stronger than ordinary plate glass yet has the same thickness and appearance. After it has been tempered, it cannot be cut or ground. Phillips, 257. Tuf-Flex, produced by Pittsburgh Plate Glass, was first used in 1942 by the U.S. Army and later in baseball stadiums. *Libbey-Owens-Ford Company Newsletter*, 1942, Libbey-Owens-Ford Glass Company Records: 1851–1991, Collection No. 066, Ward M. Canaday Center, University of Toledo, Ohio.

**28.** The majority of the plate glass manufactured in the early twentieth century was clear. By the mid-1950s tinted plate glass was widely available. Plate glass was also produced as laminated glass and insulating glass. In its most common form laminated glass consists of two sheets of glass sandwiched around and adhered to a plastic interlayer. Insulating glass consists of two sheets of glass separated by a sealed air space. In early insulating glass the glass separation and air space seal were maintained by a continuous lead strip soldered to the glass sheets at their perimeter. Today almost all flat glass is manufactured by the float process.

29. David Button and Brian Pye, eds., *Glass in Building: A Guide to Modern Architectural Glass Performance* (London: Butterworth Architecture and Pilkington Glass Ltd., 1993), 213.
30. In theory, glass products can have tensile strengths reaching 5 million pounds per square inch, based on the strength of their intermolecular bonds. In reality, glass achieves only a fraction of this strength because of the effects of Griffith flaws and stress concentration. Theodore Baumeister, Eugene A. Avallone, and Theodore Baumeister III, eds., *Marks' Standard Handbook for Mechanical Engineers,* 8th edn. (New York: McGraw-Hill, 1978), section 6–157–58.
31. American Architectural Manufacturers Association, *Structural Properties of Glass* (Des Plaines, Ill.: American Architectural Manufacturers Association, 1984), 5–10.
32. Button and Pye, 212.
33. Raymond McGrath, *Glass in Architecture and Decoration,* rev. edn. (London: The Architectural Press, 1961), 426. Alkaline paint removers used on painted surfaces adjacent to glass can cause damage if they are not removed quickly from the glass surface.
34. *Technical Bulletin ATS-104* (Toledo, Ohio: Libbey-Owens-Ford, May 21, 1987), 2.
35. McGrath, 428.
36. Martin E. Weaver, *Conserving Buildings: A Guide to Techniques and Materials* (New York: John Wiley and Sons, 1993), 233–37.
37. The techniques and cleaning compounds given here are recommended for use only on uncoated glass surfaces. Glass coatings may be damaged by these methods.
38. *Technical Bulletin ATS-118* (Toledo, Ohio: Libbey-Owens-Ford, September 26, 1988), 1.
39. When using this technique, porous building materials, such as masonry, must be protected from runoff to avoid staining. Generous amounts of water should be used to flush the glass surface to ensure that no paste residue remains.
40. Flat Glass Marketing Association, *Glazing Manual* (Topeka, Kans.: Flat Glass Marketing Association, 1980), 24–26, 32–34.

**23. PRISMATIC GLASS**

1. For more information about prismatic glass, see Dietrich Neumann, "'The Century's Triumph in Lighting': The Luxfer Prism Companies and Their Contribution to Early Modern Architecture," *Journal of the Society of Architectural Historians* 54 (March 1995): 24–53; Gordon Bock, "Glass Notes," *The Old-House Journal* 16 (July–August 1988): 35–43; Joseph Siry, *Carson, Pirie, Scott, Louis Sullivan and the Chicago Department Store* (Chicago: University of Chicago Press, 1986), 142–46.
2. "Patent of Edward Wyndus for Glasses and Lamps for Ships, Mines &c.," 1684, no. 232, British Patent Office, London; Patent of Richard Cole, no. 372, 1704, British Patent Office, London; Patent of Apsley Pellatt, no. 3,058, 1807, "Lighting the Interior of Ships, Buildings &c.," British Patent Office, London; Raymond McGrath, *Glass in Architecture and Decoration* (London: The Architectural Press, 1937), 180.
3. U.S. Patent no. 4,266, November 12, 1845; Peter Collins, *Concrete* (New York: Horizon Press, 1954), 58–60; Cecil D. Elliott, *Technics and Architecture* (Cambridge, Mass.: MIT Press, 1992), 171–72.
4. See, for example, "An Act in Relation to Buildings in the City of Providence and for Other Purposes," chap. 688, section 29, 20.
5. "The Advantages of Ribbed Glass and of Wire Glass," *Engineering News* 36 (August 13, 1896): 101; "Experiments on the Diffusion of Light through Prismatic and Ribbed Glass Windows," *Engineering News* 45 (January 10, 1901): 34–35.
6. W. Eckstein "Interior Lighting," *American Architect and Building News* 62 (1898): 3–7. Hayward Brothers and Eckstein, eds., *Trade Catalogue* (London, 1899), 4–5.
7. The first one seems to have been Robert Boughton of England, who suggested "sheets of glass with one face plane and one face with a series of parallel ridges, which are of such a form as to direct the light by simple diffusion or by refraction in the required direction." Boughton, "Daylight Diffusers," British Patent no. 139, issued on January 13, 1880. "Daylight Prism Co. v. Marcus Prism Co.," *Federal Reporter* 110 (1901): 980–85.
8. James G. Pennycuick, "Window-Glass," U.S. Patent no. 312,290, issued on February 17, 1885, and filed on August 12, 1882. Pennycuick acknowledged that "it is not broadly new to provide window-glass with reflecting ridges," but he suggested that the angle chosen in his application was more efficient.
9. Pennycuick's patent was owned by the Prismatic Glass Company, N.H., from 1887 to 1889, by the Alpha Glass and Metal Company, N.J., from 1889 to 1890, by a number of individual businessmen, and finally by Thomas W. Horn, who founded the Prismatic Glass Company of Toronto in 1896 and eventually became one of the cofounders of the Luxfer Prism Company of Chicago. In the numerous court cases in which Luxfer defended its products against competitors, Pennycuick's patent was regularly quoted as central to Luxfer's claims. The competing firms referred to Thaddeus Hyatt's patent and other unprotected patents of prismatic pavement lights as predecessors to their own products. The Luxfer Prism Patents Company et al. v. Trustees of Columbia College et al., "Demurrer Book" (1897–1911), *Case No. N-6808. U.S. Cir. S.D. New York,* 3–4.
10. "To Make Dark Places Light," *The Economist* 17 (February 17, 1897): 224.
11. Henry Crew, "On the History of Our Ideas Concerning Illumination," *American Architect and Building News* 62 (1898): 3–7. There is also a brief reference to Luxfer prisms in Henry Crew, *General Physics* (New York: Macmillan, 1908, 1919), 551.
12. L. W. Marsh, "Daylight Illumination," *Transactions of the Illuminating Engineering Society* 3 (1908): 224–32.
13. U.S. Patent no. 574,843 was issued on January 5, 1897; Winslow's patent was an improvement of an 1887 patent by Henry F. Belcher, who had joined the Luxfer enterprise. U.S. Patent no. 396,911, issued on May 11, 1887. Both patents used existing techniques of electroplating.
14. Henry Crew and Olin H. Basquin, eds., *Pocket Hand-Book of Useful Information and Tables Relating to the Use of Electro-Glazed Luxfer Prisms* (Chicago: The Luxfer Prism Company, 1898), 6.

15. Ibid., 18–27.

16. In a three-step procedure the prospective client would calculate the "Zenithtangent" from the width of the street and the height of the opposite building, choose an area to be covered by the refracted light, and find out which type of prism to use, normally one major prism (providing 80 to 90 percent within a plate) and one minor prism (10 to 20 percent). The required brilliance in the room (for desk work, fine merchandise, general merchandise, or storage) and the depth of the room would lead to the square footage needed. Nine different angles would be offered, and the corresponding letter was usually imprinted on the tile: A for protruding canopies and J, K, L, M, N, O, S, and P for ordinary sash or "forilux" installations. This procedure was complicated, but the company explained that because of its overwhelming success it had to abandon its original practice of sending experts ("lucical engineers") to every site to calculate the necessary prism angles. Crew and Basquin, 79.

17. Luxfer Prism Company, "Table of Color Corrections for Luxfer Prism Areas," n.d. I am grateful to Joe G. Jernejcic for this information.

18. One report stated that the light in the back of a room could be improved 10 to 15 times; the Luxfer Prism Company itself claimed the figure was as high as 25 times; and an independent report by the New England Mutual Insurance Companies hinted at a possible efficiency rate of up to 50 times. Marsh, 224–32; Crew and Basquin, 6; "Experiments on the Diffusion of Light," 33–34.

19. Holabird and Roche reported to the Luxfer Prism Company about its project on 84 Wabash Ave., which they had executed without a light shaft after learning about the qualities of Luxfer Prisms. For more information about the renovation of the Mandel Brothers Store in Chicago using Luxfer Prisms, which led to the omission of a light court, see "Modernizing Commercial Buildings," *The Inland Architect and News Record* 31 (September 1898): 18, 19.

20. "An Interesting Competition," *The Inland Architect and News Record* 30 (1898): 63–64.

21. "The Century's Triumph in Lighting," 39.

22. *Sweet's Catalogue of Building Construction,* 1906.

23. Crew and Basquin, 5.

24. "The Century's Triumph in Lighting," 39–40.

25. Hugh Morrison, *Louis Sullivan* (New York: W. W. Norton, 1935), 195–96.

26. Frank Lloyd Wright, "Prism Light," Design Patent no. 27.977, patented on December 7, 1897.

27. Grant Carpenter Manson, *Frank Lloyd Wright to 1910* (New York: Reinhold Publishing, 1958), 88.

28. *Sweet's Catalogue of Building Construction,* 1906.

29. "Experiments on the Diffusion of Light," 34–35.

30. *Glasses, Paints, Varnishes, and Brushes: Their History, Manufacture, and Use* (Pittsburgh: Pittsburgh Plate Glass Company, 1923), 133–37; McGrath, 573–641.

31. Knut Lönberg-Holm, "Glass," *Architectural Record* 68 (October 1930): 327–41; John W. T. Walsh, *The Science of Daylight* (London: Macdonald, 1961), 98.

**24. GLASS BLOCK**

1. The terms *block, tile, lens,* and *prism* were all used by different authors to name a heavy piece of glass embedded in a structural framework of concrete or iron.

2. There had been earlier unsuccessful attempts to produce such a product. In 1879 Friedrich von Siemens of Dresden, Germany, devised a block. "Brown Glass Bricks for Building Purposes," *Scientific American* 76 (January 2, 1897): 10.

3. Among Falconnier's patent holders were the German Adlerhuttenwerke in Penzig, Silesia, which was responsible for the glass block building at the 1893 World's Columbian Exposition in Chicago. "Glass Paving and Building Bricks," *Scientific American Supplement No. 1538* 89 (June 24, 1905): 24,643. I am grateful to Anne Laure Carre at the Sorbonne in Paris, who provided information regarding this pavilion.

4. Some buildings that used Falconnier's block were the apartment block at the Rue Franklin in Paris (1902, Auguste Perret), Castel Beranger (1899, Hector Guimard), and Villa Schwob (1914, Le Corbusier) in La Chaux-de-Fonds. The German Luxfer Prism Company included Falconnier patents in its product range.

5. See Eugene Clute, "Designing for Construction in Glass, Part I," *Pencil Points* 13 (November 1932): 741–48.

6. Hohler, "Unten offener Glasbaustein," *Zentralbatt der Bauverwaltung* 23, no. 34 (1903): 212; F. Neumeyer, "Glasarchitektur—Zur Geschichte des glasernen Steines," *Bauwelt* 73 (1982): 172–79; Knut Lönberg-Holm, "Glass," *Architectural Record* 68 (October 1930): 357.

7. Friedrich Keppler, a German-born architect, immigrated to America in 1888, joined the American Luxfer Prism syndicate, and returned to Germany to become the head of the German branch after 1900. "F. L. Keppler, Leader in Structural Glass," *New York Times,* January 8, 1940, 21; U.S. Patent no. 1,145,997, filed on July 13, 1915; this patent was based on an earlier patent of July 12, 1911.

8. Raymond McGrath, *Glass in Architecture and Decoration* (London: The Architectural Press, 1937), 184–87.

9. Ibid., 180.

10. Keppler had introduced the new European structural glass products to the United States when he returned to New York City in 1914 and founded Keppler Glass Constructions (later renamed Structural Glass Corporation). He made his patents available to the American 3-Way Luxfer Prism Company. Lönberg-Holm, 357. See also *Keppler Glass Constructions; Pavements, Floors, Roofs, Walls, Partitions, Windows and Crystal Ceilings, Translucent, but Not Transparent* (New York: F. L. Keppler, 1914).

11. The Owens-Illinois House was designed by Eloy Ruiz, measured 60 by 100 feet, and was crowned by a dynamically staggered 50-foot-high tower. At night it was lighted from within with various colors. Lorraine Welling Lanmon, *William Lescaze Architect* (Philadelphia: Art Alliance Press, 1987), 99–103, 110 n. 6.

12. "Announcing the New Pyrex Glass Construction Unit," *Architectural Record* 48 (September 1935): 15.

13. "The Pittsburgh Corning Corporation," *Pittsburgh Plate Products* 47 (March–April 1938): 3–4.
14. "The Glass Age Arrives," *Architectural Forum* 68 (February 1938): 17–28.
15. "Glass Block," *Architectural Forum* 72 (May 1940): 327.
16. The molten glass was set into a mold with compressed air injected to fill the cavity and keep the brick from collapsing. By this process the blocks tended to be poorly proportioned and were neither uniform nor structurally self-supporting. Linda Cook, "Walls of Daylight" (unpublished manuscript).
17. Sealants included solder, aluminum, resin, asphalt, grout, and portland cement. However, these tenuous bonding materials failed to accommodate the glass's natural expansion properties. Cook, 3.
18. The batch materials consisted of sand, limestone, soda ash, and nepheline syenite, from which is derived potassium, aluminum, and some additional soda. "PC Glass Block," in *Glass Manual* (Pittsburgh, Pa.: Pittsburgh Plate Glass Company, 1946), E-1, 5.
19. Ibid., E-1, 8.
20. "Glass Block," 327.
21. John Ely Burchard, "Glass in Modern Housing," *The Glass Industry* 16 (December 1935): 379–81.
22. Cook, 4.
23. See, for example, "Here Are Two New PC Glass Blocks to Meet Special Lighting Needs," *Architectural Record* 88 (February 1940): 35; and "New Insulux Prismatic Glass Block," *Architectural Record* 88 (April 1940): 119.
24. "No Windows but All Glass," *Engineering News-Record* 116 (March 26, 1936): 459–61.
25. Derek Trelstad, "The Hecht Company Warehouse," *Building Renovation,* July–August 1993, 29–33.
26. "No Window but All Glass," 459.
27. Trelstad, 30.
28. Christine Beall, "Laying Glass Block," *The Magazine of Masonry Construction,* March 1990, 131–33.

#### 25. STRUCTURAL GLASS

1. Structural glass has also been known by many other terms: *recreated rock slab, sanitary glass, rolled* or *opaque opal glass,* and *heavy obscured structural glass.* Its most common product proprietary names are Vitrolite and Carrara.
2. C. J. Phillips, *Glass: The Miracle Maker* (Chicago: Pitman, 1941), 259–60.
3. Raymond R. McGrath, *Glass in Architecture and Decoration* (London: The Architectural Press, 1937), 44.
4. Ibid., 44–46.
5. *Flat Glass and Related Glass Products,* report no. 123, 2d series (Washington, D.C.: U.S. Tariff Commission, 1937), 20.
6. "The Preservation of Historic Pigmented Structural Glass," *Preservation Brief No. 12* (Washington, D.C.: U.S. Department of the Interior, National Park Service, 1984), 1.
7. *Sweet's Catalogue of Building Construction,* 1906.
8. *Sweet's Architectural Catalogues,* 1916.
9. *Flat Glass and Related Glass Products,* 137. Structural glass had been produced abroad since the early 1900s. In 1929 imports (primarily from Belgium and Czechoslovakia) constituted less than 5 percent of the U.S. market.
10. A 1959 Carrara brochure is the last instance of prominent advertising for structural glass in *Sweet's Architectural Catalogue File.* A Pittsburgh Plate Glass brochure on curtain wall systems included in the 1963 *Sweet's Catalogue* and showing Carrara as a spandrel panel choice is the last time the material is pictured in *Sweet's.* PPG kept the trade name listed in *Sweet's* until 1969, but the material colors and information were no longer displayed. *Sweet's Catalogue File,* 1959 through 1969.
11. "Manufacture of Structural Glass," *The Glass Industry* 20 (June 1939): 215–19. See also Charles M. Gay and Harry Parker, *Materials and Methods of Architectural Construction* (New York: John Wiley and Sons, 1932), 236.
12. McGrath, 46.
13. *Glass, Paints, Varnishes and Brushes: Their History, Manufacture, and Use* (Pittsburgh, Pa.: Pittsburgh Plate Glass Company, 1923), 159.
14. McGrath, 37, 46.
15. Ibid., 47.
16. *Sweet's Architectural Catalogues,* 1929.
17. *Carrara Colorful Structural Glass* [brochure] (Pittsburgh, Pa.: Pittsburgh Plate Glass Company, 1959), 2. The maximum size of structural glass sheet permitted for use above the first floor in a curtain wall was 6 feet square. William D. Hunt, Jr., *The Contemporary Curtain Wall: Its Design, Fabrication, and Erection* (New York: F. W. Dodge Corporation, 1958), 233.
18. *Sweet's Catalogue of Building Construction,* 1906.
19. *Glass, Paints, Varnishes and Brushes,* 164.
20. Gay and Parker, 236.
21. See, for example, "Remodeling in Glass Attracts Attention and Dollars," *Architectural Forum* 72 (May 1940): 205.
22. "Glass-Faced Structural Unit Developed," *American Builder* 61 (October 1939): 70–71, 120; "Glastone: The New, Load-Bearing Colorful Masonry Unit" [brochure] (Toledo, Ohio: Libbey-Owens-Ford Company, 1939), 47.
23. Flat Glass Jobbers Association, *Glazing Manual: Specifications for Installation of Flat Glass* (Chicago: R. R. Donnelley and Sons Company, 1958), 67–68.
24. Ibid., 67–68.
25. Douglas A. Yorke, Jr., "Materials Conservation for the Twentieth Century: The Case for Structural Glass," *Association for Preservation Technology Bulletin* 13 (1981): 23.
26. Ibid., 27.
27. Geier Brown Renfrow Architects with Oehrlein and Associates, *Historic Structures Report, Department of Justice Building* (Washington, D.C.: General Services Administration, National Capital Region, 1987), 490–91.
28. Personal communication with Tim Dunn of Vitrolite Specialists, St. Louis, Mo., September 1994.

29. Geier Brown Renfrow, 490–91.
30. Ibid.
31. Personal communication with Tim Dunn, September 1994.
32. For detailed preservation information on the removal and repair of historic structural glass panels, see Yorke, 28.
33. "The Preservation of Historic Pigmented Structural Glass," 6.
34. Personal communication with Mary Oehrlein, Oehrlein and Associates, Architects, Washington, D.C., January 1993. See also Yorke, 28.

### 26. SPANDREL GLASS

1. William D. Hunt, *The Contemporary Curtain Wall: Its Design, Fabrication, and Erection* (New York: F. W. Dodge Corporation, 1958), 236.
2. *Glass Manual* (Pittsburgh, Pa.: Pittsburgh Plate Glass Company, 1946).
3. "Glass Walls Take on New Color," *Business Week,* July 9, 1955, 48.
4. *Sweet's Catalogue File,* 1958.
5. The substrate glass is softened slightly while the frit is fusing. Therefore it can stretch, sag, ripple, and bow and thus change its appearance.
6. Heat strengthening imparts compression stress in edges and surfaces and balances tension stress in the central core. Typical surface compression stresses range from 3,000 to 8,000°F for heat-strengthened spandrel glass and from 10,000 to 30,000°F for tempered spandrel glass. Usually, tong marks (dimples), bows, and ripples are considered unattractive and should not exceed limits given in ASTM's specifications. "Standard Specification for Heat-Treated Flat Glass—Kind HS, Kind FT Coated and Uncoated Glass," ASTM C-1048-91 (Philadelphia: American Society for Testing and Materials, 1991).
7. "Plate Glass Is Fired with Colored Ceramic on Back for Spandrel Covers," *Architectural Forum* 103 (September 1955): 207.
8. *Glazing Manual* (Topeka, Kans.: Flat Glass Manufacturers Association, 1980).
9. Reflective coatings are sometimes applied to spandrel glass to improve its visual match with adjacent reflective glass windows. These coatings are very thin and more vulnerable to damage than is the spandrel substrate. Some may be pitted or removed by hydrochloric as well as hydrofluoric acid.
10. Falling tools may scratch, crack, or abrade spandrel glass. Overly aggressive removal of dried compounds or paint with sharp metal or abrasive tools may damage it. Impact by steel-capped shoes or unpadded scaffolds may cause spalls. A tiny surface spall with a diameter of even 1/64 inch may mark the top of a structural flaw driven deep by a hard, sharp point. A break may originate at the flaw because of mechanical or thermal stress.
11. When ambient temperatures approach 70°F on sunny elevations, the temperature of dark spandrel glass, sealants, and framing can reach 150 to 190°F. Elevated temperatures will last 30 to 60 minutes on each hot clear day. On east elevations on clear, sunny, winter mornings following freezing nights, the daytime difference in temperature between the center and edge of the spandrel glass may exceed 50°.
12. The highest temperature differences and resultant thermal stresses may be expected when dark-colored, insulated spandrel glass is glazed in a concrete or comparable heat-sink frame. When cut, shipped, stored, and glazed as usual, annealed glass edges may resist thermal tension stresses of 1,500 pounds per square inch (psi). However, if annealed edges are damaged slightly by nipping, seaming, compression, or impact, they may break at tension stress levels below 1,000 psi. Today, all spandrel glass must be heat strengthened. At furnace temperatures in the range required for properly firing and fusing durable frits—1,000 to 2,000°F—proper soaking and quenching times soften the substrate glass slightly, cure surface and edge flaws, and add compression stress of more than 3,500 psi to surfaces and edges. *Guide Specification Glazing,* FCGS-08810, General Services Administration (n.d.).
13. The higher stress (10,000 to 30,000 psi) built into the tempered product will cause it to break into thousands of small pieces. Usually, when the glass is glazed, all will fall out of the opening immediately or at the first wind pressure or suction of 4 to 5 psi. Because of its lower compression levels, spandrel glass breaks are much simpler. Typically, only 5 to 10 runs will occur. The few large pieces are likely to stay in the opening long enough to permit temporary taping of the broken light and replacement.
14. These are nearly invisible bits of unmelted batch or process ingredients, totaling perhaps a hundred different compositions. They may occur anywhere in the end product.
15. "Glass and Glazing-Glass Failures, Fully Tempered Glass—Nickel Sulfide Inclusions," *Turner Building Digest* 8-45 (February 1978): 1. The roughly spherical diameter of nickel sulfide stone may change from, for example, 0.00100 to 0.00125 inch, causing internal stress around the stone to increase. The resultant stress increase is not sufficient to cause breakage of annealed or heat-strengthened glass. However, breaks may occur if a stone changes state while "cooking" in the tempering furnace.
16. John D. MacKenzie, "Nickel Sulfide Stones in Tempered Glass," *The Glass Industry* 57 (December 1978): 32–33.
17. To save inspection time and money when many lights are involved, it may be prudent to inspect first a small statistical sample selected by use of random numbers applied to each elevation. For many reasons, the nature and frequency of change is likely to vary significantly between elevations and sometimes between floors. Labor-management difficulties and changes in weather during glazing often significantly influence production rates and quality of installation.
18. Experienced technicians using microscopes can distinguish between mechanical bending and thermal tension stress breaks. By measuring the radius at the origin, they can determine the magnitude of the stress that caused the break. Thomas A. Schwartz, "A Pain in Your Pane?" *Building Renovation,* July–August 1993, 53–56.
19. *Recommended Techniques for Washing Glass,* Pittsburgh Plate Glass Industries (n.d.), 8.
20. If the job involves many lights, a "preproduction approval area" of perhaps fifty lights (5 feet wide by 10 feet high) should be glazed, weather tested, and color photographed to the satisfaction of all responsible parties before the successful bidder is permitted to begin

regular production glazing. "Standard Test Method for Structural Performance of Exterior Windows, Curtain Walls, and Doors by Uniform Static Air Pressure Difference," ASTM E-0330-90 (Philadelphia: American Society for Testing and Materials, 1990), and "Standard Test Method for Structural Performance of Exterior Windows, Curtain Walls, and Doors, by Cyclic Static Air Pressure Differential," ASTM E-1233-88 (Philadelphia: American Society for Testing and Materials, 1988).

## PART VI. FLOORING

### 27. LINOLEUM

1. British Patent Nos. 209 (1860) and 3,210 (1863). "History of the Wax Cloth," unpublished manuscript chronology in files of Forbo-Nairn Company, Kirkcaldy, Fife, Scotland, n.d., 4.
2. "History of the Wax Cloth," 4. Frederick Walton, *The Infancy and Development of Linoleum Floorcloth* (London: Simpkin, Marshall, Hamilton, Kent and Company, 1925), 40. Augustus Muir, *Nairns of Kirkcaldy: A Short History of the Company (1847–1956)* (Cambridge, England: W. Heffer and Sons, 1956), 86. Arthur L. Faubel, *Cork and the American Cork Industry*, rev. edn. (New York: Cork Institute of America, 1941), 102. The town was called Linoleumville; its name was changed to Travis in 1930.
3. Muir, 85–86, 112.
4. Leo Blackman and Deborah Dietsch, "A New Look at Linoleum: Preservation's Rejected Floor Covering," *The Old-House Journal* 10 (January 1982): 10. The Blabon Company apparently operated more than one factory; there was also one in Nicetown, Pa.
5. William A. Mehler, Jr., *Let the Buyer Have Faith: The Story of Armstrong* (Lancaster, Pa.: Armstrong World Industries, 1987), 11, 16.
6. Ibid., 18. The book was *The Art of Home Furnishing and Decoration* (Lancaster, Pa.: Armstrong Cork Company, 1918).
7. R. S. Morrell and A. de Waele, *Rubber, Resins, Paints, and Varnishes* (London: Baillière, Tindall and Cox, 1921), 164.
8. *Armstrong's Handbook for Linoleum Mechanics* (Pittsburgh, Pa.: Armstrong Cork Company, 1924), 56. Faubel, 102.
9. Kauri gum is a fossil gum from New Zealand.
10. U.S. Federal Specifications LLL-L 359, "Inlaid and Molded Linoleum," and LLL-L 367, "Plain, Jaspe and Marbelized [*sic*] Linoleum," state that it "shall consist of oxidized linseed oil, fossil or other resins and/or rosin or an equivalent thoroughly oxidized resinous binder intimately mixed with ground cork, and/or wood flour and pigments pressed on a backing." The foregoing description is drawn from various sources, including but not limited to Walton, 30; Muir, 72–73; *History and Manufacture of Floor Coverings* (New York: Review Publishing Company, 1899), 83–84; Bernard Berkeley and Cyril S. Kimball, *The Care, Cleaning and Selection of Floors and Resilient Floor Coverings* (New York: Ahrens Publishing Company, 1961), 33; and Edward S. Wiard, "Grading in the Cement, Coal and Cork Industries," *Metallurgical and Chemical Engineering* 62 (December 15, 1916): 687.
11. *Armstrong's Handbook*, 71, 77, 79; Faubel, 106–8; and other sources previously mentioned.
12. This is also sometimes called granular inlaid linoleum. While sources disagree about the order of the development of the geometric inlaid patterns, most put this one first, but no date was given in any of the sources consulted.
13. *Armstrong's Handbook*, 74–75; Faubel, 109; and other sources. Nairn began manufacturing this product in 1895. Armstrong was producing it by the 1910s.
14. Since the scrim incorporated into the linoleum cement dulled the requisite cutting blades, the product's development first required inventing a method for manufacturing oxidized linseed oil without the scrim.
15. *Armstrong's Handbook*, 77; Faubel, 108; and other sources. After Walton invented the process for manufacturing straight-line inlaid linoleum and an intricate machine for cutting and reassembling the individual pieces, he opened a plant in Greenwich, England, solely for the manufacture of this product, which he marketed as Walton Inlaid. Walton, 49; Muir, 115. Muir says the company was started in 1895 and called the Frederick Walton Mosaic Linoleum Company. Walton says the company was started in 1898 (one of the few dates he gives) and was called the Greenwich Inlaid Linoleum Company. Before Nairn purchased the company in 1922, his straight-line inlaid linoleum was assembled by hand. Armstrong, one of the leading manufacturers of linoleum in the United States, did not begin making straight-line inlaid linoleum until 1923. Mehler, 16.
16. Faubel, 109–10.
17. Mehler, 26–27.
18. Ibid., 90.
19. Congoleum, whose product was Gold-Seal Congoleum, was a market leader. Armstrong came out with its version, Fiberlin Rugs, in 1916 but began marketing printed linoleum in 1917 and dropped the felt-based product in 1920. However, to compete with Congoleum, Armstrong subsequently bought a felt-based rug manufacturing company and reentered the market with Quaker Rugs and Floor Coverings in 1925. Mehler, 24–25.
20. *Armstrong's Handbook*, 66–70.
21. Mehler, 19.
22. *Armstrong's Linoleum Pattern Book* (Lancaster, Pa.: Armstrong Cork Products Company Floor Division, 1935): 10; (1937): 356; "Linoleum Flooring," *Building Age and National Builder* 48 (April 1926): 326; "Linoleum and Its Proper Application," *American Architect* 119 (May 11, 1921): 565–68. Over the 16-year period of these publications, the installation method remained the same. Vinyl is installed the same way today, although the felt layer is omitted.
23. Many factories operated installation schools; Armstrong began its in 1923. Mehler, 20.
24. If a felt layer was not used in the installation or is damaged, there may be evidence of dry rot in the wood subfloor.
25. Alkali tends to damage the surface treatment, such as paint, far faster than it does the components of the linoleum itself. *Armstrong's Handbook*, 89.

**26.** Areas covered by rugs or furniture could be substantially darker than adjacent areas exposed to daylight. Ultraviolet light rays not only forestall the darkening process but may actually lighten the colors. Light from sources that do not contain ultraviolet rays will not prevent darkening to as great an extent.

**27.** *Armstrong's Pattern Book* (1935), 342, recommends two soaps still available today: Ivory and Murphy's Oil Soap. Murphy's Oil Soap has a pH of approximately 11.

**28.** Some conservation soaps available from suppliers of conservation materials have been successfully tested on linoleum. The tests were performed at the Canterbury Shaker Village in Canterbury, N.H., and Triton X-100 was selected to clean the extant linoleum. Personal communication with Susan Buck, conservator, Society for the Preservation of New England Antiquities, August 26, 1994.

**29.** Personal communication with Paul Kofoed, chemist, Erlin, Hime Division, Wiss Janney Elstner Associates, August 19, 1994.

**30.** Ibid.

**31.** In a museum setting where the floor is not exposed to much daylight or where period lighting is used as the regular illumination, exposure to fluorescent light when the room is closed to visitors may serve this purpose; however, the effect of the fluorescent light on other materials in the room should be considered. Linseed oil darkening is also a common problem in the conservation of oil paintings. Painting conservation literature or a paintings conservator may be consulted for more information about the effects of light on linseed oil.

**32.** Personal communication with Paul Kofoed, August 19, 1994. Aromatic polyurethane may not bond with the linoleum, and it yellows in sunlight.

**33.** Deutsche Linoleum Werke (DLW) still manufactures linoleum in Germany. For information regarding this product, contact Gerbert Limited, P.O. Box 4944, Lancaster, Pa. 17604-4944. Forbo-Nairn, the successor company to Michael Nairn and Company, manufactures plain linoleum in at least twelve shades and marble-pattern linoleum in more than sixty shades and combinations; contact Forbo-Nairn, P.O. Box 1, Kirkcaldy, Fife KYI 2SB, Scotland.

**34.** Personal communication with Paul Kofoed, August 19, 1994.

**35.** This design was in Armstrong's line from 1932 to 1974. Versions of it are still manufactured by Armstrong, Congoleum, and Mannington.

**36.** For a complete list of manufacturers with products suitable as linoleum substitutes, see Helene Von Rosenstiel and Gail Caskey Winkler, *Floor Coverings for Historic Buildings* (Washington, D.C.: The Preservation Press, 1988).

#### 28. RUBBER TILE

**1.** Harry L. Fisher, *Rubber and Its Use* (Brooklyn, N.Y.: Chemical Publishing Company, 1941), 101–8.

**2.** Waldemar Kaempffert, ed., "Transportation, Communication, and Power," in *A Popular History of American Invention* 1 (New York: A. L. Burt Company, 1924), 164.

**3.** Some names for *rubber* are *caoutchouc, caucho, gutta, para, borracha, goma elastica,* and *gummi.*

**4.** W. H. Reece, "Mats and Flooring," in *History of the Rubber Industry* (Cambridge, England: Institute of the Rubber Industry, 1952), 258–59.

**5.** Hubert L. Terry, *India Rubber and Its Manufacture* (London: Archibald Constable and Company, 1907), 195.

**6.** No patents were identified before 1894, when Frank Furness's patent, no. 527,961, was granted. Furness was granted a second patent, no. 565,734, in 1896. The second patent improved the tab connections by providing a recessed tab that could be mechanically fastened with a nail or screw for use on walls and ceilings.

**7.** *Rubber Tiling* (New York, N.Y.: New York Belting and Packing Company, 1918) states that the company was founded in 1846 and claims that it was the original manufacturer of interlocking tiles. Its address, 821–823 Arch Street, suggests that the company may have used Furness's patent.

**8.** While it appears that interlocking tiles were not widely advertised after World War I, this book indicates that they were still available in 1939. Victor Darwin Abel, *Floor and Wall Coverings* (Scranton, Pa.: International Textbook Company, 1939), 51.

**9.** H. E. Morse, "History of Goodyear Rubber Flooring," unpublished paper, February 27, 1929, Archives of Goodyear Tire and Rubber Company, Akron, Ohio.

**10.** "Rubber Floorings Give Satisfaction," *India Rubber World* 70 (July 1, 1924): 695.

**11.** Experiments continued to improve the durability of rubber flooring and reduce its cost; one system used reinforcement with shredded cotton fibers to produce a thinner, tougher tile, generally $\frac{3}{16}$ inch thick. James H. Stedman, "The Development of Cotton Reinforced Rubber," *India Rubber World* 69 (February 1, 1924): 289. Some companies, such as Goodyear, felt that quality rubber flooring should not contain fillers such as cotton fiber, and their manuals and advertisements clearly stated that no such fillers were used. Although rubber flooring is listed as relatively expensive, the catalogues and trade journals rarely list prices. In 1925 the manufacturing cost of rubber flooring was noted as about $.50 per square foot and its installation cost about $.95 per square foot. See Elmer Swanson, "The Installation of Rubber Flooring," *India Rubber World* 71 (February 1, 1925): 271–72. By 1935 the cost was listed as up to $8 per square yard. See Frank Orellana, "Architectural Uses of Rubber," *The Rubber Age* 31 (May 1935): 75–77.

**12.** Elmer Swanson, "Typical Rubber Floor Designs," *India Rubber World* 71 (November 1, 1924): 82–83.

**13.** *Goodyear Rubber Floors: An Architect's Reference Book* (Los Angeles: Goodyear Tire and Rubber Company of California, 1929). This book gives an overview of the types of rubber products available in addition to basic flooring.

**14.** Werner Esch, "Compounding Ingredients for India-Rubber Mixings," in *International Rubber Exhibition Catalogue* (London, 1911), 447–50.

**15.** Fisher, 60. A recipe for white tiling with a mixture of 465 units reveals that rubber is not the major component: soft clay, 275 units; rubber, 100 units; whiting, 50 units; titanium dioxide, 25 units; zinc oxide, 10 units; sulfur, 4 units; paraffin wax, 2 units; stearic acid, 1.5 units;

plasticizer, 1 unit; accelerator, 0.5 to 1 unit. Attempts were even made to combine the benefits of rubber flooring with the less costly linoleum by developing a blended product called Lino floor covering. See "Lino Floor Covering," *India Rubber World* 87 (March 1, 1933): 26; this article identifies U.S. Patent no. 1,868,787.

**16.** Harry L. Fisher, "The Origin and Development of Synthetic Rubbers," in *Symposium on the Applications of Synthetic Rubbers* (Philadelphia: American Society for Testing and Materials, 1944), 3–17. Synthetic rubber experiments were reported in journals in 1912, but it was not until the 1930s that synthetic rubber products were in general use. In the early years synthetic rubbers were made from a variety of plants and carbohydrates—for example, dandelions and potatoes—and later from chemical hydrocarbon compounds in the laboratory. For more on the stain resistance of synthetic rubber, see Harry L. Fisher, "Rubber and Its Use," *Scientific American* Supplement 2,275 (August 1, 1919): 82. For a discussion of various types of synthetic rubbers, including isoprenes and styrene-butadienes, which were often found in rubber flooring, see *Rubber Technology and Manufacture,* ed. C. M. Blow (Cleveland, Ohio: CRC Press, 1971), 20–32.

**17.** While rubber flooring as a finished material over a subfloor continued to be made in large quantities until the 1960s, rubber as a base material for a whole new line of resilient poured flooring became popular in the 1940s. These floors, called cement-latex flooring compounds or poured resilient flooring, used a base of liquid rubber latex. Aggregate chips and fillers were added to simulate a variety of monolithic floors, such as cork and terrazzo. See, for example, W. G. Wren, "Cement-Latex Flooring Compositions," *India Rubber World* 99 (March 1, 1939): 29–31; and *Architectural Engineering* (New York: F. W. Dodge Corporation, 1948), 448.

**18.** Maximilian Tosh, "The Influence of Pigments on Rubber," *The Rubber Age* 1 (August 25, 1917): 431. See also Esch, 49. This publication describes the following natural pigments for coloring rubber tiles: black consists of lamp-black, sulfite of lead, protoxide oxide of iron, and pitch; red consists of cinnabar, golden sulfide of antimony, peroxide of iron or hydrated oxide of iron; white consists of white oxide of zinc, oxide of antimony, or a mix of sulfide of zinc and a sulfate of barium known as lithopone; green consists of chrome green, which could be a light, grass-color "Victoria" green or a darker color known as Guignet. For more on the drying and storage process, see "From Latex to Manufactured Article," in *International Rubber Exhibition Catalogue,* 46.

**19.** "How Rubber Is Prepared for Tire Making," *The Rubber Age* 1 (May 10, 1917): 135–38. The process for making tires described here is the same for making flooring, although the use of fabric lining is not generally part of the rubber flooring process.

**20.** E. Morris, "Rubber Flooring in Rolls; Present Day Manufacturing Methods," *India Rubber World* 97 (October 1937): 47–49.

**21.** In 1937 the widest rolls or tables were 4½ feet wide. One description of processing rolls noted that while 15 feet was the length of roll that could be cured at one time, ten continuous sections could be wound up on a roll, thereby making a roll 4½ feet wide by 150 feet long. Care had to be taken not to overcure the edges between sections. Rolls of rubber flooring today appear to be limited to 36 inches wide, although some custom orders may be up to 48 inches wide.

**22.** Fisher, *Rubber and Its Use,* 60.

**23.** John Lipani, "Rubber in Modern Hospitals," *The Rubber Age* 38 (October 1, 1935): 21–24.

**24.** Another type of interlocking rubber tiles, also designed for use on ocean vessels, used square tiles with a spline and slot system on the edge of each tile, much like interlocking wood parquet floor tiles. These were designed by John J. Fields, Jr., of New York, and may also have found some use in architectural interiors. Fields was granted U.S. Patent no. 615,832 on December 13, 1898.

**25.** Swanson, 82.

**26.** Thinner tiles were also produced, but these were hard to cut on a true square because the rubber pulled under pressure and had a tendency to show irregularities in subfloor conditions. Swanson, 83.

**27.** In 1936 the Armstrong Cork Company purchased the Stedman Rubber Company's line of rubber tile and its Braintree, Mass., plant. William A. Mehler, Jr., *Let the Buyer Have Faith: The Story of Armstrong* (Lancaster, Pa.: Armstrong World Industries, 1987), 50.

**28.** *Goodyear Rubber Floors,* 3.

**29.** Allan Williams, "Rubber Floor Designs," *India Rubber World* 80 (May 1, 1929): 61–62.

**30.** Abel, 54.

**31.** Yvonne Shashoua and Scott Thomsen, "A Field Trial for the Use of Ageless in the Preservation of Rubber in Museum Collections," in *Saving the Twentieth Century: The Conservation of Modern Materials,* ed. David W. Grattan (Ottawa: Canadian Conservation Institute, 1993), 364.

**32.** M. J. R. Loadman, "Rubber: Its History, Composition, and Prospects for Conservation," in *Saving the Twentieth Century,* 59–60.

**33.** For more on blooms and aging, see Grattan, 65–69.

**34.** "Oil Resisting Flooring," *India Rubber World* 92 (April 1935): 44. See also "Pitchy Felt and Rubber Flooring," *India Rubber World* 79 (January 1, 1929): 62; "The Cleaning of Rubber Floors," *The Rubber Age* 10 (December 25, 1927): 297.

**35.** Grattan, 62–64.

**36.** "Maintaining Rubber Floors," *India Rubber World* 87 (December 1, 1932): 40; "Care of Rubber Flooring," *India Rubber World* 103 (February 1, 1941): 50.

**37.** Grattan, 70–71.

**38.** "International Floor Machine," *India Rubber World* 61 (October 1, 1919): 67.

**39.** "Care of Rubber Flooring," 50.

**40.** Elmer Swanson, "Laying of Rubber Floors," *India Rubber World* 70 (August 1, 1924): 744.

**41.** At Shelburne Village, Vt., Goodyear helped develop a mold to make replacement interlocking tiles out of a synthetic compound for the *Ticonderoga* in the 1980s.

**29. CORK TILE**

1. *Cork—Its Origin and Uses* (Lancaster, Pa.: Armstrong Cork Company, 1930). Cork is commercially grown in France, Italy, Morocco, Tunisia, and Algeria, but Spain and Portugal are usually considered to produce the largest amount and the best quality.
2. For the physical properties of cork, see Arthur L. Faubel, *Cork and the American Cork Industry,* rev. edn. (New York: Cork Institute of America, 1941), 5–9.
3. See, for example, Gilbert E. Stecher, *Cork: Its Origin and Industrial Uses* (New York: Van Nostrand, 1914).
4. The range of products created from these leftovers included wall and ceiling insulation and acoustical materials, among others.
5. William A. Mehler, Jr., *Let the Buyer Have Faith: The Story of Armstrong* (Lancaster, Pa.: Armstrong World Industries, 1987), 10–11.
6. Faubel, 25. Faubel states that around 1900 cork floor tiles began to be manufactured on a commercial basis, but the David E. Kennedy firm described itself in the 1925 *Sweet's Architectural Catalogues* as the "pioneer" in cork floors, "having installed the first cork tile floor in 1899."
7. Mehler, 11; Faubel, 87, 35. Other early producers of cork tiles included the Beaver Tile Company of Cranston, R.I., and the Johns-Manville Company in New York City. Regarding the Beaver Tile Company, see "600 Per Cent Increase in Production after Fuel Change," *Gas Age Record* 52 (December 22, 1923): 771–72. For more on Johns-Manville, see F. B. Green, "Laying Cork Tile Flooring," *Concrete-Cement Age* 5 (September 1914): 132–33. The Dodge Cork Company began producing cork tile in 1947 and produced its first vinyl cork tile in 1949. Personal communication with Arthur B. Dodge, Jr., September 15, 1994.
8. Faubel, 91. Yearly cork tile production during the 1930s was approximately one-third that of 1927, a decrease that may be partially explained by the economy's downturn and competition with other flooring materials, such as rubber and asphalt.
9. Faubel, 87–89. Originally the kilns were electric, but the introduction of gas around 1923 resulted in a great improvement in the quality of the cork tiles.
10. Stanley C. Taylor, "Linoleum and Cork Composition Flooring Materials," *Architectural Forum* 49 (October 1928): 577–82.
11. Arthur B. Dodge, Jr., to Thomas C. Jester, July 15, 1994.
12. Faubel, 89. Beveled tiles did not require sanding because the beveled edges compensated for irregularities in the subfloor.
13. James H. Longshore, "Library Floors," *Library Journal* 62 (September 1, 1937): 634.
14. Because of its resiliency, cork flooring was also promoted by manufacturers as being especially useful as a retrofit material for all existing floor types. This versatility no doubt contributed to its initial popularity as a replacement floor covering for existing floors in older buildings.
15. Longshore, 634.
16. Victor Darwin Abel, *Floor and Wall Coverings* (Scranton, Pa.: International Textbook Company, 1938), 49.
17. Faubel, 124.
18. Taylor, 577.
19. Thomas S. Hines, "Sources and Precedents, Southern California, 1920–1942," in *Blueprints for Modern Living: History and Legacy of the Case Study Houses* (Cambridge, Mass.: MIT Press, 1989), 91.
20. Abel, 49.
21. Some cork tiles were made with a coved base so that a cork tile floor could be extended upward as a wall covering.
22. *Sweet's Catalogue of Building Construction,* 1915.
23. Green, 132–33.
24. Many companies stipulated that their own brand of mastics and glues be used in specifications they provided for laying cork flooring and cautioned that only if company installers or authorized agents carried out the work could the product be guaranteed.
25. Bernard Berkeley and Cyril S. Kimball, *The Care, Cleaning and Selection of Floors and Resilient Floor Coverings* (New York: Ahrens Publishing Company, 1961), 42–43.
26. *Floors and Stairways* (Chicago: Time-Life Books, 1978), 28–29.
27. Replacement tiles were installed in the Gropius House in the late 1980s. See Thomas Fisher, "Updating Mass-Produced Parts," *Progressive Architecture,* April 1989, 111–12.

**30. TERRAZZO**

1. Louis Del Turco, "The Historical Background of Terrazzo Floors, Part 1," *Mosaics and Terrazzo* 2 (February 1931): 11.
2. Ibid., 11
3. Ibid., 12.
4. Ibid., 14.
5. Ibid., 32.
6. Louis Del Turco, "The Historical Background of Terrazzo Floors, Part 2," *Mosaics and Terrazzo* 2 (March 1931): 16.
7. "How to Construct Terrazzo Floors," *Building Age* 40 (February 1918): 98.
8. It is very important for the designer to communicate to the structural engineer the type of terrazzo floor to be installed, because it may affect the structural slab. For instance, if a thick-set system with a sand bed is to be used, the structural system may have to be dropped 2 inches in the vicinity of the terrazzo floor so the finish elevation matches that of other flooring. If a monolithic terrazzo is to be used, the structural engineer must take into account both the modulus of elasticity and the coefficient of expansion of the terrazzo topping. Both these factors should be considered in all cases of terrazzo installation.
9. Resinous toppings can be catalyst-cured resins, latex (emulsion) resins, and polyester materials, such as latex-resins, polyvinyl chloride, neoprene, and acrylic matrixes. Harold B. Olin, John L. Schmidt, and Walter H. Lewis, *Construction: Principles, Materials, and Methods,* 8th edn. (Chicago: The Institute of Financial Education and Interstate Printers and Publishers, 1980), 453–54.
10. Daniel W. Kessler, Arthur Hockman, and Ross E. Anderson, "Physical Properties of Terrazzo Aggregates," *Building Materials and Structures Report BMS98* (Washington, D.C.: National Bureau of Standards, May 20, 1943), 2.

11. H. S. Wright, "Ornamental Concrete Floor Surfacings," *Engineering and Contracting* 65 (July 1926): 22.
12. The National Terrazzo and Mosaics Association publishes standards for color and size selection.
13. For chemically bonded and monolithic systems, panels should be separated at 15- to 20-inch intervals. While the strips are not required for resinous bonded systems, saw-cut control joints must be provided at a maximum distance of 15 to 20 inches. Divider strips would then be placed at the control joint.
14. Del Turco, Part 2, 16.
15. *Sweet's Catalogue of Building Construction,* 1906.
16. Personal communication with Derek Hardy, executive director, National Terrazzo and Mosaic Association, to the author, February 1995.
17. By the mid-1920s preassembled brass strip units were available. Pascal Sylvester, a Chicago installer of terrazzo, patented such a technique, which is described in "The Business of Brass Strip Design," *Mosaics and Terrazzo* 2 (June 1931): 17.
18. Derek Hardy to the author, September 1994 and January 1995.
19. "Terrazzo for Heavy-Duty Industrial Floors," *Engineering and Contracting* 6 (January 1924): 216.
20. Kenneth Reid, "Terrazzo for Wall Decorations," *Pencil Points* 10 (July 1929): 481–84.
21. *Modern Mosaic and Terrazzo Floors* (Harrison, N.J.: L. Del Turco and Brothers, 1924), 6.
22. Olin, Schmidt, and Lewis, 453–54. The most recent development in the terrazzo industry is the use of resinous, rather than cementitious, terrazzo toppings. These can be applied in very thin coats, often ¼ to ½ inch thick, which bond directly to the substrate. Resinous, thin-set toppings can be applied to concrete, wood, or metal subfloors. This method was developed for use in applications where greater strength, chemical resistance, and abrasion resistance are required. The resinous material is more expensive than the traditional portland cement mixtures, but the thin-set method involves shortcuts intended to reduce labor costs. Resinous thin-set terrazzo has become reasonably priced and keeps terrazzo competitive with other nonresilient floor systems. The topping is finished in the same way as cementitious toppings.
23. "Proposed NTMA Specifications for Epoxy Terrazzo," in *NTMA Performance Specifications* (Chicago: National Terrazzo and Mosaic Association, 1963), 7a.
24. This method is used primarily when an existing concrete slab has a smooth finish. It is believed that a bonding agent will strengthen the bond between substrate and topping.
25. J. E. Tufft, "Architect Tells Why Some Terrazzo Cracks," *Mosaics and Terrazzo* 2 (October 1931): 11.
26. Foundation movement and cement expansion and contraction cause cracks that will follow the line of least resistance. Pasquale Galassi, *Suggesting a Standard Specification for Terrazzo Work* (Hartford, Conn.: Pasquale Galassi, 1923), 5.
27. Tensile stresses must be relieved by providing joints in slabs, and cracking of terrazzo adjacent to metal strips must be controlled. "Recommendations for Monolithic Terrazzo," in *Terrazzo Information Guide* (Des Plaines, Ill.: National Terrazzo and Mosaic Association, 1990), 43–44.
28. Acids and crystalline cleaners will cause pitting and chipping of the surface. Elliot C. Spratt, "Timeless Terrazzo," *Skyscraper Management* (August 1957): 12.
29. Kessler et al., 2, 12.
30. "Custodians' Guide to the Proper Maintenance of Terrazzo," in *Terrazzo Information Guide,* 1.
31. "Insight to Cleaning Terrazzo Floors," Technical Bulletin No. 41, National Terrazzo and Mosaic Association, March 1980, 1.
32. "Custodians' Guide to the Proper Maintenance of Terrazzo," 1.
33. "The Proper Treatment of Hard Floors," *Hillyard Floor Treatments Informational Pamphlet* (St. Joseph, Mo.: The Hillyard Company, n.d.), 3.
34. Stain removers can be divided into three general categories: solvents, absorbents, and bleaches. Grease stains are most common and are difficult to remove. Kitchen grease, chewing gum, and lipstick may be dissolved using solvents such as carbon tetrachloride. Fresh grease and food stains, such as ketchup and mustard, may be removed using absorbents such as chalk, talcum powder, blotting paper, or cotton. Difficult stains may be bleached with household ammonia, hydrogen peroxide, acetic acid, or lemon juice. Olin, Schmidt, and Lewis, 453.
35. Other effective removal solutions for such ink stains are Javelle water (a solution of calcium of sodium hypochlorite), ammonia water, chlorinated lime poultice, and soaps. *Terrazzo: Specifications, Details, Technical Data* (Alexandria, Va.: National Terrazzo and Mosaic Association, 1970), 66.
36. "Shop Talk," Technical Bulletin No. 57, National Terrazzo and Mosaic Association, July 1984, 1.

**31. VINYL TILE**

1. L. H. Griffiths, "The Advent of Use of the Decorative Thermoplastic Floor Tile," *Chemistry and Industry* (May 17, 1952): 435; *Modern Plastics Encyclopedia: 1947,* 3 vols. (New York: Plastics Catalogue Corporation, 1947), 1:40, 135–38, 342.
2. Jeffrey Meikle, "Plastics," in *Formica and Design,* ed. Susan G. Lewin (New York: Rizzoli, 1991), 43.
3. Robert Lanzillotti, *The Hard-Surface Floor-Covering Industry* (Pullman: State College of Washington Press, 1955), 27. Coumarone-indene in the highest grade is a clear solid that melts at a temperature above 130°C and has a pale amber color. It is derived from coal tar and by-products of steel production obtained from the fraction of coal tar naphtha, chemically known as the unsaturates in the presence of catalysts, or by a process employing heat and sulfuric acid. Carleton Ellis, *The Chemistry of Synthetic Resins,* 2 vols. (New York: Reinhold Publishing, 1935), 1:123; Griffiths, 435.
4. Irving Skeist, *Plastics in Building* (New York: Reinhold Publishing, 1966), 246; "Vinylite House," *Architectural Record* 75 (January 1934): 36.
5. "The Spectacular Growth of Plastics," in *New Horizons in Construc-tion Materials* (Los Altos, Calif.: Hudson Publishing Company, 1962); "Application Trends in 1951," *Modem Plastics* 29 (January 1952): 87–91.

6. D. V. Rosato, *Asbestos: Its Industrial Applications* (New York: Reinhold Publishing, 1959), 88; *Plastics Catalog: 1944 Encyclopedia of Plastics* (New York: Plastics Catalogue Corporation, 1943), 888; Lanzillotti, 21.

7. S. J. Sencer, "If I Could Meet Your Next Customer for Vinyls . . . ," *Floor Covering Profits* (February 1954): 48; "Application Trends in 1951," *Modern Plastics* 29 (January 1952): 91.

8. *Modern Plastics Encyclopedia* (New York: Plastics Catalogue Corporation, September 1956), 21.

9. Lanzillotti, 33, 39. Major manufacturers included Armstrong Cork Company (Corlon); Kentile; Azrock (Cortina); Pabco (Floron); Johns-Manville (Terraflex); Robbins Floor Products (Geometile, Monogram tile); American Biltmore Rubber Company (Amtico); Flintkote (Vinylcraft, Vincor, Flexachrome); Sloane-Delaware (Flor-Ever); and Dodge Cork Company (Vinyl-Cork).

10. Lanzillotti, 29.

11. Vinyl tiles could also be embossed to create various patterns. The two embossing processes were mechanical and chemical. Mechanical embossing involved the heat and pressure applied by a pattern roller during the calendering. Chemical embossing involved the use of chemical reagents in the plastic composition, which cause differential expansion when heat is applied. W. V. Titow, *PVC Technology,* 4th edn. (London: Elsevier Applied Science Publishing, 1984), 1,092–93.

12. Caleb Hornbostel, *Materials for Architecture* (New York: Reinhold Publishing, 1961), 400.

13. The proportion of vinyl chloride to vinyl acetate was typically 85:15. W. Mayo Smith, *Vinyl Resins* (New York: Reinhold Publishing, 1958), 191; dioctyl phthalate, butyl benzyl phthalate, and dipropylene glycol dibenzoate were common plasticizers. Skeist, 246.

14. A longer fiber would not be satisfactory since it would result in a rougher surface on the finished product. The great bulk of the asbestos that went into the early manufacture of tiles came from Canadian mines. Griffiths, 433.

15. Smith, 193. Dodge Vinyl-Cork Tile is an example of a backed vinyl, in which a cork base and vinyl top are thermowelded together to form a homogeneous unit. Vinyl-Cork Tile, Standard Cork Tile: Catalog No. 55 (Lancaster, Pa.: Dodge Cork Company, 1954).

16. Pure PVC resin is generally used in these tiles because of its superior physical properties, while the vinyl chloride–vinyl acetate copolymers are generally used in vinyl asbestos tiles. Skeist, 251.

17. In 1966 the price ranged from $0.35 to $3.50 per square foot, depending largely on the amount of resin used. Skeist, 251.

18. Because some thermoplastic tiles (for example, vinyl asbestos) are less affected by rising damp than other flooring materials, they may be applied directly to concrete, which is in direct contact with earth, with less likelihood of breakdown. Griffiths, 434.

19. Vinyl asbestos tile contains an asbestos mineral fiber component embedded within the polymeric tile binder, typically polyvinyl chloride or polyvinyl chloride copolymer–based resins. Adhesives that also contain asbestos mineral fillers may also be used in conjunction with vinyl asbestos tile. The asbestos fiber constituent allows water (liquid and vapor) to migrate through the adhesive and fosters movement of moisture through the entire cross section of the installed tile. This breathability, which is not characteristic of other types of vinyl tile, makes vinyl asbestos tile more durable under aggressive, moist environments. Vinyl composition tile contains, in addition to polyvinyl chloride binder resin, other inert mineral fillers such as a calcite filler component but no asbestos component. It is significantly less permeable than vinyl asbestos tile and, in effect, can act as a water barrier. Solid vinyl tile primarily consists of the polymeric binder, such as polyvinyl chloride, and is substantively void of a purposeful addition of an inert filler component such as asbestos or calcite.

20. A vinyl floor can be well suited for a specific, existing environment, but if conditions change at any time during the service life of the floor, long-term durability concerns may arise.

21. K. J. Saunders, *Organic Polymer Chemistry* (London: Chapman and Hall, 1973), 92–96.

22. F. E. Condon and Herbert Meislich, *Introduction to Organic Chemistry* (New York: Holt, Rinehart and Winston, 1960), 328. Vinyl tiles or adhesive plasticizing agents such as a phthalate ester may degrade in a moist, alkaline environment into a soap—metal salt of a long-chain fatty acid (carboxylic acid)—and an alcohol, both of which would reduce bond adhesion of the installed tiles, which have oily or soapy characteristics. This degradation process is called alkaline saponification or hydrolysis depending on existing conditions.

23. The chemical formula for ettringite is $3CaO{\bullet}Al_2O_3{\bullet}CaSo_4{\bullet}31H_2O$. Frederick M. Lea, *The Chemistry of Cement and Concrete* (New York: Chemical Publishing Company, 1970), 223; William G. Hime, "Gypsum: Not for Moist Environments," *Construction Specifier* 46 (July 1993): 41–42.

24. *Standard Test Method for Indicating Moisture in Concrete by the Plastic Sheet Method,* ASTM D4263-83 (Philadelphia: American Society for Testing and Materials, 1988), 325. The plastic sheet test involves taping and sealing a small, rectangular section of plastic sheeting—4 millimeters (0.004 inch) polyethylene—over the substrate and then leaving the sheet in place overnight (a minimum of 16 hours). Formations of moisture condensates on the underside of the plastic sheet indicate that moisture conditions are excessive.

Electronic moisture meters provide "percent moisture" values, but these are not absolute values. Without an extensive standardized procedure using materials of identical composition and identical physical characteristics to that of the substrate being evaluated, the moisture meter values obtained are useful only as relative comparisons of existing lesser and greater moisture levels at localized areas tested in the same project.

25. A procedure used by the author is to grind down test areas to expose the bare concrete substrate using a diamond wheel grinder. Moisture levels are measured by placing a preweighed petri dish containing pellets of calcium chloride inside a plastic chamber of known surface area (approximately 0.48 square feet). The chamber to the on-grade concrete slab surface is then sealed for a finite period (60 hours). Calcium chloride is a hygroscopic salt, a compound that has an affinity

for moisture. The weight gain of the calcium chloride pellets is then determined using an analytical balance.

**26.** Laboratory studies can also be used to identify the type and degree of substrate surface preparation, the condition of the adhesive, and the percentage of surface contact of the adhesive to the tile and substrate bonding planes.

**27.** Application of curing compounds and sealers can inhibit the drying of newly placed concrete slabs, thereby contributing to higher moisture. They can also lessen the initial adhesion of installed vinyl tiles.

**28.** Bayard S. Johnson, "Recommended Procedures for Maintenance," in *Installation and Maintenance of Resilient Smooth-Surface Flooring* (Washington, D.C.: National Academy of Sciences, National Research Council, Building Research Institute, 1955), 105.

**29.** Bernard Berkeley and Cyril S. Kimball, *The Care, Cleaning, and Selection of Floors and Resilient Floor Coverings* (New York: Ahrens Publishing Company, 1961), 23.

**30.** When undertaking remedial work, specify known or proven materials and techniques in accordance with manufacturers' recommendations and trade specifications, confirm that the chosen technique has a proven track record for its intended use and environment, and provide on-site consultants or quality control personnel to ensure a successful installation. In some cases laboratory or on-site test mock-ups of the chosen repair scheme may be required.

**31.** *Test Method—Interim Method for the Determination of Asbestos in Bulk Insulation Samples,* EPA-600/M4-82-020 (Washington, D.C.: U.S. Environmental Protection Agency, December 1982), 1, 6; Federal Register 55, No. 246, Department of Transportation, 49 CFR part 107, *Performance-Oriented Packaging Standards; Changes to Classification, Hazard Communication, Packaging and Handling Requirements Based on UN Standards and Agency Initiative; Final Rule* (Friday, December 21, 1990), 76.

**32.** *Evaluation of Workers' Exposure to Airborne Fibers during the Removal of Resilient Floor Coverings and Asphaltic Cutback Adhesives using Recommended Work Practices* (Arlington, Va.: Environ Corporation, May 1, 1992), n.p. Recommended work practices for removal of vinyl tile flooring, including asbestos-containing material, can be obtained from manufacturers and the Resilient Floor Covering Institute. *An Update on Recommended Work Practices and Exemption from OSHA Initial Monitoring Requirement* (Rockville, Md.: Resilient Floor Covering Institute, November 1992), 2–4; *Recommended Work Practices for the Removal of Resilient Floor Coverings* (Rockville, Md.: Resilient Floor Covering Institute, 1990), 1–2, 6–26.

**33.** *Managing Asbestos in Place—A Building Owner's Guide to Operations and Maintenance Programs for Asbestos-Containing Materials* 20T-2003 (Washington, D.C.: U.S. Environmental Protection Agency, July 1990), viii, 2–3.

**34.** Asphaltic mastic (cutback) is in wide use in the industry today. It has limitations but holds up well under dry or ideal conditions. If moisture levels are excessive, this material can begin to emulsify, resulting in the mastic oozing through the vinyl tile flooring joints and staining the finish tile surface. This type of adhesive can become embrittled with age and is, therefore, less tolerant of aggressive conditions and long-term shrinkage or expansion of adjacent strata. Finally, its adhesion performance tends to be less than those of other adhesive types.

A latex emulsion adhesive that contains polyvinyl acetate has inherent risks, especially when tiles are installed on a slab-on-grade concrete substrate. High moisture can also cause this adhesive to revert to an emulsified state. Polyvinyl acetate is also susceptible to chemical deterioration known as hydrolysis and saponification. Hydrolysis deterioration requires a moist environment, whereas saponification occurs under a moist, alkaline environment. The latter chemical breakdown results in the formation of an alcohol and a metal salt of a carboxylic acid, both of which act as bond-inhibiting agents. For these reasons do not consider using any adhesive containing polyvinyl acetate or polyvinyl acetate-copolymer without knowing its limitations and the inherent risk involved.

A styrene-butadiene-based latex adhesive can often be used with success, but under certain conditions the butadiene component can break down (oxidize), resulting in an embrittled, powdery, aged adhesive.

Aside from potential re-emulsion under excessive moisture, a 100 percent acrylic-based latex adhesive is least vulnerable to long-term degradation under normal environments. It is therefore ideally suited for vinyl tile installations.

Another type of adhesive well suited for vinyl tile installations is products that contain a Neville resin, a synthetic hydrocarbon. Similarly, epoxy- and polyurethane-based adhesives are used for some types of resilient flooring. However, they too can fail under high moisture conditions.

**35.** Factors influencing bonding include proper use and mixing of the specified or chosen materials, application of the adhesive using the recommended notched trowel and in accordance with the manufacturer's directions, placement and heavy rolling of the newly installed tile to maximize surface contact of adhesive to tile backside and substrate, and proper precautions to limit traffic for a designated amount of time to allow for adequate drying and curing of materials.

**36.** These surface imperfections are evident from reflection, diffraction, and scattering of incident light on the untreated floor. Thorough cleaning of freshly installed tiles should be postponed until the adhesive has set and cured (a minimum of five days). *Standard Practice for Application of Floor Polishes to Maintain Vinyl Asbestos Tile or Flooring,* ASTM 3364–88 (Philadelphia: American Society for Testing and Materials, 1988), 364-65; *Standard Practice for Application of Floor Polishes to Maintain Multilayer Composite Tile or Flooring,* ASTM D4386-89 (Philadelphia: American Society for Testing and Materials, 1989), 477–78.

**37.** *Standard Practice for Preparation of Substrate Surfaces for Coefficient of Friction Testing,* ASTM D4103.90 (Philadelphia: American Society for Testing and Materials, 1990), 444.

## PART VII. ROOFING, SIDING, AND WALLS

### 32. ASPHALT SHINGLES

1. The terms *asphalt* and *bitumen* are sometimes used interchangeably. Bituminous materials are "mixtures of natural or pyrogeneous origin." Asphalt is a cementitious material whose constituents include bitumens obtained by refining petroleum. Edwin J. Barth, *Asphalt Science and Technology* (New York: Gordon and Breach Science Publishers, 1962), 1.
2. Asbestos was added to the saturated felts of some products as a way of increasing their fire resistance.
3. Herbert Abraham, *Asphalt and Allied Substances: Their Occurrence, Modes of Production, Uses in the Arts, and Methods of Testing*, 1st edn. (New York: Van Nostrand, 1918), 398.
4. John N. Vogel, Theodore J. Karamanski, and William A. Irvine, *One Hundred Years of Roofing* (Rosemont, Ill.: National Roofing Contractors Association, 1986), 33–34.
5. "The Man with One Idea—and Courage," *Grand Rapids Herald*, October 21, 1923. Vogel et al., 90.
6. Numerous patents exist for special production methods and patterns for asphalt shingles, but the basic idea of the individual rectangular asphalt shingle was never patented. See, for example, Abraham, 3d edn. (1929), 526–41.
7. *Sweet's Catalogue of Building Construction*, 1906.
8. Vogel et al., 64. In 1926 this organization's name was changed to the Asphalt Shingle and Roofing Association.
9. National Board of Fire Underwriters, *Shingle Roofs as Conflagration Spreaders: An Appeal to the Civil Authorities and Civic and Commercial Bodies* (New York: National Board of Fire Underwriters, 1916), 59. Asphalt shingles are currently marketed with three levels of fire resistance.
10. National Board of Fire Underwriters, 59.
11. For examples of pattern variations, see Abraham, 3d edn., 526–41.
12. As early as 1918, a similar multitab shingle was available from the Flintkote Company in a slightly smaller size (32⅛ inches by 10 inches). Abraham (1st edn.), 414.
13. Ralph N. Traxler, *Asphalt: Its Composition, Properties, and Uses* (New York: Reinhold Publishing, 1961), 222–23.
14. Ibid., 232.
15. Although the manufacturing process could be intermittent, whereby the felt was cooled before being coated, most processes were continuous. For mineral fillers, see Traxler, 227. Modern stabilizers have included silica, slate dust, talc, micaceous materials, dolomite, and trap rock. J. L. Strahan, *Manufacture, Selection, and Application of Asphalt Roofing and Siding Products* (New York: Asphalt Roofing Industry Bureau, 1964), 5.
16. Traxler, 234.
17. Ibid., 229.
18. A substantial body of information provided by the roofing manufacturers and installers associations details the established and correct procedures for installing asphalt roofing shingles. National Roofing Contractors Association, *Residential Steep-Slope Roofing Materials Guide* (Rosemont, Ill.: National Roofing Contractors Association, 1981).
19. Strahan, 45.
20. Abraham (1963 edn.), 416.
21. Strahan, 42.
22. Abraham (1963 edn.), 302.
23. Traxler, 240.
24. Barth, 594.
25. The asphalt-saturated felts used to construct shingles perform poorly in temperatures below 40° and above 80°F. When heated above 80°, they become relatively soft; in temperatures below 40° they can become brittle. Consequently, they often suffer physical damage when tampered with during the wrong weather cycle.
26. Asphalt Roofing Manufacturers Association, "Algae Discoloration of Roofs," *Technical Bulletin* (Rockville, Md.: Asphalt Roofing Manufacturers Association, 1993).
27. At the Harry Truman Home in Independence, Mo., the National Park Service duplicated the shingles that the Trumans had installed in the 1960s. The particular pattern was not available, but shingles of the same color blend and weight were available in a multitab cut. The Park Service was ultimately able to purchase a run of shingles without tabs, which were hand cut in the field.

### 33. PORCELAIN ENAMEL

1. Porcelain enamel is sometimes confused with synthetic and baked-on enamels, which are usually composed of resins dissolved in solvents and applied at temperatures around 300°. See H. D. McLaren, "Porcelain Enamel as a Protective Finish," *Enamelist* 26 (1949): 8.
2. E. E. Howe, "Physical Characteristics of Porcelain Enamel," in *Porcelain Enamel in the Building Industry* (Washington, D.C.: National Academy of Sciences, Building Research Institute, 1954), 14.
3. A. I. Andrews, *Porcelain Enamels* (Champaign, Ill.: Garrard Press, 1961), 6.
4. "Porcelain Enameled Steel Buildings," *Iron Age* 134 (September 6, 1934): 19.
5. C. A. W. Vollrath, "Early History of Enameling in the Vollrath Company," *Journal of the American Ceramic Society* 6 (January 1923): 239. See also J. E. Hanson, ed., *A Manual of Porcelain Enameling* (Cleveland, Ohio: The Enamelist Publishing Company, 1937), 10–11.
6. The Enamel Division of the American Ceramic Society was formed in 1919. See H. F. Staley, "Developments in the Enameling Technology during the Past Twenty-Five Years," *Journal of the American Ceramic Society* 6 (January 1923): 243.
7. *Enameling iron*, as the new iron was called, ranged from 18 to 28 gauge; it could be welded, stamped, cast, punched, drawn, or cut before enameling to form grilles, louvers, steps, clips, and flanges. Enameling iron is characterized by a coefficient of thermal expansion that is close to the enamel coating, which reduces stresses. Andrews, 125.
8. *Ceramic Products Cyclopedia* (Chicago: Industrial Publications, 1930), 664; Hanson, 170–90.
9. Andrews, 128–29.

**10.** "Vitreous Enamels for Aluminum," *Finish* 6 (January 1949): 42–43, 68–70.

**11.** To create textured finishes, several stippling and mottling techniques were used. One method created a bumpy appearance by spraying the final enamel coat (sometimes in a different color). Another method was to mix more than one color for the final coat. William Scarlet to Thomas Jester, July 26, 1993. See also Roy E. Dybvig, "Is Architectural Porcelain Enamel a Logical Building Material?" *Finish* 6 (March 1949): 42.

**12.** D. H. Grootenboer and Don Graf, "A Material of Versatility," *Pencil Points* 20 (March 1939): 183.

**13.** Porcelain enamel has also been used for industrial applications. In 1926 the Republic Enameling Company experimented with corrugated porcelain enamel roofing for warehouses. Republic manufactured its corrugated roofing under the trade name V-Corr (later Vitric Steel); a similar product, Porcelok, was manufactured by the Davidson Enamel Company in Lima, Ohio. Most manufacturers used 20-gauge corrugated steel. See "Premium Roofing at Competitive Prices," *Ceramic Industry* 65 (July 1955): 3, 59; "Porcelok Porcelain Enameled Roofing Promotion Draws Attention," *Enamelist* 19 (February 1942): 8–12; and "Vitric Steel," Box 2, Luther Ray Archive, Library of Congress.

**14.** L. W. Ray, "An Early Headache Becomes a Practical Movable Building," *Finish* 5 (June 1948): 20. Joints were covered with 2-inch-wide metal trim screwed to furring. According to Ray, this White Castle "probably was the first installation of this kind in the country." The author could not locate any evidence that would suggest architectural sheets or panels existed before this date.

**15.** "White Castles Constructed of Porcelain Enameled Sheets," *Steel* 95 (October 29, 1934): 28. The Porcelain Steel Building Company Division of White Castle was later known as the Porcelain Steel Builders.

**16.** Moses P. Handy, *The Official Directory of the World's Columbian Exposition* (Chicago: Conkey Co., 1893), 280. Bergmann's factory, Eisenwerke, displayed Metal Enamel.

**17.** Columbian sold its shingles under the trade name Porcelite. LeRoy W. Allison and Malcolm B. Catlin, "The Place of the Porcelain Enameled Steel House," *Iron Age* 134 (September 13, 1934): 18. At least one house from this period in Terre Haute, located at 1027 South Center Street, is extant with green porcelain enamel shingles; it is unclear at this time whether this is Columbian's first roofing installation. See *Vigo County Interim Report: Indiana Historic Sites and Structures Inventory* (1984). This inventory was conducted by the Historic Landmarks Foundation of Indiana. I am indebted to Scott Zimmerman for this information.

**18.** *Porcelain Enamel in the Building Industry* (Washington, D.C.: The Building Research Institute, National Research Council, National Academy of Sciences, 1955), 56. "Porcelain Enamel Tile Makes Appearance in Detroit," *Ceramic Industry* 5 (1925): 532–33. This shingle was invented by roofing contractor Neil Burgett. Burgett's patent, no. 1,868,845, was granted on July 26, 1932, and assigned to Joseph Hoehl, who later became Wolverine's vice president. Wolverine apparently changed its name from the Glassiron Products Company, which offered identical shingles in the 1929 edition of *Sweet's Architectural Catalogues.*

**19.** Allison and Catlin, 14–19. The earliest known U.S. tile patent, no. 1,535,301, was granted to James Francis of London on April 28, 1925.

**20.** "Porcelain Enamel Steel Tile," *Enamelist* 12 (1935): 5–7, 29–31. The first Porcelain Tile Company patent, no. 1,594,614, was granted to George D. Haines on August 8, 1926.

**21.** In 1952 the Porcelain Enamel Products Corporation, a division of the Bettinger Corporation, purchased the rights to the Veos tile, presumably from the Davidson Enamel Company, which advertised the Veos tiles in the 1942 issue of *Sweet's Catalogue File.* These tiles were originally manufactured by the Porcelain Tile Company. See "This Porcelain-on-Steel Tile Sells at Competitive Prices," *Enamelist* 64 (April 1955): 103–5.

**22.** See "Columbian Develops New Wall Tile," *Enamelist* 9 (March 1932): 16–17, and *Sweet's Architectural Catalogues,* 1932. Columbian's patent, no. 1,917,388, was granted on July 11, 1933; *Sweet's Architectural Catalogues,* 1932, 1933, and *Sweet's Catalogue File,* 1942.

**23.** Before choosing porcelain enamel tiles for his restaurant designs, Howard Johnson first painted asphalt shingles orange, but the effect was a dull and unacceptable finish. The color-fast qualities of porcelain enamel meant that replacement was seldom necessary. See Philip Langdon, *Orange Roofs, Golden Arches* (New York: Alfred Knopf, 1986), 48.

**24.** Initially manufacturing costs made it difficult for porcelain enamel tiles to compete with clay tiles. After several years the Louisville Enameled Products Company took over this tile business and lowered production costs with an automatic spraying unit. "Porcelain Enameled Steel Tile," 7. By 1942 only a few manufacturers were manufacturing tile. *Sweet's Catalogue File,* 1942. R. B. Schall and K. M. Gibbs, "The Porcelain Enamel Shingle in Residential and Special Use Construction," *Enamelist* 26 (1949): 3–7.

**25.** For a list of early manufacturers of porcelain enamel sheets and panels, see *Enamel Trade Directory* (Newark, N.J.: Ceramic Publishing Company, 1937).

**26.** The original design called for porcelain enamel panels, but was changed to 36-inch-long Ferro-Clad shingles. Presumably, the original design was flawed, or the capability to execute the design was lacking. See "Porcelain Enamel for House Exterior," *American Builder* 53 (June 1932): 49. The revised plan is described in "House to Have Porcelain Enamel Siding on Insulated Sheets," *Iron Age* 132 (July 14, 1932): 74.

**27.** "Enamel Building at Great Lakes Exhibition Is Co-operative Display," *Steel* 99 (July 20, 1936): 49, 64.

**28.** Between 1937 and 1939 the number of architectural enamel manufacturers increased considerably. For a list of frit manufacturers, suppliers of enameling sheets, architectural enamel panels, pointing compounds, and attachments devices, see Grootenboer and Graf, 195–96.

**29.** Custom panels were also used in combination with stock sizes, which ranged in width from 24 to 36 inches and in length from 4 to 8 feet. "Architectural Porcelain Enamel," *Architectural Forum* 66 (May 1937): 460.

**30.** Panels developed by the Maul Macotta Corporation in Detroit were concrete backed and may have been the only load-bearing system.

Stainless steel battens covered exposed joints. To keep the panels flat, producers spot-welded clips to the backs of panels, over which concrete was poured.

**31.** Later panels with cores were also known as sandwich panels. See, for example, "How a Sandwich Panel Is Made," *Ceramic Industry* 64 (May 1955): 71.

**32.** Other early flanged panels were installed on gas stations in grout before the advent of clip assemblies. Carl F. Block, "Attaching Porcelain Enamel Panels," *Sheet Metal Worker* 39 (January 1948): 1,106–9; see also "Architectural Porcelain Enamel," 459.

**33.** R. M. King, "Number One Headache for Architectural Enamel Manufacturers," *Finish* 1 (December 1944): 14–16, 52.

**34.** Problems with inexperienced contractors forced some manufacturers to offer installation services. Specialty contractors also appeared. Howard Michel, "The Future of Architectural Porcelain Enamel," *Finish* 1 (December 1944): 18.

**35.** As the curtain wall systems gained acceptance with nationally recognized building codes, obstacles to porcelain enamel use diminished. See "Full Steam Ahead for Porcelain Enamel in Construction," *Ceramic Industry* 64 (March 1955): 47; B. L Wood, "Building Codes and the Porcelain Enamel Industry," *Enamelist* 21 (January 1944): 42.

**36.** "Steel in 'Glass' Coating," Baltimore Porcelain Steel Corporation, Box 2, Luther Ray Archive, Library of Congress.

**37.** "Kawneer Enamels Aluminum for Profit," *Ceramic Industry* 55 (November 1950): 55–56.

**38.** John M. Clancy, "Architectural Ceramics in the Future of the Building Industry," *Ceramic Age* 77 (October 1961): 57.

**39.** See Tom Wolf and Leonard Garfield, "A New Standard of Living: The Lustron Home, 1946–1950," in *Perspectives in Vernacular Architecture 2,* ed. Camille Wells (Columbia, Mo.: University of Missouri Press, 1986), 51–60.

**40.** "Porcelain Enamel and Curtain Walls," *Enamelist* 27 (1950): 2, 27. See also William D. Hunt, Jr., *The Contemporary Curtain Wall* (New York: F. W. Dodge Corporation, 1958), 199–222.

**41.** The 16-gauge spandrel panels were laminated to a paper honeycomb core. See William H. Scarlet, "Smooth Skin," *Construction Specifier* 44 (June 1991): 67.

**42.** Properly designed panels were rigid to control wind stresses, insulated to offer thermal stability and dampen sound transmission, and sealed to prevent water and vapor transmission. "How Should the Curtain Wall Perform?" *Ceramic Industry* 65 (December 1955): 53.

**43.** William Lescaze, "An Architect's Viewpoint of Porcelain Enamel Used Alone and in Combination with Other Materials," in *Porcelain Enamel in the Building Industry,* 68–69. See also Clancy, 42.

**44.** "What Is Offered in Panels," *Ceramic Industry* 64 (May 1955): 69.

**45.** Andrews, 580.

**46.** G. H. Spencer Strong, "Chemical Properties of Porcelain Enamel Coatings," in *Porcelain Enamel in the Building Industry,* 8.

**47.** David C. Clark, Carlo G. Pantano, Jr., and Larry L. Hench, *Corrosion of Glass* (New York: Magazines for Industry, 1979), 40.

**48.** Roy Newton and Sandra Davidson, *Conservation of Glass* (London: Butterworth, 1989), 150–53.

**49.** Hydrogen diffusion through exposed steel from condensation is believed to create enough pressure to cause spalling. See Dwight G. Moore and William N. Harrison, "Fifteen-Year Exposure Test of Porcelain Enamel," *Building Materials and Structures Report BMS148* (Washington, D.C.: National Bureau of Standards, June 28, 1957), 3.

**50.** The nature of the stresses and strain between the enamel and metal is reflected by differences in each material's coefficient of thermal expansion, which can cause residual stresses and strains. Andrews, 67–68.

**51.** A number of tests are available to evaluate specific porcelain enamel properties and may be useful in researching the durability of porcelain enamel and potential conservation treatments. See W. N. Harrison, "Methods of Testing Architectural Porcelain Enamels," in *Porcelain Enamel in the Building Industry,* 45–52. A complete list of American Society for Testing and Materials tests for porcelain enamel is found in *Metals Handbook* (Metals Park, Ohio: American Society for Metals, 1982), 509–31.

**52.** One treatment frequently asked about is refiring porcelain enamel. Although in-process repairs are made by manufacturers, these complex techniques are neither practical for weathered porcelain enamel nor reversible. *Metals Handbook,* 524.

**53.** In some instances calcium carbonate has also been used as a mild abrasive. Margaret A. Baker, "1939 Exposure Test of Porcelain Enamel on Steel 30-Year Inspection," *National Bureau of Standards Building Science,* series 38 (August 1971): 5.

**54.** For a discussion of urethanes and epoxies, see National Bureau of Standards, *Development of Interim Performance Criteria for Restoration Coatings for Porcelain Enamel Surfaces,* NBSIR 82-2553 (Washington, D.C.: National Bureau of Standards, 1982).

**55.** Woodrow Carpenter to Thomas Jester, April 18, 1994.

**56.** Newton and Davidson, 183.

**57.** Personal communication with James D. Connolly, Erlin, Hime and Associates, August 19, 1994.

**58.** Robert A. Mitchell tested several auto body repair materials, which are similar to fillers used by companies that repair bathroom fixtures, to infill chipped enamel. However, the weathering characteristics of these materials is unknown. See Robert A. Mitchell, "What Ever Happened to Lustron Homes?" *Association for Preservation Technology Bulletin* 23 (1991): 44–53. See also James Stedge, Steven Greenburg, Jon Myers, and Duane Mosch, "On the Spot Repair of Chipped Porcelain Enamel" (bachelor's thesis, College of Ceramics, State University of New York at Alfred, May 1973).

**59.** William Scarlet to Thomas Jester, July 26, 1993.

**34. ACOUSTICAL MATERIALS**

**1.** Emily A. Thompson, "'Mysteries of the Acoustic': Architectural Acoustics in America, 1800–1932" (Ph.D. diss., Princeton University, 1992), 193.

**2.** Sabine's theory of architectural acoustics revolved around five elements: loudness; interference and resonance (associated with live and dead spots in a room); reverberation (the time it takes for a sound to die away within a room); echo; and extraneous noises.

**3.** Most acoustical concerns in general building applications focused on extraneous noise and control of reverberation. Extraneous noise meant street noise, equipment noise, and noise from other parts of a building. By the 1920s excessive noise was recognized as an impediment to business activities; it interfered with concentration and caused fear, fatigue, and irritation. The Celotex Corporation, *Sound Conditioning—Quiet Comfort for School and College* (Chicago: The Celotex Corporation, 1946). Celotex called its acoustical services "sound conditioning," which it defined as "the art and practice of treating interiors so as to minimize the annoyance of 'unwanted sound.'"

**4.** Thompson, 207.

**5.** Ibid., 208.

**6.** During the 1920s and 1930s the Keaseby and Mattison Company listed itself as "Miners of Chrysotile Asbestos and Manufacturers of Asbestos and Magnesia Products."

**7.** *Sweet's Architectural Catalogues,* 1924.

**8.** A similar system was used before 1923 to correct reverberation in the auditorium at the University of Illinois. The felt was placed on furring, creating a second air space in addition to the one between the felt and the canvas, an approach determined by F. R. Watson, one of Sabine's followers, to be more efficient than applying the felt directly to the plaster. F. R. Watson, *Acoustics of Buildings* (New York: John Wiley and Sons, 1923), 55.

**9.** *Sweet's Architectural Catalogues,* 1935.

**10.** *Sweet's Architectural Catalogues,* 1932.

**11.** Silent Ceal, another variant of the felt-and-membrane system devised by the Acoustical Corporation of America, consisted of perforated sheet metal that held a thick felt layer in place at the ceiling. The metal was then covered with texture-painted canvas. Vern O. Knudsen, *Architectural Acoustics* (New York: John Wiley and Sons, 1932), 250–51.

**12.** *Sweet's Architectural Catalogues,* 1924. The Mazer Acoustile Company's product Acoustile solved that problem, but the solution involved applying canvas to panels of Acoustifelt. Installations included the U.S. Supreme Court Chamber and the House of Commons (1916–27, John A. Pearson) in Ottawa. Mazer also indicated that it could do acoustical installations in the "Old Method," that of felt and membrane.

**13.** *Sweet's Architectural Catalogues,* 1935. The Fels Planetarium (1933, John T. Windrim) at the Franklin Institute in Philadelphia was finished in these products, the tile on the dome and the Holorib on the walls. The Holorib was cut out into a silhouette of the Philadelphia skyline to shield indirect lighting at the base of the dome.

**14.** Nashkote is not listed in the Johns-Manville literature after 1938. Keaseby and Mattison, who developed the early felt-and-membrane method, devoted their energies to sound-absorbing plaster in the later 1920s and 1930s, and no mention can be found of the Acoustical Corporation of America after 1933. *Sweet's Architectural Catalogues,* 1936, 1937, 1939, and 1940.

**15.** *Sweet's Architectural Catalogues,* 1924, 2,132. The patent for Sabinite was held by the Riverbank Laboratories, Wallace Clement Sabine's lab.

**16.** The United States Gypsum Company may have changed the formula to eliminate the asbestos.

**17.** Knudsen, 240–41.

**18.** *Sweet's Catalogue File,* 1945.

**19.** *Sweet's Catalogue File,* 1947.

**20.** Kalite Company, *Sound Control with Kalite Sound Absorbing Products* (New York: Certain-Teed Products Corp., 1937).

**21.** *Sweet's Catalogue File,* 1947. A similar plaster used at the Nebraska State Capitol (1928, Bertram Goodhue) was described by Andy Ladygo at the Interiors Conference for Historic Buildings II in 1993.

**22.** An acoustical plaster derivative that developed after World War II was the spray-applied acoustical coating. In 1945 the Sprayo-Flake Company offered Spray-Acoustic in three varieties: X, mineral wool; R, asbestos fiber; and XR, a combination of mineral wool and asbestos fiber. These materials were combined with a fireproof binder and sprayed with a special air gun over a plaster or lath substrate. They had a deep-fissured appearance and could be tinted if the natural off-white color was not acceptable. By 1947 the company had developed Spray Acoustic Pigmented Surface Coating for redecorating. *Sweet's Catalogue File,* 1945, 1947.

**23.** *Sweet's Catalogue,* 1945.

**24.** Minnesota and Ontario Paper Company, *Universal Insulite in Building Construction* (International Falls, Minn.: Minnesota and Ontario Paper Co., 1922). The material was said to contain "90%, by volume, of minute, hermetically sealed, air cells." The catalogue shows an office with the burlap-covered boards secured with thin wood battens.

**25.** *Sweet's Architectural Catalogues,* 1924.

**26.** *Sweet's Architectural Catalogues,* 1926. The company also manufactured Tri-Mount Acoustical Doors at this time.

**27.** *Sweet's Architectural Catalogues,* 1927.

**28.** *Sweet's Architectural Catalogues,* 1926. Acoustex was installed at the walls in the Hall of Representatives of the South Carolina state capitol around 1925. Its condition today indicates that it was not a highly durable material and was not suitable for use as a wainscot.

**29.** *Sweet's Architectural Catalogues,* 1931, 1932. Acoustex was available in twelve standard colors, custom colors, and stenciled designs.

**30.** *Sweet's Catalogue File,* 1935. By 1939 Acoustex was a product of National Gypsum Company, which advised that it could be cemented or nailed to a backup material or suspended on the Gold Bond Acoustical Suspension System, a concealed spline system based on the use of metal Ts hung from the structural ceiling. *Sweet's Catalogue File,* 1939.

**31.** *Sweet's Catalogue File,* 1935.

**32.** By 1937 Celotex, one of the giants in the acoustical products field, had begun producing Calistone and Calicel Cast Plaster Ornament. *Sweet's Catalogue File,* 1937.

**33.** *Acousti-Celotex (Used in Buildings of All Types for Acoustical and Sound Quieting Purposes)* (Chicago: The Celotex Company, 1925).

**34.** *Sweet's Architectural Catalogues,* 1927. Acousti-Celotex also came in twenty predecorated stencil designs; for additional decorative possibilities, it could be stained or dyed.

**35.** By 1935 Celotex had acquired the Thermax Company, one of the insulation companies that had developed acoustical materials, and offered Thermax Absorbex, a cement wood fiberboard that could be attached to room surfaces or laid in formwork for concrete so that it was exposed when the forms were removed. They were also able to report that "by a new exclusive process, all Acousti-Celotex Cane Fiber Tile is Ferox treated. The Ferox process is a proven method whereby the individual fibers . . . are coated with a chemical complex which is toxic to fungi, termites, and other cellulose destroying organisms." *Sweet's Catalogue File,* 1935. Some noteworthy Celotex applications of the 1920s and 1930s are the Convention Hall (1929, Lockwood and Green) in Atlantic City, N.J., and the Calhoun College Dining Hall (Absorbex) (c. 1930, James Gamble Rogers) at Yale University, New Haven, Conn.

**36.** Knudson, 125.

**37.** In a direct shot at Acousti-Celotex, United States Gypsum noted that Quietile had "a pleasing appearance unmodified by puncturing." *Sweet's Catalogue File,* 1936.

**38.** Ibid.

**39.** Armstrong Cork and Insulation Company advertisement, *The American Architect* 138 (June 1930): 11.

**40.** *Sweet's Catalogue File,* 1937.

**41.** William Jordy, "PSFS: Its Development and Its Significance in Modern Architecture," *Journal of the Society of Architectural Historians* 21 (May 1962): 53, as cited in Thompson, 255, n. 154.

**42.** Advertisement for Owens-Corning Fiberglas, *Architectural Record* 114 (November 1953): 311.

**43.** Advertisement for the Edwin F. Guth Company, *Architectural Record* 114 (November 1953): 65.

**44.** Owens-Corning Fiberglas Corporation, *Sound Control Products* (Toledo, Ohio: Owens-Corning Fiberglas Corp., 1954).

### 35. GYPSUM BOARD

**1.** *Gypsum Construction Handbook* (Chicago: United States Gypsum Company, 1978), 10.

**2.** U.S. Patent no. 520,123 was issued on May 22, 1894.

**3.** Richard F. Stone, "Gypsum Wall Board: A Twentieth Century Building Product from a Jurassic Period Material," *The Building Official and Code Administrator* (March–April 1994): 13.

**4.** Multi-ply gypsum plasterboard consists of one or more layers of gypsum plaster and alternate layers of an open-surface, unsized paper to provide a bond. J. Miller Porter, "Gypsum Plaster and Wall Boards," *ASTM Proceedings 22* (1922): 358; *Gypsum Drywall Industry Newsletter* (October–November 1971): 26; George L. Lincoln "Quarrying and Milling Gypsum," *Engineering and Mining Journal* 129 (February 24, 1930): 183.

**5.** In 1910 Kelley started the American Plaster and Plaster Board Company at Paterson, N.J. He also started the Kelley Plaster and Plaster Board Company, although the date is not known. U.S. Patent no. 905,951 was issued to Stephen Kelley. Raymond Ives Blakeslee, *John Schumacher et al.: Plaintiffs v. Buttonlath Manufacturing Company et al.* (Los Angeles: Parker and Stone, 1921): 151.

**6.** Utzman secured a patent in 1920 for a machine to make gypsum board with folded paper edges; U.S. Patent no. 1,330,413 was issued to him on February 10, 1920.

**7.** U.S. Patent no. 1,176,322 was issued on March 21, 1916, to John and Joseph E. Schumacher. Blakeslee, 151.

**8.** "United States Government Master Specification for Gypsum Wall Board," Department of Commerce, Bureau of Standards, Circular No. 211 284 (April 30, 1925): 2; Lincoln, 183.

**9.** Stone, 13.

**10.** "Glut in Gypsum?" *Barron's* 36 (October 22, 1956): 11.

**11.** Sheetrock tile board was impressed to resemble ceramic tile and could be enameled to further the illusion. Gyp-lap was an insulating sheathing material, ⅜ inch thick, used as a substitute for shiplap wood sheathing and insulating building paper. "Gyp-Lap—A New Fire Resistive Insulating Sheathing Material," *Engineering and Contracting* 62 (July 1924): 250.

**12.** "Patents + Prices = Trouble," *Business* Week, June 17, 1950, 62.

**13.** "Gypsum in the War Effort," *Rock Products* 47 (March 1947): 68; "Versatile Gypsum Finds Application in New Wallboards and Roofs," *Scientific American* 168 (July 1943): 13.

**14.** This created minuscule air pockets throughout the slurry that enhanced the growth of strong gypsum crystals and made the product much lighter than the previous wallboard, which had a 100 percent gypsum core. Stone, 14.

**15.** John P. Hayes, *National Gypsum Company: The Power of Balance* (New York: The Newcomen Society of the United States, 1985), 11.

**16.** Stone, 14. This paper had highly absorbent interior and exterior layers for better bonding, and less absorbent layers in the middle.

**17.** Gypsum alone is a barrier to the passage of heat, fire, and smoke for long periods of time largely because of its water content (50 percent by volume), which acts as an integral sprinkler system under high temperature. "Fire Resistant Plaster Board Offers Apartment Builder Dry-Wall Economies," *Architectural Forum* 95 (August 1951): 218. Many fires attain a temperature of 2,000°. At 1,700° a steel column can bear only its own weight. When the steel is protected by gypsum, the heat penetrates the plaster at such a slow rate that in fires of ordinary duration the metal would hardly get warm. "Gypsum as a Fireproof Material," *Municipal Engineering* 48 (January 1915): 73.

**18.** U.S. Patent no. 2,526,066 was issued on October 17, 1950, to Michael Croce of Bestwall for ½-inch, fire-rated board that provided 45 minutes of fire resistance or added 30 minutes protection to existing assemblies. A second U.S. Patent, no. 2,681,863, was issued to Michael Croce and Clarence G. Shuttleworth in July 1954 for a ⅝-inch board providing 1-hour fire resistance and consisting of glass fibers and unexpanded vermiculite additives. C. E. Abbey, "Type X Wallboard, What It Is, and Where and When It Originated," in *Gypsum in the Age of Man* (Evanston, Ill.: Gypsum Association, 1974), 7.

19. As the glass fibers act as a strengthening binder, the vermiculite aggregate expands under heat to balance the shrinkage of the gypsum as it loses water. U.S. Patent no. 2,744,022 was issued in May 1956 to Michael Croce and Clarence Shuttleworth. "Fire Resistant Plaster Board Offers Apartment Builder Dry-Wall Economies," 218.
20. Other varieties of the same material are known as gypsite and alabaster, while anhydrite is a variety containing no water. "Gypsum: Its Properties, Definitions, and Uses," *Engineering and Industrial Management* 5 (April 17, 1921): 413.
21. "Glut in Gypsum?" 11.
22. The term *plaster of paris* is often applied to all calcined gypsum because of the large number of gypsum rock beds found near Paris, France. "What You Should Know about Gypsum," *Building Age* 42 (July 1920): 61–62. The rotary, or Cummer, process has the advantage over the kettle method in lower power consumption. W. A. Felsing and A. D. Potter, "Gypsum and Gypsum Products," *Journal of Chemical Education* 7 (December 1930): 2,801–2.
23. "Newest Plaster Board Plant on West Coast," *Rock Products* 50 (September 1947): 75.
24. "British Gypsum Wallboard Plant with Latest American Equipment," *Rock Products* 37 (December 1934): 31–32.
25. "Labor Requirements for Gypsum Wall Plaster and Board," *Monthly Labor Review* 65 (October 1947): 455.
26. "The Gypsum Industry: Its Growth and Innovations," in *Gypsum in the Age of Man*, 4.
27. "New Gypsum Tile Board," *Engineering and Contracting* 62 (August 1924): 483.
28. United States Gypsum Company, *Sheetrock: The Fireproof Wallboard* (Chicago: R. R. Donnelley and Sons, 1937), 44.
29. H. Leslie Simmons, *Repairing and Extending Finishes*. Part I, *Plaster, Gypsum Board, Ceramic Tile* (New York: Van Nostrand Reinhold, 1990), 135–36.
30. *Manual of Gypsum Drywall Construction* (Chicago: Gypsum Drywall Contractors International, 1960), 8–10.
31. Ibid., 8–10.
32. Stone, 18.

**36. BUILDING SEALANTS**

1. *Adhesion* describes the sealant's ability to bond to adjacent materials. *Cohesion* describes the sealant's ability to resist internal cracking and tearing. *Elasticity* is the property by which the sealant accommodates joint movement. *Weathering ability* is the sealant's ability to retain its characteristics when exposed to the environment.
2. Julian R. Panek and John P. Cook, *Construction Sealants and Adhesives*, 3d edn. (New York: John Wiley and Sons, 1991), 181–84.
3. Asbestos fiber was used in at least one oil-based sealant available in 1961. It is described as a component of Vulcatex elastic caulking compound, manufactured by the A. C. Horn Companies as described in their product literature, copyrighted in 1960. *Sweet's Catalogue File*, 1961. "Elastic" glazing compound was manufactured by H. B. Fred Kuhls. *Sweet's Catalogue File*, 1944. D-P white lead putty was manufactured by Dicks-Point Company in Dayton, Ohio. *Sweet's Catalogue File*, 1950.
4. Polysulfide polymer is a mercaptan-terminated, long-chain aliphatic containing disulfide linkages.
5. In two-component products one container held the Thiokol rubber polymer, filler, and plasticizer, while the other held lead dioxide paste and plasticizer. In 1961 one-component polysulfide sealants that did not require mixing became available.
6. Panek and Cook, 107–17. The popularity of polysulflde-based sealants grew rapidly; during the 1970s approximately thirty-five different formulations were available. These formulations, made by various manufacturers, were performance tested and approved for use by Thiokol. In the early 1980s polysulfide sealants lost out to better performing urethane and silicone sealants. However, since polysulfide sealants can last up to twenty years, many buildings still have joints containing them. Polysulfide sealants also remain in some joints underneath newer sealants. Today polysulfide sealants are still used for submerged applications and to a limited extent for cladding joints.
7. Panek and Cook, 118–28.
8. R. B. Seymour, ed., *Plastic Mortars, Sealants, and Caulking Compounds* (Washington, D.C.: American Chemical Society, 1979), 113–27.
9. Panek and Cook, 118–28.
10. Ibid., 175–80.
11. Ibid., 159–63.
12. Urethane sealant is based on a polymeric material prepared by the isocyanate-addition-polymerization-reaction between di- or polyisocyanates and di- or polyfunctional hydroxyl compounds.
13. Panek and Cook, 118–28.
14. If a sealant is installed at very cold temperatures, frost may interfere with proper bonding and curing. If the sealant is installed at very hot temperatures, it may flow, and its durability may be adversely affected; this is especially likely with dark-colored sealants in locations exposed to direct sunlight.
15. Silicones that are more than 13 percent fluid can stain vulnerable stones. Staining of paints or elastomeric coatings can occur when they are applied over a previously caulked crack in the field of a building panel because of the migration of the plasticizer from the sealant bead into the coating over it. When light-colored sealants are applied over neoprene gaskets, the oil in the gasket may also stain the sealants.
16. Some sealants have been found to contain hazardous substances. Some polysulfide sealants applied during the 1960s have been found to contain polychlorinated biphenyls (PCBs,) which are a suspected health hazard. Also, asbestos fiber and lead were used as components of oil-based sealants available in the 1950s. Therefore, special handling and disposal procedures are required for any conservation program. It is necessary to conduct laboratory testing to determine whether a particular sealant has PCBs. Also, sealant samples should be transported to the laboratory in aluminum foil; some plastic bags have oils that can contaminate sealants.

# BIBLIOGRAPHY


## PART I. METALS

### 1. ALUMINUM

Adams, Chris, and David Hallam. "Finishes on Aluminum—A Conservation Perspective." In *Saving the Twentieth-Century: The Conservation of Modern Materials*, edited by David W. Grattan. Ottawa: Canadian Conservation Institute, 1993.

Brimelow, E. I. *Aluminum in Building.* London: Macdonald, 1957.

Brown, Hiram. *Aluminum and Its Applications.* New York: Pitman, 1948.

*Care of Aluminum,* 4th edn. Washington, D.C.: The Aluminum Association and the Architectural Aluminum Manufacturers Association, 1977.

Carr, Charles C. *ALCOA: An American Enterprise.* New York: Rinehart, 1952.

Hobbs, Douglas Brown. *Aluminum: Its History, Metallurgy, and Uses.* Milwaukee, Wisc.: Brace Publishing, 1938.

Van Horn, Kent R., ed. *Aluminum.* Vol. 3, *Fabrication and Finishing.* Metals Park, Ohio: American Society for Metals, 1967.

Weidlinger, Paul. *Aluminum in Modern Architecture*, 2 vols. Louisville, Ky.: Reynolds Metals Company, 1956.

### 2. MONEL

*Architectural and Ornamental Monel Metals.* New York: International Nickel Company, n.d. [c. 1930].

Betteridge, Walter. *Nickel and Its Alloys.* New York: Halsted Press, 1984.

Culberson, Laura J. "The Deterioration of 'Non-Corroding' Nickel Alloys in Architectural Applications: A Study of Nickel Silver, Monel Metal, and Chromium-Nickel Stainless Steels." Master's thesis, Columbia University, May 1994.

*Polishing Monel Metal and Pure Nickel.* New York: International Nickel Company, n.d. [c. 1930].

*Practical Design in Monel Metal for Architectural and Decorative Purposes.* New York: International Nickel Company, 1931.

Stansfield, A. "Monel Metal." *Canadian Society of Civil Engineering, Transactions* 23 (1909): 302–9.

Thompson, John F., and Norman Beasley. *For the Years to Come: A Story of International Nickel of Canada.* New York: G. P. Putnam's Sons, 1960.

Uhlig, Herbert H. *The Corrosion Handbook.* New York: John Wiley and Sons, 1948.

*Welding, Brazing, and Soldering Monel Metal and Pure Nickel: No. 1 Working Instructions.* New York: International Nickel Company, 1931.

Williams, Hugh R. "Monel Metal: Points of Superiority of This New Natural Alloy in All Fields for Non-Corroding Steel," *Scientific American* Supplement 88 (August 16, 1919): 98–99.

### 3. NICKEL SILVER

Curry, D. M. "Producing Castings of Nickel Silver." *The Foundry* 68 (June 1940): 46–47, 123–27.

Everhart, John L. *Engineering Properties of Nickel and Nickel Alloys.* New York: Plenum Press, 1971.

Gayle, Margot, David W. Look, and John G. Waite. *Metals in America's Historic Buildings.* Rev. edn. Washington, D.C.: U.S. Department of the Interior, 1992.

Howard-White, F. B. *Nickel, an Historical Review.* Princeton, N.J.: D. Van Nostrand Company, 1963.

Malmstrom, L. L. "This Trend to White Metals." *Metalcraft* 8 (March 1932): 128–33.

McGinnis, Joseph P. "Melting German Silver Alloys." *The Metal Industry* 28 (December 1930): 556.

*Nickel and Its Alloys.* Circular No. 100, Bureau of Standards, U.S. Department of Commerce. Washington, D.C.: Government Printing Office, 1924.

*Nickel Silver in Architecture.* New York: International Nickel Company, 1935.

"Nickel Silvers." *Machine Design* 18 (August 1946): 159–62.

Perry, E. "White Metals, Brasses, and Bronzes." *The Metal Industry* 28 (November 1930): 512–13.

Simmons, A. L. "Tarnishing of Nickel-Silver." *Metal Progress* 58 (September 1950): 345–47.

### 4. STAINLESS STEEL

American Iron and Steel Institute. *Cleaning and Descaling Stainless Steels.* Washington, D.C.: American Iron and Steel Institute, 1983.

*The Architectural and Domestic Uses of Stainless Steel.* New York: Electro Metallurgical Company, 1937.

"Composition Ranges of Standard Stainless Steel Type Numbers." *Iron Age* 49 (February 26, 1942): 68.

*Curtain Walls of Stainless Steel.* Princeton, N.J.: American Iron and Steel Institute, 1955.

Eberhard, Ernest. "Architectural Uses of Chrome Nickel Steel, the Metal Used for the Chrysler Tower and the Empire State Building." *American Architect* 138 (August 1930): 42–47.

Gayle, Margot, David W. Look, and John G. Waite. *Metals in America's Historic Buildings.* Rev. edn. Washington, D.C.: U.S. Department of the Interior, 1992.

Zapffe, Carl A. *Stainless Steels.* Cleveland: The American Society for Metals, 1949.

**5. WEATHERING STEEL**

Boyer, H. E., and T. L. Gall, eds. *Metals Handbook.* 10th edn. Metals Park, Ohio: ASM International, 1990.

Dinkaloo, John. "Bold and Direct Using Metal in a Strong Basic Way." *Architectural Record* 133 (July 1964): 135–42.

——. "The Steel Will Weather Naturally." *Architectural Record* 131 (August 1962): 148–50.

Larrabee, C. P., and S. K. Coburn. "The Atmospheric Corrosion of Steels as Influenced by Changes in Chemical Composition." In *First International Congress on Metallic Corrosion.* London: Butterworth, 1961.

Okada, H., et al. "The Protective Rust Layer Formed on Low-Alloy Steels in Atmospheric Corrosion." In *Fourth International Congress on Metallic Corrosion.* Houston: National Association of Corrosion Engineers, 1969.

Sakai, T., and S. Tokunaga. "Maintenance-Free Service of Weathering Steel by Rust Stabilization Accelerating Treatment." In *Corrosion/79.* Houston: National Association of Corrosion Engineers, 1979.

Scott, John C. "Conservation of Weathering Steel Sculpture." In *Saving the Twentieth Century,* edited by David W. Grattan. Ottawa: Canadian Conservation Institute, 1991.

Storad, A. S. "Coating Systems for High Strength Low Alloy Steel: Are They Necessary?" In *Corrosion/83.* Houston: National Association of Corrosion Engineers, 1983.

Wurth, L. A. "The Role of Chromium in Weathering Steel Passivation." *Materials Performance* 3 (January 1991): 62–63.

## PART II. CONCRETE

**6. CONCRETE BLOCK**

Beall, Christine. *Masonry Design and Detailing for Architects, Engineers, and Contractors.* 3d edn. New York: McGraw-Hill, 1993.

Bell, Joseph. *From Carriage Age to Space Age: The Birth and Growth of the Concrete Masonry Industry.* Herndon, Va.: National Concrete Masonry Association, 1969.

Grant, William. *Manufacture of Concrete Masonry Units.* 2d edn. Chicago: Concrete Publishing Corporation, 1959.

National Concrete Masonry Association. "Maintenance of Concrete Masonry Walls." TEK No. 3. Herndon, Va.: National Concrete Masonry Association, 1972.

——. "Removal of Stains from Concrete Masonry Units." TEK No. 45. Herndon, Va.: National Concrete Masonry Association, 1972.

Newbury, S. B. "Hollow Concrete Block Building Construction in the United States." *Concrete and Constructional Engineering* 1 (May 1906): 118.

Simpson, Pamela H. "Cheap, Quick and Easy: The Early History of Rockfaced Concrete Block." In *Perspectives in Vernacular Architecture* 3, edited by Thomas Carter and Bernard L. Herman. Columbia, Mo.: University of Missouri Press, 1989.

Torrance, William. "Types of Hollow Concrete Blocks Used in the United States and Their Patents." *Concrete and Constructional Engineering* 1 (July 1906): 206–14.

**7. CAST STONE**

Baker, Ira Osborn. *A Treatise on Masonry Construction.* New York: John Wiley and Sons, 1909.

Childe, H. L. *Manufacture and Uses of Concrete Products and Cast Stone.* London: Concrete Publications, 1930.

Cowden, Adrienne B. "The Art of Cast Stone: The History of the Onadaga Litholite Company and the American Cast Stone Industry." Master's thesis, Cornell University, August 1995.

Prudon, Theodore H. M. "Confronting Concrete Realities." *Progressive Architecture* 62 (November 1981): 131–37.

Stauffer, Sara. "Cast Stone: History and Technology." Master's thesis, Columbia University, 1982.

Whipple, Harvey, ed. *Concrete Stone Manufacture.* Detroit: Concrete-Cement Age Publishing Company, 1918.

Wiegel, Fred. "Testing Cast Stone." *Journal of the American Concrete Institute* 3 (September 1931): 33–36.

**8. REINFORCED CONCRETE**

*ACI Manual of Concrete Practice.* Detroit: American Concrete Institute, 1994.

Collins, Peter. *Concrete, the Vision of a New Architecture.* New York: Horizon Press, 1959.

Draffin, Jasper O. "A Brief History of Lime, Cement, Concrete, and Reinforced Concrete." *Journal of the Western Society of Engineers* 48 (March 1943): 14–47.

Emmons, Peter H. *Concrete Repair and Maintenance Illustrated.* Kingston, Mass.: R. S. Means Company, 1994.

Hool, George S., and Nathan C. Johnson. *Concrete Engineers' Handbook.* New York: McGraw-Hill, 1918.

Kozsmatka, Steven H., and William C. Panarese. *Design and Control of Concrete Mixtures.* 13th edn. Skokie, Ill.: Portland Cement Association, 1988.

Newlon, Howard, Jr., ed. *A Selection of Historic American Papers on Concrete, 1876–1926.* Detroit: American Concrete Institute, 1976.

Pfeifer, Donald W. "Part I. Steel Corrosion Damage on Vertical Concrete Surfaces." *Concrete Construction,* February 1981, 91–93.

——."Part II. Steel Corrosion Damage on Vertical Concrete Surfaces." *Concrete Construction,* February 1981, 97–101.

Ransome, Earnest L., and Alexis Saurbrey. *Reinforced Concrete Buildings: A Treatise on the History, Patents, Design, and Erection of the Principal Parts Entering into a Modern Reinforced Concrete Building.* New York: McGraw-Hill, 1912.

**9. SHOTCRETE**

ACI Committee 506R-85. "Guide to Shotcrete." In *ACI Manual of Concrete Practice*. Part 5. Detroit: American Concrete Institute, 1985.

Crom, Theodore R. "Discussion on Cement Gun Mortar." *Proceedings: Journal of the National Association of Cement Users,* vol. 8. Detroit: American Concrete Institute, 1912.

———. "Dry-Mix Shotcrete Practice." In *Shotcreting,* ACI SP-14. Detroit: American Concrete Institute, 1965.

Gebler, Steve, P. Koll, George Seegebrecht, Albert Litvin, and J. Vincent. "Shotcrete: Surveillance for Durable Structures—Case Histories." In *Proceedings of the Second Canmet/ACI International Conference on Durability of Concrete,* vol. 2, ACI SP-126. Montreal: American Concrete Institute, August 1991.

Glassgold, I. Leon. "Evaluation and Design of Concrete Structures and Innovations in Design." In *Proceedings, ACI International Conference, Hong Kong, 1991,* vol. 1, ACI SP-128. Detroit: American Concrete Institute, 1991.

Prentiss, G. L. "The Use of Compressed Air in Handling Mortar and Concretes." In *Proceedings: Journal of the National Association of Cement Users,* vol. 7. Detroit: National Association of Cement Users, 1911.

Reading, T. J. "Shotcrete as a Construction Material." In *Shotcreting,* ACI SP-14. Detroit: American Concrete Institute, 1965.

Rodriguez, Louis. "Samuel W. Traylor, the Cement Gun, and the Process of Innovation." Master's thesis, Lehigh University, 1989.

Seegebrecht, George W., Albert Litvin, and Steve H. Gebler. "Durability of Dry-Mix Shotcrete." *Concrete International* 11 (October 1989): 47–50.

**10. ARCHITECTURAL PRECAST CONCRETE**

*Architectural Precast Concrete.* MNL 122. Chicago: Precast/Prestressed Concrete Institute, 1973.

Fisher, Hugo C. "The Navy's New Ship Model Testing Plant." *Journal of the American Concrete Institute* 35 (April 1939): 317–36.

Ironman, Ralph, and Richard S. Huhta. "Shocked Concrete Comes to the States." *Concrete Products* 63 (December 1960): 29–32.

*Manual for Quality Control for Plants and Production of Architectural Precast Concrete.* MNL 117. 3d edn. Chicago: Precast/Prestressed Concrete Institute, 1996.

Morris, Anthony E. J. *Precast Concrete in Architecture.* London: George Goodwin, 1978.

Peterson, J. L. "History and Development of Precast Concrete in the United States." *Journal of the American Concrete Institute* 50 (February 1954): 477–96.

**11. PRESTRESSED CONCRETE**

*First United States Conference on Prestressed Concrete.* Cambridge, Mass.: MIT Press, 1951.

Lin, T'ung-yen. *Design of Prestressed Concrete Structures.* New York: John Wiley and Sons, 1955.

Magnel, Gustave. *Prestressed Concrete.* London: Concrete Publications, 1948.

Nasser, George D., ed. *Reflections on the Beginnings of Prestressed Concrete in America.* Chicago: Prestressed Concrete Institute, 1981.

## PART III. WOOD AND PLASTICS

**12. FIBERBOARD**

"Insulation Board for Home Building." *American Building* 61 (December 1939): 43–71.

Kollman, F. P., Edward W. Kuenzi, and Alfred J. Stamm. *Principles of Wood Science Technology.* Vol. 2, *Wood-Based Materials.* New York: Springer-Verlag, 1975.

Muench, Carl G. "An Outline of the Insulation Board Industry." *Paper Trade Journal* 125 (July 31, 1947): 48–51.

Rossman, Joseph A. "History of Laminated Wall Board Patents." *Paper Trade Journal* 86 (January 19, 1928): 45–47.

Suchsland, Otto, and George E. Woodson. *Fibre-Board Manufacturing Practices in the United States.* Handbook No. 640, Forest Service, U.S. Department of Agriculture. Washington, D.C.: Government Printing Office, 1987.

"Wall Board Patent History." *Paper Trade Journal* 86 (May 31, 1928): 50–53.

**13. DECORATIVE PLASTIC LAMINATES**

Crosmans, George J., Jr. "Laminated Ureas for the Building Industry." *Modern Plastics* 12 (June 1935): 13–15, 61–63, 70.

Duffin, D. J. *Laminated Plastics.* New York: Reinhold Publishing, 1958.

Grattan, David W., ed. *Saving the Twentieth-Century: The Conservation of Modern Materials.* Ottawa: Canadian Conservation Institute, 1993.

Lewin, Susan Grant, ed. *Formica and Design.* New York: Rizzoli, 1991.

*Modern Plastics Encyclopedia.* 4th edn. New York: Plastics Catalogue Corporation, 1956.

Morgan, John. "Cleaning and Care of Plastic Artifacts." *Polymer Preprints* 33 (1992): 643–44.

———. *Conservation of Plastics.* London: Museums and Galleries Commission, Conservation Unit, and Plastics Historical Society, 1991.

"Plastics for Interior Design." *Modern Plastics* 17 (March 1940): 38–39, 74.

"Standard Practice for Determining the Effect of Heat on Plastics, Designation D794–82." *Annual Book of American Society for Testing and Materials Standards,* vol. 08.01. Philadelphia: American Society for Testing and Materials, 1982.

**14. PLYWOOD**

Cour, Robert. *The Plywood Age: A History of the Fir Plywood Industry's First Fifty Years.* Portland, Ore.: Douglas Fir Plywood Association, 1955.

Haines, Charles M. "The Industrialization of Wood." Ph.D diss., University of Delaware, 1990.

Meyer, Louis H. *Plywood: What It Is, What It Does.* New York: McGraw-Hill, 1947.

Mora, Alexander. *Plywood: Its Production, Use, and Properties.* London: Timber and Plywood, 1932.

Perry, Thomas D. *Modern Plywood.* New York: Pitman, 1942.

———. *Modern Wood Adhesives.* New York: Pitman, 1944.

U.S. Department of Commerce, Forest Products Division. *American Douglas Fir Plywood and Its Uses.* Washington, D.C.: U.S. Department of Commerce, 1937.

Wood, Andrew D., and Thomas G. Linn. *Plywoods: Their Development, Manufacture, and Application.* London: W. and A. K. Johnston, 1950.

**15. GLUED LAMINATED TIMBER**

American Institute for Timber Construction. *Timber Construction Manual.* Rev. edn. New York: John Wiley and Sons, 1994.

American Society of Civil Engineers. *Evaluation, Maintenance, and Upgrading of Wood Structures.* New York: American Society of Civil Engineers, 1982.

*Designing Structural Glued Laminated Timber for Permanence.* Technical Note No. 12. Englewood, Colo.: American Institute for Timber Construction, November 1986.

Freas, A. D., and M. L. Selbo. *Fabrication and Design of Glued Laminated Wood Structural Members.* Technical Bulletin No. 1069. Washington, D.C.: U.S. Department of Agriculture, 1954.

"Injecting New Life into a Veteran Structure." *Technology and Conservation* 7 (Summer 1982): 10–12, 14.

*In-Service Inspection, Maintenance and Repair of Glued Laminated Timbers Subject to Decay Conditions.* Technical Note No. 13. Englewood, Colo.: American Institute of Timber Construction, December 1986.

Selbo, M. L., and A. C. Knauss. "Glued Laminated Wood Construction in Europe." *Journal of the Structural Division: Proceedings of the American Society of Civil Engineers* 7 (November 1958): 1840-1–1840-19.

Steinhaus, Max. "Historic Notes on Wood-Laminating in the United States." *Forest Products Journal* 14 (January 1964): 49.

*Use of Epoxies in Repair of Structural Glued Laminated Timber.* Technical Note No. 14. Englewood, Colo.: American Institute for Timber Construction, January 1990.

Wibbens, Russell P. "Structural Glued Laminated Timber." In *Composite Engineering Laminates,* edited by Albert G. H. Dietz. Cambridge, Mass.: MIT Press, 1969.

Wilson, T. R. C. *The Glued Laminated Wooden Arch.* Technical Bulletin No. 691. Madison, Wisc.: U.S. Department of Agriculture, 1939.

**16. FIBER REINFORCED PLASTIC**

Dietz, Albert G. H. "Paper and Plastics in Buildings." *Technical Association of the Pulp and Paper Industry* 39 (December 1956): 38A–50A.

Koehler, Charles, ed. *Plastics in Building.* Washington, D.C.: Building Research Institute, 1955.

Lubin, George. *Handbook of Fiberglass and Advanced Plastics Composites.* New York: Reinhold Publishing, 1969.

McNeill, Ian. "Fundamental Aspects of Polymer Degradation." In *Polymers in Conservation.* Manchester, England: Royal Society of Chemistry, July 1991.

Morgan, Phillip. *Glass Reinforced Plastics.* London: Iliffe Books, 1961.

Parkyn, Brian, ed. *Glass Reinforced Plastics.* London: Butterworth, 1970.

Skeist, Irving, ed. *Plastics in Building.* New York: Reinhold Publishing, 1966.

Sonneborn, Ralph. *Fiberglas Reinforced Plastics.* New York: Reinhold Publishing, 1954.

## PART IV. MASONRY

**17. STRUCTURAL CLAY TILE**

Freitag, Joseph K. *Architectural Engineering.* New York: John Wiley and Sons, 1895.

Plummer, Harry C., and Edwin F. Warner. *Principles of Tile Engineering: Handbook of Design.* Washington, D.C.: Structural Clay Products Institute, 1947.

*Proceedings: National Workshop on Unreinforced Hollow Clay Tile.* Oakridge, Tenn.: Center for Natural Phenomena Engineering, 1993.

Wermiel, Sara. "Structural Hollow Tile." *Building Renovation,* Spring 1994, 41–44.

White, Charles E. *Hollow Tile Construction.* New York: John Wiley and Sons, 1895.

**18. TERRA COTTA**

American Institute of Architects. *Public Works Specifications: Ceramic Veneer.* AIA File No. 9. Washington, D.C.: American Institute of Architects, 1961.

*Architectural Terra Cotta: Standard Construction.* New York: National Terra Cotta Society, 1914.

Geer, Walter. *The Story of Terra Cotta.* New York: Tobias A. Wright, 1920.

Hunderman, Harry J., and Deborah Slaton. "Diagnostics: Terra Cotta." *Building Renovation,* November–December 1992, 63–66.

———. "Terra Cotta: Analysis and Repair." *The Construction Specifier* 42 (July 1989): 50–57.

Mack, Robert C. "The Manufacture and Use of Architectural Terra Cotta in the United States." In *The Technology of Historic American Buildings: Studies of the Materials, Craft Processes, and the Mechanization of Building Construction,* edited by H. Ward Jandl. Washington, D.C.: Foundation for the Association for Preservation Technology, 1983.

Prudon, Theodore H. M. *Architectural Terra Cotta and Ceramic Veneer in the United States Prior to World War II: A History of Its Development and an Analysis of Its Deterioration Problems and Possible Repair Methodologies.* Ph.D. diss., Columbia University, 1981.

———. "Looking Back: Architectural Terra-Cotta." *Building Renovation* (May–June 1993): 57–60.

Stockbridge, Jerry G. "Evaluation of Terra Cotta on In-Service Structures." In *Special Technical Publication No. 691.* Philadelphia: American Society for Testing and Materials, 1980.

*Terra Cotta: Standard Construction.* New York: National Terra Cotta Society, 1927.

Thomasen, Sven E., and Carolyn L. Searls. "Diagnosis of Terra Cotta Glaze Spalling." In *Masonry: Materials, Design, Construction, and Maintenance. Special Technical Publication No. 992.* Philadelphia: American Society for Testing and Materials, 1988.

Tiller, deTeel Patterson. *The Preservation of Historic Glazed Architectural Terra-Cotta.* Preservation Brief No. 7. Washington, D.C.: U.S. Department of the Interior, 1979.

**19. GYPSUM BLOCK AND TILE**

American Society for Testing and Materials. *Standard Specifications for Gypsum Partition Tile or Block.* ASTM Designation C 52-54. Philadelphia: American Society for Testing and Materials, 1925; rev. edns. 1927, 1933, 1941, and 1954.

Newland, David H. *Gypsum Resources and the Gypsum Industry of New York.* Albany: New York State University, 1929.

"Novel Building Method Utilizes Precast Gypsum Units." *American Builder and Building Age* 51 (August 1931): 62–64.

Porter, John M. *The Technology of the Manufacture of Gypsum Products.* Washington, D.C.: U.S. Government Printing Office, 1926.

Schweim, Henry J. *Gypsum Partition Tile.* Washington, D.C.: Gypsum Association, 1926.

Stone, Ralph W. *Gypsum Products—Their Preparation and Uses.* Technical Paper No. 155, U.S. Bureau of Mines. Washington, D.C.: U.S. Government Printing Office, 1917.

United States Gypsum Company. *Red Book.* Chicago: United States Gypsum Company, 1936.

U.S. Bureau of Mines. *Gypsum, Its Uses and Preparation.* Circular No. 6163. Washington, D.C.: U.S. Government Printing Office, 1929.

U.S. Department of Commerce, National Bureau of Standards. *The Technology of the Manufacture of Gypsum Products.* Circular No. 281. Washington, D.C.: U.S. Government Printing Office, 1926.

**20. THIN STONE VENEER**

Beall, Christine. 3d edn. *Masonry Design and Detailing.* New York: McGraw-Hill, 1993.

Consiglio, Antonio. *A Technical Guide to the Rational Use of Marble.* Rome: Italian Marble Industry Association, 1970.

*Marble Engineering Handbook.* Mount Vernon, N.Y.: Marble Institute of America, 1962.

Siebert, J. S., and F. C. Biggin. *Modern Stone-Cutting and Masonry.* New York: John Wiley and Sons, 1986.

**21. SIMULATED MASONRY**

"Faux Past: When Formstone Made a Home a Castle." *(Baltimore) City Paper,* May 5, 1989, 11.

"Formstone: Love It or Loathe It." *The (Baltimore) Sun,* November 15, 1978, B1.

"Formstone's Glorious Past." *The (Baltimore) Evening Sun,* February 24, 1992, B2.

Knight, Lewis Albert. "Process of Making Artificial Stone Wall Facings." U.S. Patent no. 2,095,641, October 12, 1937.

*Modernizing Magic.* Columbus, Ohio: Perma-Stone Company, 1954.

Peffer, Harry C., Richard L. Harrison, and David E. Ross. "Structural Material and Process of Making Same." U.S. Patent no. 1,852,672, April 5, 1932.

Prudon, Theodore H. M. "Simulating Stone, 1860–1940: Artificial Marble, Artificial Stone, and Cast Stone." *Association for Preservation Technology Bulletin* 21 (1989): 79–91.

Ross, David E., Floyd P. Wymer, and Richard L. Harrison. "Composite Structural Material." U.S. Patent no. 1,852,676, April 5, 1932.

"Rostone—A New Industry." *American Builder* 55 (June 1933): 40–41.

"Saving the City's Monumental Illusion—Formstone." *(Baltimore) Evening Sun,* April 29, 1991, C1.

## PART V. GLASS

**22. PLATE GLASS**

American Architectural Manufacturers Association. *Structural Properties of Glass.* Des Plaines, Ill.: American Architectural Manufacturers Association, 1984.

Button, David, and Brian Pye, eds. *Glass in Building: A Guide to Modern Architectural Glass Performance.* London: Butterworth Architecture and Pilkington Glass, 1993.

Davis, Pearce. *The Development of the American Glass Industry.* Cambridge, Mass.: Harvard University Press, 1949.

Flat Glass Marketing Association. *Glazing Manual.* Topeka, Kans.: Flat Glass Marketing Association, 1980.

Lutz, Ernst. "The Manufacture of Rolled Plate." *The Glass Industry* 11 (October 1930): 227–34, 255–61, 277–83.

McGrath, Raymond. *Glass in Architecture and Decoration.* London: The Architectural Press, 1961.

Newton, Roy, and Sandra Davison. *Conservation of Glass.* London: Butterworth, 1989.

*Technical Bulletin ATS-104.* Toledo, Ohio: Libbey-Owens-Ford, May 21, 1987.

*Technical Bulletin ATS-133.* Toledo, Ohio: Libbey-Owens-Ford, October 6, 1986.

**23. PRISMATIC GLASS**

American Luxfer Prism Company. *Luxfer Prisms: Description with Illustrations.* Chicago: American Luxfer Prism Company, 1897.

"American Luxfer Prism Company." *The Economist* 20 (August 13, 1898): 199.

"Architectural Possibilities of the Luxfer Prisms." *The Inland Architect and News Record* 29 (March 1897): 18.

Basquin, Olin H. "Daylight Illumination of Stores." *The Illuminating Engineer* 2 (1907–8): 805–13.

"The Century's Triumph in Lighting." *The Inland Architect and News Record* 34 (January 1900): 39–40 (supplement).

Crew, Henry, and Olin H. Basquin, eds. *Pocket Hand-Book of Useful Information and Tables Relating to the Use of Electro-Glazed Luxfer Prisms.* Chicago: Luxfer Prism Company, 1898.

"Experiments on the Diffusion of Light through Prismatic and Ribbed Glass Windows." *Engineering News* 45 (January 10, 1901): 34–35.

Neumann, Dietrich. "'The Century's Triumph in Lighting': The Luxfer Prism Companies and Their Contribution to Early Modern Architecture." *Journal of the Society of Architectural Historians* 54 (March 1995): 24–53.

**24. GLASS BLOCK**

Beall, Christine. "Laying Glass Block." *The Magazine of Masonry Construction,* March 1990, 131–33.

"The Glass Age Arrives." *Architectural Forum* 68 (February 1938): 17–28.

"Glass Block Building Featured at Century of Progress." *The Glass Industry* 14 (August 1933): 91.

McGrath, Raymond. *Glass in Architecture and Decoration.* London: The Architectural Press, 1937.

"PC Glass Block." In *Glass Manual.* Pittsburgh, Pa.: Pittsburgh Plate Glass Company, 1946.

Phillips, C. J. *Glass: The Miracle Maker.* New York: Pitman, 1941.

Trelstad, Derek H. "From Beer to Buildings: The Curious History of Glass Masonry." In *Preserving the Recent Past,* edited by Deborah Slaton and Rebecca Shiffer. Washington, D.C.: Historic Preservation Education Foundation, 1995.

——. "The Hecht Company Warehouse." *Building Renovation,* July–August 1993, 29–33.

**25. STRUCTURAL GLASS**

Dyson, Carol J., and Mansberger, Floyd. "Structural Glass: Its History, Manufacture, Repair and Replacement." In *Preserving the Recent Past,* edited by Deborah Slaton and Rebecca A. Schiffer. Washington, D.C.: Historic Preservation Education Foundation, 1995.

*Flat Glass and Related Glass Products: A Survey Covering the Principal Producing and Trading Countries, with Particular Reference to Factors Essential to Tariff Considerations.* Report no. 123, 2d series. Washington, D.C.: U.S. Tariff Commission, U.S. Government Printing Office, 1937.

Flat Glass Jobbers Association. *Glazing Manual: Specifications for Installation of Flat Glass.* Chicago: R. R. Donnelley and Sons Company, 1958.

*Glass, Paints, Varnishes, and Brushes: Their History, Manufacture, and Use.* Pittsburgh, Pa.: Pittsburgh Plate Glass Company, 1923.

"Manufacture of Structural Glass." *The Glass Industry* 20 (June 1939): 215–19.

McGrath, Raymond. *Glass in Architecture and Decoration.* London: The Architectural Press, 1937.

*The Preservation of Historic Pigmented Structural Glass (Vitrolite and Carrara Glass).* Preservation Brief No. 12. Washington, D.C.: U.S. Department of the Interior, 1984.

Yorke, Douglas A., Jr. "Materials Conservation for the Twentieth Century: The Case for Structural Glass." *Association for Preservation Technology Bulletin* 13 (1981): 18–29.

**26. SPANDREL GLASS**

Flat Glass Marketing Association. *Glazing Manual.* Topeka, Kans.: Flat Glass Manufacturing Association, 1980.

Hunt, William D. *The Contemporary Curtain Wall: Its Design, Fabrication, and Erection.* New York: F. W. Dodge Corporation, 1958.

"Plate Glass Is Fired with Colored Ceramic on Back for Spandrel Covers." *Architectural Forum* 103 (September 1955): 207, 212.

Schwartz, Thomas A. "Pain in Your Pane?" *Building Renovation,* July–August 1993, 53–56.

## PART VI. FLOORING

**27. LINOLEUM**

*Armstrong's Handbook for Linoleum Mechanics.* Pittsburgh, Pa.: Armstrong Cork Company, 1924.

*Armstrong's Linoleum Pattern Book.* Lancaster, Pa.: Armstrong Cork Products Company, Floor Division, 1915.

Berkeley, Bernard, and Cyril S. Kimball. *The Care, Cleaning and Selection of Floors and Resilient Floor Coverings.* New York: Ahrens Publishing Company, 1961.

Blackman, Leo, and Deborah Dietsch. "A New Look at Linoleum: Preservation's Rejected Floor Covering." *The Old-House Journal* 10 (January 1982): 9–12.

Faubel, Arthur L. *Cork and the American Cork Industry.* Rev. edn. New York: Cork Institute of America, 1941.

*History and Manufacture of Floor Coverings.* New York: Review Publishing Co., 1899.

"How Linoleums and Oilcloths Are Made." *Scientific American* 97 (July 13, 1907): 1, 28–30.

"Linoleum and Its Proper Application." *American Architect* 119 (May 11, 1921): 565–68.

Mehler, William A., Jr. *Let the Buyer Have Faith: The Story of Armstrong.* Lancaster, Pa.: Armstrong World Industries, 1987.

Muir, Augustus. *Nairns of Kirkcaldy: A Short History of the Company (1847–1956).* Cambridge, England: W. Heffer and Sons, 1956.

Parks, Bonnie W. "Floorcloths to Linoleum: The Development of Resilient Flooring." In *The Interiors Handbook for Historic Buildings,* vol. 2, edited by Michael J. Auer, Charles E. Fisher III, Thomas C. Jester, and Marilyn E. Kaplan. Washington, D.C.: Historic Preservation Education Foundation, 1993.

Von Rosenstiel, Helene, and Gail Caskey Winkler. *Floor Coverings for Historic Buildings: A Guide to Selecting Reproductions.* Washington, D.C.: The Preservation Press, 1988.

Walton, Frederick. *The Infancy and Development of Linoleum Floorcloth.* London: Simpkin, Marshall, Hamilton, Kent, and Company, 1925.

**28. RUBBER TILE**

Blow, C. M., ed. *Rubber Technology and Manufacture.* Cleveland: CRC Press, 1971.

"Care of Rubber Flooring." *India Rubber World* 103 (February 1, 1941): 50.

Fisher, Harry L. *Rubber and Its Use.* Brooklyn, N.Y.: Chemical Publishing Co., 1941.

*History of the Rubber Industry.* Cambridge, England: The Institute of the Rubber Industry, 1952.

"The Influence of Pigments on Rubber." *The Rubber Age* 1 (August 25, 1917): 431.

"Laying Rubber Floors." *India Rubber World* 70 (August 1, 1924): 744.

Loadman, M. J. R. "Rubber: Its History, Composition and Prospects for Conservation." In *Saving the Twentieth Century: The Conservation of Modern Materials,* edited by David W. Grattan. Ottawa: Canadian Conservation Institute, 1993.

"Maintaining Rubber Floors." *India Rubber World* 87 (December 1, 1932): 40.

"Typical Rubber Floor Designs." *India Rubber World* 71 (November 1, 1924): 82–83.

**29. CORK TILE**

Abel, Victor Darwin. *Floor and Wall Coverings.* Scranton, Pa.: International Textbook Company, 1938.

Berkeley, Bernard, and Cyril S. Kimball. *The Care, Cleaning, and Selection of Floors and Resilient Floor Coverings.* New York: Ahrens Publishing Company, 1961.

*Cork: Its Origins and Uses.* Lancaster, Pa.: Armstrong Cork Company, 1930.

Faubel, Arthur L. *Cork and the American Cork Industry.* Rev. edn. New York: Cork Institute of America, 1941.

Green, F. B. "Laying Cork Tile Flooring." *Concrete-Cement Age* 5 (September 1914): 132–33.

Howard, E. H. "Modern Floor Coverings." *Architectural Forum* 35 (July 1921): 29–30.

Longshore, James H. "Library Floors." *Library Journal* 62 (September 1, 1937): 634–36.

Mehler, William A., Jr. *Let the Buyer Have Faith: The Story of Armstrong.* Lancaster, Pa.: Armstrong World Industries, 1987.

Taylor, C. Stanley. "Linoleum and Cork Composition Flooring Materials." *Architectural Forum* 49 (October 1928): 577–82.

**30. TERRAZZO**

Del Turco, Louis. "The Historical Background of Terrazzo Floors, Part 1." *Mosaics and Terrazzo* 2 (February 1931): 11–14, 32; "The Historical Background of Terrazzo Floors, Part 2." *Mosaics and Terrazzo* 2 (March 1931): 15–16, 30.

Del Turco, L., and Brothers. *Modern Mosaic and Terrazzo Floors: A Handbook on the Improved Method of Laying Terrazzo Floors with Metal Dividers.* Harrison, N.J.: L. Del Turco and Brothers, 1924.

Galassi, P. "Some Specifications for Terrazzo." *Mosaics and Terrazzo* 2 (March 1931): 17–28.

"How to Construct Terrazzo Floors." *Building* Age 40 (February 1918): 98.

Kessler, David W., Arthur Hockman, and Ross E. Anderson. "Physical Properties of Terrazzo Aggregates." *Building Materials and Structures Report BMS98.* Washington, D.C.: National Bureau of Standards, May 20, 1943.

Reid, Kenneth. "Terrazzo for Wall Decorations." *Pencil Points* 10 (July 1929): 481–84.

Spratt, Elliot C. "Timeless Terrazzo." *Skyscraper Management* (August 1957): 12–13.

*Terrazzo Information Guide.* Des Plaines, Ill.: National Terrazzo and Mosaic Associations, 1990.

Wright, H. S. "Ornamental Concrete Floor Surfacings." *Engineering and Contracting* 65 (July 1926): 19–23.

**31. VINYL TILE**

*Addressing Moisture-Related Problems Relevant to Resilient Floor Coverings Installed over Concrete.* Rockville, Md.: Resilient Flooring Covering Institute, 1985.

Berkeley, Bernard, and Cyril S. Kimball. *The Care, Cleaning, and Selection of Floors and Resilient Floor Coverings.* New York: Ahrens Publishing Company, 1961.

Griffiths, L. H. "The Advent of Use of the Decorative Thermoplastic Floor Tile." *Chemistry and Industry,* May 17, 1952, 432–36.

Harriman, Lew. "Drying and Measuring Moisture in Concrete, Part 1." *Materials Performance* 34 (January 1995): 34–36.

———. "Drying and Measuring Moisture in Concrete, Part 2." *Materials Performance* 34 (February 1995): 55–59.

*Installation and Maintenance of Resilient Smooth-Surface Flooring.* Washington, D.C.: National Academy of Sciences, National Research Council, Building Research Institute, 1955.

*Kentile Resilient Flooring Technical Handbook.* 14th edn. Chicago: Kentile, n.d.

Lanzillotti, Robert. *The Hard-Surface Floor-Covering Industry.* Pullman: State College of Washington Press, 1955.

*Modern Plastics Encyclopedia.* 1944. Reprint, New York: Plastics Catalogue Corporation, 1956.

Skeist, Irving. *Plastics in Building.* New York: Reinhold Publishing, 1966.

Smith, W. Mayo. *Vinyl Resins.* New York: Reinhold Publishing, 1958.

*Standard Practice for Application of Floor Polishes to Maintain Vinyl Asbestos Tile or Flooring,* ASTM 3364-88. Philadelphia: American Society for Testing and Materials, 1988.

"Vinylite House." *Architectural Record* 75 (January 1934): 2–3, 35–36.

## PART VII. ROOFING, SIDING, AND WALLS

**32. ASPHALT SHINGLES**

Abraham, Herbert. *Asphalt and Allied Substances: Their Occurrence, Modes of Production, Uses in the Arts, and Methods of Testing.* 1918. Reprint, New York: D. Van Nostrand Company, 1963.

Asphalt Roofing Manufacturers Association. *Residential Asphalt Roofing Manual.* Rockville, Md.: Asphalt Roofing Manufacturers Association, 1984.

Barth, Edwin J. *Asphalt Science and Technology.* New York: Gordon and Breach Science Publishers, 1962.

"The Man with One Idea—and Courage." *Grand Rapids Herald,* October 21, 1923.

Strahan, J. L. *Manufacture, Selection, and Application of Asphalt Roofing and Siding Products.* New York: Asphalt Roofing Industry Bureau, 1964.

Traxler, Ralph N. *Asphalt: Its Composition, Properties, and Uses.* New York: Reinhold Publishing, 1961.

Vogel, John N., Theodore J. Karamanski, and William A. Irvine. *One Hundred Years of Roofing in America.* Rosemont, Ill.: National Roofing Contractors Association, 1986.

Watson, John H. *Roofing Systems: Materials and Applications.* Reston, Va.: Reston Publishing Company, 1979.

**33. PORCELAIN ENAMEL**

Allison, LeRoy W., and Malcolm B. Catlin. "The Place of the Porcelain Enameled Steel House." *Iron Age* 134 (September 13, 1934): 14–19.

Andrews, A. I. *Porcelain Enamels.* Champaign, Ill.: Garrard Press, 1961.

Block, Carl F. "Attaching Porcelain Enamel Panels." *Sheet Metal Worker* 39 (January 1948): 106–9.

Grootenboer, D. H., and Don Graf. "A Material of Versatility." *Pencil Points* 20 (March 1939): 181–96.

Hanson, J. E., ed. *A Manual of Porcelain Enameling.* Cleveland: The Enamelist Publishing Co., 1937.

McLaren, H. D. "Porcelain Enamel as a Protective Finish." *Enamelist* 26 (1949): 8–13.

Mitchell, Robert A. "What Ever Happened to Lustron Homes." *Association for Preservation Technology Bulletin* 23 (1991): 44–53.

National Bureau of Standards. *Development of Interim Performance Criteria for Restoration Coatings for Porcelain Enamel Surfaces.* NBSIR 82-2553. Washington, D.C.: National Bureau of Standards, Center of Building Technology, Building Materials Division, 1982.

Newton, Roy, and Sandra Davidson. *Conservation of Glass.* London: Butterworth, 1989.

*Porcelain Enamel in the Building Industry.* Washington, D.C.: The Building Research Institute, National Research Council, National Academy of Sciences, 1954.

Ray, L. W. "An Early Headache Becomes a Practical Movable Building." *Finish* 5 (June 1948): 18–22, 70, 73.

**34. ACOUSTICAL MATERIALS**

Collins, George R. "The Transfer of Thin Masonry Vaulting from Spain to America." *Journal of the Society of Architectural Historians* 27 (October 1968): 176–201.

Knudson, Vern O. *Architectural Acoustics.* New York: John Wiley and Sons, 1932.

Knudson, Vern O., and Cyril M. Harris. *Acoustical Designing in Architecture.* Rev. edn. New York: John Wiley and Sons, 1978.

Sabine, Hale J. "Manufacture and Distribution of Acoustical Materials over the Past 25 Years." *Journal of the Acoustical Society of America* 26 (September 1954): 657–61.

Thompson, Emily A. "'Mysteries of the Acoustic': Architectural Acoustics in America, 1800–1932." Ph.D. diss., Princeton University, 1992.

**35. GYPSUM BOARD**

"British Gypsum Wallboard Plant with Latest American Equipment." *Rock Products* 37 (October 1934): 30–32.

"The Gypsum Industry: Its Growth and Innovations." In *Gypsum in the Age of Man.* Evanston, Ill.: Gypsum Association, 1974.

Porter, J. Miller. "Gypsum Plaster and Wall Boards." *ASTM Proceedings* 22 (1922): 358.

Stone, Richard F. "Gypsum Wallboard: A Twentieth Century Building Product from a Jurassic Period Material." *The Building Official and Code Administrator,* March–April 1994, 13–18.

Stone, R. W. *Gypsum Deposits of the United States.* Washington, D.C.: U.S. Government Printing Office, 1920.

United States Gypsum Company. *Sheetrock: The Fireproof Wallboard.* Chicago: R. R. Donnelley and Sons, 1937.

**36. BUILDING SEALANTS**

*Adhesives and Sealants in Building.* Washington, D.C.: The Building Research Institute, National Research Council, National Academy of Sciences, 1958.

Damusis, Adolfas, ed. *Sealants.* New York: Reinhold Publishing, 1968.

Panek, Julian R., and John P. Cook. *Construction Sealants and Adhesives.* 3d edn. New York: John Wiley and Sons, 1991.

*Sealants for Curtain Walls.* Washington, D.C.: The Building Research Institute, National Research Council, National Academy of Sciences, 1959.

Scheffler, Michael J., Deborah Slaton, and James D. Connolly. "Building Sealants." *Building Renovation,* September–October 1993, 49–52.

Seymour, R.B., ed. *Plastic Mortars, Sealants and Caulking Compounds.* Washington, D.C.: American Chemical Society, 1979.

# SOURCES FOR RESEARCH

**Editor's note:** Names and addresses given below were current at the time of the book's initial publication in 1995.

## LIBRARIES, ARCHIVES, AND RESEARCH INSTITUTIONS

### LIBRARIES AND ARCHIVES

American Ceramic Society
Ceramic Information Center
735 Ceramic Place
Westerville, OH 43081-8720
*Holdings include more than 47,000 book titles, 600 periodical titles, and an extensive collection of product and company literature, U.S. government reports on microfilm, standards manuals, and handbooks.*

American Concrete Institute
P.O. Box 19150 Redford Station
Detroit, MI 48219
*An independent, nonprofit research organization with a library containing 800 volumes on concrete, 450 bound periodicals, and 3,000 technical reports. Collection focuses on concrete—its structural applications, maintenance, manufacture, and use.*

American Institute of Architects
Library and Archives
1735 New York Avenue, N.W.
Washington, DC 20006-5292
*Holdings include more than 30,000 books, nineteenth- and twentieth-century journals, AIA reports, and professional newsletters. Collection also includes unpublished manuscripts and privately published works of various architects, a photographic collection, artifacts, drawings, and many historical periodicals.*

Armstrong World Industries, Inc.
Technical Center and Information Services
2500 Columbia Avenue
Lancaster, PA 17604
*Focuses on chemistry, polymer science, chemical engineering, physics, and material science. Holdings include 10,000 books and bound periodicals, 9,500 research reports, U.S. patents on microfilm, and numerous reference guides.*

The Athenaeum of Philadelphia
East Washington Square
Philadelphia, PA 19106-2688
*Focuses primarily on American architecture and building technology. Maintains a large collection of about 2,000 trade catalogues.*

Avery Architectural and Fine Arts Library
Columbia University
New York, NY 10027
*Holdings include more than 250,000 volumes, including a large collection of American trade catalogues from the eighteenth to the twentieth century. Archives contain two major building trades collections: the Guastavino Collection and selected company files from the New York Architectural Terra Cotta Company.*

Center for Glass Research and Advanced Ceramic Technology
Alfred University
New York State College of Ceramics
Alfred, NY 14802
*More than 75,000 volumes on pottery, glass art, and glass and ceramic engineering. Focus of research is basic and applied information on glass properties, manufacturing processes, and performance behavior.*

Chicago Historical Society
Hedrich-Blessing Collection
Clark Street at North Avenue
Chicago, IL 60614-6099
*Photographs, negatives, slides, transparencies, and other visual records relating primarily to the history and development of the Chicago area. The majority of the collection dates from between 1930 and 1970.*

Corning Museum of Glass
Juliette K. and Leonard S. Rakow Library
One Museum Way
Corning, NY 14830-2253
*Devoted primarily to the art and history of glass and glassmaking. Includes more than 10,000 books, 15,000 bound periodical volumes, an extensive collection of internal company technical reports, correspondence, annual reports, patents, and more than 5,000 microfiche of early trade catalogues.*

Curt Teich Postcard Archives
Lake County Museum
27277 Forest Preserve Drive
Wauconda, IL 60084
*Extensive postcard collection of the Curt Teich Company (operated in Chicago from 1898 through 1974). Includes copies of every image printed and production materials, including original artwork and layout drawings, product samples, photographic prints, negatives, and client letters.*

Dow Chemical Collection
Post Street Archives
205 Post Street
Midland, MI 48640
*Holdings include material about a wide range of plastic products.*

Forest History Society
Carl A. Weyerhaeuser Library
701 Vickers Avenue
Durham, NC 27701
*Periodicals, newsletters, and books, in addition to a large photographic collection. Archives include holdings of the American Forest Council, American Forestry Association, National Forest Products Association, Society of American Foresters, and Western Timber Association.*

Forest Products Laboratory and Library
1 Gifford Pinchot Drive
Madison, WI 53705-2398
*More than 150,000 books and journals relating to wood and chemistry, including information on wood-related topics such as pulp and paper products, glued laminated timber, wood composite panels, wood engineering systems, chemical adhesives, preservatives, and product development and testing.*

General Electric Company Archives
Hall of History
I River Road
Schenectady, NY 12345
*Information on GE products and GE development research on building materials.*

Getty Conservation Institute
1200 Getty Center Drive, Suite 700
Los Angeles, CA 90049
*Collection of more than 12,000 books, periodicals, and video cassettes on the conservation and restoration of all types of cultural property.*

B. F. Goodrich Company Archives
Pierce Library
University of Akron
Akron, OH 44325-1072
*Holdings include the company history and information on rubber and laminated products.*

Hagley Museum and Library
P.O. Box 3630
Wilmington, DE 19807
*Devoted to the documentation of America's commercial, industrial, and technological heritage. Repository for the Conference Board and the National Association of Manufacturers. Archives contain business records for the Du Pont Company, Alan Wood, and Bethlehem Steel. Holdings also include more than 18,000 trade catalogues.*

Institute of Paper Science and Technology
W. R. Hasleton Library
500 10th Street, N.W.
Atlanta, GA 30318-5794
*World's largest collection of information related to pulp and paper manufacturing; founded in 1929. Holdings include 27,000 monographs, 28,000 bound journals, 850 journal subscriptions, more than 230,000 patents, and a vertical file collection of 41,000 items.*

Library of Congress
Washington, DC 20540
*Extensive runs of historic periodicals, government publications on materials, trade catalogues, and numerous indexes. The Science and Technology Division includes approximately 3.2 million technical research reports issued by many U.S. government agencies and a standards collection of more than 500,000 titles.*

Linda Hall Library
5109 Cherry Street
Kansas City, MO 64110-2498
*Independent research library with extensive historical and current holdings in science, engineering, and technology, including standards, specifications, journals, and conference proceedings. Recently acquired the collection of the Engineering Society Library (founded in 1908).*

Massachusetts Institute of Technology
77 Massachusetts Avenue
Cambridge, MA 02139
*Extensive collections in the fields of engineering, materials science, technology, and architecture. Large holdings of historic periodicals, trade catalogues, architectural drawings, archival collections, and technical literature.*

National Agricultural Library
10301 Baltimore Boulevard
Beltsville, MD 20705
*Holdings include more than 2 million books and 22,000 periodicals. Its computerized database (AGRICOLA) contains 3.1 million citations to agricultural literature.*

National Association of Home Builders
Library and Information Center
1201 15th Street, N.W.
Washington, DC 20005-2800
*Focuses on the latest developments in materials, construction methods, and housing design. Subscribes to more than 200 industry-related periodicals.*

National Building Museum
Pension Building
401 F Street
Washington, DC 20001
*Records of the Stewart Construction Company, the Northwestern Terra Cotta Company, and the Benedict Stone Company. Other holdings include 10,000 prints by the Wurts Brothers, a postcard collection dating from 1904 to the present, and Kress Company records (1906–85).*

National Institute of Standards and Technology
Research Information Center
U.S. Department of Commerce
Gaithersburg, MD 20899
*More than 200,000 books and bound periodicals and 2,000 periodical subscriptions, as well as reports of NIST and the National Bureau of Standards. Research focuses include chemistry, physics, and material science and engineering. Also located at NIST is the Building and Fire Research Laboratory, established in 1971. Its holdings consist of 50,000 national and international research reports, books, journals, and conference proceedings.*

National Museum of American History
Smithsonian Institution
14th Street and Constitution Avenue, N.W.
Washington, DC 20560
*Holdings in the Dibner Library relate to science, technology, natural history, and anthropology and date from the fifteenth to the twentieth centuries. The Warshaw Collection consists of 1,020 linear feet with thousands of images of trade cards, posters, trade catalogues, pamphlets, lithographs, and advertising artifacts on more than 565 subjects, including building materials and American culture from the late 1800s to the 1950s. The Columbia Collection consists of approximately 350,000 items, including trade catalogues. The manuscript collection includes the Leo H. Baekeland Papers—journals, diaries, patents, technical papers, biographical materials, photographs—and other information about Baekeland and the Bakelite Company.*

National Park Service
Harpers Ferry Center
Office of Library, Archives, and Graphics
P.O. Box 50
Harpers Ferry, WV 25425-0050
*More than 1,000 original trade catalogues, most dating from 1870 to 1940, as well as a large collection of original patent labels dating from 1872 to 1940.*

National Trust for Historic Preservation Library Collection
University of Maryland at College Park
Architecture Library
College Park, MD 20742
*Holdings include 12,000 books, 300 periodicals, 18,000 postcards, trade catalogues, and unpublished historic structures reports.*

North Carolina State University
College of Textiles
P.O. Box 8301
Raleigh, NC 27695
*Contains 20,000 volumes on textiles and allied subjects, plus 700 journal subscriptions on technology and chemistry of textiles, including basic and applied studies of natural and manmade fibers, fiber formation, polymer science, fiber processing and product development, and color instrumentation and control.*

Northwest Architectural Archives
University of Minnesota
826 Berry Street
St. Paul, MN 55114
*More than 7,000 trade catalogues and 250 stock books, in addition to collections on various architects, engineers, and contractors.*

Portland Cement Association
5420 Old Orchard Road
Skokie, Il 60077
*Holdings focus on literature from 1900 to the present and include 8,100 books; 15,000 bound periodical volumes; 1,650 U.S. patents; 22,000 government, university, and foreign reports; and 720 bibliographies.*

University of California, Santa Barbara
Department of Special Collections
Santa Barbara, CA 93106
*Holdings include the Lawrence B. Romaine collection of trade catalogues, which is computerized on the RLIN database and contains more than 50,000 late-nineteenth- to mid-twentieth-century trade catalogues, many on building materials.*

University of Illinois at Urbana–Champaign
Ricker Architectural Library
608 East Laredo Taft Drive
Champaign, IL 61820
*Second largest architectural library in the United States with extensive book and bound periodical collection. The Small Homes Council is an independent research center affiliated with the architecture department and contains an eclectic collection on residential design from the 1930s. The engineering library contains technical books and reports relating to materials testing done by the engineering department since the turn of the century. The main library contains a complete run of historic* Sweet's *catalogues.*

U.S. Patent and Trademark Office
2110 Jefferson Davis Highway
Arlington, VA 22202
*Maintains patent and trademark indices and the actual patents on both hard copy and microfilm. Patents and trademarks can be researched by name, state, number, assignor, and invention.*

The Ward M. Canaday Center
University of Toledo
Toledo, OH 43606
*An important resource on the history of the American glass industry. Holdings include 22,000 books, 2,500 linear feet of archives and manuscripts, and 25,000 photographs. Included are records for the Toledo glass industry dating to the early nineteenth century and more than 80 linear feet of archival material on the Libbey-Owens-Ford Glass Company spanning the years 1851 to 1991.*

Henry Francis du Pont Winterthur Museum, Inc.
The Louise du Pont Crowninshield Research Building
Route 52, Kennett Pike
Winterthur, DE 19735
*More than 70,000 books and 500,000 manuscripts and visual images on American art and material culture dating to 1920, including approximately 4,000 American and British trade catalogues. The Joseph Downs collection of manuscripts and printed ephemera contains 100,000 manuscripts, visual images, and printed matter relating to American architecture and decorative arts.*

**RESEARCH INSTITUTIONS**

Advanced Steel Processing and Products Research Center
Colorado School of Mines
Golden, CO 80401
*A university-industry cooperative research organization operated through the department of metallurgical engineering, with research focusing on the evaluation of steel processing and new products, including stainless steel– and nickel-based alloys. Conducts studies on the formability and mechanical properties of steel, including fatigue and fracture testing.*

Architectural Conservation Laboratory
University of Pennsylvania
407 Meyerson Hall
Philadelphia, PA 19104-6311
*Dedicated to advanced training and technical research in the conservation of the built environment. Research focuses on technical examination, materials characterization, and treatment of historic buildings and monuments. Also conducts a broad range of professionally sponsored research and training programs, including the preparation of bibliographies and review of technical literature, building materials analyses, and the development and execution of conservation treatments.*

Asphalt Institute
Research Park Drive
P.O. Box 14052
Lexington, KY 40512-4052
*An independent, nonprofit research and engineering organization with affiliated centers in 17 states. The library contains 5,000 volumes focusing on the uses and testing of asphalt, as well as research data developed by government agencies and member companies.*

Buildings Engineering and Architectural Research Center
New Jersey Institute of Technology
323 Martin Luther King, Jr., Boulevard
Newark, NJ 07102
*Contains more than 99,000 books, 33,000 bound periodical volumes, and 33 reels of microfilm. Research focuses on building technology and design, including planning, materials, building systems, construction, and evaluation and education.*

Carleton Laboratory
Department of Civil Engineering
Columbia University
500 West 120th Street
New York, NY 10027
*Research focuses on materials and structures, including strength, material properties, creep, fatigue, vibration, and other conditions of masonry, concrete, and steel.*

Center for Advanced Cement-Based Materials
Northwestern University Department of Civil Engineering
Technological Institute
Evanston, IL 60208
*Conducts studies of cement properties and processes and principles for the design of cement-containing materials. Research focuses on fracture and fiber-reinforcement, long-term performance and prediction in the design of stronger, lighter, less expensive, and more energy-efficient materials for new construction and the restoration of existing infrastructure.*

Center for Building Performance and Diagnostics
Carnegie Mellon University
Department of Architecture
Pittsburgh, PA 15213-2683
*Carnegie Mellon University contains within its system several centers that focus on materials research, including the ASTM Test Monitoring Center (the testing facility for the American Society for Testing and Materials), the Center for Iron and Steel Making Research, the Center for Materials Production, the Engineering Design Research Center, and the Materials Characterization Center. The Institute of Building Sciences focuses on the study of design sciences and building performance. The Materials Characterization Center conducts research on ceramics, metal, adhesives, cements, minerals, polymers, plastics, electronics, and composites. Projects include surface and corrosion studies, fracture analysis, thin film studies, pigment analysis, and development of new instrumentation and analytical techniques. The engineering and science library contains more than 272,000 books, 91,000 bound periodical volumes, and CMU theses.*

Center for Ceramic Research
College of Engineering
Rutgers University
Brett and Bowser Roads
P.O. Box 909
Piscataway, NJ 08855-0909
*A university-industry cooperative research center operating under the Institute for Engineered Materials. Research focuses on ceramic processing, structural ceramics, and ceramic films and coating.*

Center for Preservation Research
Graduate School of Architecture, Planning, and Preservation
Columbia University
400 Avery Hall
New York, NY 10027
*Research focuses on materials and technology and developing the art and science of conserving the built environment, while providing the most conducive environment for scholars and practitioners to carry out advanced studies in historic preservation.*

Center for Public Buildings
Economic Development Institute
Georgia Institute of Technology
Atlanta, GA 30332-0390
*Concerned with all facets of building conservation technology and diagnostics, with emphasis on existing buildings, facilities management, architectural methods, materials composition, energy management, and building performance. Conducts research on all aspects of construction technology. Maintains the Cultural Resource Assistance Information Network (CRAIN) database.*

Colorado State University
Science Laboratory, Department of Forest and Wood Sciences
Fort Collins, CO 80523
*Research focuses on engineering and chemical properties of wood and wood-based products. The laboratory conducts product development and evaluation and responds to consumer requests for information, providing consulting and technical assistance programs.*

Conservation Analytical Laboratory
Smithsonian Institution
Washington, DC 20560
*A specialized unit dedicated to research and education in an interdisciplinary area uniting the arts and humanities with the physical and natural sciences. Research focuses on collections of cultural or scientific significance. Research projects are directed toward improving and perfecting knowledge and technical skills needed to ensure that the most up-to-date methods are used in the preservation and conservation of cultural resources and historic building materials.*

Construction Industry Institute
University of Texas at Austin
3208 Red River, Suite 300
Austin, TX 78705-2650
*Research addresses management, planning, and design aspects of construction, as well as methods and materials of construction and techniques. The institute collects information from engineering and construction projects and collates information to develop a relevant information bank.*

Fibrous Materials Research Center
Department of Materials Engineering
Drexel University
31st and Chestnut Street
Philadelphia, PA 19014
*Research addresses textile structural composites, structural toughening of composites by 3-D fiber, thermoplastic and thermoset composites, fiber reinforced composites, and materials and structural characterization and analysis.*

Industrial Research Institute
University of Windsor
Windsor, Ontario, Canada N9B 3P4
*Conducts contract research on various topics, including materials testing, civil engineering, noise and vibration, and the deterioration of concrete.*

Industrial Research Institute, Inc.
1550 M Street, N.W.
Washington, DC 20005
*Formed in 1938 as a consortium of about 270 industrial companies with technical research departments. Seeks to generate industrial and academic research collaboration as well as industry-government cooperation in research matters.*

Institute for Research in Construction
National Research Council of Canada
Building M-20, Montreal Road
Ottawa, Ontario, Canada KIA OR6
*Conducts research to provide information needed by the Canadian construction industry. The library acts as a national source of general technical information on construction and general building documentation.*

Institute of Wood Research
Michigan Technological University
Houghton, MI 49931
*Conducts research on wood engineering, particle board, composite products, wood fiber, wood waste utilization, wood chemicals, pulping, adhesive systems, and wood preservatives.*

Massachusetts Institute of Technology
77 Massachusetts Avenue
Cambridge, MA 02139
*Includes many centers and laboratories involved in the study of material science and engineering. The Laboratory of Architecture and Planning (LAP) conducts research on energy and buildings, environmental management, architecture, and urban design. The Materials Processing Center conducts research on materials processing and engineering and economic studies in both ferrous and nonferrous metals, ceramics, polymers, electronic materials, composites, and other materials. The Industry Composites and Polymer Processing Program conducts projects such as the cost-effective manufacture of composite components, fiber reinforced plastics, and thermoplastic composites.*

Materials Research Center
Civil Engineering Laboratory
Lehigh University
Bethlehem, PA 18015
*Conducts research on materials, structural mechanics, industrial testing projects, fatigue and fracture of building systems, structural concrete, and structural connections and stability. The library contains 2,000 books and 12,000 research reports on steel and concrete fatigue and environmental civil engineering.*

Materials Research Laboratory
Case Western Reserve University
Olin Building
Cleveland, OH 44106
*The Materials Research Laboratory is involved in the improvement and development of new materials such as glass, ceramics, metals, and polymers. Conducts studies on surfaces, deformation, internal friction in metals and polymers, transformation in ceramics, polymers and metals, and fabrication.*

National Academy of Sciences
2101 Constitution Ave, N.W.
Washington, DC 20418
*Operated by the National Research Council to investigate, examine, experiment, and report on any matters relating to art, science, or technology according to the requests of any government department.*

National Center for Asphalt Technology
Auburn University
211 Ramsay Hall
Auburn, AL 36849
*Focuses on asphalt technology, chemistry, and material properties. Databases include experimental features, new products, and literature in asphalt technology.*

National Center for Preservation Technology and Training
Northwestern State University
P.O. Box 5682
Natchitoches, LA 71497
*Develops and promotes preservation and conservation skills and technologies for identification, evaluation, treatment, monitoring, and interpretation of cultural resources; develops and facilitates training for the preservation field; and applies technological benefits from research by other agencies, institutions, and private industry.*

Paint Research Associates
430 West Forest
Ypsilanti, MI 48197
*Founded in 1943 to undertake research on paints and coatings, including paint formulation and testing, resin and latex synthesis, specialty product development, and literature and patent searches. Also maintains a database on resin and latex formulations.*

Plastics Institute of America
Stevens Institute of Technology
Castle Point Station
Hoboken, NJ 07030
*Founded in 1961 to conduct research on polymer science and its practical application in evaluation of properties of polymeric materials and in fabrication, processing, and use of plastics as engineering materials.*

Plastics Research Center
Department of Industry and Technology
Ball State University
Muncie, IN 47306
*Maintains a collection of plastics periodicals dating from 1943 and provides information on plastics processing and testing. Research focus is the accelerated aging of plastics. Also provides contract research services and consultation on material and production problems.*

Portland Cement Association
5420 Old Orchard Road
Skokie, IL 60077
*Founded in 1916 to conduct research on concrete technology and durability, as well as building and structural uses of concrete. Operates a construction technology laboratory that conducts research and technical services in construction materials, products, and applications.*

Pulp and Paper Research and Education Center
Auburn University
242 Ross Hall
Auburn, AL 36849-5128
*Collection focuses on pulp and paper technology and by-product utilization.*

Structural Research Laboratory
Iowa State University
Ames, IA 50011
*Founded in 1904 to conduct research on structural engineering, including studies of prestressed and reinforced concrete, steel and timber structures, sandwich walls, materials fatigue, and behavior of structural members and connections.*

## INDEXES, BIBLIOGRAPHIES, DATABASES, AND REFERENCE MATERIALS

### INDEXES

*Agricultural Index.* New York: H. W. Wilson Company.
*A cumulative subject index of agricultural periodicals. Except for the first volume, which covered the period 1916–18, it has been issued yearly. In 1964, beginning with volume 19, the name was changed to* Biological and Agricultural Index.

*Aluminum Industry Abstracts.* Materials Park, Ohio: ASM and Institute of Metals.
*A monthly compendium of information on aluminum and its alloys derived from the business and technical literature published worldwide. It continues and enlarges on the earlier journal,* World Aluminum Abstracts, *which covered the period 1968–91. Each issue is divided into five sections: business, environmental issues, production, research, and a comprehensive subject listing.*

*25th Anniversary Index to Volumes I–XXIV of the APT Bulletin.* Fredericksburg, Va.: Association for Preservation Technology International.
*An index of APT Bulletins from 1969 to 1992. Entries are listed alphabetically by subject and author. The index was published in 1994 as volume 25, numbers 1–2, in the Bulletin series.*

*Art and Archaeology Technical Abstracts (AATA Online).* Los Angeles, Calif.: The Getty Conservation Institute and the International Institute for Conservation of Historic and Artistic Works.
*Surveys more than 2,500 sources per year relating to the technical examination, investigation, analysis, restoration, conservation, and documentation of architectural, artistic, and archeological resources. Each volume includes an author index, an annual cumulative subject index, and bibliographical supplements. The first five volumes were published between 1955 and 1966 by the IIC as* IIC Abstracts.

*The Art Index.* New York: H. W. Wilson Company.
*An annual cumulative index dating from 1929 to the present of selected periodicals, museum bulletins, and annuals covering fine arts, archeology, architecture, arts and crafts, ceramics, decoration and ornament, graphic arts, industrial design, interior decoration, landscape architecture, painting, and sculpture.*

*Avery Index to Architectural Periodicals.* New York: Avery Architectural and Fine Arts Library at Columbia University.
*A photographic reproduction of all cards in Columbia University's Avery Architectural Library. Includes not only the Avery collection but also all architectural and art books on the Columbia University campus. A printed catalogue of the collection first appeared in 1895; the first photoreproduction of the catalogue cards was published in 1958 in six volumes. The present catalogue, with three supplements, represents books and periodicals catalogued through May 1977.*

*Bibliography of Agriculture.* Phoenix: Onyx Press.
*A monthly index, begun in 1941, of the literature of agriculture and the allied sciences maintained in the National Agricultural Library. Indexes publications in 17 languages. The bibliography is divided into five sections: a main entry section, checklist of new government publications, list of books recently acquired by the library, subject index, and author index.*

*Comprehensive Dissertation Index.* Ann Arbor, Mich.: University Microfilm International.
*Consists of citations of dissertations accepted for doctoral degrees by American educational institutions in all fields of study. Citations are listed by subject and key words. The first set of abstracts spans the period 1861–1972; the second set, 1973–82. Thereafter, the index was published each year.*

*Dissertation Abstracts International.* Ann Arbor, Mich.: University Microfilms International.
*Consists of worldwide dissertation abstracts available on microfilm or as xerographic reproduction. The abstracts are arranged alphabetically by subject group. Key word, author, and title indexes are also included. Also available online.*

*Engineered Materials Abstracts.* Metals Park, Ohio: ASM International.
*Provides comprehensive coverage of the world's published literature concerning the science of polymers, ceramics, and composite materials intended for use in the design, construction, and operation of structures, equipment, and systems; and the practices of materials science and engineering as they relate to these materials.*

*The Engineering Index (Ei).* New York: J. R. Dunlap.
*Focuses primarily on topics relevant to civil and mechanical engineering yet also includes publications from other related fields. In addition to bibliographic information, each entry includes a brief description of the article. Published since 1884, it is international in scope and indexes more than 2,700 serials. The name was changed to* Ei *in 1968. This index or a portion of it is available online.*

*Guide to Architectural Trade Catalogs from Avery Library, Columbia University.* Frederick, Md.: UPA Academics Editors.
*A reference to all the trade catalogues on microfiche at the Avery Library. The microfiche collection is organized by the modern* Sweet's Catalogue *arrangement, except for the first section, which has incorporated those catalogues under the "General" heading. The guide is divided into six sections: a guide to the subjects in the 16* Sweet's *categories; a table of contents listing all the catalogues alphabetically with each of the 16 categories; a single alphabetical listing by company; a geographical guide; a chronological guide; and a subject index.*

*The History and Art of Glass: Index of Periodical Articles.* Compiled by Corning Museum of Glass. Boston: G. K. Hall.
*Available in two volumes: 1956–79 (vol. 1) and 1980–82 (vol. 2), which list more than 14,000 articles in many languages drawn from periodicals, conference proceedings, annuals, and yearbooks. Both volumes are arranged in three sections: general publications; technological publications (including preservation); and historical publications. The indexes are compiled from the annual checklists published in the museum's* Journal of Glass Studies.

*Index of Patents Issued from the U.S. Patent Office.* Washington, D.C.: Government Printing Office.
*An alphabetical listing of patentees and inventions. The early years tend to include such information as city and state of patentee, invention name, date the patent was issued, official gazette volume, and page number. Inventor and patent numbers are always included.*

*Index of Trademarks Issued from the U.S. Patent Office.* Washington, D.C.: Government Printing Office.
*Contains an alphabetical listing of the trademarks issued providing trademark numbers, assignor, and dates. Before 1927 trademark listing was included in the* Index of Patents.

*Industrial Arts Index.* New York: H. W. Wilson Company.
*A cumulative subject index to a select list of engineering, trade, and business periodicals as well as books and pamphlets. First published in 1914, it has continued to be issued yearly. In 1958 (vol. 46) it was divided into two indexes: the* Applied Science and Technology Index, *which carried on the same numbering, and the* Business Periodicals Index.

*The Institute of Mining and Metallurgy (IMM) Abstracts.* London: The Institute of Mining and Metallurgy.
*A survey of world literature on economics, geology, mining, mineral dressing, extraction, metallurgy, and allied subjects. About 3,000 abstracts are included in each issue. The abstracts do not have indexes; they are available on microfiche and microfilm and online through IMM.*

*Metals Abstracts.* Woking, England: American Society for Metals and the Institute of Metals.
*Originally entitled* Metals Review. *In 1945 the American Society for Metals compiled the monthly lists into a reference volume representing a survey of metallurgical literature published in 1944. Thereafter, the* ASM Review of Metal Literature *was produced yearly as a reference work. Contains annotations of articles and technical papers from foreign and domestic engineering, scientific, and industrial journals and books. In 1969 the index was incorporated into* Metals Abstracts.

*National Union List of Catalogues, Pre-1956 Imprints.* New York: H. W. Wilson Company.
*Lists select portions of the catalogued collections of major American and Canadian research libraries, as well as rare items held in the collections of smaller libraries. The index includes entries for books, pamphlets, maps, atlases, and music published before 1956 and also contains periodicals and other serials that have been catalogued by the Library of Congress. Entries are arranged alphabetically by author and include a full citation and libraries known to have the source material. Works dated 1956 and later are to be found in* The National Union Catalogue.

**BIBLIOGRAPHIES**

*Durability of Building Materials Bibliography.* Published in *Durability of Building Materials* 2 (May 1984).
*Includes approximately 450 references. To facilitate retrieval, numbers are assigned to many key words; these numbers accompany the entries and help categorize the contents of the articles listed.*

*Historic Concrete: An Annotated Bibliography.* Compiled by Adrienne B. Cowden. Washington, D.C.: National Park Service, U.S. Department of the Interior, 1993.
*Provides an overview of the published literature relating to the history, manufacture, deterioration, repair, and applications of concrete from 1900 to 1950. The references are divided into five sections: history and evolution of concrete; concrete manufacture; concrete failure, deterioration, and repair; applications; and reference materials.*

*Master's Theses in the Pure and Applied Sciences.* New York: Plenum Press.
*An annual list of master's theses completed at accredited colleges and universities in the United States and Canada. Covers theses completed since 1955.*

*Pure and Applied Science Books: 1876–1982.* New York: R. R. Bowker Company.
*A listing of more than 220,000 titles in science and technology published and distributed in the United States. It spans publication dates from before 1800 through 1982 and covers all aspects of the physical and biological sciences and their applications, including technology, engineering, agriculture, domestic arts and science, and manufacturers.*

*Twentieth-Century Building Materials: 1900–1950: An Annotated Bibliography.* Compiled by George M. Bleekman III et al. Washington, D.C.: National Park Service, U.S. Department of the Interior, 1993.
*Consists of information on building materials introduced or significantly developed during the first half of the twentieth century. The bibliography is organized into four major sections: general building materials; classification of materials; construction systems; and additional resources.*

*World List of Scientific Periodicals.* London: Butterworth.
*Consists of more than 60,000 titles of periodicals concerned with the natural sciences published between 1900 and 1960. Periodicals first published at any time before 1900 are listed if they continued publication into the twentieth century. Because of the growth in science publication in the years 1951–60, titles appearing during this period account for about a quarter of the present edition. Arrangement is alphabetical by title with standard abbreviations, locations, dates of publication, and coverage given where known. Holdings of participating British libraries are given. Later continued as the* British Union-Catalog of Periodicals Incorporating World List of Scientific Periodicals.

**DATABASES**

The Conservation Information Network
Client Services Canadian Heritage Information Network
365 Laurier Avenue West, 12th Floor
Ottawa, Ontario, Canada KIA 0C8
Internet: CAN-SERVICE@immedia.ca
*An international computerized information and communication network that is a joint effort of the Getty Conservation Institute, the Canadian Heritage Information Network, the Canadian Conservation Institute, the Smithsonian Institution Conservation Analytical Laboratory, the International Council of Museums (ICOM), the International Council on Monuments and Sites (ICOMOS), and the International Centre for the Study of the Preservation and Restoration of Cultural Property (ICCROM). Its main service is the access to continuously maintained databases, such as the bibliographic reference database (BCIN), which consists principally of a merger of the bibliographic references of the ICCROM library together with the Getty Conservation Institute's* Art and Archaeology Technical Abstracts. *Most references contain an abstract, and services of CIN also include e-mail and a bulletin board.*

The Historic Twentieth-Century Building Products Database—National Park Service
Preservation Assistance Division
National Park Service
P.O. Box 37127
Washington, DC 20013-7127
*A catalogue of twentieth-century building products and materials, as well as appropriate preservation treatments, suitable substitute materials, research sources, and techniques. Entries include information about a product's dates of production, appearance, manufacturing process, composition, and uses. Under development.*

AECNET—The Electronic Resource Network of Architecture, Engineering and Construction
Environmental Dynamics Design, Inc.
Long Island, N.Y.
Internet: AECNET@saecnet.cpbx.net
*An online information and communications resource primarily for architectural, engineering, and construction professionals. AECNET will bring together information and services from professional organizations, building product manufacturers, trade associations, industry publications, software publishers, and other related sources. Under development.*

**REFERENCE MATERIALS**

*American Society for Testing and Materials Annual Book of Standards.* Philadelphia: American Society for Testing and Materials.
*Consists of all current formally approved ASTM standards and tentative test methods, definitions, recommended practices, classifications, specifications, and other related material, such as proposed methods.*

*Materials Handbook: An Encyclopedia for Managers, Technical Professionals, Purchasing and Productions Managers, Technicians, Supervisors, and Foremen.*
New York: McGraw-Hill.
*Consists of the important characteristics and economics of more than 14,000 commercially available materials. Entries are alphabetical by general category of materials. The handbook also includes a brief section on the nature and properties of materials with tabular data and a detailed index. First published in 1929.*

*Modern Plastics Encyclopedia.* Bristol, Conn.: Plastics Catalog Corporation.
*Issued as part of the periodical* Modern Plastics *and organized into four sections: a textbook with brief chapters on materials, chemicals, additives, fillers, property enhancers, reinforcements, processing, tooling, testing, fabricating, and finishing; a design guide; a data bank with tables; and a section on suppliers. First published in 1941, with subsequent annual editions still being published.*

*Preserving the Recent Past.* Washington, D.C.: Historic Preservation Education Foundation, 1995.
*Proceedings from this 1995 conference. Includes chapters on resource evaluation, preservation and reuse strategies, and conservation of materials, as well as a bibliography.*

*Saving the Twentieth Century: The Conservation of Modern Materials.* Ottawa: Canadian Conservation Institute.
*Proceedings from the 1991 symposium addresses topics of major importance to museums that had begun to collect a wide range of twentieth-century material culture. Contains articles on the deterioration of artifacts made of such materials as plywood, aluminum, weathering steel, rubber, and plastic.*

*Sweet's Architectural Catalogue File.* New York: McGraw-Hill.
*First published in 1906. Made up of various building products and manufacturers' catalogues. Each volume contains an alphabetical index of all products in the file, as well as a listing of proprietary names. For each product heading, firm names and the catalogue codes are listed. Many libraries have a recent set, while some libraries have a microfiche version of the historic issues from 1906 to 1949.*

*Thomas Register of American Manufacturers.* Woodbridge, Conn: Research Publications, Inc.
*A series of annual volumes beginning in 1905 listing companies, products, and services. With several volumes per issue, the information is divided into three sections: an alphabetical listing of thousands of products and services, with sources of each product and service listed geographically under its subject heading; profiles of U.S. companies in alphabetical order, with addresses, phone numbers, branch plants, names of executives, and asset ratings; and catalogue file with information from numerous catalogues; catalogues are bound alphabetically by company name and cross-referenced in the first section.*

## PROFESSIONAL AND TRADE ASSOCIATIONS

**GENERAL**

American Architectural Manufacturers Association
1540 East Dundee Road, Suite 310
Palatine, IL 60067-8322

American Institute for Conservation
1717 K Street, N.W., Suite 301
Washington, DC 20006

American Society for Testing and Materials
1916 Race Street
Philadelphia, PA 19103-1187

Association for Preservation Technology International
P.O. Box 3511
Williamsburg, VA 23187

DoCoMoMo International
Eindhoven University of Technology
BPU Postvak 8, P.O. Box 513
5600 MB Eindhoven
The Netherlands

Materials Research Society
9800 McKnight Road
Pittsburgh, PA 15237

Modern Architecture Preservation League
P.O. Box 9782
Denver, CO 80209

Plastics Historical Society
c/o Plastics and Rubber Institute
11 Hobart Place
London SWIW OHL
UK

Society for Architectural Historians
1365 Astor Street
Chicago, IL 60610-2144

Society for Commercial Archeology
National Museum of American History,
Room 5010
Washington, DC 20560

Society for the History of Technology
Department of Social Sciences
Michigan Technological University
1400 Townsend Drive
Houghton, MI 49931-1295

**METALS**

The Aluminum Association, Inc.
900 19th Street, N.W., Suite 300
Washington, DC 20006

Aluminum Extruders Council
1000 North Rand Road, Suite 214
Wauconda, IL 60084

American Institute of Steel Construction
I East Wacker Drive, Suite 3100
Chicago, IL 60601-2001

American Iron and Steel Institute
1101 17th Street, N.W.
Washington, DC 20036

Copper Development Association
260 Madison Avenue, 16th Floor
New York, NY 10016

Iron and Steel Society
410 Commonwealth Drive
Warrendale, PA 15086

Metal Construction Association
1767 Business Center Drive, Suite 302
Reston, VA 22090

National Association of Architectural Metal Manufacturers
600 South Federal Street, Suite 400
Chicago, IL 60605

National Ornamental and Miscellaneous Metals Association
804–10 Main Street, Suite E
Forest Park, GA 30050

United Steelworkers of America
5 Gateway Center
Pittsburgh, PA 15222

**CONCRETE**

American Concrete Institute
P.O. Box 19150
22400 West Seven Mile Road
Detroit, MI 48219

Concrete Reinforcing Steel Institute
933 North Plum Grove Road
Schaumburg, IL 60173-1206

Gunite-Shotcrete Association
12306 Van Nuys Boulevard
Lake View Terrace, CA 91342

International Concrete Repair Institute
1323 Shepard Drive, Suite D
Sterling, VA 20164-4428

National Concrete Masonry Association
Research and Development Laboratory
2302 Horse Pen Road
Herndon, VA 22070

National Precast Concrete Association
10333 North Meridian, Suite 272
Indianapolis, IN 46290

Portland Cement Association
5420 Old Orchard Road
Skokie, IL 60077

Precast/Prestressed Concrete Institute
175 West Jackson Boulevard, Room 1859
Chicago, IL 60604

**WOOD AND PLASTICS**

American Hardboard Association
1210 West Northwest Highway
Palatine, IL 60067

American Institute of Timber Construction
7012 South Revere Parkway, Suite 140
Englewood, CO 80112

American Laminators Association
P.O. Box 11700
Tacoma, WA 98411

American Pulpwood Association
600 Jefferson Plaza, Suite 350
Rockville, MD 20850

American Wood Preservers Institute
1945 Old Gallows Road, Suite 150
Vienna, VA 22182

APA—The Engineered Wood Association
P.O. Box 11700
Tacoma, WA 98411

Composites Fabricators Association
1735 North Lynn Street, Suite 950
Arlington, VA 22209-2022

Decorative Laminate Products Association
1100 South LaSalle Street, Suite 1400
Chicago, IL 60605

Hardwood Plywood Manufacturers Association
P.O. Box 2789
1825 Michael Faraday Drive
Reston, VA 22090-2789

Laminated Timber Institute of Canada
P.O. Box 2699
Station D
Ottawa, Canada KIP 5W7

National Forest Products Association
1111 19th Street, N.W., Suite 700
Washington, DC 20036

National Particle Board Association
18928 Premiere Court
Gaithersburg, MD 20879

National Plastics Center Museum
210 Lancaster Street
P.O. Box 639
Leominster, MA 01453

Plywood Research Foundation
P.O. Box 11700
Tacoma, WA 98411-0700

Resilient Floor Covering Institute
966 Hungerford Drive, Suite 12-B
Rockville, MD 20850

Society of the Plastics Engineers
14 Fairfield Drive
Brookfield, CT 06804-0403

Society of the Plastics Industry
1275 K Street, N.W., Suite 400
Washington, DC 20005

SPI Composites Institute
355 Lexington Avenue
New York, NY 10017

Western Wood Products Association
Yeon Building
522 Southwest Fifth Avenue
Portland, OR 97204

**MASONRY**

American Ceramic Society
735 Ceramic Place
Westerville, OH 43081-8720

Brick Institute of America
11490 Commerce Park Drive, Suite 300
Reston, VA 22091-1532

Friends of Terra Cotta
771 West End Avenue, Apt. 10E
New York, NY 10025

Gypsum Association
810 First Street, N.E., Suite 510
Washington, DC 20002

International Masonry Institute
823 15th Street, N.W.
Washington, DC 20005

Masonry Contractors Association of America
1550 Spring Road, Suite 320
Oak Brook, IL 60521

Masonry Institute of America
2550 Beverly Boulevard
Los Angeles, CA 90057

The Masonry Society
3775 Irois Avenue, Suite 6
Boulder, CO 80301-2043

**GLASS**

American Ceramic Society
735 Ceramic Place
Westerville, OH 43081-8720

Glass Association of North America
3310 Southwest Harrison Street
Topeka, KS 66611-2279

National Glass Association
8200 Greensboro Drive, Suite 302
McLean, VA 22101

Society of Glass and Ceramic Decorators
1627 K Street, N.W., Suite 800
Washington, DC 20006

**FLOORING**

American Ceramic Society
735 Ceramic Place
Westerville, OH 43081-8720

American Floorcovering Association
13-154 Merchandise Mart
Chicago, IL 60654

National Association of Floor Covering Distributors
401 North Michigan Avenue
Chicago, IL 60611-4267

National Terrazzo and Mosaic Association
3166 Des Plaines Avenue, Suite 132
Des Plaines, IL 60018

National Tile Contractors Association
626 Lakeland East Drive
P.O. Box 13629
Jackson, MS 39236

Resilient Floor Covering Institute
966 Hungerford Drive, Suite 12B
Rockville, MD 20850

Rubber Manufacturers Association
1400 K Street, N.W., Suite 900
Washington, DC 20005

Tile, Marble, Terrazzo, Finishers, Shopworkers and Granite Cutters International Union
101 Constitution Avenue, N.W.
Washington, DC 20001

Wood and Synthetic Flooring Institute
4415 West Harrison Street, Suite 242-C
Hillside, IL 60162

**ROOFING, SIDING, AND WALLS**

Acoustical Society of America
500 Sunnyside Boulevard
Woodbury, NY 11797

Adhesive and Sealant Council
1627 K Street, N.W., Suite 1000
Washington, DC 20006

Adhesive Manufacturers Association
401 North Michigan Avenue
Chicago, IL 60611

Asphalt Institute
P.O. Box 14052
Lexington, KY 40512-4052

Asphalt Roofing Manufacturers Association
6000 Executive Boulevard, Suite 201
Rockville, MD 20852-3803

Association of the Wall and Ceiling Industry
307 East Annandale Road, Suite 200
Falls Church, VA 22042-2433

National Council of Acoustical Consultants
66 Morris Avenue, Suite IA
Springfield, NJ 07081-1409

National Roofing Contractors Association
206 E Street, N.E.
Washington, DC 20002

Porcelain Enamel Institute
102 Woodmont Boulevard, Suite 360
Nashville, TN 37205

Roof Consultants Institute
7424 Chapel Hill Road
Raleigh, NC 27607

Roofing Industry Educational Institute

14 Inverness Drive East

Building H, Suite 110

Englewood, CO 80112

Sealant, Waterproofing, and Restoration Institute

3101 Broadway, Suite 585

Kansas City, MO 64111

# AUTHORS AND CONTRIBUTORS

**Editor's note:** Affiliations given below were current at the time of the book's initial publication in 1995.

FLORA A. CALABRESE is a structural engineer with Wiss, Janney, Elstner Associates in Northbrook, Illinois.

IRENE J. COHEN is an architect with McDonald's Corporation in Oak Brook, Illinois.

JAMES D. CONNOLLY, a chemist, is manager of Erlin, Hime Associates in Northbrook, Illinois.

ADRIENNE B. COWDEN is a historic sites surveyor with P. A. C. Spero and Company in Baltimore.

CAROL J. DYSON is an architectural historian with the Illinois Historic Preservation Agency in Springfield.

SUSAN M. ESCHERICH is a historian with the Preservation Assistance Division of the National Park Service, Washington, D.C.

DAVID C. FISCHETTI is a structural engineer with DCF Engineering in Cary, North Carolina.

SIDNEY FREEDMAN is director of architectural services at the Precast/Prestressed Concrete Institute in Chicago.

REBECCA GALLAGHER is a design consultant in Mt. Pleasant, South Carolina.

PAUL E. GAUDETTE is an engineer with Wiss, Janney, Elstner Associates in Chicago.

EDWARD A. GERNS is an architect with Wiss, Janney, Elstner Associates in Chicago.

CAROL S. GOULD is a historian with the Preservation Assistance Division of the National Park Service, Washington, D.C.

ANNE E. GRIMMER is an architectural historian with the Preservation Assistance Division of the National Park Service, Washington, D.C.

WILLIAM G. HIME, a chemist, is a principal of Erlin, Hime Associates in Northbrook, Illinois.

HARRY J. HUNDERMAN is an architect with Wiss, Janney, Elstner Associates in Chicago.

MIKE JACKSON is chief architect of the Division of Preservation Services of the Illinois Historic Preservation Agency in Springfield.

THOMAS C. JESTER is an architectural historian with the Preservation Assistance Division of the National Park Service, Washington, D.C.

WALKER C. JOHNSON is a principal of Johnson-Lasky Architects in Chicago and a Fellow of the American Institute of Architects.

BRUCE S. KASKEL is an architect and structural engineer with Wiss, Janney, Elstner Associates in Chicago.

STEPHEN J. KELLEY is an architect and structural engineer with Wiss, Janney, Elstner Associates in Chicago.

PAUL D. KOFOED is a chemist with Erlin, Hime Associates in Northbrook, Illinois.

KIMBERLY A. KONRAD is a preservation planner with the Boston Landmarks Commission.

ANN MILKOVICH MCKEE is assistant professor of architecture in the Historic Preservation Program at Ball State University.

ROBERT W. MCKINLEY, formerly manager of technical services at PPG Industries, is a management consultant based in Hancock, New Hampshire.

ANDREW MCNALL is a Ph.D. candidate in philosophy at the University of Wisconsin.

KATHLEEN CATALANO MILLEY is an architectural historian with the Mid-Atlantic Regional Office of the National Park Service, Philadelphia.

DIETRICH NEUMANN is assistant professor of art history at Brown University.

HOWARD NEWLON, JR., is adjunct professor in the Historic Preservation Program at the University of Virginia.

WILLIAM J. NUGENT is a structural engineer with Wiss, Janney, Elstner Associates in Northbrook, Illinois.

SHARON C. PARK is senior historical architect with the Preservation Assistance Division of the National Park Service, Washington, D.C.

CONRAD PAULSON is a structural engineer with Wiss, Janney, Elstner Associates in Chicago.

MICHAEL J. SCHEFFLER is a professional engineer with Wiss, Janney, Elstner Associates in Chicago.

ROBERT SCORE is an architect with Muller and Muller in Chicago.

JOHN C. SCOTT is an architectural and fine arts conservator with the New York Conservation Center, New York City.

CAROLYN L. SEARLS is an engineer with Wiss, Janney, Elstner Associates in Emeryville, California.

PAMELA H. SIMPSON is professor of art history at Washington and Lee University.

AMY E. SLATON is a doctoral fellow in the Department of the History of Science, Harvard University.

DEBORAH SLATON is an architectural conservator with Wiss, Janney, Elstner Associates in Chicago.

BONNIE WEHLE PARKS SNYDER is a principal with P.S. Preservation Services and an environmental planner with the California Department of Transportation, Sacramento, California.

JERRY G. STOCKBRIDGE is an architect, structural engineer, and president of Wiss, Janney, Elstner Associates in Northbrook, Illinois.

NICOLE L. STULL is a graduate of the Historic Preservation Planning Program, Cornell University.

ANNE T. SULLIVAN is an architect with Johnson-Lasky Architects in Chicago and is an adjunct professor in the Historic Preservation Program, School of the Art Institute of Chicago.

MICHAEL A. TOMLAN is director of the Historic Preservation Planning Program, Cornell University.

DEREK H. TRELSTAD is senior editor of *Building Renovation Magazine* and adjunct assistant professor in the Historic Preservation Program, Columbia University.

ANTHONY J. T. WALKER is an architect with Damond, Lock, Grabowski Partners in London.

ANNE E. WEBER is an architect with Ford, Farewell, Mills and Gatsch in Princeton, New Jersey.

DAVID P. WESSEL is an architectural conservator with Architectural Resources Group in San Francisco.

KENNETH M. WILSON recently retired as director of collections and preservation at the Henry Ford Museum and Greenfield Village in Dearborn, Michigan. He is the author of *New England Glass and Glassmaking and American Glass: 1760–1930.*

# INDEX

**Note:** *Italicized* page numbers refer to illustrations.


A. O. Smith Corporation Research and Engineering Building, 14, *97*
AAA Building, *174*
Aberene Stone Company, 139
abrasion, 29–30, 37–38, 43
acids, 16, 17, 23, 28, 57, 154–55, 177, 206
Acme Cement Plaster Company, 131
Acoustex, 233–34
Acoustical Corporation of America, 232
acoustical materials, 4, *230,* 231–35, *232, 233*
Acousti-Celotex tile, *xxxi,* 234
Acoustifibroblock, 233
Acoustone, 234, *235*
acrylic sealants, 241–42
adhesives: for cork tile, 199, 200; in decorative plastic laminates, 98; in fiberboard, 90; in glued laminated timber, 105, 106, 108; for linoleum, 188–89; in plywood, 101; for rubber tile, 193; for shotcrete repair, 74; for vinyl tile, 212–13
admixtures, in shotcrete, 71
advertising, 6, 19–21, *20,* 33–34, 47, 92, *92, 182,* 183, *220*
Agasote Millboard Company, 89
aggregates, 47–48, 54–55, 57, 67, 71, 77, 78
airport construction, 8
Akeley, Carl E., 71
Akoustolith, 231, *233*
Alcoa (Aluminum Company of America), 13, *15,* 16
algae, on asphalt shingle, 221
alkalines, 16, 17, 23, 65, 67, 153–54, 177, 187, 212
alkyd resins, 111
Allegheny Steel Company, 33–34
Allentown Pneumatic Gun Company, 71
alloying elements, 41
alum, 13
aluminum, 8, *12,* 13–18, 34, 223
aluminum alloys, 8, 14–16, 18
American 3-Way Prism, 160
American Institute of Timber Construction, 106–7
American Linoleum Manufacturing Company, 183
American Plaster Company, 131
American Red Cross Building, *235*
American Rolling Mill Company, 33, 223
American Society for Testing and Materials (ASTM), 2, 5, 39. *See also* product standards
Amtico Permalife tile, *xxiv*
anchorage, 60, 80–81, 127–28, *129,* 137–40, *138,* 141, 142, 167
Anderson, Arthur, 86
Anderson House, 192, *195*
annealed glass, 154
anodic coatings, 14, 17
arches, *104,* 105, 106, 109
architectural acoustics, 231
architectural precast concrete, *76,* 77–81, 131
architecture: cast stone in, *52, 54,* 57; concrete block in, 49, *50*; glass block in, *165*; Monel in, 21, *22*; nickel silver in, 27–28; porcelain enamel in, 223, 224–25; prismatic glass in, *159,* 159–60; reinforced concrete in, 63–64, *64, 66*; shotcrete in, 72–73; stainless steel in, 33–36, *35*; terrazzo in, 205; weathering steel in, 41, *41*
argentan, 25
Ar-ke-tex tile, 121–22
Armco-Ferro House, 225
Armstrong Cork Company, *xxiii, 2,* 183, 185, 197, 234
Art Deco, 27, *28,* 34, 127, *164,* 171
Art Moderne, 171, 224
asbestos, 4, 8, 210, 213, 231
asphalt shingles, *216,* 217–21, *218*
asphalt tile, 209
Association of Cast Stone Manufacturers, 53
Atlantic Gypsum, 234
atmospheric aging, of rubber tile, 193
Auburn Automobile Company, *xxiii*
austenitic stainless steels, 31
automobiles, 101, 151

backed vinyl tile, 210
Baekeland, Leo, 95
Baha'i Temple, 77, *80*
baked-on enamels, for aluminum, 14
Bakelite, 95, *208,* 209
balloon-frame structures, in U.S., 1
bar stock Monel, 22
Basquin, Olin H., 157
batts, for thermal insulation, 4
Bayonne Casting Company, 19
Beaver Board Company, 89, *91*
bedding orientation, of thin stone veneer, 140–41
Benedict Stone Company, 55, *56,* 57
Bergmann, Theodor, 223, 224
Berzelius, Jöns Jacob, 111
Bessemer ore, 19
Bestwall Firestop, 238
Beth Sholom Synagogue, *xxix*
billets, 31
Biltrite Rubber Company, *xxiv*
Bird and Son, *216,* 217
blasting, 17, 29–30, 44, 68
blowholes, in nickel, 27
Blumer's Bakery, *xxvi*
board-and-tile materials, 232, *232,* 232–34
Boeing Company Development Center, 86
bolting, of weathering steels, 41, 43
bonded terrazzo, 205
bonding agents, for architectural precast concrete, 80
Boston Acoustical Engineering Company, 233–34
brads, for cork tile, 199
breakage. *See also* impact damage
breakage, glass, 153, 154, 177
Brearley, Harry, 31
bridges, 42, 63, 83–84
Bruning Residence, *165*
Brussels World's Fair (1958), 112, *113*
Building Board, 93
building cycles, *4*
building sealants, *240,* 241–44
Butyl, development of, 8
butyl sealants, 241, *242*

Cabot's Quilt, 231
calcined gypsum, 131, 238–39
calendering, 184–85, 192
Calicel Company, 234
Calistone, 234
capitalism, 1
Carbide and Carbon Chemicals Corporation, 209
carbon, 31, 39
carbonation, 65, 73
carpentry, in concrete construction, 9
Carrara, 169, 170, 171, *171*
Cary High School, *108*
cast aluminum, *xx,* 14, 17, 18
cast iron, porcelain enamel on, 223
cast Monel, 21–22, 24
cast stainless steel, 38
cast stone, *52,* 53–60
Cast Stone Institute, 53
casting, of architectural precast concrete, 78
Cathedral of Learning, 14
cathodic protection, of reinforced concrete, 68–69
caulk, 241
Celotex Corporation, *xxviii, xxxi,* 89, 93
Celotex fiberboard, *xxviii*
*Cement and Concrete Reference Book, xxxii*
Cement Gun Company, 71
Cemesto, 93
Century of Progress Exposition (Chicago), 6, *92,* 93, 143, 164, 209, 224–25
ceramic veneer, 125–26, 127, *128*
cerium oxide, 155
Certain-Teed Products Corporation, 131, 232
chalking, of rubber tile, 193, *195*
charcoal, 27
chemical processing, of fiberboard, 90
Cheshire Glass Company, 151

Chicago Daily News Building, 27, *28*
Chicago Foundry Company, 224
Chicago Terra Cotta Company, 125
Chicago World's Fair (1933), *212*
China, nickel silver in, 25
chloride, in reinforced concrete, 65–67
chromium, in stainless steel, 31, 36–37
Chrysler Building, 14, *33*, 34
churches, glued laminated timber in, 106
cinder blocks, 47–48
City Bank Farmers Trust Company Building, 27
Civic Center Building, *40*, 41
cladding, *32*, 115, 116, 126–27, 137, 141, *141*, 142
clay. *See* structural clay tile; terra cotta
Clay, Henry, 89
cleaning. *See* conservation
Cleveland Gypsum Company, 232
Club Moderne, *171*
coal tar, 217
coatings: for aluminum, 14, 17–18; for architectural precast concrete, 80, 81; for cast stone, 58; for concrete block, 51; for fiber reinforced plastic, 115; for linoleum, 188, 189; for Monel, 24; for nickel silver, 30; for porcelain enamel, 229; for rubber tile, 194; for shotcrete, 74; for weathering steels, 39, 44
Cobb, Judd, 89
coke dust, 27
Cold Spring Granite Corporation, 139
cold-finished aluminum, 17
cold-pressed plywood, 102
cold-rolled Monel, 21
cold-rolled stainless steel, 33
colors: of aluminum coatings, 14; of architectural precast concrete, 80; of asphalt shingle, 219, 221; of concrete block, 49; of cork tile, 197–98, *199*; of decorative plastic laminates, 95; of linoleum, 186, *186*, *187*; of nickel silver, 27; of porcelain enamel, 229; of simulated masonry, 146–47, 148; of structural glass, 170; of terrazzo, 204; of vinyl tile, *211*
Columbian Enameling and Stamping Company, 224
columns, reinforced concrete, 61–63
composition roofing, 217
compounding, of rubber, 192
concrete, 9. *See also specific types*
Concrete Block Machine Manufacturers Association, 47
Concrete Block Manufacturers Association, 47
Concrete Producers Association, 47
Concrete Products Company of America, 84
Concrete Technology Corporation, 86
conservation: of aluminum, 16–18; of architectural precast concrete, 78–81; of asphalt shingles, 219–21; of cast stone, 57–60; of concrete block, 50–51; of cork tile, 200–201; of decorative plastic laminates, 97–99; of fiber reinforced plastic, 115–16; of glass block, 166–67; of glued laminated timber, 107–9; of gypsum, 134–36; of linoleum, 187–89; of Monel, 23–24; of nickel silver, 28–30; of plate glass, 153–55; of porcelain enamel, 225–29; of reinforced concrete, 64–69; of rubber tile, 193–95; of sealants, 242–44; of shotcrete, 73–75; of spandrel glass, 177–80; of stainless steel, 36–38; of structural clay tile, 122–23; of structural glass, 172–73; of terra cotta, 127–30; of terrazzo, 205–7; of thin stone veneer, 140–42; of vinyl tile, 211–14; of weathering steels, 43–44
Considère, Armand, 63
construction industry, 6–9
contact molding, of fiber reinforced plastic, 111–12
continuous annealing lehr, 151–52, *152*
control joints, in concrete block, 51
Cooper Union Building, 119
copper, 8, 19, 25, 27, 29, 39
copper electroglazing, 158
copper steel. *See* weathering steel
Coral Court Motel, *120*
cork, *xxii*, 4, 183–84, *196*, 197–201
corkboard, 197
Corkoustic, 234
Corning Glass Works, 164
corrosion: of aluminum, 16–17; of gypsum, 134; of Monel, 22, 23; of nickel silver, 27, 28, 29; of plate glass framing, 154, 155; of porcelain enamel, 227, 229; of reinforced concrete, 65–67, 68–69; of shotcrete, 73; of spandrel glass, 179; of stainless steel, 31, 36–38; of terra cotta, *129*; of thin stone veneer, 141; of weathering steels, 41, 42, *42*, 43, 44
corrugated fiber-reinforced translucent sheets, 112
corrugated glass, 157
Cor-Ten Steel, 39, 41
cracking: of architectural precast concrete, 80; of cast stone, 57, 58; of concrete block, 50, 51; of fiber reinforced plastic, 114; of gypsum, 134–35; of reinforced concrete, *67*; in sealants, 242–43; of shotcrete, 73–74, 75, *75*; of spandrel glass, 178; of structural clay tile, 122; of structural glass, 172–73; of terra cotta, 128; of terrazzo, 204, 205–6, 207, *207*; of thin stone veneer, 140, 142
cranes, for plate glass, 151
crazing, 57, 58, 114
Crew, Henry, 157
Cronstedt, Aksel Fredrik, 25
cupronickel metal, identifying, 28
curing, 55, 68, 78, 79–80, 191, 192, 206, 243
curtain walls: aluminum, *15*, 16, *16*; glass, 9, 153; porcelain enamel, 225, *228*; sealants for, 241, *242*; spandrel glass, *174*, *176*, *178*; stainless steel, *32*, 36
Cushocel, 232

Dantsizen, Christian, 31
David E. Kennedy Company, 197
David W. Taylor Model Testing Basin, 77
Davy, Humphrey, 13
Daylight Prism, 160
debonding. *See* delamination
decorative plastic laminates, *94*, 95–99
Deere Company Administrative Center, 41, *41*
deformation, 98, 153
degreasing, 24, 29
dehydrochlorination, 211
delamination, 57, 58, 60, *60*, 73–74, *75*
Delaware and Hudson Railroad Company Building, *54*, 57
Demonstration Home no. 2, *xvii*
Demonstration Home no. 4, *150*
deoxidation, in nickel manufacture, 27
Department of Justice, *xxviii*
Deville, Henri Saint-Claire, 13
Dextone Company, 77
Dick, G. A., 14
Dietz, Albert G. H., 114
Dill, Richard E., 83
Disney World, *82*, 86
Disneyland, *110*
divider strips, in terrazzo, *202*, 204, 205
domes, 63, 72, 112, *113*
Doric Building, 84
Douglas Fir Plywood Association, 101
Dow Corning, 241
draining, for weathering steels, 43–44
Dri-Bilt houses, 102
dry process, for fiberboard, 91
dry-mix shotcrete, 71, *72*
dry-tamp process, 54–55, 57
Dunn, Alan, *9*
dyes, for aluminum coatings, 14

Earley, John J., 77
Edwin F. Guth Company, 234
efflorescence, 50–51, *123*
elastomeric sealants, 241
Electricians Union Building, *178*
electricity, 5, 160
electrolytic process, for aluminum, 13
electroplating, 25, 27
electropolishing, of stainless steel, 33
Elkinton and Company, 25
Ellis, Carleton, 111
embossed molded inlaid linoleum, 185
Empire State Building, 14, 34, 152
environment, in weathering steel corrosion, 41
epoxy, 60, 74, 80, 109, 229
Equitable Building, 16
erosion, 16–17, 42, 58, 67, 115
expositions, during Great Depression, 6
extruded aluminum, 14, 17, 18
extrusion, 14, 125

Faber, Herbert, 95, 96
facing tile, 121
fairs, during Great Depression, 6
Falconnier, Gustave, 163
Fallingwater, *67*, 198
Farnsworth House, 152–53
faying surfaces, of weathering steels, 43
federal government, 6, 9
Federal Housing Administration, 6
Federal Reserve Bank, 137
felt, 187, 219
felt-and-membrane systems, 231–32, *235*
felted roofs, 217
ferritic stainless steels, 31
Ferro Enamel Corporation, 224
fiber reinforced plastic, *110*, 111–16
fiberboard, 89–93
fiber-reinforced plastic, corrugated, *xxix*
Field Museum of Natural History, 72
fillers, for decorative plastic laminates, 96
finishes: for acoustic materials, 234; for aluminum, 14, 17–18; for architectural precast concrete, 77, 78; for cast stone, 55, *59*; for cork tile, 197, 198; for decorative plastic laminates, 96; for fiber reinforced plastic, 112; for fiberboard, 93; for gypsum block, 133; for gypsum board, 238; for linoleum, 184; for Monel, 21, *22*; for plywood, 103; for porcelain enamel, 223; for rubber tile, 194; for shotcrete, 73; for simulated masonry, *145*, 146, 147–48; for stainless steel, 31–33, 38; for structural clay tile, *120*, 121–22; for structural glass, 170; for thin stone veneer, 137; for vinyl tiles, 213
fireproofing, 72, 121
fire-resistance, 119, 238
firing processes, 120, 175
flashing, 51, 220
flaws, in plate glass, 153
Flintkote Company, *218*
float process, for plate glass, 152, 155
floor arches, *119*, 120, *121*, 123
flooring: gypsum block, 133–34. *See also specific types*

fluorides, in structural glass, 169–70
Fogg Museum, 231
folded plate form, for fiber reinforced plastic, 114
Ford, John B., 151
Ford Motor Company, *152*
forged Monel, 21, 22
Formica, *xxix*, 4, *94*, 95, 96–97
Formstone, 143, 145, *145*, 147
foundations, for weathering steel, 43
Fourier transform infrared (FTIR), 213
Fowler, S. T., 61
fracture, 129, 153, 154
Frear Stone, 53
freeze-thaw cycling, 57, 67, 78–79, 122, 128, 140
Fresnel, Augustin-Jean, 157
Freyssinet, Eugène, 83
frits, for spandrel glass, 175
Fuller, Buckminster, 112
Furness, Frank, 191

Gage Building, 159
galvanic corrosion, 16, 17, 23, 37
gaskets, 154, 155, 179, *242*
Geitner, Maximilian Werner von Bausman, 25
Gelbman, Louis, 48
General Bakelite Company, 95
General Electric Company, 111, 241
General Electric experimental house, 97
General Electric Turbine Building, 36
General Motors Technical Center, 225, *228*
geodesic radomes, 112, *113*
German Evangelical Church, 14
German Silver, 25
glass. *See also specific products*
glass, with steels, *35, 36*, 43
glass block, *xxvi, xxvii, 7, 162*, 163–67
glass fiber-reinforced felt, 219
glass walls, 9, 16, 152–53
glazed structural clay tile, *xix*
glued laminated timber, *104*, 105–9
glues. *See* adhesives
Glyptal, 111
Goelet Building, *26*, 27
Gold Bond, 238
Goodyear Tire and Rubber Company, *xxiv, xxv, 190*, 191
graffiti, 29, 30, 37, 115
Gratelite ceilings, 234
Great Depression, 6, 34–36, 57
Great Lakes Exposition (1936), 225
Greyhound Bus Depot, *xxvii*
Griffith, A. A., 153
grinding, 31, 204
Gropius, Walter, *xxii*, 165, 198
grout, for structural clay tile, 122
Guastavino, Raphael, 231
Gunite, *70*, 71
Gun-Stone, 72
gypsum, and World War I, 3, 4
gypsum block, 131–36
gypsum board, 8, *236*, 237–39. *See also* Sheetrock
gypsum partition tile, 131, *132*
gypsum recrystallization, 212
gypsum tile, 131–36
gypsum wallboard, 131

H. H. Robertson Company, 133
Hale, W. E., 89
Hall, Charles Martin, 13
Hallidie Building, 152, *152*
Hamilton, Alexander, *29*
Hanisch, Max C., 105
Harbor Plywood Corporation, 102
hard tile, 119
hardboard, 89, *90*, 90–91, 93
hardwood veneers, 101
Harry Truman Home, *220*
Hayden Planetarium, 72
Haydite, *46*
heat. *See* temperature
heat-absorbing glass, adoption of, 6
Hecht Company Warehouse, *xxvi, 164*
Henninger Brothers, 25
Herculite, 153
Héroult, Paul T., 13
Hetzer, Otto, 105
Hiarni, 25
high-rise buildings, 9, 63, 152
Hilton Hotel, 78
Hollow Building Block Company, 47
hollow glass blocks, 164
Homasote, 89
Home Owners Loan Corporation, 6
horizontal shear, in glued laminated timber, 108
hotels, *29*
hot-pressed plywood, 102
hot-rolled aluminum, assessment of, 17
hot-rolled Monel, 21
housing construction, 6, 90, 92, 102, *103*, 112, 152–53, 225
Howard Johnson's, 224, *226*
Hushkote Sound Absorbent Plaster, 232
Hyatt, Thaddeus, 61, 157
hydraulic extrusion, 14
hypochlorites, and Monel, 23

impact damage, 98, 115, 172, *172*, 221, 227, *229*
industrial research, 1, 5, 9, 96, 105–6
infill walls, structural clay tile in, 121
Ingalls Building, 63
ingots, 31
inlaid linoleum, *184*, 184–85
Inland Steel, 223
Inland Steel Building, *35*
installation: of acoustic materials, 231–35; of aluminum, 14–16; of architectural precast concrete, 78; of asphalt shingles, 219, *220*; of cast stone, 57; of concrete block, 49; of cork tile, 198–99; of decorative plastic laminates, 96–97; of fiber reinforced plastic, 112–15; of fiberboard, 92–93; of glass block, 165–66; of glued laminate timber, 106–7; of gypsum, 133–34, *134, 135*; of gypsum board, 239; of linoleum, *185*, 186–87, *188*; of Monel, 21–23; of nickel silver, 27–28; of plate glass, 152–53; of plywood, 102–3; of porcelain enamel, 223–25, *224, 225*; of prestressed concrete, 84–86; of prismatic glass, 158–61, *159, 161*; of reinforced concrete, 61–64; of rubber tile, 192–93; of sealants, 242; of shotcrete, 72–73; of simulated masonry, 145–48; of spandrel glass, *176*; of stainless steel, 33–36; of structural clay tile, 120–22; of structural glass, *170*, 170–72; of terra cotta, 126–27; of terrazzo, *204*, 204–5; of thin stone veneer, 137–40; of vinyl tile, 210–11; of weathering steel, 41–42
insulated plate glass, 153
insulation board, 89, 91–93
Insulux glass block, *xxvi*, 164
intergranular corrosion, 36
International Nickel Company (Inco), 19, *26*
International Style, 6
iron, 19, 31, 39
Jackson, P. H., 83, *83*
Jacquet, Pierre, 33
Jenney, William LeBaron, 125
John Breuner Company Building, *128*
John F. Kennedy Center for the Performing Arts, 137, *141*
John F. Kennedy Space Center, 112
Johns-Manville, 231
Johnson, George H., 119
Johnson Wax Administration Building, 63, *64*
joining, of Monel sheets, 21
joints: in fiber reinforced plastic, 115; in glass block, 166, 167; in glued laminated timber, 106; in gypsum board, 239; in porcelain enamel, 225, 229, *229*; sealants in, 242, 243; in structural glass, 172; in thin stone veneer, 142
Joltcrete machine, 48–49
Junkers, Hugo, 14

Kahn, Julius, 63
Kalite, 232
Kalwall panels, 112
Kawneer Company, 225
Keaseby and Mattison, 231, 232
Keller, Robert, 112
Kelley, Stephen, 237
Kelley Island Lime and Transfer Company, 232
Kenflex, *208*
Kennedy Center for the Performing Arts, 137, *141*
Kennedy Space Center, 112
Keppler, Friedrich, 163
Kilnoise Acoustical Plaster, 232
Kimball, Francis H., 125
Kocher, Lawrence, 102
Kreischer, Balthasar, 119
Kresge Auditorium, *xxi*
Kress Company, *153*
Kroll, William J., 31

lacquer solvent, and Monel, 24
lacquering, 17, 30, 229
lag screws, for glued laminated timber, 109
laminated glass, wartime use of, 4
laminations, 106
lanolin coatings, for nickel silver, 30
Lasting Products Company, 143
latex sealants, 242
leaks, in asphalt shingle, 220
Lenox Plate Glass Company, 151
Lescaze, William, 164, 225
leveling-course material, for vinyl tile, 212
Lever House, 21, 36, *36*, 176
Lewis and Clark Centennial Exposition, 101
Libbey-Owens-Ford Company, 152, 169, *174*, 175
Liberty Silver, 25
light, 98, 115, 177, 189, 200, 206, 211
lighting, 5, 157, *158*, 163, 165–66
linoleum, *xxiii, 2, 182*, 183–89
linoxyn, 183, 184
linseed oil, 183, 184, 187–88
load-bearing walls, 121, 122, 166
Lockwood Paper Company, 237
Loring, Sanford, 125
Los ore, 25
lumber, 3. *See also specific products*
Lustron houses, *xxxi, 227*
Luxfer Prism Company, *156*, 157–58, 159, 160
Lyman, Azel Storrs, 89

MacBeth-Evans Glass Company, 164
machine molding, of fiber reinforced plastic, 111–12
Macoustic, 232

Magnel, Gustave, 83–84
magnetic testing, of Monel, 23
Maillart, Robert, 61
Maison de Verre, 163
manufacturing processes: for aluminum, 14; for architectural precast concrete, 77–78; for asphalt shingles, 219; for cast stone, 54–55, *55*; for concrete block, 48–49, *49*; for cork tile, 197–98, *198*; for decorative plastic laminates, 96; for fiber reinforced plastic, *111*, 111–12; for fiberboard, *90*, 90–92; for glass block, *164*, 164–65; for glued laminated timber, 106; for gypsum board, *238*, 238–39; for gypsum tile, 133, *134*; for linoleum, *184*, 184–86; for Monel, 21; for nickel silver, 25–27; for plate glass, 151–52; for plywood, 102; for porcelain enamel, 223; for prestressed concrete, 84; for prismatic glass, 157–58; for rubber tile, 191–92; for shotcrete, 71; for simulated masonry, 143–45; for spandrel glass, 175; for stainless steel, 31–33; for structural clay tile, 119–20; for structural glass, 169–70; for terra cotta, 125–26; for terrazzo, 203–4; for thin stone veneer, 137, *139*; for vinyl tile, 209–10, *210*; for weathering steels, 39–41
marbleized rubber tile, *xxiv*, 192
marble, 204. *See also* thin stone veneer
Marietta Manufacturing Company, 169
Marina Building, 114
Marina City, *66*
martensitic stainless steels, 31
Masonite, *88*, 89, 90–91, *91*, *92*
masonry. *See specific products*
Massachusetts Institute of Technology, *xxi*
mastic, 171–72, 173, 199
materials science, as discipline, 10
Maurer, Eduard, 31
Mayer, Ward, 106–7
Mayo, John K., 101
Mayo Clinic, 198
McKeown Brothers Company, 105
mechanical processing, for fiberboard, 90
medium-density fiberboard, 89, 93
Melan, Josef, 63
metal, 1–2, 221, 223. *See also specific types*
metal alloys, 1, 8, 14–16, 18, 25–30, *26*
metal reinforcement, of cast stone, 57, 58, 59
metal skeleton structures, in U.S., 1
metals industries, during war, 3, 6–8
Micarta, 4, *97*
Midland Terra Cotta Company, 127
military training facilities, 8, 106
mining, of gypsum, 131
Minnesota and Ontario Paper Company, 89
Miracle Blocks, *48*
Mirawall, 225
mock-ups, of repairs, 68
models, for terra cotta, 125, *126*
Modern Diner, *xxix*
Modernism, structural glass in, 171
moisture. *See also* water infiltration; water infiltration and moisture
moisture emission testing, 212
molds, 54–55, 157–58, 165, 197, *198*
Monel, *xx*, 19–24, 28
Monel Nickel Company, 21
Monell, Ambrose, 19
Monel-Plymetyl, 22–23
Monnartz, Philipp, 31
Monsanto House of the Future, *110*, 114
mortar, 122, 137, 166, 167, 205
Mo-Sai Institute, *76*, *77*, *79*
mosaic, 203
mosaic terra cotta tile, *xviii*
mottled porcelain enamel panels, *xxvi*
Muench, Carl G., 89
multi-tab strip shingles, 217–19, *218*
munitions manufacture, in World War I, 2
Museum of Modern Art, 165, 176, *177*
Museum of the City of New York, *29*

Nashkote, 231–32
National Bank of Commerce, 27
National Bureau of Standards, 6
National Fire Protection Association (NFPA), 5
National Fireproofing Corporation, *118*, 122
National Gallery of Art, 198
National Gypsum Company, 6, 8, 232, 238
NEG Industries, 173
NeoClad, 173
Neoprene, development of, 8
Neponset Twin, 217
Neusilber. *See* nickel silver
Neutra, Richard, 102, 198
New York Belting and Packing Company, 191
New York Trust Company Building, *12*
New York World's Fair (1939), 6, 102, *150*
New York World's Fair (1964), 96–97, 114
Niagara-Hudson Building, *xx*
nickel, 19, 25, 27, 31
nickel arsenide, 25
nickel silver, 25–30, *26*
nickel sulfide stones, 177–78
Norcross, Orlando W., 61
North British Rubber Company, 191
North Star Granite Corporation, 139
Northern Trust Company, *235*
Northwestern Terra Cotta Company, *124*, 125, *126*
Norton Building, 86
Notre Dame du Raincy, 77
Novus Sanitary Structural Glass, 169, 170

O'Conor, Daniel, 95
Oersted, Hans Christian, 13
oil coatings, for nickel silver, 30
oil-based caulk, 241
opal glass, 169
*opus sectile*, 203
*opus tesselatum*, 203
*opus vermiculatum*, 203
ore, 19, 25
Orford Nickel and Copper Company, 19
ornamentation, 57, *58*, 126–27
Owens-Corning, 234
Owens-Corning Fiberglas, 111
Owens-Illinois Company, *xxvi*, *7*, 163–64
oxidation, 193, 219–20
oxide films, 16, 17, 21, 28, 36
oxidized linseed oil, 183, 184, 187–88
oxidizing acids, 29
oxidizing salts, 23
oxygen, and weathering steels, 39–41

Pacific Stone and Concrete Company, 53
Pacific Telephone and Telegraph Building, *126*
paint, 17–18, 44, 58, 231, 233
Palmer, Harmon S., 47
Panel Structures, 112
Paramount Theater, *xviii*
Parsons, Frank, 183
patching, 51, 59–60, 74, *75*, 99, 129, 189, 200, 206–7
patents, 31–33, 89, 147
patina, 17, 23, 28, 39, 43, 44
patterns, *xxv*, 99, 187, 189, *194*, *201*, *211*
PC Block, 164
Penguin Coffee Shop, *97*
Penn-American Plate Glass Company, 169, 170
Pennsylvania Railroad Station, 234
Pennsylvania Station, 19
Pennycuick, James, 157
Perfatile, 234
perforated tile, 4, *230*, 232, 234
Perkins Glue Company, 101
Perma-Stone, *xix*, 143, *144*, 145, 146–47, *147*, *148*
Peshtigo High School, 105, *107*
Peterson, Fred A., 119
petrographic examination, 129
petroleum products, research in, 9
phenolic resin laminate, 4, 95
Philadelphia Savings Fund Society Building, 234
Pilkington Brothers, 152
pins, for terra cotta repair, 129
Pittco, *5*
pitting, 36–37, 154
Pittsburgh Corning, 164, 167
Pittsburgh Plate Glass Company, *5*, *150*, 151, 152, 169, 175
Pittsburgh Reduction Company, 13
plainface concrete block, 49
plastering, 4, 8, 232
plastics, 10, 17, *94*, 95–99, *110*, 111–16
plate glass, *xxvi*, *150*, 151–55
plates, in glued laminated timber repair, 109
Plexiglas, heat-absorbing glass, 6
Plyshield, *100*
plywood, *xvii*, 4, 22–23, *100*, 101–3
pointing, 51, 129, 166
Police Administration Building, 78, *81*
polishing, 21, 31–33, 155, 169, 194
polyester laminates, 95
polysulfide polymer, 241
pooling, with weathering steels, 43–44
porcelain enamel, *xxvi*, *xxxi*, 14, *222*, 223–29
Porcelain Tile Company, 224
porous tile, 119–20
Porstelain, 224
portland cement, 2, 47, 53
Portland Cement Association, *xxxii*
Post, George B., 125
posttensioning, 84
powdered filling, for insulation, 4
precast concrete, *xxviii*, 64, *76*, 77–81, 131, 140
precast gypsum tile, 131
precipitation-hardened stainless steels, 31
Precision Built Home system, 90
prefabricated concrete, 9
prefinished plywood, 103
Preload Corporation, 83, 84
Presdwood, 89, *92*
pressed glass block, 163–64
pressed-sheet aluminum, spandrel panels, 14
pressure treatments, for glued laminated timber, 108
prestressed concrete, 9, *82*, 83–86
pretensioning, 84
prismatic glass, *156*, 157–61
product standards. *See* standards
professional organizations, 5
pulp, in fiberboard, 90
Pyrex, 164
Pyrobar, 131, *132*, 133–34, *134*, *135*

quality control, for cast stone, 53
*Queen Mary*, 96
Quietile, 234
quilted stainless steel, 36
Quonset Hut, *92*

R. Guastavino Company, 232
Radiating Light Company, 157
Radio City Music Hall, 232
Ransome, Ernest, 61
Ransome, Frederick, 53
Ray, L. W., 224
realkalization, of reinforced concrete, 69
Réaumur, René, 111
Reconstruction Finance Corporation, 6
reinforced concrete, 61–69
reinforcement, 57, 58, 59, 73, 106, 109
Reliance Building, 137
replacement. *See* conservation
Republic National Bank, *13*
Republic Steel, 33, *222*
resins, 95, 101, 111, 197, 200
resurfacing, of fiber reinforced plastic, *114*
retrofit, with weathering steels, 43
reuse of materials, 167, 214
Reynolds, Herbert M., 217
Richfield Oil Building, *193*
Rigidized Metals Corporation, *33*
rigidized stainless steel, 36
Rigid-Tex, *33*
Rockefeller Center, 93
rockfaced concrete block, 49
rod stock Monel, 22
Rogers Brothers, 25
rolled Monel, 24
Roman mosaic, 203
roofs, 9, 16, *16,* 17, 19, 21, 93, 133, 134, 217–21
Rookery Building, 126
Rossman, Joseph, 89
Rostone, 143–46
rubber, during World War II, 6–8
rubber tile, *xxiv, xxv, 190,* 191–95, *192*
Rule-Page Building, 137
Rumford Tile, 231
rust, 42, 43, 44. *See also* patina

S. M. and C. M. Warren Company, 217
Saarinen, Eero, *xxi,* 41, *65,* 225
Sabine, Wallace Clement, 231
Sabinite, 232
Sackett, Augustine, 237
Sackett Wall Board Company, 237
St. Ignatius Loyola Church, 21–22
St. John's Abbey Church, 72–73
St. Leonard Catholic Church, 106
St.Thomas's Church, 231
salvaging, of glass block, 167
sample cleaning, of aluminum, 17
Samson Plaster Board Company, 237
San Vitale Church, 119
Sanacoustic tile, 232
sand-bed systems, 205
sanding, of cork tile, 200
sandwich panels, 112, 115, 116
Sani-Onyx, 169
Sanpan panels, 112
Schine's Theater, *99*
Schokbeton process, 78
Schumacher, John, 237
Schumacher, Joseph, 237
scratches, on stainless steel, 38
seal failure, in glass block, 166
sealants, 43, 81, 154, 155, 179, 206, 227
Searchlight Prism, 160
Sears, Roebuck and Company, *62*
Seattle Art Museum, *xx*
Seattle World's Fair, 86
*seminato,* 203
semiporous tile, 119
Sendzimer cold reduction process, 33, 38
7-Up Pavilion, 114
seven-wire unit, 84
shear reinforcement, of glued laminated timber, 109
sheet aluminum, 14
sheet flooring, 191
sheet Monel, 24
sheet stainless steel, 31
Sheetrock, *xxx, 236,* 237
shelf aging, of rubber tile, 193, *195*
shells, 63, 114
shop drawings, 108
shotcrete, *70,* 71–75
shredded filling, for thermal insulation, 4
shrinkage, 50, 125
siding, aluminum, *16*
Silen-Stone, 233–34
Siliceous Stone, 53
silicone sealants, 241, 242, 243, 244
silver plating, 25
simulated masonry, 143–48
sinks, Monel, *20*
skyscrapers, 9, 63, 152
slabs, 31, 61–63, *62, 63*
slag wool, 4
slush method, 48
Smith, John T., 197
Solar Prism, 160
Soldier Field, *56,* 57, *57*
solid vinyl tile, 210–11
solvents, 30, 98, 194
sound projection, 5
spalling, 50, 51, 128, 140, *141*
spandrel glass, *174,* 175–80
spandrel panels, aluminum, 14
Spandrelite, 175
spatter linoleum, 185
Speedtile, 122
"Spirit of Light," *xx*
Squibb Building, 27
staining, 37, 43, 154–55, 206, 243
stainless steel, 21, 31–38, *32, 34, 35*
stainless steel sculpture, *xx*
stallboard light, 157
stamped stainless steel, *34*
Standard Oil, *226*
standards: for asphalt shingles, 217; for cast stone, 53; for concrete block, 47; for fiber reinforced plastic, 115; for glued laminated timber, 107, 108; during Great Depression, 6; for gypsum, 133; for metals, 1–2; for plywood, 102; for prismatic glass, 158; in private sector, 5–6; for reinforced concrete, 64; for shotcrete, 74; for terra cotta, 127; for thin stone veneer, 139–40; for vinyl tiles, 213; during World War I, 3–4
Stanley, Robert Crooks, 19
steam cleaning, 17, 23, 43
Stearns Manufacturing Company, 48
Stedman Naturalized Flooring, 193
steel, 8, 22, 106, 109, 223, 225
steel frame, thermal insulation for, 4
Sterling Streamliners, *xxix*
stock terra cotta, 127
storefronts, *5, 33, 153, 160,* 171
Stran-Steel House, 225
Straub, F. J., 47–48
Streamline Moderne, 34
stress corrosion, 28
Stria tile, 234
stripping, of linoleum, 188
structural clay tile, *118,* 119–23
structural glass, *168,* 169–73
Structural Glass Corporation, 163
substrates, 175, 214, 225
sulfate salts, and reinforced concrete, 67
Sullivan, Louis, 127, 159–60
Super Harbord, 102
Supply Priorities and Production Board, 8
surface deposits, on nickel silver, 29
suspended acoustical ceilings, 234–35
synthetic rubber, 6–8, 191

T. Eaton Department Store, *22*
Tampa Bay Bridge, 84
tarnish, of nickel silver, 28, *29*
Taut, Bruno, 163
technical societies, 5
Temcoustic, 234
temperature, 97–98, 177, 243
tempered plate glass, 153, 154, 175
Tempered Presdwood, 89, 93
tensile strength, of plate glass, 153
termites, 107–8
terra cotta, *124,* 125–30
terrazzo floors, *xxiii, 28, 202,* 203–7
testing: of architectural precast concrete, 79; of concrete block, 47, 50; of decorative plastic laminates, 98; of glued laminated timber, 106, 108–9; of gypsum, 134; of Monel, 23; of nickel silver, 29; of plate glass, 154; of plywood, 102; of porcelain enamel, 228–29; of prismatic glass, 159; of reinforced concrete, 68; of rubber tile, 193; of sealants, 243–44, *244*; of shotcrete, 74, 75; of structural clay tile, 119, 122; of terra cotta, 128–29; of terrazzo, 206; of thin stone veneer, 141, *141,* 142; of vinyl tile, 212
Textured Ceiling Board, 234
thermal insulation, 4, 9, 153, 164
Thermopane, 153
thin stone veneer, 137–42
Thiokol, *240,* 241
thiourea-formaldehyde resins, 95
Thompson, Robert, 19
Tic Toc Cafe, *171*
Ti-Namel, 223
titanium steel, 223
trade associations, 5
Tremco Manufacturing Company, 241
trial mixes, for architectural precast concrete repair, 79
trim, stainless steel, 36
tuck-pointing, 122
Tuf-Flex, 153
Turner, C. A. P., 61
TWA Terminal, *65*
Twindow, 153
230 Park Avenue, 34

unbonded terrazzo, 205
underbed, for terrazzo, 205
Underwriters' Laboratory (UL), 5–6
Union Fibre Company, 233
Union Terminal, 21
Union Trust Building, *22*
Unit Structures, *104,* 105, 106
United Nations General Assembly Building, 21, *24*
United States Gypsum Company, *xxx,* 8, 131, 133–34, 232, 234, *236,* 237–39
Unitrave system, 131
Unity Temple, *73, 75*
Universal hydraulic testing machine, *3*
Universal Insulite, 233
University of Connecticut at Storrs, 78
urea-formaldehyde resins, 95
urethane foam insulation, 43

urethane sealants, 242, 243
U.S. Air Force Academy, 16
U.S. Courthouse, 232
U.S. Plywood, 102
U.S. Steel, 39, 41
Utzman, Clarence, 237

valley linings, 220
Van Benshoten, William H., 112
Van der Leeuw House, 198
varnishing, of aluminum, 17
veins, in cast stone, 55
Venice, 203
Veos, 224
Vinoy Park Hotel, *136*
vinyl asbestos tile, 210, 211, 213
vinyl composition tile, *xxiv*
vinyl flooring, 189
vinyl tile, *208,* 209–14
Vinylite, 209, 210
visual examination: of aluminum, 17; of architectural precast concrete, 79; of asphalt shingle, 220; of concrete, 50; of glass block, 166; of reinforced concrete, 68; of sealants, 243; of shotcrete, 73–74; of spandrel glass, 178; of structural clay tile, 122; of structural glass, 172; of terra cotta, 129; of terrazzo, 206
Vitrified Metallic Roofing Company, 224
Vitrolite, *xxvii,* 169, *170,* 171
Vitrolux, *174,* 175, *176*
Von Bunsen, Robert, 13
Vorländer, Daniel, 111
vulcanization, 191, 192

Waldorf-Astoria Hotel, 27
Wallace, Robert, 25
wallboard. *See* fiberboard; gypsum wallboard
Walnut Lane Bridge, 83–84, *85*
Walton, Frederick, 183, 185
Wanamaker Building, 133
War Industries Board, 3–4
War Production Board, 8
warping, of stainless steel, 37
Washburn Park Tank, *85*
washing, of Monel, 24
Washington Monument, 13
Wasmuth, R., 31
water infiltration and moisture: and architectural precast concrete, 78–79, 81; and asphalt shingle, 219, 220; and cast stone, 60; and concrete block, 50, 51; and cork tile, 200; and decorative plastic laminates, 98; and fiber reinforced plastic, 115; and glued laminated timber, 107–8; and gypsum, 134; and linoleum, 187; and plate glass, 153–54; and porcelaand enamel, 227; and reinforced concrete, 68; and shotcrete, 73, 74; and structural clay tile, 122, *123*; and terra cotta, 127–28; and thand stone veneer, 140; and vinyl tile, 211–12, 213, 214; and weathering steels, 43
waterproofing, of plywood, 102
waxing, 17, 30, 194, 200
wear, on linoleum, 187, *189*
weathering, 67, 115
weathering steel, 39–44, *40*
weld stains, on stainless steel, 37
welding, 18, 41, 43, 44
Werkbund Exhibition (1914), 163
Westinghouse Electric and Manufacturing Company, 95
wet process, for fiberboard, 90–91
wet-cast process, 54–55
wet-mix shotcrete, 71, *72, 75*
Wheeling Tile Company, 122
White Castle, 223–24, *226*
white copper, 25
White Horse Barn, 77
White House, 93
Whitey's Cafe, *xxi*
Wieboldt-Rostone House, 143, *146, 148*
Will Rogers Theater, *50*
window glass, 152
Winslow, William H., *158*
wire brushing, 17, 30
wire glass, *xxix*
Wolverine Porcelain Enamel, 224, *226*
woods. *See specific products*
Woolworth Building, 127, 152
World War I, 2–4, 101, 105, 237
World War II, 6–8, 14–16, 27–28, 57, 95, 102, 106, 111, 237
World's Columbian Exhibition (1893), 163, 224
World's Fair (1933), *212*
World's Fair (1939), 6, 102, *150*
World's Fair (1964), 96–97, 114
World's Fair House, 96–97
Wright, Frank Lloyd, *xxix,* 63, *64, 67, 158,* 160, 198
Wrigley Building, 127
wrought aluminum, welding of, 18
Wyche, E., 31

Youngstown Pressed Steel Company, 224

zinc, in nickel silver, 25, 27

# ILLUSTRATION CREDITS

Front cover, page viii: See fig. 1.3

Back cover: Photo: Kyle Normandin © J. Paul Getty Trust.

Plates 1, 11, 22, 25, 27; frontispiece, figs. 6.1, 12.1, 22.1, 27.3, 28.4, 29.1: Richard Longstreth

Plate 2: Library of Congress, Prints & Photographs Division, HABS CAL,1-OAK,9--41 (CT)

Plate 3; figs. 21.3, 21.5: Eldon Hambel

Plate 4; figs. 17.1, 18.1, 20.1, 20.2: Sweet's Architectural Catalogues, 1932

Plates 5, 7: Randy Juster

Plate 6; figs. 2.5: Derek H. Trelstad

Plate 8; p. ii: Library of Congress, Prints & Photographs Division, Korab Collection. LC-KRB00-209

Plate 9; fig. 32.5: Lake County (IL) Discovery Museum, Curt Teich Postcard Archives

Plate 10: Courtesy of Historic New England

Plates 12–14; fig. 17.3: Bob Barrett

Plates 15, 17; figs. 6.2, 6.3, 7.1, 7.6, 7.7, 12.3, 21.1, 25.1, 25.2, 29.5, 30.1, 30.4, 31.3: Richard Cheek

Plate 18: Library of Congress, Prints & Photographs Division, HABS DC-862-2 (CT)

Plate 19: Mike Jackson

Plate 20: Library of Congress, Prints & Photographs Division, HABS SC,40-COLUM,16–11 (CT)

Plate 21; figs. 21.6, 33.7: Library of Congress, Prints & Photographs Division, HABS

Plate 23: Library of Congress, Prints & Photographs Division. LC-DIG-highsm-12032

Plate 24: Laszlo Regos Photography

Plate 26: Library of Congress, Prints & Photographs Division, HABS IND,64-CHEST,1–14 (CT)

Fig. I.1: *Cement and Concrete Reference Book*, Portland Cement Association (Chicago: Portland Cement Association, 1941)

Figs. I.2, 27.4, 27.5, 29.4: Armstrong World Industries, Inc.

Fig. I.3: *Materials Testing: Theory, Practic and Significance*, Herbert Gilkey, Glenn Murphy, and Elmer Bergen (New York: McGraw-Hill, 1941)

Fig. I.4: *Journal of the American Statistical Association*, June 1933

Figs. I.5, 11.4, 19.3, 26.4, 36.1: Smithsonian Libraries, National Museum of American History Library

Fig. I.6: Chicago History Museum (HB18509) Photograph: Hube Henry, Hedrich Blessing

Fig. I.7: *The Last Lath*, Alan Dunn (New York: F. W. Dodge Corporation, 1947)

Figs. 1.1, 2.1, 23.6, 24.5, 26.5, 30.5: Thomas C. Jester

Fig. 1.2: *Aluminum in Modern Architecture*, Paul Weidlinger (Louisville, KY: Reynolds Metal Company, 1956)

Figs. 1.3, 7.3–7.5, 18.2, 22.4: Courtesy the National Building Museum

Fig. 1.4: *Architectural Forum*, July 1955

Fig. 1.5: Stephen Senigo

Fig. 2.2: Library of Congress Prints and Photographs Division, HABS MICH, 82-DETRO,42–13

Fig. 2.3: *Practical Design in Monel Metal for Architectural and Decorative Purposes*, International Nickel Company (New York: International Nickel Company, 1931)

Fig. 2.4: Avery Architectural and Fine Arts Library

Fig. 3.1: *Architectural Record*, September 1938

Fig. 3.2: Chicago History Museum (ICHi-19848) Photograph: Kaufmann & Fabry

Figs. 3.3, 3.4: *Nickel Silver in Architecture*, International Nickel Company (1955)

Fig. 3.5: © Nilda Rivera

Fig. 4.1: *Architectural Forum*, 1955

Fig. 4.2: *Sheet Metal Worker*, July 1945

Fig. 4.3: Timothy A. Clary/Getty Images

Fig. 4.4: American Diner Museum, Providence, Rhode Island

Fig. 4.5: Chicago History Museum (HB-21235-B). Photograph: Heidrich-Blessing

Fig. 4.6: *Curtain Walls in Stainless Steel*, American Iron and Steel Institute (Princeton AISI, 1955)

Fig. 4.7: National Park Service, Preservation Assistance Division

Fig. 5.1: *Progressive Architecture*, June 1967

Fig. 5.2: Library of Congress, Prints and Photographs Division, Korab Collection, LC-DIG-krb-00599

Figs. 5.3, 5.4: Carolyn L. Searls

Fig. 6.4: Kristin and Corry H. Kenner

Fig. 6.5: Portland Cement Association

Fig. 7.2: Albany Institute of History and Art

Fig. 7.8: Architectural Resources Group

Fig. 8.1: *Architectural Forum*, June 1928

Figs. 8.2, 8.3: *Concrete Engineers Handbook*, George A. Hool and Nathan C. Johnson (New York: McGraw Hill, 1918)

Fig. 8.4: Library of Congress Prints and Photographs Division, LC-HS503-2979 (ONLINE) [P&P]

Fig. 8.5: Library of Congress, Prints and Photographs Division, Korab Collection, LC-DIG-krb-00599

Fig. 8.6: Chicago History Museum (HB-23215-D) Photograph: Heidrich-Blessing

Figs. 8.7–8.9, 17.5, 18.5, 20.3, 20.4, 36.3: Wiss, Janney, Elstner Associates

Figs. 9.1, 9.5–9.7: Robert A. Bell Architects

Fig. 9.2: *Journal of the National Association of Concrete Users*, 1911

Figs. 9.3, 9.4: *Shotcreting*, SP-14 (Detroit: American Concrete Institute, 1965)

Fig. 10.1: Mo-Sai, Precast Concrete Facing and Curtain Wall, 1963 Catalogue

Figs. 10.2, 10.4, 11.1, 11.3: Precast/Prestressed Concrete Institute

Fig. 10.3: Baha'i House of Worship

Figs. 11.2, 28.2: United States Patent and Trademark Office

Fig. 12.2: *The Story of Hardboard*, American Hardboard Association (Chicago: American Harboard Association, 1961)

Fig. 12.5: Courtesy Masonite Corporation

Fig. 12.6: Chicago History Museum (ICHi-25470) Photograph: Kaufmann & Fabry

Fig. 13.1: *Architectural Record*, August 1943

Fig. 13.2: Chicago History Museum (HB-20953-A) Photograph: Bill Engdahl, Heidrich-Blessing

Fig. 13.3: *American Architect and Building Age*, May 1932

Fig. 13.3: Armet and Davis

Fig. 13.3: Chicago History Museum, (HB-01035-T) Photograph: Ken Hedrich, Hedrich-Blessing

Fig. 13.4: Kimberly Konrad Alvarez

Fig. 14.1: *Progressive Architecture*, April 1950

Fig. 14.2: *Plywoods: Their Development, Manufacture and Application*, Andrew Wood and Thomas Linn (New York: Chemical Publishing, 1943)

Figs. 14.3, 15.2: Forest Products Laboratory, U.S. Department of Agriculture

Figs. 15.1, 15.3: Sentinel Structures, Inc. Courtesy Andreas Jordahl Rhude

Fig. 15.4: David C. Fischetti

Fig. 16.1: Richard Cheek, *Popular Science Magazine*, April 1956

Figs. 16.2, 16.5: Scott Bader

Fig. 16.3: *Glass Reinforced Plastics*, Phillip Morgan(New York Interscience Publishers, 1961)

Fig. 16.4: Courtesy the Estate of R. Buckminster Fuller

Fig. 16.6: Kalwall Corporation

Fig. 17.2: *Architectural Construction Vol 2. An Analysis of Structural Design in American Buildings, Book Two:Steel Construction* (New York: Voss and Varney, John Wiley and Sons, 1927)

Fig. 17.4: *Building Renovation Magazine*, Spring 1994

Fig. 18.3: California State Library

Fig. 18.4: Mary Swisher

Fig. 19.1: *Pencil Points*, October 1925

Fig. 19.2: *Gypsum Products, Perparation and Uses*, R. W. Stone (Washington, DC: Government Printing Office, 1917)

Figs. 19.4, 35.2: United States Gypsum Company

Fig. 19.5: *Architectural Graphic Standards*, Charles G. Ramsey and Harold R. Sleeper (New York: John Wiley and Sons, 1932)

Fig. 19.6: *Building Renovation Magazine*, January–February 1993

Fig. 20.4: Library of Congress Prints and Photographs Division, LC-HS503-3130

Fig. 21.2: US Patent Office

Fig. 21.4: Chicago History Museum (HB-0163-F) Photograph: Ken Hedrich, Hedrich-Blessing

Fig. 22.3: Library of Congress, Prints & Photographs Division, HABS CAL,38-SAN-FRA,149–1

Figs. 23.1–23.4: Dietrich Neumann

Fig. 23.5: *Architectural Forum*, August 1925

Fig. 21.1: *Pencil Points*, April 1940

Fig. 24.2: *Modern Glass Practice*, Samuel Scholes (Chicago: Industrial Publications, 1941)

Fig. 24.3: Library of Congress, Prints and Photographs Division, HABS DC-862-1

Fig. 24.4: Chicago History Museum (HB-04382-C) Photograph: Hedrich-Blessing

Fig. 24.4: Chicago History Museum (HB-04382-N)

Fig. 25.3: University of Louisville, Ekstrom Library

Fig. 25.4: Library of Congress, Prints & Photographs Division, HAER

Figs. 25.5, 25.6: Carol Dyson

Figs. 26.1, 26.2: Ward Canaday Center, University of Toledo, Libbey-Owens-Ford Collection

Fig. 26.3: Digital Image © The Museum of Modern Art/Licensed by SCALA/Art Resource, NY

Fig. 27.1: *Ladies' Home Journal*, March 1924

Fig. 27.2: Forbo-Nairn

Fig. 27.6: *Pencil Points*, January 1935

Fig. 27.7: *Architectural Forum*, December 1930

Fig. 27.8: Bonnie Wehle Parks Snyder

Figs. 28.1, 32.3: Sweet's Architectural Catalogues: 1931

Fig. 28.3: Goodyear Tire and Rubber Company

Fig. 28.5: Sharon C. Park

Figs. 29.2, 29.3: *Cork and the American Cork Industry,* Arthur Faubel (New York: Cork Institute of America, 1941)

Fig. 30.2: Lawrence Di Filippo, The Venice Art Terrazzo Co. Inc.

Fig. 30.3: *Pencil Points*, March 1936

Fig. 31.1: *Architectural Record*, October 1955

Fig. 31.2: Kentile, Inc.

Fig. 31.4: *Modern Plastics*, June 1933

Fig. 32.1: Bird & Son Catalogue

Fig. 32.2: *Asphalt and Allied Substances*, Herbert Abraham (New York: Reinhold Press, 1963 ed)

Fig. 32.6: Alan O'Bright

Fig. 33.1: Library of Congress, Prints and Photographs Division, Luther Ray Archive

Fig. 33.2: *Pencil Points*, March 1939

Fig. 33.3: *Architectural Forum*, May 1937

Fig. 33.4: White Castle Systems, Inc.

Fig. 33.6: Chicago History Museum (HB-05559-C) Photograph: Hedrich-Blessing

Fig. 33.8: Library of Congress, Balthazar Korab Collection, Prints & Photographs Division, LC-KRBOO- 81

Fig. 33.9: Richard Ryan

Fig. 33.10: William H. Scarlet and Associates

Figs. 34.1, 34.2: Owens-Corning Fiberglass Corporation

Fig. 34.3: Thomas C. Jester, *Architectural Forum*, June 1928

Fig. 34.4: Chicago History Museum (HB-12809-G) Photograph: Hedrich-Blessing

Fig. 34.5: David Bell

Fig. 35.1: Catalogue of US Gypsum Company, 1937

Fig. 35.3: Chicago History Museum (HB-09513-V4) Photograph: Hedrich-Blessing

Fig. 36.2: *Adhesives Age*, October 1959